Mana·   ·ers,
Sci·   ·ogists

Second Edition

Management for Engineers, Scientists and Technologists

Second Edition

John V. Chelsom
Andrew C. Payne
Lawrence R. P. Reavill

John Wiley & Sons, Ltd

Copyright © 2005   John Wiley & Sons Ltd, The Atrium, Southern Gate, Chichester,
West Sussex PO19 8SQ, England

Telephone   (+44) 1243 779777

Email (for orders and customer service enquiries): cs-books@wiley.co.uk
Visit our Home Page on www.wileyeurope.com or www.wiley.com

Reprinted July 2005, April 2006

This publication is designed to provide accurate and authoritative information in regard to the subject matter
covered. It is sold on the understanding that the Publisher is not engaged in rendering professional services. If
professional advice or other expert assistance is required, the services of a competent professional should be
sought.

Other Wiley Editorial Offices

John Wiley & Sons Inc., 111 River Street, Hoboken, NJ 07030, USA

Jossey-Bass, 989 Market Street, San Francisco, CA 94103-1741, USA

Wiley-VCH Verlag GmbH, Boschstr. 12, D-69469 Weinheim, Germany

John Wiley & Sons Australia Ltd, 33 Park Road, Milton, Queensland 4064, Australia

John Wiley & Sons (Asia) Pte Ltd, 2 Clementi Loop #02-01, Jin Xing Distripark, Singapore 129809

John Wiley & Sons Canada Ltd, 22 Worcester Road, Etobicoke, Ontario, Canada M9W 1L1

Wiley also publishes its books in a variety of electronic formats. Some content that
appears in print may not be available in electronic books.

Library of Congress Cataloging-in-Publication Data

Chelsom, John V.
    Management for engineers, scientists, and technologists / John V. Chelsom,
Andrew C. Payne, Lawrence R. P. Reavill. – 2nd ed.
        p. cm.
    Rev.ed. of: Management for engineers / Andrew C. Payne, John V. Chelsom,
Lawrence R. P. Reavill. c1996.
    Includes bibliographical references and index.
    ISBN 0-470-02126-8 (pbk. : alk. paper)
    1. Engineering – Management. I. Chelsom, John V. II. Reavill, Lawrence R. P.
III. Payne, Andrew C. Management for engineers. IV. Title.
    TA190.C485 2004
    658'.002'462 – dc22
                                                                2004013224

British Library Cataloguing in Publication Data

A catalogue record for this book is available from the British Library

ISBN 10: 0-470-02126-8 (P/B)
ISBN 13: 978-0-470-02126-2 (P/B)

Typeset in 10/12 pt Photina by Laserwords Private Limited, Chennai, India.
Printed and bound in Great Britain by Antony Rowe Ltd, Chippenham, Wiltshire.
This book is printed on acid-free paper responsibly manufactured from sustainable forestry
in which at least two trees are planted for each one used for paper production.

With thanks to Jo for all she gave me, especially our family, Joanna and Tim with their Sophie and Robert John, and John J and Angela with their Jay Harold, and our friends.

J. V. C. 2004

To the memory of Dorothy, her love, forbearance and courage.

A. C. P. 2004

To Anne, for her continuing patience and help, especially in periods of overload and stress.

L. R. P. R. 2004

Contents

Acknowledgements   ix
Introduction to the second edition   xi
Preface   xiii

PART I   BUSINESS BASICS   1
1   Business basics   3
2   The business environment   14
3   Management styles: From Taylorism to McKinsey's 7Ss   32
4   Management of quality   43
5   Materials management   63
6   Managing design and new product development   76
7   Organizations   103
8   Managing to succeed   122

PART II   MANAGING ENGINEERING RESOURCES   147
9   Human resource management – the individual   149
10   Groups of people   173
11   Communication   204
12   Work study   235
13   Costing and pricing   250
14   Measuring financial performance   268
15   Project investment decisions   296
16   Maintenance management   324
17   Project management   335
18   Networks for projects   364
19   Project management – managing construction procurement   376
20   Inventory management   388
21   Management of the supply system   410
22   Marketing   433
23   A case study in starting an SME   466
Appendix 1   A guide to writing a business plan   487
Appendix 2   Quality management tools   496
Appendix 3   Case study: Developing a network   512
Appendix 4   DCF tables   522
Index   535

Acknowledgements

This edition owes a considerable debt to all those involved in the development and writing of the first edition of *Management for Engineers* published in 1996. We especially wish to thank those who have made a further contribution to this edition: Tony Curtis of the Plymouth Business School, assisted by Elly Sample of the Peninsular Medical School, for revising Chapter 22; Fred Charlwood for revising Chapter 16; Diane Campbell for revising Chapter 14. We remain fully responsible for any errors in these and other chapters.

We would like to acknowledge help from the following organizations: ABB (Asea Brown Boveri); BAe (British Aerospace); British Steel (now part of Corus); CSW Group; DaimlerChrysler AG; Ford Motor Company; IBM; Ingersoll Milling Machine Co. USA; Lamb Technicon; Nissan; the Office of Official Publications of the European Community; PA Consulting Group; Siemens AG; The British Library of Political and Economic Science at the London School of Economics; The Royal Statistical Society.

John V. Chelsom gratefully acknowledges the contribution of two outstanding individuals to the success of the start-up company (OURCO) that provides the case study in Chapter 23. Sir Martin Wood founded and developed his own successful high-technology company, Oxford Instruments. Sir Martin used some of his rewards to establish The Oxford Centre for Innovation, whose facilities were vital to OURCO's early growth. Mr Stewart Newton founded and developed a successful financial organization, Newton Investment Management. His applied business acumen and substantial financial support were vital to OURCO's survival and eventual success. These two gentlemen did more for OURCO Group Ltd than all the DTI's 183 schemes for SMEs put together.

Introduction to the Second Edition

In the light of the continuing success of *Management for Engineers*, published in 1996, it has become clear that, while the needs of undergraduate and graduate engineers were being met, the very similar needs of scientists and technologists were not being addressed. The purposes of this new edition are to bring to the attention of a much wider readership the fundamentals of management, and to bring the text more up to date.

While there have been significant changes in the business environment since 1996 there has been no really important addition to the basic management skills and knowledge required by engineering, science and technology students about to start their working careers, nor for those who wish to move from being specialist practitioners into 'management'.

Where there have been developments in some aspects of management, we considered that these could be incorporated without major change to the book's structure. And where sections were out of date, such as the chapter on employment law or the appendix on the Single European Market, we felt they could be removed without much loss. Thanks to the Internet, current information on such topics is readily available elsewhere.

There are two new chapters, Chapters 19 and 23, and one new appendix, Appendix 1. Chapter 19, which originally concerned itself with the project management of large projects, has been rewritten to consider the especial circumstance in which an engineer, scientist or technologist might find himself on the project management team for the procurement of the design and construction of a new facility, whether it be new buildings, a new manufacturing or processing facility or a new laboratory. There have been some major new initiatives in managing construction procurement and these are addressed in this chapter. The only all new chapter is Chapter 23, which relates to small and medium enterprises (SMEs). It takes the form of a case study, based on close observation of a start-up new technology company that was founded by a young engineer and that now includes several scientists and technologists in key positions. There is a related appendix, Appendix 1, providing guidelines to developing a business plan. Other new material has been integrated with the updating of each of the original chapters, and new references have been added to help locate material that had to be excluded because of space constraints. Where possible, we have given Internet references.

With these changes we believe we have made the book a useful 'primer' for all students of management. From our experience and feedback we know that, as well as engineers, the first edition was used by business studies students and management students – and even by managers. We hope that in its updated form scientists and technologists can be added to its readership, and that all readers and users will enjoy it.

John Chelsom   Andrew Payne   Lawrie Reavill

Preface

Management skills make engineers, scientists and technologists better at their jobs. This is true even if they are working as specialists in the field for which they have trained. A study of 30 major UK government-funded collaborative scientific research and development projects [1] found that almost 90% of them failed to meet their objectives. Many projects foundered, overrunning their cost and timing objectives, due to conflicting objectives, perspectives and expectations, and lack of project management, teamworking and communication skills. Studying management can overcome such shortcomings. This book is designed to assist such learning.

The need for management-trained scientists extends far beyond collaborative R&D projects. It exists in commerce, industry, government and education. The *Sunday Times* (13 June 1993) said: 'British management's biggest defect is its lack of technocrats: managers combining technical know-how with financial acumen.' This defect is not peculiar to Britain, and it persists more than 10 years on. The defect can be remedied more readily by management training of those with the technical know-how than by trying to provide accountants, lawyers and economists with an understanding of science and technology.

In business, the need for technocrats is now even greater. The top business issues of the 1990s, identified by PA Consulting [2] and shown in Figure P.1, continue to be important, but the search for winning products now extends deeper into the science base. Management of the costs and risks of new technology is therefore more demanding, and the time-to-market issue now focuses on harnessing market-driven science to accelerate the introduction of top-quality products featuring technological advances. More recent research [3] puts 'strategic planning for technology products' top among management of technology issues. When combined with the 'advanced materials revolution' [4] these developments take management deep into the 'supply system' (see Chapter 21) and create many opportunities for those technocrats with scientific and business know-how. Career prospects for 'techies' who are skilled managers have never been better.

The skills and knowledge required by those who wish to be successful managers were summarized by W. Edwards Deming [5] as:

- appreciation for a system;
- some knowledge about variation;
- some theory of knowledge; and
- some knowledge of psychology.

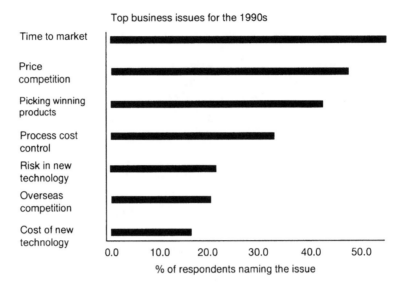

Figure P.1   Top business issues

This book makes a start in the direction that Deming would have us travel. It treats business as a system, takes a systemic approach and describes ways of dealing with variation. The implicit theory of knowledge is the traditional inductive/deductive scientific method, which was translated by Deming into his Plan–Do–Study–Act cycle. We incorporate some psychology in Chapters 9 and 10. The book was not deliberately based on Deming's philosophy, but the coincidence is not altogether surprising. When the first edition was written, the authors were in the Systems Science Department of City University, London, teaching management to undergraduate and postgraduate engineers, and 'management and systems' to other undergraduates. They all came to City University after careers spanning a total of more than 80 years as managers in companies with multinational systems and organizations, and before that they all had a science-based education. So their systemic thinking had common roots with Deming's – he trained initially as a physicist. This all suggests that scientific training is a good basis for becoming a manager or, for the really talented, as in Deming's case, a world-class management expert.

The book is in two parts: Part I is a series of chapters on management applications and concepts, starting with basic issues such as 'What is a business?' and 'What is management?', continuing through management of quality, materials and new product development, and concluding with examples of companies who provide models of good management.

Part II starts with chapters on human resources management and communication, and goes on to provide some tools and techniques, such as critical path networks, discounted cash flow and inventory control – with exercises and worked examples. These examples relate mainly to large organizations, but the lessons apply also to small and medium enterprises. 'Know your customer' is good advice for any company, and large companies, directly or indirectly, are the customers of small companies. To balance this emphasis on large organizations, this second edition includes a new chapter (Chapter 23), dealing with the challenges facing small start-up companies. The main body of Part II continues to unveil some of the mysteries of activities such as finance, marketing and purchasing – a greater understanding of these functions is a precondition for greater cooperation with them.

To assist those who may wish to pursue a particular subject further, references and bibliographies are provided throughout. These make specific recommendations for further reading, identifying useful chapters and passages or outlining the contents of the referenced material.

We hope that the book will be useful to students and to teachers, that it will be occasionally enjoyable as well as informative, and that it may inspire and assist many young engineers, scientists and technologists to become successful managers. There are opportunities vacant and awaiting you.

REFERENCES

1. Dennis, J. (1993) *The Application of Systems Science Methodologies to the Assessment of Collaborative R&D*, Project for the MBA in Engineering Management, City University, London.
2. Bone, S. (1992) *Chief Executives' Attitudes to Innovation in UK Manufacturing Industry*, PA Consulting, London.
3. Scott, G. M. (1996) *Management of New Product Development*, Research study, University of Connecticut, Storrs, CT.
4. Chelsom, J. V. and Kaounides, L. (1995) 'The advanced materials revolution', David Bennett and Fred Steward (eds), *Proceedings of the European Conference on Management of Technology*, Aston University, Birmingham.
5. Deming, W. Edwards (1993) *The New Economics for Industry, Government, Education*, MIT, Cambridge, MA.

BIBLIOGRAPHY

'W. Edwards Deming (1900–1993): The man and his message', the 1995 Deming Memorial lecture delivered by Dr Henry Neave to the First World Congress on Total Quality Management, traces the development of Deming's philosophy of management from Statistical Process Control to the Theory of Profound Knowledge. The Congress proceedings were published by Chapman & Hall, ISBN 0 412 64380 4.

'Getting equipped for the twenty-first century', an article by John V. Chelsom in *Logistics Information Management*, Volume 11, Numbers 2 and 3, 1998, published by MCB University Press, deals with changes in the focus of competition and the concept of total quality. The article adapts some of Deming's teaching as presented by Henry Neave, and suggests modifications to other management models presented by Professors N. Kano and Y. Kondo, to accommodate the impact of the materials revolution and other new technologies. Articles in *Logistics Information Management* are available online at http://www.mcb.co.uk.

Part I

Business Basics

1

Business Basics

Don't skip this chapter! It may be basic = simple, but it is also basic = provides a framework or base for your better use of the rest of the book.

1.1 INTRODUCTION

This chapter identifies the basic business functions and shows how they relate to each other to form a 'system' that leads from the idea – the concept of a product or service – to satisfying the customers for that idea. It identifies the need to find funds – that is, money – to get the business started, and to generate more funds by operating the system so that the business can be sustained and expanded. The first part of the chapter describes 'management' and the various roles of managers. The basic business functions are then introduced, and the way they relate to each other to form the 'business chain' or business system is outlined. The chapter concludes with a description of some of the ways in which the performance of business systems may be measured.

1.2 ABSOLUTE BASICS

In absolutely simple terms, businesses, and many other organizations, are concerned with obtaining reward from an idea. How this is done is the task of management, as shown in Figure 1.1.

What is management? This question can be answered by considering first what it is that managers manage – that is, the inputs to the business system – and secondly the roles of managers as they help to transform the inputs to outputs, products or services, through the business process.

So, what are the inputs? Through the nineteenth century it was normal to consider three inputs to a business system:

- land;
- labour;
- capital.

This has its roots in agriculture-based economies, where the capital was used to provide tools for the labour to work the land, plus 'material' in the form of livestock, feed and seed. As economies became more industrialized, the emphasis moved away from land and capital became a more important element, to provide

Figure 1.1   From the idea to the rewards – management's task is to find the most efficient, most effective route

more equipment for the labour as well as a wider range of material inputs. Even in manufacturing, labour was the most important cost element from the late nineteenth century until the mid-twentieth century. At this point, mechanization was supplemented by automation and labour's share of total costs fell rapidly. Mechanization had increased labour productivity through devices such as the moving assembly line, which was introduced in the automotive industry in 1913 by Henry Ford I, initially for the assembly of magnetos in the Highland Park plant in Detroit and subsequently for the assembly of the complete vehicle. Automation, through numerically or computer-controlled machines for cutting, shaping and joining materials, through materials handling equipment and through reprogrammable universal transfer devices (UTDs) – better known as robots – has accelerated this decline in labour's share of cost. As a result, material and equipment costs have become progressively more significant.

Managing materials and equipment requires a lot of data about them and about the companies that supply them. Managers also need to know what is going on inside their own organization, and what is happening outside in the marketplace and general environment. So, by the 1980s a new input became a vital part of many businesses: *information*. Developments in information technology (IT) and the advent of the Internet have increased the importance of information management as a means of optimizing business performance. Some organizations, for example Asea Brown Boveri (ABB), have used their IT skills as a source of competitive advantage (see Chapter 8).

Thus there are now four inputs or factors to be managed: land, labour, capital and information. But what does it mean, 'to manage'?

From the 1950 edition of *Chambers' Dictionary*, 'to manage' means: 'To have under command or control; to bring round to one's plans; to conduct with great carefulness; to wield; to handle; to contrive; to train by exercise, as a horse', or the intransitive form, 'To conduct affairs.'

Some of these terms are rather militaristic or dictatorial, but most of the elements of later definitions of management are there. Note that the manager has 'plans' and brings others round to them, and that 'training' is included. These are elements that have grown in importance.

The 1993 edition of *Chambers* adds: 'to administer, be at the head of; to deal tactfully with; to have time for; to be able to cope with; to manipulate; to bring about.' This suggests that the world outside business and industry has detected little change in management in more than 40 years, and still sees it as a form of constraint or control. Some authorities closer to the action share this view. Stafford Beer, a deep thinker and prolific writer on the subject, described management as 'the science and profession of control' [1]. As shown briefly below and in more detail in Chapter 3, these definitions are too restrictive – *good management entails more positive features, such as initiative and leadership.*

Mintzberg [2] quotes a definition from 1916 by a French industrialist, Henri Fayol, who said that the manager 'plans, organizes, coordinates and controls' but goes on to show that managers actually do rather different things most of the time. Mintzberg defines a manager as a 'person in charge of an organization or one of its subunits'. From his own observations and from studies by others in the US and the UK, Mintzberg concluded that managers spend their time in ways that can be grouped into three separate roles: interpersonal, informational and decisional. Elements of each role are shown in Figure 1.2.

While some of the old dictatorial terms from the dictionary definition are still there in Mintzberg's analysis – 'figurehead', 'monitor', 'handler', 'allocator' – there are some important softer additions – 'disseminator', 'negotiator' *and, most important, 'leader' and 'entrepreneur'.*

Within the role of leader, Mintzberg notes, 'Every manager must motivate and encourage his employees, somehow reconciling their individual needs with the goals of the organization.' This is more like the style that most organizations aim for today. It recognizes that employees are individuals, with needs that may sometimes conflict with corporate goals, and that corporations do have goals.

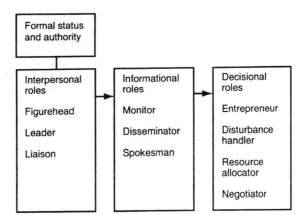

Figure 1.2   Mintzberg's three roles of management

Setting corporate goals, and encouraging and enabling all employees to share them and work towards their achievement, is one of top management's major tasks.

Mintzberg also states, 'As entrepreneur, the manager seeks to improve his unit, and adapt it to changing conditions in the environment ... is constantly on the lookout for new ideas ... as the voluntary initiator of change.' Many organizations are still striving to realize this image of the manager at all levels, and to create a working environment where constant improvement and new ideas are encouraged by involvement and empowerment of all employees, not just managers, to the limits of their abilities.

So, today's manager is enabler, coach and counsellor, as well as leader, entrepreneur, communicator, planner, coordinator, organizer and controller. The manager performs these roles within the business system, and in some cases in setting up the business system.

1.3   THE BUSINESS SYSTEM

The manager performs within an environment and manages resources to achieve some end or objective. Whatever the resources and the objective, there are some features common to the route from the idea or concept to the end result. The sequence of processes and the functions to be performed are similar whether the product is a dynamo or a doughnut, software or a

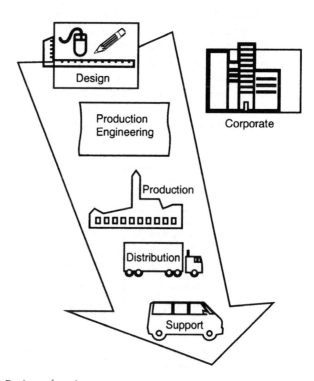

Figure 1.3   Business functions

symphony, a car or a cure. A chart identifying the major functions is shown in Figure 1.3.

What the organization does (i.e. what is to be produced, where it is to be sold, where the facilities are to be located) is largely determined in the 'corporate' box.

'Design' is concerned with *what* the product or service or system contains, what its dimensions are, what it is made from and how it performs to meet the market requirement. 'Production engineering' is concerned with developing *how* – how the components of the product or service or system are made and assembled. The production, distribution and support functions are concerned with *when* operations are performed – when material, labour or information is brought in, when production, distribution and service activities are performed. A more detailed list of the decisions and actions in each of these groups of functions is shown in Figure 1.4.

The collection of processes and functions can be regarded as a system, or a business or a business system, through which the idea is turned into a design, which is turned into a product, which is made, sold, distributed, serviced and eventually replaced and scrapped or recycled. Figure 1.5 represents such a system.

Within the system, the core functions of design, sales and production are supplemented by analysts, advisers and scorekeepers concerned with financial, legal and personnel matters. In Figure 1.3 these are contained in the remote 'corporate' box, which is sadly realistic – one of the most difficult management tasks is to close the gap between advisers and monitors in one group, and 'doers' in other parts of the organization.

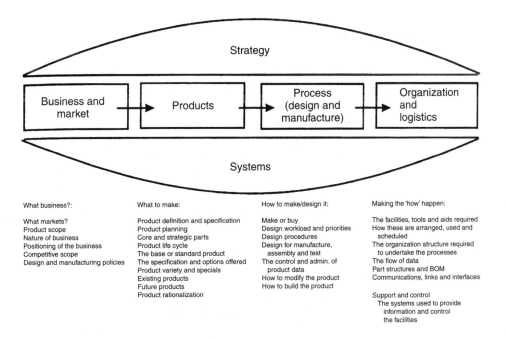

Figure 1.4   Business chain

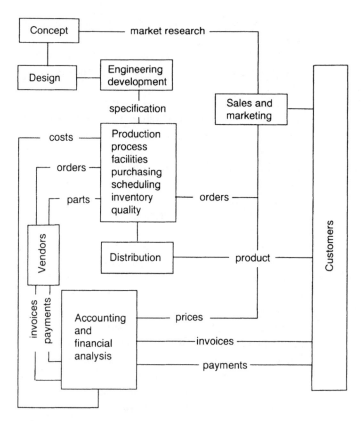

Figure 1.5   A corporate system

1.4   HOW THE SYSTEM WORKS

The entrepreneur, or research team, with a brilliant idea may find that a great deal of waste is avoided if the appeal of the idea is checked first with potential users or buyers. This may be done by market research through specialists, or by the entrepreneur, or by the organization's own sales and marketing activity. This is not an infallible process. One of the most famous market failures was the Ford Edsel, a car introduced for the US market in the 1950s, which was also one of the most expensively researched. As at election times, what the pollsters think the public say they will do is not always what they in fact do. Internal committees may be no better – the video cassette recorder was invented in the Victor company of the USA, but its management thought it had no market. The idea was only brought to market by Matsushita, through its ownership of the Japanese Victor Company (JVC). The Sony Walkman, on the other hand, was the result of logical and lateral thinking rather than third-party market research, and was enormously successful.

Market tested or not, the idea will need design and development to turn it into something fit for production. The production processes have to be developed, production personnel put in place with any necessary equipment, materials procured and the completed product delivered to the customer. The customer in most cases has to be charged, and suppliers of goods and services paid. Some products need support or service after they have been sold and taken into use, and disposal or recycling at the end of the product life have to be considered.

Many of these activities have to be performed, and paid for, before payments are received from customers, so the idea needs backing with money to bring it to market or into use. This may be the entrepreneur's own money, or it may be borrowed – in which case the lender will require a return in the form of interest – or it may be subscribed in exchange for a share in the business. Shareholders will require either interest payments or dividends – that is, a share of profits – as reward for risking their capital. In Figure 1.6, the shaded lines represent flows of information, product or funds, or actions that have to be completed before money starts to flow into the organization as payments from customers.

The first few months in the life of any business are critical, as the owners wait for the inflow of payments for sales to overtake the outflow of payments for costs. Of course these 'cash flows' have to be carefully considered at all times, not just at start-up. Cash-flow management through the 'working capital cycle', as described in Chapter 14, is vital to a company's survival. Lack of control in this activity is the most frequent cause of failure in start-up companies – the enterprise runs out of money before it is firmly established in the marketplace. Cash-flow problems

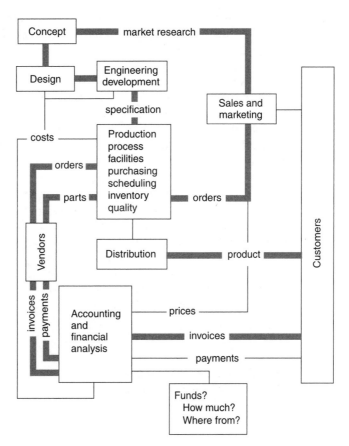

Figure 1.6   A corporate system with cash flow out (shown by shaded areas)

can also create difficulties for large, well-established concerns – see the example of ABB in Chapter 8.

For short-term success, the entrepreneur has to meet all the organization's costs from sales of the product, and have something to spare to make it all worthwhile. Longer term, the 'something to spare' has to cover continual generation of more and better ideas, the maintenance and renewal of facilities, development of personnel, and reaction to competitors and other external factors. Long-term success comes from doing all these things better than other organizations in the same or a similar business – that is, better than the competition.

The more successful the original idea turns out to be, and the more successful the company becomes, the greater the likelihood that competitors will appear. A management tool for dealing with competitive threats – Porter's Five Forces analysis – is shown in Chapter 22.

1.5   MEASURING PERFORMANCE

Managers need to measure several aspects of company performance. Where 'other people's money' is involved, those other people will want to know how 'their' company is doing.

Some measures come from the financial accounts that most companies are required by law to keep. The accounts comprise two major elements – the profit and loss (P&L) accounts and the balance sheet (described in more detail in Chapter 14).

A very simple indication of what appears in P&L and balance sheets is shown below. The P&L accounts show revenues – the value of sales – and the costs incurred in making those sales. The difference between revenue and costs is the profit (if revenue exceeds costs) or loss (if costs exceed sales revenue). The trend of profit or loss in successive accounting periods is one important performance measure.

The balance sheet shows the company's assets on one side and its liabilities on the other. The total value of the assets equals the total of the liabilities – hence the name 'balance sheet'. Growth in the balance sheet total is normally seen as 'a good thing', but much depends on why the figure has grown, and what use has been made of the investment or assets.

Two of the most common performance measures that can be derived from the accounts and give an indication of how well the assets have been used are return on capital employed (ROCE) and return on sales.

Return on capital employed is calculated as

$$\frac{\text{profit before tax}}{\text{capital employed}}$$

The normal target is 10% or more.

Return on sales is calculated as

$$\frac{\text{profit before tax}}{\text{sales revenue}}$$

The normal target is 5% or more.

Some of those terms may need explanation: profit before tax (PBT) is the difference between sales revenue and costs. The following simple P&L account shows £5m PBT.

Sales revenue (£m)		100
Costs (£m)		
Labour	10	
Material	60	
Overhead	25	
		95
Profit before tax (£m)		5

'Overhead' includes (but may not be limited to):

- Depreciation – the portion of fixed assets written off each year, recognizing that they wear out and will need to be replaced.
- Utilities – the costs of gas, electricity, fuel oil, water.
- Rates – taxes on property paid to local government.
- Central staff etc. – these costs are usually allocated to 'operations', so much per employee in each operating unit.

(See Chapter 13 for more information on overheads.)

'Assets' comprise:

Fixed assets
such as land, buildings, machinery, equipment

Current assets
cash and near cash (e.g. bank balances)
debtors (amounts owed to the company)
inventory

less
creditors (amounts the company owes its suppliers)
short-term loans (e.g. overdrafts)

Total assets are the sum of fixed assets and current assets.

It is easy to understand why companies wish to make more than 10% annual return on their assets – the money could earn between 4% and 7% invested in government bonds. This is a lot easier and, for most governments' bonds, less risky than creating and running a business.

Another performance measure normally available from the accounts is 'sales per employee'. This is a rather crude indicator of labour productivity, and not very helpful for interfirm comparisons. It can be useful to indicate productivity changes over successive accounting periods, but, as already stated, labour is a reducing element of cost, and there are other more powerful indicators of a company's overall performance.

For manufacturing, one such indicator is 'inventory turnover', which is the ratio of total cost of materials during a year to average stock or inventory. This measures the frequency of inventory 'turns' in a year. In the very simple P&L account above, material costs were £60m for the year. If average inventory was £6m, it would have 'turned over' 10 times a year. This is a fairly typical rate, but not a good one. Manufacturers such as the car producers Nissan Manufacturing UK and General Motors Saturn company in Tennessee claim turnover rates of 200 times a year – that is, not much more than one day's stock. Such rates can only be achieved with 'perfect' quality within the plant and from suppliers, plus balanced production facilities, well-trained labour, and careful management of incoming suppliers' material and distribution of the assembled product. *That is why inventory turnover is such a good business performance indicator.* An article in *The TQM Magazine* [3] describes how inventory turnover can not just be used to measure past management performance, but can also predict future performance of manufacturing organizations.

Another powerful measure is the percentage of sales derived from products less than two years (or one year) old. This shows how innovative the company is and can be applied to many types of business. As shown in the Preface, new product 'time to market', or innovativeness, was the top business issue of the 1990s. This measure cannot be derived from the financial accounts, but some progressive companies are including such information in their reports. It is an *output* measure, and therefore a better indicator of the effectiveness of a company's research and development (R&D) effort than the alternative *input* measure, R&D expenditure as a percentage of sales, which may be high because R&D is inefficient or misdirected.

These financial and nonfinancial performance indicators are very much measures of management performance, and should be seen as a package to be used together, rather than as separate indicators for shareholders or managers. This was neatly expressed by one of the UK's most successful managers, Gerry Robinson. His summary of the overall aim of businesses and the role of managers is:

'You are in business to manage the corporate affairs of the company in which people put their money. I have never seen a dilemma between doing what shareholders want and what is good for the company and the people in it. The two go hand in hand' [4].

1.6   SUMMARY

In this chapter a general form of business system was described and an indication given of the major elements or functions that make up the system, and the processes that turn inputs to the system into outputs. Changes in the relative importance of the land, labour and capital inputs were indicated, and the newer, significant input – information – was introduced. Some definitions of 'management' and the manager's task were given, including a reference to Mintzberg's 'three roles of management' – interpersonal, informational and decisional. The chapter concluded with outlines of some of the ways in which the performance of the business and its managers is measured by managers themselves, and by outside observers and analysts.

REFERENCES

1. Beer, S. (1959 and 1967) *Cybernetics and Management*, Unibooks English Universities Press, London.
2. Mintzberg, H. (1975) 'The manager's job: Folklore and fact', *Harvard Business Review*, July/August 1975.
3. Davidson, A. R., Chelsom, J. V., Stern, L. W. and Janes, F. R. (2001) 'A new tool for assessing the presence of total quality', *The TQM Magazine*, Volume 13, Number 1. (MCB editors voted this the 'Outstanding Article of 2001'. *TQM Magazine* can be viewed via www.emerald-library.com.)
4. Quoted in *The Times*, December 1992, during an interview following his appointment as head of Granada television. Robinson was a millionaire by the time he was 40 as a result of leading Compass, the corporate catering group, after its management buyout from Grand Metropolitan. He is not an engineer or scientist by training, but an accountant – with an approach that engineers and scientists can learn from!

2

The Business Environment

2.1 INTRODUCTION

As soon as a business is established, it creates its own contacts with the business environment through its customers, employees and suppliers. The more successful it is, the more contacts it makes – competitors appear, 'authorities' become interested, and new products and markets have to be explored. Soon, the company is involved in the 'business game', surrounded by other players on a huge pitch that covers the globe, all governed by common factors – the rules: economics, demographics, politics, technology, ecology – and all affected by and contributing to change. Surprisingly, even at the start, in its own corner of the field the new company is influenced by events and developments on the other side of the pitch – or, realistically, as far as the other side of the world. This chapter reviews the major elements or factors that make up the business environment, and considers how their effects may be grouped into opportunities or threats to influence the strategy of business organizations seeking competitive advantage.

Variations in the significance of the external factors, depending on the type of business and its location in the supply chain, are described. Common business objectives are set against this background of external factors, and the ways in which the objectives interact with each other and with the environment are shown, with special reference to the role that engineers, scientists and technologists can play in achieving each objective.

2.2 BUSINESS OBJECTIVES: THE GAME

Almost all companies and organizations have the same objectives:

- greater customer satisfaction;
- higher-quality products and services;
- lower operating costs;
- lower capital costs;
- shorter lead times – quicker to market; and, of course,
- survival.

For businesses, an implicit, superordinate objective is to make profit. This is essential for survival, and will follow from lower costs, greater customer satisfaction and the other objectives listed.

Today these objectives are being pursued in a business environment that features:

- slow economic growth;
- relentless cost competition;
- government and consumer pressures;
- changing consumer expectations;
- increasing market complexity;
- faster technological change;
- globalization of key industries.

Managers who have trained as engineers or scientists are uniquely well equipped to help companies succeed in achieving these aims in this environment, but before considering the ways in which they can contribute, two more influences on their role are introduced. Different business functions, such as design, manufacturing, distribution and service, which all involve engineers, scientists and technologists in different ways, have already been mentioned. The two further influences on their tasks to be reviewed here are:

(1) Location of the organization in the 'supply chain.'
(2) The type of organization (product or project based).

2.3   THE SUPPLY CHAIN

The requirement for a particular management skill, the importance of particular objectives and the influence of particular external factors depend somewhat on where the manager's organization is located in the supply chain (Figure 2.1). Environmental factors may be more important to extractive industries that supply basic raw materials, and to process industries, such as power generation or oil refining. Demographic factors – that is, the size and age structure of populations – may be more important to service or assembly industries with relatively high labour requirements.

The supply chain stretches from holes in the ground or seabed where metals, minerals, fuels and feedstuffs are extracted from mines, quarries and boreholes, through refining of basic materials, the processing and forming of parts and components, the stages of subassembly, final assembly, distribution, service and maintenance, and, eventually, disposal – possibly into more holes in the ground or seabed. Along the way, all manner of services and equipment are required. At every stage there are supplier/customer relationships – internally, between different functions within the organization, and externally, between the organization and its suppliers and customers. It is these internal and external relationships that occupy most of a manager's time. They may also be the source of a new employee's first business experience, through participation in preparation of an offer to supply a product or service, or through helping to review and analyse such an offer from within the receiving organization, or through 'chasing' a supplier organization to expedite delivery of information or the product or service.

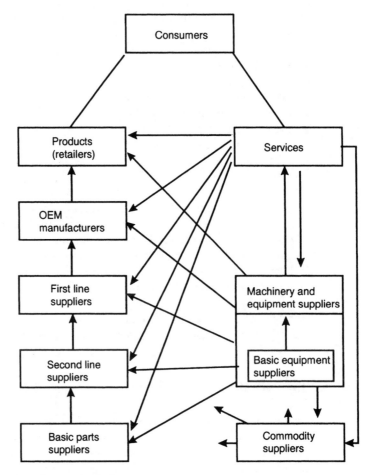

Figure 2.1   The supply chain

2.4   PRODUCT-BASED COMPANIES AND PROJECT-BASED COMPANIES

2.4.1   Product-based companies

A product-based company is one that makes a range of products such as hardware, light fittings, motor cars or television sets, which are designed to meet a general market need. A series of models and options may be necessary to cover the majority of customers' requirements and, together with a mix of technologies, this can result in a high level of complexity in designing, manufacturing and distributing the product. The products are made in relatively high volumes, which leads to a high percentage of management and engineering effort being devoted to manufacturing and distribution. This is illustrated in Figure 2.2.

The figure indicates that production schedules for product-based companies are based on sales forecasts, which is still generally the case. Some product-based companies, though, such as those in the car industry or personal computer makers, are seeking to become more responsive in their manufacturing operations

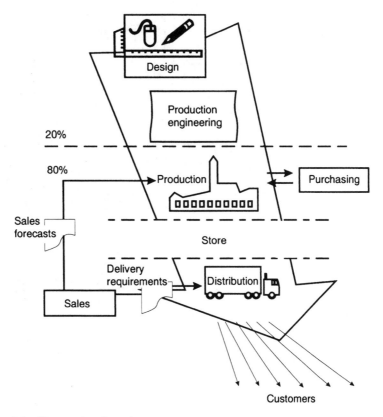

Figure 2.2 The product-based company

and to supply within a few weeks or even days, to fill specific orders. Another change affecting product-based companies is the growing emphasis on design and production engineering. This stems from the realization that manufacturing concerns to improve quality and increase responsiveness are often better resolved in these early phases of the business cycle. In turn, this means that suppliers, and hence purchasing, are also more involved in the design stages than solely with production as shown in the illustration. The figure was provided by PA Consultants in 1991, immediately after it had been used in a workshop for a leading UK engineering company. It has been reproduced as then used to make the point that many aspects of engineering and business are continually changing.

2.4.2 Project-based companies

Project-based companies design and manufacture complex items such as aircraft or locomotives in relatively low volumes, or design and construct such things as bridges, processing plant, ships or oilrigs, which may be unique, one-off projects. Some of these products, such as aircraft, have extended working lives of 20 years or more, so that aftersales support such as maintenance, servicing and upgrading are more important than for product-based companies. A greater proportion of management and engineering and technological effort is therefore absorbed by

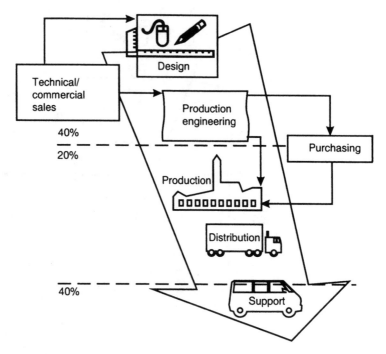

Figure 2.3   The project-based company

such support functions. Equally, more effort goes into design (which may include scientists designing materials at the molecular level) and into production or construction engineering compared with product-based companies (Figure 2.3).

Selling and buying a 'project' is a more complicated task than selling or buying a product like a domestic washing machine, and is done by teams rather than individuals. The teams contain engineers and scientists of many kinds, who have to work together with other professionals such as lawyers, accountants and buyers to secure the best contract for their company. Learning to work as a member of a team is particularly important for engineers and scientists in project-based companies (see Chapter 17), and is increasingly important for engineers in product-based companies. Project-based or product-based, the companies and teams have to work in a business environment with similar external factors.

2.5   EXTERNAL FACTORS

To determine a business strategy, or any other strategy, it is necessary to consider the environment within which the manager's unit is performing. This consideration should identify the 'OT', opportunities and threats, of 'SWOT' analysis arising from the external factors that are considered below. ('SW' stands for strengths and weaknesses, which are identified by internal review to determine how the organization is placed to handle the opportunities and threats.) SWOT analysis is useful at any level in business (see also Chapter 22, section 22.6) and in many other aspects of life – such as preparing for a sporting encounter, or even an unsporting encounter – so engineers and other 'techies' will benefit in many ways if they use the approach.

Like most management tools, SWOT is not new – a much cheaper version with a different mnemonic, 'SMEAC', was used at the very low level of the British infantry section at least 50 years ago. It may be easier to remember and use than SWOT analysis. SMEAC covers:

S	– Situation.	What is the operating environment?
M	– Mission.	What are we aiming to do?
E	– Execution.	How are we going to do it?
A	– Administration.	What do we need to get it done?
C	– Communication.	Who needs to know what? When? How?

A review of the situation or business environment, to identify the factors that are important to the organization's mission and the ways in which the mission is to be pursued, is the initial management task. This applies whether the unit being managed is a small team within an office, or a whole department, or a manufacturing plant, or an entire company.

External factors can be classified in two ways:

(1)   Groups of people, sometimes labelled stakeholders, such as shareholders, suppliers, customers, competitors, unions, the media, governments, financial institutions. (These are not separate groups!)
(2)   Abstract concepts, such as economics, politics, technology, ecology, culture. (Again, the borders overlap.)

Figure 2.4 illustrates the way in which these factors 'surround' the organization, with the stakeholders in the inner ring having more direct, short-term influence than the factors in the outer ring.

In this chapter, only the outer general or global factors will be considered. Some of the stakeholders and inner factors are covered elsewhere – suppliers in Chapter 21 and competitors in Chapter 22, for example. Chapter 11 aims to assist engineers, scientists and technologists when they face the task of communicating with (and within) the groups shown in the inner circle.

Some factors coming from the outer group that affect many managers are as follows.

Economics. Worldwide economic growth, measured by Gross National Product (GNP), was slow in the 1980s and even slower in the 1990s – less than 2% annually over the decade. Even the 'driver' economies of Japan, Germany and the USA slowed in the first few years of the 1990s. At the turn of the century, Japan and Germany were still in the economic doldrums, but the USA was growing faster. By the start of the twenty-first century, in addition to growth in the USA, impetus to global growth was coming from new quarters such as China, India and Russia. (See Table 2.2.)

Japan became used to double-digit growth rates from the 1960s and was shocked to find itself struggling with annual rates of 4% or less in the early 1990s – although 4% is the sort of increase that many western economies would be glad to reach. Germany consistently grew at 4 or 5% for many years, but struggled after reunification at the levels that are more familiar to Britain, Italy

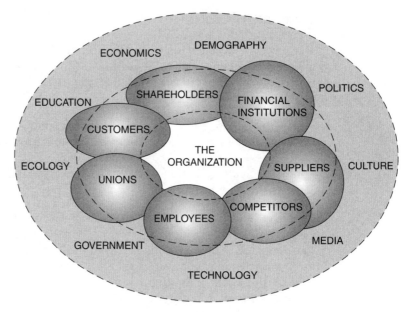

Figure 2.4   External factors influencing the organization

Table 2.1   Real GDP % change per annum in G7 countries 1991–1996

	1991	1992	1993	1994	1995	1996
USA	−0.7	2.6	3.1	3.9	2.7	3.6
Japan	4.0	1.3	0.1	1.0	1.9	3.4
France	1.7	1.9	−1.1	2.8	1.8	1.1
Germany	0.7	1.4	−1.0	2.2	1.7	0.8
Italy	1.3	0.9	−0.7	2.2	2.9	1.1
UK	−2.2	−0.5	2.0	3.5	2.9	2.6
Canada	−1.7	0.7	2.2	4.1	2.8	1.6
G7	0.4	1.7	1.3	2.8	2.4	2.7
Total OECD*	0.5	1.7	1.3	2.8	2.6	3.0

*The Organization for Economic Cooperation and Development (OECD) countries comprise over 20 of the world's more advanced economies. They are dominated by the seven largest 'free world' economies of the 'G7' countries: USA, Japan, Germany, France, UK, Italy and Canada.
Source: *OECD Economic Outlook*, Vol. 2003/1, No. 73, June 2003

and France. From 1993, for the rest of the decade, both Japan and Germany experienced growth rates between zero and 2% (see Table 2.1).

This slow growth led to surplus capacity in many industries – automotive, aerospace, shipbuilding, steel and so on. Despite the surplus, new capacity was added as part of national or company policy, and the effectiveness of existing capacity was increased by efficiency improvements. This led to intense competition, with national and company efforts enhanced or frustrated by fluctuating exchange rates. For example, the value of the Japanese yen doubled versus the US dollar in the 1980s, completely changing the economics of shipping products from Japan to the USA. Japanese companies therefore set up assembly plants in the USA for motorcycles, televisions, copiers, cars and more, and their major suppliers followed. Some of the American states with limited industrial development, such as Tennessee, came up with their own economic policies to attract this investment and became alternative sources for the US market. With further exchange rate changes in the early 1990s, and partly due to political influences, these new, efficient, low-cost, high-quality American assembly plants were in some cases used as a base for exporting to Europe, or back to Japan.

The United States (with Canada in close association) took over as the driver of the world economy in the closing years of the twentieth century, but a combination of political, cultural, demographical and technological influences began to change the shape of global economics in the twenty-first. GDP growth in the early years is shown in Table 2.2, where China and Russia have been added to the G7 countries, and some of the factors affecting recent and later years are discussed below.

Demography. Changes in the age structure of populations – which are almost inevitable, barring some form of global catastrophe – have important effects on the threats and opportunities for countries, companies, products and services – and people. Figures 2.5 and 2.6 (taken from an internal Ford publication distributed in 1985) show projections of the age structure of the populations of the USA and western Europe.

These projections were made in 1985, but this is one area where forecasts are very reliable and, as expected, the projections held good. The chart shows, for example, that the number of Americans in the 35 to 44 age group was expected to increase by 68% between the years 1980 and 2000. Over the 30-year period 1970 to 2000, in both Europe and North America, the projections showed a middle age bulge (not an anatomical feature, but an increase in the proportion of the population and in the absolute numbers, in the 35 to 55 age range). There was also a growing old age tail, as lifespans increased with better diet, better health care and better education and, in Europe particularly, as the effects of loss of lives in World War II diminished.

Table 2.2   Real GDP % change per annum in G7 countries 1997–2004 (plus two of the new growth economies, China and Russia 2001–2004)

	1997	1998	1999	2000	2001	2002	Forecast 2003	Forecast 2004
USA	4.4	4.3	4.1	3.8	0.3	2.4	2.5	4.0
Japan	1.8	−1.1	0.1	2.8	0.4	0.3	1.0	1.1
France	1.9	3.5	3.2	4.2	1.8	1.2	1.2	2.6
Germany	1.4	2.0	2.0	2.9	0.6	0.2	0.3	1.7
Italy	2.0	1.8	1.7	3.1	1.8	0.4	1.0	2.4
UK	3.4	2.9	2.4	3.1	2.1	1.8	2.1	2.6
Canada	4.2	4.1	5.4	4.5	1.5	3.4	2.7	3.4
G7	3.1	2.7	2.8	3.6	0.8	1.5	1.8	2.9
China					7.3	8.0	7.7	7.1
Russia					5.0	4.3	5.0	3.5
Total OECD	3.5	2.7	3.1	3.8	0.8	1.8	1.9	3.0

Source: OECD Economic Outlook, Vol. 2003/1, No. 73, June 2003

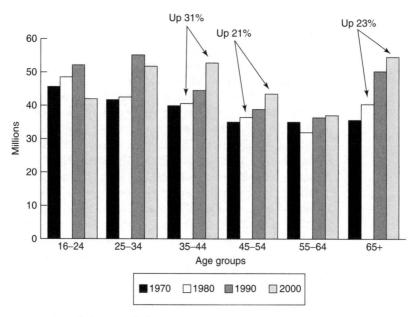

Figure 2.5   Population age shifts – Europe

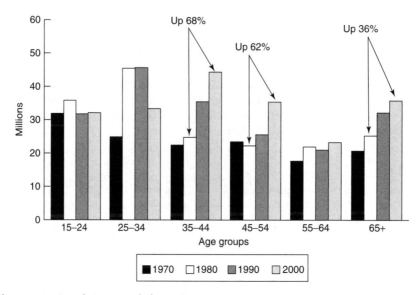

Figure 2.6   Population age shifts – USA

The way these expanded middle age and old age groups use their relatively high disposable income has a profound effect on the demand for services and consumer goods. They tend to be better informed and more demanding customers, and their priorities and values are changing compared with their counterparts 20 or even 10 years ago. This shifts the edge of the intense competition generated by economic factors into quality and product features that particularly appeal to the 'golden oldies', and to competition on convenience as well as cost. These older groups tend to be impatient, so they want the improved products and services *now*, which introduces time competition – shorter times from idea to availability, and shorter times from order to delivery. So, products become more complex and tailored to segments of the market, and production and delivery systems have to be adjusted to deal with this complexity.

While demographic trends in the closing decades of the twentieth century had similar economic effects in Europe and North America, in the opening decades of the twenty-first century they 'will transform the economic importance of different areas of the world', as described in an article by Christopher Smallwood [1]. Through a combination of differential birth and immigration rates, the population of the United States will grow from some 250 million – that is, 100 million less than Europe in 2003 – to catch up with Europe by 2040 and exceed Europe by 40 million by 2050. The working population of Europe will decline, and America's will increase. Consequently, the US share of global GDP will grow from 23% in 2000 to 26% in 2050, while Europe's will shrink from 18% to 10%. A later article by David Smith [2] shows that despite this growth, US GDP will be overtaken by China's by 2041, and possibly by India's later in the century – as a result of political and technological developments, as well as demographic factors.

These shifts in economic power have to be recognized in the business planning and management of companies engaged in engineering and technology. Managers have to predict what products and services will be required, where they can be sold most profitably, and where they can best be sourced.

Governments. Local, national and supranational governments all take on greater responsibilities in areas such as health, safety, care for the environment, consumer protection and much else. Keeping up with the resultant legislation, and complying with it, is an increasing part of a manager's job. Designers and researchers in the pharmaceutical, automotive and aircraft industries in particular need to be aware of existing and expected legal requirements affecting the content, production and use of their industries' products in all the countries where they are to be made and used. Managing the associated information is almost an industry in itself, and complying with the laws is an important driver of technological change.

Governments can be credited or blamed for the rapid economic growth of China and Russia, the sustained strength of the USA and the decline of 'Euroland' – the countries using the euro as their common currency. China's decisions to join the World Trade Organization in 2001 and to peg its currency to the US dollar have made it a low-cost manufacturer for many of the consumer products required by 'the West', enabling it to overtake Japan as America's biggest trading partner and generating a $100 bn trade surplus with the US. These decisions also helped make China an increasingly important market for equipment and engineering companies that can help modernize China's manufacturing base and infrastructure – and for makers of consumer goods for China's new wealthy elite. In contrast, the European Monetary Union policy of retaining a 'strong' euro has made it difficult for European companies to retain their share of US markets, and has drawn increased imports from the USA, Japan – and China.

Culture. This is shown as an external factor in Figure 2.4. It was culture that kept China out of the international trade community, and extreme cultural differences could be cited as a root cause of the attack on the World Trade Center in September

2001. Immediate effects of the attack were a plunge in business confidence and the US stock market, a cessation of air travel, and diversion of resources to a 'war on terror', which together knocked US GDP growth from about 4% p.a. to zero, overnight – see Table 2.2. Thus culture, though a somewhat vague concept, has significant real-world influence and has to be considered in long-term planning.

Technological change. The pace of technological change continues to increase. Information technology (IT) and electronics are two of the fastest-changing areas, and these changes themselves generate more potential for change by putting powerful tools at the disposal of designers. A complete car, aircraft or oil refinery can be designed and shown in three dimensions on screen in sufficient detail to 'sit in' or 'walk through', and components or processes can be tested by computer simulation in a fraction of the time it would take to build and test a model or a prototype. The designs and test results can be flashed across the world instantly. Materials with new properties are being developed, bringing the possibility of new products and processes – and new competition. Bringing these new materials and technologies to market ahead of competition is becoming, or has become, the most important source of competitive advantage (see Chapter 22).

Globalization. Because of the huge costs of developing new products in many industries, such as aerospace, pharmaceuticals, computers and cars, it is becoming increasingly common for design to take place in one centre only, and for manufacturing or assembly facilities to be established in several locations to minimize production and delivery costs. Readily transportable products like aeroplanes can be built wherever the necessary skills are available and flown or floated to the customer, but products like cars and consumer electronics (TV and hi-fi, for example) tend to be assembled closer to major markets. Centralized design and dispersed production of 'end products' by the OEMs (original equipment manufacturers) has implications all the way down (or is it up?) the supply chain. The technological changes in IT mentioned above make 'global' companies of high-street banks, utility companies and many others – call centres, design offices, 'back-office' operations and so on can be located in countries across the world with low labour costs.

A more complete picture of the external factors at work in the business environment is shown in Figure 2.7, from the Department of Trade and Industry's book *Manufacturing in the Late 1990s* [3]. It is worth spending time looking at the threats and opportunities, and considering how they work their way through to the strategies of companies close to home – wherever that may be.

At first it may be difficult to see the links between, say, the expanding Pacific Rim countries, exchange rate fluctuations and the affairs of a small UK supplier to one of the long-established European vehicle manufacturers. However, such a supplier has to consider the effects of the arrival of the Japanese as vehicle producers in Europe (because currency movements and political pressures preclude reliance on continued growth in the shipment of cars from Japan). Will the European vehicle manufacturer – that is, the UK supplier's present customer – lose a large part of its market share to the Japanese? Will the customer perhaps even go out of business? Will it be demanding price cuts as part of its battle to survive? Should the supplier begin to court the newly arrived producers? If it does, what will be the reaction of its current major customer? What do the Japanese expect from their suppliers? Will this mean new equipment, retraining of staff and involvement with designers in Japan? Suddenly, what is happening on the other side of the world becomes

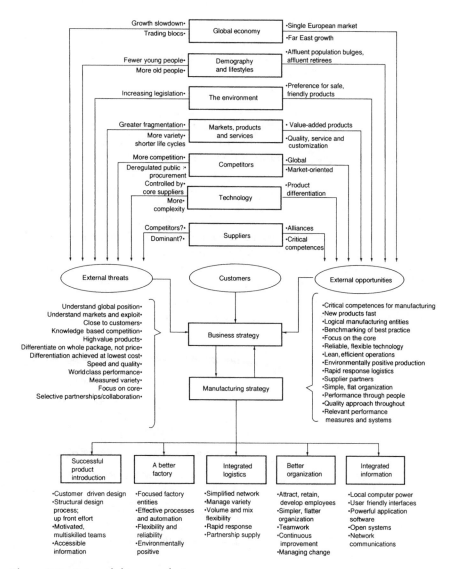

Figure 2.7   External drivers or factors

an important factor in the development of the UK supplier's business plan, and it has to review its strengths and weaknesses (the internal factors) to determine its response.

2.6   THE OPPORTUNITY FOR ENGINEERS, TECHNOLOGISTS AND SCIENTISTS

2.6.1   Natural strengths

Engineers, technologists and scientists probably have better basic strengths than other professional groups to handle the pursuit of those common business goals listed at the start of this chapter, against the background of external factors described above. This can be inferred from a look at the corporate objectives from a 'techie's' point of view. They are considered in turn below.

2.6.2   Greater customer satisfaction

Think about customers in the internal sense, as shown in Figure 2.8, from a Ford Simultaneous Engineering team presentation. The major internal customer and supplier functions are engineering and technological activities, so it should be the case that engineers and technologists in the supplier functions will readily understand the needs of engineers and technologists in the customer functions, and will be motivated to satisfy them. With product development increasingly starting in the science base (see Chapter 6, section 6.6), scientists, too, become part of the design team, and have to satisfy customer colleagues.

In the wider, external context, customers – as shown in Figures 2.5 and 2.6 – are getting older. This is unavoidable and should not depress young engineers and technologists. We all get older one year at a time, which means that from a population's present age structure it is possible to predict its future age structure with some precision. This is one case where making predictions, even about the

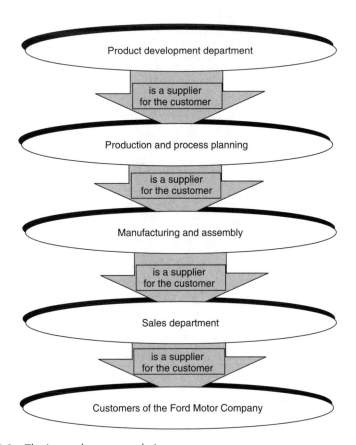

Figure 2.8   The internal customer chain

	Under $40k			Over $40k			
	1985	2000	PP change	1985	2000	PP Change	Total PP Change
Under 25	9%	6%	−3	2%	1%	−1	−4
25–34	19	11	−8	9	9	0	−8
35–44	10	9	−1	11	17	6	5
45–54	6	7	1	8	13	5	6
over 55	17	14	−3	9	13	4	1
Total	61%	47%	−14	39%	53%	14	0

19% { 35–44, 45–54 } 30%

Figure 2.9 Increasing real affluence – USA
Source: DRI population trends applied to NNCBS age/income distribution

future, is not so dangerous. By combining the expected age structure with some rather less secure forecasts about social and income patterns, it is possible to outline customer profiles some years ahead. Figure 2.9 shows such a forecast of age and income distributions in the USA.

It can be seen that potential customers were expected to become richer. This is not so certain as the prediction of their life expectation, but the general trend should be an encouragement to young engineers and technologists. The chart reflects an expected increase in real affluence. From a mid-1980s base when $40 000 p.a. was a comfortable level of earnings, it was expected that the 35 to 54 age group earning more than $40 000 (in real terms) would increase from 19% to 30% of the population by the year 2000. The increased number of middle-aged Europeans probably expected a similar comfortable state of affluence.

So, customers are getting older and wealthier. Maybe they are getting wiser, but we should not rely on it. They are also becoming better educated, more discerning and more demanding.

In both senses, internal and external, customers are changing and their requirements are changing as well. This is where engineers and others with scientific training can score, because they are used to dealing with change. Change results from potential difference, temperature difference, or a force or chemical reaction. Inertia and impedance slow down change. A control system listens, measures and adjusts so that the desired performance or result is achieved despite variations. The same analytical approach can be used to recognize the forces creating change and the obstacles to change in business, and to measure and hear 'the voice of the customer', which is the key to quality and the start of new product and process development (Figure 2.10).

2.6.3 Higher quality

Customer satisfaction and improved quality are inseparable. Compliance, performance, reliability, durability, availability and disposability are all aspects of quality that affect customer satisfaction.

Engineers and technologists help to meet all these quality aspirations through design, both of the product and the process. At the same time, they can, by using tools such as QFD, FMEA, Pokayoke and the concept of the internal customer, resolve the 'design dilemma'. (See Chapter 4 for an explanation of these tools and Chapter 6 for an illustration of the design dilemma.)

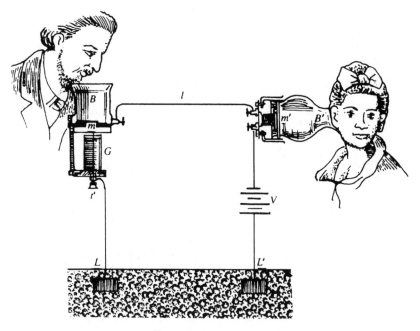

Figure 2.10   The listening engineer – hearing the customer

2.6.4   Lower operating costs

The key to lower operating costs is higher quality, and doing things right first time every time. Redoing things, sorting, dismantling, correcting, scrapping, replacing, compensating, inspecting, checking and chasing are all unnecessary wasteful activities that add 20% or more to cost. (See Chapter 4.)

Engineers and scientists can eliminate these wasteful activities, through design for quality, design for manufacture, design for assembly, design for mainte-nance – and design for customer satisfaction and delight. They are therefore key people in the drive for lower operating costs.

2.6.5   Lower capital costs

Quality is again the key to lower capital costs. Capacity utilized in making defects is lost capacity. The way this can be changed by engineers and technologists is illustrated by achievements in Ford's paint shop at Halewood, which at one time had a first-time OK rate of about 65%, meaning that 35% of painted bodies needed rework before they met the levels of quality required for the next phase of production. With problem-solving techniques, minor investment and improved labour relations this was raised to over 90% – similar to the best European and Japanese paint shops, and equivalent to almost 50% more capacity, with a similar increase in labour productivity. Production control, maintenance planning and other activities for which engineers and technologists are responsible can have similar effects – but to influence investment levels, these aspects have to be considered in the planning phases and included in the QFD analysis – another province of engineers and technologists. This is the sort of factor that cuts investment costs, rather than 'squeezes' by accountants.

Another spectacular example of savings that engineers can achieve in this field also comes from the motor industry. A 12-year-old transfer line for machining engine blocks was the bottleneck in production of the engine whose availability was restricting sales. For every extra engine produced, another car could be sold. The line ran only just over 50% of the time. One engineer, working with the shop-floor operators and using simple analytical methods, found out why (the main reason was waiting for parts) and devised ways to increase production by over 50%, and at the same time reduce manning levels by 22 employees per shift. The changes in the way the line was managed were valid for other lines – including the next transfer line of that type, which could be designed with 20% fewer stations – at a saving of more than £3m.

An example that is not from the motor industry, but from close to it, is the British Steel Llanwern works. It was reckoned that its total quality performance programme helped raise plant utilization from 80 to 90%, saving £20m a year. However, it is not clear whether this success can be claimed by engineers and technologists – in fact, like all successful quality programmes, it was the result of involvement of the whole workforce (which included many engineers, scientists and technologists!).

2.6.6   Shorter lead times

The main cause of programme delays is change – design change leading to component change, process change and equipment change. The next most important cause is the development process itself, including the sourcing of bought-out materials, parts and equipment, which generates a series of delays if performed sequentially.

Design engineers may argue that they are not the source of changes but innocent victims of the whims of the customer or senior management. This is not very persuasive – it is really an admission of poor communication and an inadequate new product development process for which they themselves are largely responsible.

By working together in teams, using the simultaneous engineering approach to new product development (see Chapter 6) and seeking to satisfy internal customers as described in section 2.6.2, scientists and engineers can eliminate – or at least reduce by significant margins, such as 80% – the changes to product and process that cause delay in new product development. They are the drivers of 'faster to market'.

2.6.7   Government, pressure groups and technological change

This review of the engineers' and scientists' natural strengths in relation to business objectives has mentioned most of the global influences that were listed, but rather understated the scientists' and engineers' roles in:

- government and consumer pressures;
- technological change.

The government and consumer pressures always translate into new or modified products or processes – the scientists' and engineers' territory. Technological change is generated largely by scientists and engineers, so handling it is also their responsibility.

Government and consumer pressures are most keenly felt in the areas of health and safety, and in environmental protection. Most developed countries have legislation relating to health and safety at work, as well as safety in the home, in all forms of transportation, in public buildings, and in connection with products such as foods and pharmaceuticals. These all have important implications for researchers, technologists and engineers. It is a small step from product safety to product quality and consumer protection. The consumer movement probably began with safety issues raised by Ralph Nader in the USA in his book *Unsafe at Any Speed* [4], which criticized the design of some of the then recently introduced cars. Now there are regulations on product liability that go beyond safety, to function and durability, and on to environmental effects in use and after use. Processes, too, may be required to comply with standards relating to noise, emissions, effluents and residues, and meeting all these standards involves scientists and engineers.

Environmental issues were highlighted by Rachel Carson in 1963, in her book *The Silent Spring* [5]. A wide range of environmental pressure groups has built up since – some private, some funded by governmental bodies like the United Nations. Recognition of public concerns as expressed by such bodies can have an influence on corporate image, and for this reason alone the environmental lobby is yet another factor for scientists, engineers and managers to consider. There can also be a direct effect on business performance: it was an environmental issue (asbestos-related disease) that almost ruined ABB and dominated top management attention for over three years – see Chapter 8.

2.7 SUMMARY

This chapter identified in general terms the business objectives that are being pursued by almost all organizations, and the major factors in the external business environment that have to be considered by managers working towards achievement of these corporate goals. The concept of the 'business chain' was introduced, followed by a classification of engineering and technology companies into product-based and project-based organizations. Together with the business functions described in Chapter 1, this should give some idea of the wide range of business roles for scientists, technologists and engineers. The chapter concluded with an indication of the outstanding opportunities for such 'techies' in all of these roles to contribute to the achievement of corporate objectives, and to respond to the threats and opportunities in the global business environment.

REFERENCES

1. Smallwood, Christopher (2003) 'People power rings changes', *The Sunday Times*, August 10. The article highlights points from Christopher Small-wood and Andrew Jones (2003) *Demographic Trends and Economic Growth: Positioning Companies for the Long Term*, Makinson Cowell. London.

2. Smith, David (2003) 'World tilts towards the east', Economic Outlook, *The Sunday Times*, October 10.
3. PA Consulting (1993) *Manufacturing in the Late 1990s*, Department of Trade and Industry, HMSO, London. Available from `www.dti.gov.uk/publications`.
4. Nader, R. (1965) *Unsafe at Any Speed*, Grossman, New York.
5. Carson, R. L. (1963) *The Silent Spring*, Hamilton, London.

3

Management Styles: From Taylorism to McKinsey's 7Ss

3.1 INTRODUCTION

This chapter describes the major changes in the ways that successful organizations were managed in the twentieth century. It shows how today's successful companies have come to practise a style of management designed to maximize the contribution of individual employees who, at the same time, work as team members rather than isolated stars. Terms such as participative management and employee involvement are introduced, which are aspects of this style of management behaviour. Attention is drawn to the coincidence between early 1980s findings by McKinsey from analysis of successful American companies and the search for explanations of the spectacular success of the Japanese in many western markets. Both found the same emphasis on teamwork, respect for the individual, shared values and goals. An earlier, transitional period is described, when the most successful companies were those with the best financial controls and marketing skills. The chapter starts with a look at management styles when attention was concentrated on labour productivity and manufacturing systems, and notes that, even then, product innovation was a key success factor. The chapter ends with a look at more recent developments, and finds encouraging signs of more systemic management using many of the basic tools recommended throughout this book.

3.2 LABOUR DAYS

The major engineering companies that grew from the Industrial Revolution were typically in heavy industry – mining, steelmaking, shipbuilding, railways, chemicals, textiles and so on – requiring huge sites, huge investments and huge labour forces. Technical skills were important for product and process development, but the dominance of labour in the cost structure drove management's attention to labour productivity or low wage rates as the key to success. Table 3.1, taken from an article by Nyquist [1], shows how direct labour cost (that is, the wages of employees directly concerned with production), as a proportion of total cost, has declined from about half at the end of the nineteenth century to one-tenth by the 1980s.

Table 3.1   Cost structure models

	1890–1920	1920–1980	1980–1990	Forecast 2000
Direct labour	0.50	0.30	0.10	0.01
Materials	0.30	0.45	0.60	0.70
Overhead	0.20	0.25	0.30	0.29
Cost of goods	1.00	1.00	1.00	1.00

It is easy to see why so much effort went into reducing labour costs in the past, and why management attention should now be focused on other aspects of cost and performance. Nyquist's forecast for the year 2000 was not fulfilled – there was little change in labour's 10% share from 1990. Many organizations reduced overhead by outsourcing services such as data processing, distribution, security and catering, and the model could be updated by grouping materials with purchased services, totalling about 70%, and reducing overhead to about 20%. Nevertheless, the forecast made an important point: in order to control costs, management should focus on purchased materials and services and on overhead.

Early efforts to cut labour costs in manufacturing concentrated on the performance of the individual. Some payment systems were devised to make employees work harder by tying their pay to the number of pieces produced, called 'piecework'. Alternatively, workers were paid an hourly rate. Under both systems, work simplification and standard rates of performance for each task were seen as ways to increase output. Adam Smith in 1776 [2] noted the benefits of specialization: the 'division of labour' in a pin factory. He observed that when operations such as 'cut to length', 'form point' and 'form head' were performed separately, operators developed specialist skills and dexterity, and production soared. More sophisticated versions of the same approach were the basis of 'work study', 'time and motion study' and 'industrial engineering', developed by Taylor, Gilbreth and others in the USA. This analysis of operator actions, speeding up of production through more efficient workplace layout, and the development of individual manual skills somehow acquired the label 'scientific management'. Although highly regarded at the time, it is now less admired and has been accorded the somewhat dismissive labels of 'Taylorism' (after Frederick Taylor its main proponent) or 'Fordism' (as a result of Henry Ford I applying Taylor's methods in his manufacturing and assembly plants in the early half of the twentieth century). Even towards the century's end, some faint effects were still discernible in Ford plants and management attitudes.

Management structures in the nineteenth and early twentieth centuries were hierarchical and militaristic, reflecting the social structures of the day. Employees were expected to obey orders, particularly in manufacturing where most of them worked. They had few rights, so there was little need for an elaborate personnel function, nor much need for other staff. Financial work was little more than bookkeeping, and sales, particularly for the railway and shipbuilding industries, was almost a branch of diplomacy or government. Engineering and manufacturing were the pillars supporting these huge companies.

Some of the most significant changes in manufacturing processes, and eventually in management, were first developed in the automotive industry around the beginning of the twentieth century. After the spread of Taylorism, another of these significant changes, with effects still being felt, was the introduction of the moving assembly line. This was an early example of the power of lateral thinking, being introduced in 1913 by Ford for the assembly of magnetos as a result of seeing carcass processing in the Chicago meat factories, where the operators had their work brought to them on the hooks of overhead conveyors.

Frederick Taylor's methods were applied to each task on the assembly line, and their effectiveness was enhanced by the improved precision of machining operations. This meant that successive parts from the machine shops were interchangeable and could be fitted into assemblies without further attention from skilled workers. The effect of the moving assembly line on productivity was so enormous that its use quickly spread through the automotive industry and outside. However, it did little for engineering management, other than to reinforce the emphasis on high volume, the deskilling of labour and the subordination of employees to the needs of the production line. Companies continued to be monolithic, hierarchical and, the dominant ones, highly integrated – that is, they processed or even extracted their own raw materials, made their own parts and subassemblies, and managed their own distribution channels. In the case of railways, where development peaked in the nineteenth century, it is possible to detect a form of integration that was reintroduced and seen as novel in the late twentieth century – design, build, operate and maintain (DBOM) contracts. (See Chapter 19.)

Where outside suppliers were involved, their role was, like that of the labour, subordinate to the production line. The accident of history and geography that concentrated the motor industry on Detroit had a profound effect on this aspect of management. The scale of assembly operations, the importance attached to volume, the priority given to keeping the assembly line going and the dominance of the Michigan peninsular by the car companies led to another set of master/servant relationships – the car company's buyer was the master, and suppliers, however big, were the servants.

Other early events in the motor industry sowed the seeds of a style of engineering management that prevailed for almost 60 years. In 1917, DuPont bought 24% of General Motors (GM) and began to introduce their financial management expertise. In 1918, GM bought the United Motors Corporation, a conglomerate of suppliers headed by Alfred P. Sloan, who became a GM director. Sloan and the DuPont financial control systems proved a formidable combination. By 1921 he was beginning to dominate the company, which was suffering 'conflict between the research organization and the producing divisions, and ... a parallel conflict between the top management of the corporation and the divisional management' [3]. Sloan's solutions to these conflicts shaped the future of General Motors. He became chief executive officer and president in 1923, and implemented plans that he had developed in the prior two years. He had already introduced a technique that is still one of the key marketing tools – market segmentation – by focusing effort on six 'grades' of cars. (See Chapter 22 for more details of segmentation and other marketing concepts.) He labelled these 'a' to 'f', by price range: $450–$600, $600–$900 and so on. (The prices have changed, but the grades are still used in the car industry – for example Micra, Fiesta and Clio are 'B Class'; Focus, Astra and Golf are 'C Class' etc.).

Sloan developed these grades as part of an overall plan to rationalize GM's range of cars, in order to reduce development complexity and increase component volume by sharing subassemblies between different end products. The rationalization enabled GM to enter the low-cost segment of the market, which until then had been dominated by Ford to such an extent that Ford took more than 50% of the total US market. Sloan's plan was to differentiate the GM product in each grade, and to charge premium prices for products that were relaunched each year with novel features. GM's ability to offer new features, such as starter motors, quick-drying coloured paint and chrome trim, and to charge premium prices for them, owed much to market-oriented scientific research and development.

Many modern engineering companies still struggle to make such effective links between their scientists, technicians and customers. Segmentation, premium pricing and annual model change were all new marketing practices when GM introduced them in the early 1920s. Combined with instalment buying (hire purchase) and the concept of the trade-in, these practices were a response to changed consumer needs. These needs were no longer being met by Ford's policy of supplying only one model at ever lower prices – a policy that had been enormously successful for 19 years, thanks to the perfect fit of the original, innovative 1908 Model T with the market requirement of that time, and to the subsequent labour-saving assembly line methods.

The change to market-led rather than supply-led product and production policies was a turning point in the evolution of modern management, but though GM's contribution to marketing management was remarkable, it was its (or Sloan's) general approach to the management of large businesses that was most significant. The key features were decentralized organization and financial controls.

'It was on the financial side that the last necessary key to decentralization with co-ordinated control was found. That key, in principle, was the concept that, if we had the means to review and judge the effectiveness of operations, we could safely leave the prosecution of those operations to the men in charge of them' [3].

The basic elements of financial control by which the operations were reviewed and judged were cost, price, volume and rate of return on investment. Control of these elements is still a good way to run a business, as long as control does not stifle initiative. It was a doctrine in General Motors that 'while policy may originate anywhere, it must be appraised and approved by committees before being administered by individuals' [3]. This doctrine recognizes the possibility of an individual originating new ideas, as well as the role of individuals in implementing policy decisions, so that, like many of Sloan's ideas from the 1920s, it is consistent with late twentieth-century views of good management practice. It is also worthy of note that GM's four control elements – cost, price, volume and rate of return – do not specifically mention labour productivity. Sloan's approach saw the beginning of the end of labour as the prime concern of management.

According to Sloan, this type of organization – coordinated in policy and decentralized in administration – 'not only has worked well for us, but also has become standard practice in a large part of American industry'. He could have said 'world industry'.

Since Sloan wrote those words in the 1960s his wonderful concept has become less effective. One reason for this is that personal success has become almost independent of corporate success.

Within companies managed on these lines, the reward systems related status and benefits to the number of levels controlled and the number of 'direct reports' (people) coordinated. This encouraged complex structures and systems – by adding levels of supervision, managers became directors and directors became vice-presidents. It also encouraged the growth of separate divisions held together only by the centrally exercised financial controls, with the exercise of those controls becoming more dominant than the divisions' efforts to run their businesses. Also, the numerous committees became very powerful, and an individual's performance as a member of, or standing before, one of these committees had implications for his/her career. Consequently, great effort went into preparing for committee meetings, and staff reviews and pre-meeting meetings added to the growth of bureaucracy.

For a long time this was not all bad: Sloan retired as CEO of the world's biggest and most profitable organization. He was, at that time, described as the world's highest-paid executive.

3.3   THE GIANT KILLERS

In the years following World War II, i.e. after 1945, many companies managed in the GM style achieved great success. Ford deliberately copied GM and became the world's second largest company. International Telephone and Telegraph (ITT) under Harold Geneen earned a reputation similar to GM's, and it was only late in the twentieth century that Geneen's appreciation of the importance of quality was put alongside his skills in financial control. American companies dominated many of the most important industrial sectors, such as automotive, aerospace, oil and petrochemicals, and computers. In almost every part of the world, and in almost every industry, overwhelming influence was exercised by financially controlled, committee-managed business giants.

Again, it was events in the automotive industry that demonstrated that another significant change in corporate management had been taking place. The oil crises of the 1970s accelerated realization that US automotive companies were not competitive with the Europeans or Japanese in terms of price, quality or economy in use. The Japanese had already captured shipbuilding, motorcycles, cameras, copiers, TV and radio, but their invasion of the US car market was a blow to American national pride. It may have been this that prompted the search by McKinsey for 'excellence' among 62 of America's largest and most successful companies, though they track it back to a seminar on innovation that they ran for Royal Dutch/Shell in Europe. McKinsey [4] found that the most successful US companies put more emphasis on what they called the *four soft Ss – style, skills, staff* and *shared values* (or goals in some versions) – than on the *three hard Ss – strategy, systems* and *structure*. In total there are seven factors, all interrelated, all with the initial letter 'S', so a diagram linking them became known as 'McKinsey's 7S framework'. This framework had been developed in the late 1970s and was used as an analytical tool in the 'search for excellence'. The originators claim that each factor has equal weight, but many practising managers believe that 'shared values' is a more significant factor than the others. A brief explanation of the four soft Ss is given below, based on the introduction to Peters and Waterman's *In Search of Excellence* [5].

Shared values – these were originally called 'superordinate goals' – represent the set of values or aspirations that underpin what a company stands for and believes in. It is only when these goals are known and shared throughout an organization that individual and sectional efforts support each other. The concept should be familiar to electrical engineers from Ewing's experiment with electromagnets. (In this experiment, dozens of small pivoted magnets are set out in rows on a table with a magnetic field passing underneath. Initially, the magnets align themselves in groups, but each group points in a different direction. As the strength of the field is increased, the groups come more and more into line, until eventually they are all pointing in the same direction. The magnets retain this alignment even when the field is later reduced or removed.) Shared values are like the superimposed field – they bring everyone's efforts to bear in the same direction.

Style is the pattern of top management action that establishes the corporate culture. The style in the most successful companies was found to be participative – that is, employees and junior managers were involved in making changes and decisions that affected them or the performance of their job.

Staff is obviously the 'people' part of organizations but, less obviously, includes their development as individuals and as more effective contributors to corporate goals.

Skills are the company's core competencies, vested in individuals and departments, and sustained by learning and training.

These four soft Ss are vital to the encouragement of personal initiative and responsibility, which, combined with teamwork and customer focus, underpin excellent quality and rapid product development.

By a completely different route, several western companies found the same key characteristics in successful Japanese companies. The different route was through study of competitors or associates in Japan. Teamwork between product development and manufacturing, and between producers and their suppliers, was identified as a significant difference between the Japanese and the visiting companies from the USA and Europe. This was the main reason for the Japanese companies' more rapid development of new products, and in part the explanation of the product's superior quality. The other part of the quality explanation was customer focus – really understanding what the customer wanted or would want in the future. The greater effectiveness and efficiency of the Japanese companies were generally ascribed to the way they involved employees and suppliers in pursuit of shared goals.

Drawing these two sets of conclusions together resulted in 'action plans' in those giant companies that had seen their dominant market positions eroded. Those plans that were based on the hard Ss were in some cases still being actioned 10 years later. Good new systems and sensible new structures were not the answer without the support and commitment of all employees, and these do not come without attention to the soft Ss. Even those companies that realized this had great difficulty undoing the influence of years of bureaucracy and complex control systems, and the persistence of low cost as an overriding objective.

In later chapters covering management of quality (Chapter 4) and product innovation (Chapter 6), the importance of employee involvement will be yet more apparent. However, involvement in learning or management is not a new idea, as the following quotation shows:

'Tell me . . . I'll forget
Show me . . . I'll remember
Involve me . . . I'll understand'
   K'ung Fu Tze (551–479 BC)

There was one marked difference between the characteristics of McKinsey's selection of excellent American companies and most of the successful Japanese companies studied in Japan. McKinsey's excellent companies concentrated on their core businesses, from which McKinsey concluded that, to be successful, companies should 'stick to the knitting' [5] and grow organically rather than by acquisition of companies in different fields of business. Acquisitions, mergers and diversification were widespread in the USA and Europe during the 1960s and 1970s. For example, ITT expanded in automotive components and plumbing fittings, Gulf and Western added film studios and machine tools to their oil interests, British American Tobacco moved into insurance and Ford and GM bought into aerospace. An example from the construction industry is the UK firm Beazer, which bought local companies such as M.P. Kent and French-Kier (itself a merger), followed by a major acquisition in the USA, and was itself bought by Hanson, a diverse holding company. In retrospect, many of these actions have been criticized as short-termism and in the 1990s some conglomerates improved performance by selling off their noncore businesses. (Beazer, the housebuilder, was a management buyout from Hanson, for example.)

In contrast, many of the successful Japanese companies were, and still are, members of huge corporations or business houses (*Keiretsu*) with subsidiaries sometimes closely linked as buyers and suppliers, but in other cases having little obvious connection. Mitsubishi, for example, makes motor cars and car components, machinery, electrical and electronic products for domestic and industrial use, has a metals and minerals division concerned with materials extraction, processing and development, and is Japan's biggest brewer. Toyota makes cars, car components and machinery, and has links that provide financial, shipping and marketing services. Sony has widely applied its skills in electronics and miniaturization, and has 'vertically integrated' into film production to secure raw material for its range of video products.

For students of management, the significance of the contrast is that what appear to be similar strategies can be criticized in one environment and praised in another. This calls for an explanation. One possible explanation is that in the unsuccessful mergers and acquisitions there were insufficient transferable management skills, and a mismatch between the management and technological expertise required to run the parent and the purchased companies. (Poor research, which failed to identify the skills gap, may have contributed to some of the unsuccessful acquisitions. The experience of ITT supports this: it has continued to prosper through many years of acquisition and divestment, *because the company has sound management systems*.) Another possible explanation is that there were insufficient financial resources, in that some of the American mergers were funded on the basis of projected earnings from the acquired or diversified company that did not materialize.

An explanation that was offered in the 1970s is that Japanese successes, and their interfirm cooperation, are based on their domestic environment and 'culture'. This has been disproved by many examples of overseas investment by Japanese companies, such as the 'transplant' car, TV and electronics factories in the USA and the UK, using local labour and, to an increasing extent, local suppliers. These counter-examples support the first possible explanation given above and reinforce the McKinsey findings: an effective balance of the seven Ss will lead to successful management in many different industries in many different environments. Robert Lutz, when he was chief operating officer of the Chrysler Corporation, described the importance of this balance as the 'overriding lesson from our experience' – see Chapter 8, section 8.4.

3.4   WHAT'S NEW, COPYCAT?

A search for a significant, lasting change of management style in the past decade reveals no significant lasting change.

There have been attempts to sell concepts such as total quality management (TQM) or business process reengineering (BPR) as *the* key to a step change in business performance and enduring success, but the initial enthusiasms passed. Both can make an important contribution, but neither has provided a significant change of management style. Reavill [6] concluded that 'TQM is alive and well and living as a fully established and valued technique'. Chelsom *et al.* [7] identified weaknesses in BPR and other transformation models and presented some alternatives.

Hamel [8] claims that, to be successful in the twenty-first century, organizations will have to constantly reinvent themselves through 'business concept innovation'. This calls for revolutionary new ways of doing things, and doing revolutionary new things. Unfortunately, one company cited as a good example of successful application of business concept innovation was Enron. In December 2001, shortly after publication of the first edition of Hamel's book, Enron collapsed into what was then the world's biggest bankruptcy. At the end of 2003 Enron's former chief financial officer was awaiting trial on more than 100 charges of fraud, insider trading, money laundering and filing false income tax returns. The company's apparent success owed more to creative accounting than to creative leadership.

This does not invalidate Hamel's proposition, nevertheless. Nor does the case of Sony, where the development of PlayStation and PlayStation 2 is also used by Hamel as an example of concept innovation. Sony's new division based on PlayStation grew from nothing to the source of more than 40% of the group's operating profit in the space of the five years to 1999. But total operating profits at Sony almost halved by 2002, and fell again in the first half of

2003. In October 2003 the group announced a $3 bn restructuring programme and 20 000 job cuts (*The Times*, October 29), while its rival Matsushita was recording sales and profit growth. Matsushita's success was based on 'brisk sales' of DVDs and flat-screen televisions, which Sony had been slower to develop. This makes Hamel's point that corporations have to *repeatedly* reinvent themselves.

Hamel also argues that 'continuous improvement is an industrial-age concept, and while it is better than no improvement at all, it is insufficient in the age of revolution', and that 'the foundation for radical innovation must be a company's core competencies (what it knows) and its strategic assets (what it owns)'. A corporate capability for innovation should be developed in the same way as a capability for quality. The way in which this is to be done has much in common with the way that successful quality programmes were developed: harness the innovative capacity of *all* employees; use multifunctional teams; use the Plan–Do–Study–Act cycle (in this case called Imagine–Design–Experiment–Assess–Scale, giving the mnemonic IDEAS). So, even implementation of Hamel's new ideas needs a touch of 'back to basics'.

It seems that the success of leading organizations comes not from a novel style of management, but from consistent application of well-established tools, techniques, policies and practices *to do something different*. Far from being a disappointment, this should be reassuring. Engineers, scientists and technologists planning to become managers in successful companies do not have to learn the secrets of some new management style, they just have to acquire some basic management skills. Some examples of application of 'the basics', with links to chapters where they can be studied, are shown below.

From 'BMW revs up for a sales drive in Asia' (*Sunday Times*, October 19, 2003):

- (Chairman) Helmut Panke 'hopes that new markets in Asia will account for a growing proportion of sales of new models and the revamped 5- and 7-series'. (Basic skill: portfolio analysis, Chapter 22.)
- BMW has opened a factory in China, already the company's third-largest market for the 7-series: 'Manufacturing follows the markets because we then have lower logistics costs'. (Basic skill: management of the supply system, Chapter 21.)

(N.B. Herr Panke is a former nuclear scientist.)

From 'Poor management controls at C&W "caused downfall"' (*The Times*, April 3, 2003):

- Richard Lapthorne (C&W's new chairman) said that when he arrived in January 'there was no system of financial control that a manager at a company like Mars, Ford, Unilever or IBM could understand. Poor management control was the more likely cause of the company not delivering its strategy rather than the strategy itself'. (Basic skills: measuring financial performance, Chapter 14; project management, Chapter 17.)

(N.B. C&W appointed Francesco Caio as chief executive 'to get down and do what needs to be done'. Signor Caio has an Electronic Engineering degree from Milan and an MBA from INSEAD.)

From the statement of Ford Motor Company Business Principles, 2002:

- **Products and customers**
  We will offer excellent products
  and services

  We will achieve this by:
  - Focusing on customer
    satisfaction and loyalty, and
    keeping our promises
  - Using our understanding of the
    market *to anticipate customer needs*
  - Delivering innovative products
    and services that offer high value
    in terms of function, price,
    quality, safety and environmental
    performance

(Basic skills: customer focus, management of quality, total customer care, Figure 4.12, Chapter 4.)

From 'Boffins who can communicate' ('The Message', David Yelland, *The Times*, November 7, 2003):

- On Wednesday Intel informed *The Wall Street Journal* that a technical paper produced by its R&D gurus had 'removed the biggest roadblock to increasing the performance of computer chips over the next decade'.[1] In reality, I should imagine the techies at Intel and Texas Instruments – and possibly IBM – are about on a par. But Intel has an extra competitive advantage – it knows how to communicate. (Yelland's point: capitalizing on brilliant science needs brilliant publicity. Basic skill: communication, Chapter 11.)

3.5 SUMMARY

Management styles changed from dirigiste, through control-centric to inclusive or participative during the twentieth century, but there have been no significant new developments in the past ten years. Creativity and innovation, to take advantage of new technologies, have become more important, but their achievement is based on well-established tools and techniques. Would-be managers can learn these from other chapters in this book.

REFERENCES

1. Nyquist, R. S. (1990) 'Paradigm shifts in manufacturing management', *Journal of Applied Manufacturing Systems*, 3 (2) Winter, University of St Thomas, St Paul, MN.
2. Smith, A. (1776) *The Wealth of Nations*.
3. Sloan, A. P. Jr (1967) *My Years With General Motors*, Pan Books, London.

---

[1] The 'roadblock' was that no way had been found to prevent the small leakages of electricity from silicon chips which were holding back their performance. Intel's chip's capacity to carry transistors had reached a ceiling of 55 m.

4. Waterman, R. H., Peters, T. J. and Phillips, J. R. (1980) (all McKinsey employees at the time) 'Structure is not organization', *Business Horizons*, June.

5. Peters, T. and Waterman, R. (1982) *In Search of Excellence*, McGraw-Hill, New York.

6. Reavill, L. R. P. (1999) 'What is the future direction of TQM development?' *The TQM Magazine*, Volume 11, Number 5. The journal archive can be viewed online at www.emerald-library.com.

7. Chelsom, J. V., Reavill, L. R. P. and Wilton, J. T. (1998) 'Left right, left right, wrong!' *The TQM Magazine*, Volume 10, Number 2. The journal archive can be viewed online at www.emerald-library.com.

8. Hamel, G. (2002) *Leading the Revolution*, Plume/Penguin, London.

BIBLIOGRAPHY

'Managing people: Lessons from the excellent companies', an article by Julien R. Phillips in the *McKinsey Quarterly*, Autumn 1982, gives several examples of the '7Ss of management' in practice.

'Leadership, management and the seven keys', an article by Craig M. Watson in the *McKinsey Quarterly*, Autumn 1983, describes the differences between leadership and management, using the 7Ss.

Leading the Revolution, by Gary Hamel, published in paperback by Plume/Penguin, is an easy read, rather like something from the 'For Dummies' series. It gives advice about how individuals with an idea can 'make a difference'. This can be by doing something revolutionary from within an existing organization or by starting something completely new.

'Getting equipped for the 21st Century', an article by John V. Chelsom in *Logistics Information Management*, Volume 11, Numbers 2 and 3, 1998, has a message similar to *Leading the Revolution*, delivered in a more prosaic style and fewer pages. Like Hamel, Chelsom chose as an example a company that was successful at the time, but is no longer – in this case FIAT Auto. The message is still valid. *Logistics Information Management* can be accessed at www.mcb.co.uk/lim.htm.

4

Management of Quality

4.1 INTRODUCTION

The concept of 'quality' has changed enormously in the past 50 years, from one that meant compliance with a specification to versions that cover the whole design, production, sales, distribution, use and disposal cycle. Management of quality has similarly changed – in an active way to spread the new ideas, and in a reactive way because of the demands of the new concepts and the use of quality as a competitive weapon. Immediately after World War II, the term 'quality' usually meant compliance with a specification, and often allowed a 'small' percentage of noncompliance. Responsibility for compliance was assigned to specialists who checked quality by inspection. This was costly and, worse, it was ineffective. Major changes began (mainly in Japan) with the work of people like W. Edwards Deming and Joseph Juran (both Americans), who spread the ideas of defect prevention through control of production and business processes, and elimination of causes of major variation to bring never-ending improvement. Some causes were found to be in the design of the process or product itself, so preventive actions moved back in the product cycle to the design phase. By the 1980s, in most industries and markets, quality had become whatever the customer perceived it to be, so customer needs were considered at the design phase and a range of new 'tools' was developed to take account of these needs. Internal customers were included in these reviews and their quality requirements were recognized, creating 'total quality' throughout the organization and beyond – forward into sales and service activities, and backward through the supply chain.

Now, 'Total Quality is about total business performance . . . It is about turning companies around, and achieving step change improvement' [1].

This chapter traces the changes in quality concepts and quality management, and introduces some of the tools and techniques for improvement.

4.2 QUALITY BY INSPECTION

Immediately after World War II and into the 1950s, there was such demand for consumer goods in the western economies that almost anything could be sold. In Britain, for example, there were two-year waiting lists for cars, and one year after delivery the car could be sold second-hand for more than its new price. As a result, production quantity was more important than quality in the eyes of most managers.

Control of quality was by mass inspection and inspectors were under pressure to pass a product as 'good enough' to sell or use. End customers, who are the final inspectors, were often so grateful to get the product that they put up with less than perfect quality.

Mass inspection was extremely inefficient. Every item in every batch made or received was checked against its specification, at several stages of production (see Figure 4.1). In consumer goods manufacture, such as vehicles or refrigerators, this meant that inspection departments – often known as 'receiving inspection' at goods inward, where parts from suppliers were received – had to keep up-to-date records of the specifications of hundreds, or thousands, of components and the procedures for testing them. These procedures required a wide range of gauges and test equipment, and a correspondingly wide range of skills to use them. Not surprisingly, inspectors were an elite, highly paid group, and there were lots of them. It was not unusual for companies, even Japanese companies, to boast that 'One in ten of our employees is an inspector – this is your guarantee of quality'. Of course, it was not a guarantee at all – defects still got through. In addition to the costs of the army of inspectors and their extensive records and expensive equipment, there were huge costs in inventory. Material and components could not be used until they were inspected, and assemblies could not be shipped until they had been checked. The stages of production where inspection took place, and where delays could occur, are shown in Figure 4.1.

The figure shows inspection taking place at suppliers prior to despatch to the assembler, and being repeated on receipt by the assembly company. Later, the assembled product was inspected before despatch to the customer, who completed a more rigorous inspection by putting the product to use.

These practices took a long time to die out. For example, in the early 1970s, managers from Ransome, Sims and Jeffries (RSJ), a leading supplier of agricultural equipment, visited Ford to exchange ideas about the use of computers in materials management. Probing for explanations of RSJ's low inventory turnover rates

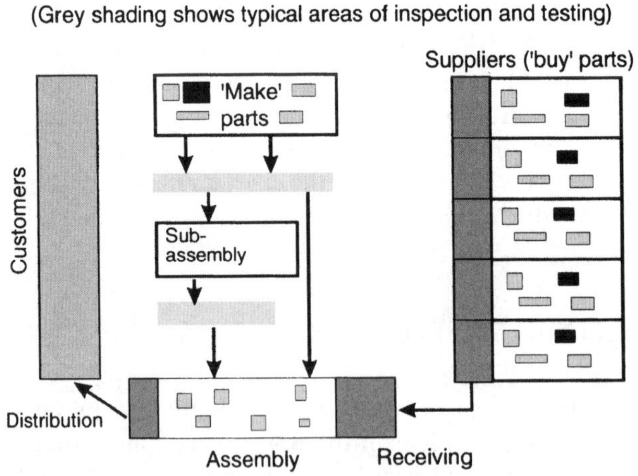

Figure 4.1   Quality management by inspection

revealed that the average time material was held in receiving inspection was one month! Even then, it was known that Japanese producers had turnover rates above 12 times a year; that is, total inventory of less than one month. In another example from the late 1970s, British Leyland (BL), to reassure customers about the quality of a new car, announced proudly that 20 000 square feet of factory space had been allocated for final inspection and rectification. Rover (BL's successor), having learnt from its partner and competitor, Honda, to reach the forefront of total quality management, would not tolerate such wasteful use of facilities – it has eliminated the need for it.

The costs of space and inventory carried by firms like RSJ and BL were not the end. What happened when a fault was found? If it was found by the inspectors in receiving, it meant that the whole batch had to be sorted, or sent back to the supplier for sorting, rework and replacement, which generated a whole series of other costs for accounting, record adjustments, shipping and handling. A defect found during or after assembly would in addition require an interruption of the assembly process while the defective part was extracted and replaced. If a defect was not found until the product was in use, there were further costs under warranty (guarantee), plus, possibly, other actions to pacify the customer, and the immeasurable cost of lost sales as the word spread that the product was faulty.

The way in which costs pile up, according to when a fault is found, is shown in Figure 4.2. The number and size of the 'x' gives an indication of the relative cost of rectifying the defect. While the value of each 'x' is unknown to most producers, it is generally reckoned that in an organization without a good quality management system, the total cost of quality defects is in the range of 15 to 30% of turnover.

The pressure of all these costs led to some helpful changes in quality control, but the system still relied on end-product inspection. One change was to move the inspection process back down the supply chain. If suppliers made more thorough checks prior to shipment, fewer defects reached receiving and some of the costs there were avoided, along with the costs of returning and replacing batches containing defective parts. Another change was the introduction of sample inspection. If checking a 10% sample revealed no faults, it might be assumed (sometimes wrongly) that the whole batch was OK. If one or two faults were found, another 10% sample would be taken. If this was OK, the batch might be accepted. If not, the batch would be returned – or if there was an urgent need

Action required	Where the fault was found					
	As a part	In sub assy	In assy	In transit	At sale	In use
Inspect	X	X	X	X	X	X
Sort	X	X	X	X	X	X
Return	X	X	X	X	X	X
Disassemble		X	XXX	XXX	XXX	XXX
Rework	X	XX	XXX	XXX	XXX	XXX
Scrap	X	XX	XXX	XXX	XXX	XXX
Reschedule	X	XX	XXX	XXX	XXX	XXXX
Replace	X	XX	XXX	XXX	XXXX	XXXX
Compensate					XXXX	XXXX

Figure 4.2   The escalating cost of nonquality

for the part, the batch would be 100% checked and the faulty parts returned. Practices such as these throughout manufacturing might still result in defect rates around 5% in the end product, and this may have been accepted indefinitely had it not been for the quality revolution in the Far East – led by Americans. Engineers and technologists faced with an inspection-based quality system should do all they can to spark a similar revolution, and move to defect prevention through process control, rather than partial defect detection by inspection.

4.3   QUALITY BY PROCESS CONTROL

Deming and Juran are generally recognized as the leaders of the quality revolution. It is not surprising that their philosophies have much in common, since they both worked for Western Electric Company in the late 1920s and were influenced by that company's approach to people management, as well as its quality management. The common elements of their views of quality management are:

- At least 80% of defects are caused by 'the system', which is management's responsibility, not by operators.
- Quality is defined by the customer.
- Quality improvement can be continuous and is never-ending.
- Quality improvement involves everyone in the organization.

A third American, Philip Crosby, shared many of Deming and Juran's views, but his approach was based on the cost of quality – or nonquality, since he maintained that 'Quality is free' [2].

During World War II, Deming applied his expertise to the American manufacturing base to increase output through defect reduction. After the war he offered his services to individual companies, but they were so busy concentrating on volume to meet pent-up demand that he was rejected, and became a government employee as a statistician with the US Census. It was in this capacity that he was sent to Japan in 1950. His presence became known to the Japanese, and he was invited to present his ideas about quality management to a group of top industrialists. He was soon in demand to address audiences of thousands, which launched the *Quality Revolution*. His role in transforming the Japanese car industry was so significant that, much later, he became sought after by US car companies to help combat the surge of Japanese imports. Ford, which had earlier rejected him, retained Deming as a consultant in 1980 at a daily rate higher than his monthly pay had he been hired in 1950.

Deming's work stressed most strongly the reduction of variance in processes as the key to continuously improving quality, using what he called 'simple statistical tools'. The most important tool is statistical process control (SPC), based on the fact that, once 'special' causes of variation are removed, variation will follow the normal, or Gaussian, distribution. Figure 4.3 shows some distributions before and after removal of special causes.

In metal cutting, a skew distribution might be due to faulty tool setting; a bi-modal distribution might be caused by drawing raw material from two sources; and the normal distribution would be achieved when all such special causes of variation have been removed.

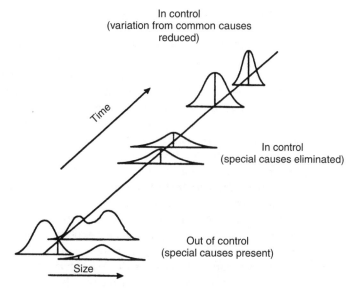

Figure 4.3   The results of improving process control

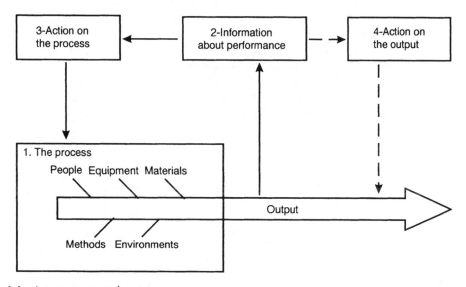

Figure 4.4   A process control system

The search for special causes, or root causes of faulty output, can be made in a systematic way using cause-and-effect analysis, illustrated in Figure 4.4.

Measurement of output data may indicate a defect, and the need to take action to sort and correct faulty product, and to take action on the process to prevent further defects. But what action? On which input? Headings under which the search should be made are shown in the box labelled 'the process' in Figure 4.4. For example, is the material to specification? Is the operator fully trained? Is the equipment working properly? The answer 'No' leads to further study, and to 'twigs' on the input 'branches'. As the questions and answers are entered on the chart, it takes a form that many people prefer to call a 'fishbone diagram'.

However, it is not necessary to wait for a defect to identify the need for action on the process. For each key parameter of the process it is possible to measure the range, R, within which acceptable variations are likely to fall, and to establish upper and lower control limits (UCL and LCL) either side of the mean (x-bar). By continuously sampling the key parameters and plotting the mean value of each sample on a chart (called an x-bar/R chart), trends can be detected, indicating that a defect is likely to occur if the process is not adjusted. Figure 4.5 shows an on-screen control chart for a process at a British Steel plant.

The point outside the UCL on the chart would normally mean that corrective action was necessary, but in this particular case it was the result of changing the target value of the parameter for a different customer. The four values for the moving mean indicate that four readings of the variable are made in each sample. (This may have been a BS productivity improvement measure, as the general practice is to sample five values.) The screen messages to the process controller in the box headed 'out of control' flash when the graph crosses a limit. The operator then has to touch the PF10 key to acknowledge that the condition has been noted. Corrective action would then be taken according to a set of instructions, and completion of this action would be confirmed by touching the PF9 key.

At its integrated works at Llanwern and Port Talbot, British Steel used SPC monitors at every operation from unloading materials at the dockside, through the sinter plant to the blast furnaces, the continuous casting process, the rolling mills, coiling, annealing, packing and despatch. Control charts were displayed at consoles in the plants, and each of them could be viewed in the central SPC room. The displays in the plants showed corrective action messages if adverse trends were detected, and the x-bar plot changed to red if it strayed outside the control

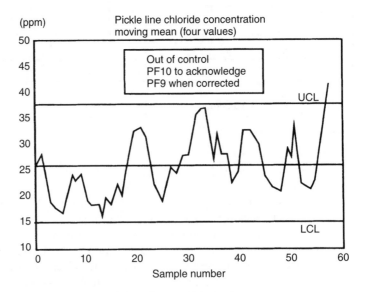

Figure 4.5   A process control chart from British Steel

limit. This degree of automation cannot always be justified, and there is much to be said for manual completion of the control charts as a way of engaging operators more deeply in quality management. (This is what British Steel did at the outset, as part of the training programme.) British Steel's use of SPC helped to make it fully competitive on quality and one of the world's lowest-cost producers. Despite its dedication to quality, British Steel had a sign at the entrance to its Llanwern Works that read 'Cost Quality Delivery'. Ford tried for years to have it put quality first on the sign, as it did in its operations. The sign had acquired the status of a war memorial, and maybe served to remind everyone entering Llanwern just how far BS had gone down the quality road. A paper by Dr Bernard Hewitt entitled 'Total Quality at Port Talbot Works' gives an account of the introduction of SPC [3].

The importance of involving all employees in SPC has become widely recognized. One of the best examples in the UK is at Nissan Manufacturing's unit shop (the engine plant) at Washington, near Sunderland. A display, about four feet high, spells out the merits of SPC, as well as the steps to implement it. This is reproduced in Figure 4.6.

Process control can be taken beyond the x-bar/R chart to consider the capability of the process. A process that has been well designed and implemented will result in only small variations in results, and the average (mean) result will be close to the centre of the tolerance range. *Capability* considers both the spread of results and their relation to the specification range. The usual measure of spread is six standard deviations (sigmas). This covers 99.73% of the normal distribution; that is, less than three items in a thousand can be expected to fall outside this range. Dividing this measure of variation by the tolerance gives an indication of the process potential, labelled Cp. The formula is:

$$Cp = \frac{\text{specification width}}{\text{process width (six sigma)}}$$

By relating the position of the distribution of parameter values to the centre of the tolerance range, a capability index Cpk is calculated. This is done by comparing half the spread (from the mean to the upper or lower three sigma value) to the distance to the nearer of the UCL or LCL. The formula is:

$$Cpk = \frac{\text{distance from the mean to nearest specification limit}}{\text{half the process distribution width (i.e. three sigma)}}$$

Illustrations of process potential and process capability are shown in Figure 4.7.

In the first chart, there is a fairly tight distribution of the measured parameter, with the six sigma spread equal to half the width of the specification range. The capability potential Cp is therefore 2.0. However, the distribution is centred towards the upper end of the specification range, and the process would generate an unacceptable level of defects.

In the second chart, the distribution is shown relative to the target (the middle of the specification range). In this case, the distance of the distribution mean from the nearest specification limit (in this instance the upper limit) is equal to half the process width (i.e. three sigmas), so the process capability index Cpk is 1.0. This is not an acceptable value (more than 5% of production is likely to be outside specification) and corrective action would be taken, such as adjusting the position of a fixture, to centre the process spread closer to the target.

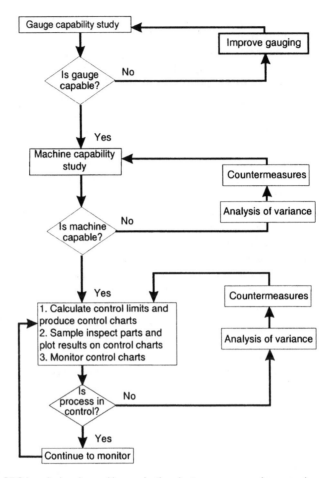

Figure 4.6   Implementation of SPC in Nissan's Sunderland unit shop

The third chart shows the process distribution mean on target. The distance to the nearest specification limit is now double the three sigma value, so the Cpk is 2.0. With a distance of six sigma to the specification limits, this process would generate about three defects in a million. Pursuit of this level of control gave rise to Six Sigma quality management systems, described in section 4.5 below.

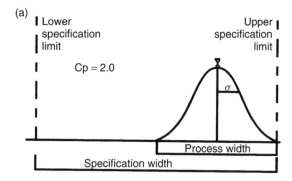

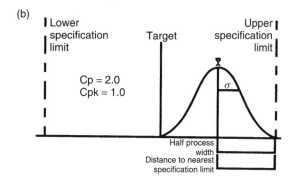

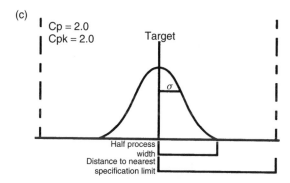

Figure 4.7   Improving process capability

The value of Cpk can be increased by continuously reducing the variations in the process, and by moving the centre of the spread closer to the centre of the tolerance. Cpk can also be increased by widening the tolerance, which is sound practice if the greater tolerance is the result of robust design (see Chapter 6). Increasing the tolerance just to get a bigger Cpk is pointless, but is not unknown when plants are judged by the percentage of processes with Cpks greater than 2.0. Many manufacturing processes aim for a Cpk value greater than 2.0 – which means that only a few parts per million (ppm) will be outside the tolerance. So, by controlling the process, defects can be prevented and inspection of the output becomes unnecessary.

With in-process gauging and computer control, Cpk, like x-bar/R, can be calculated and displayed automatically, but again it is preferable to start with manual calculation to aid understanding and to secure the involvement of the operators.

Whether performed manually or automatically, defect prevention by process control is a far more effective way of assuring the quality of output. Costs of appraisal and prevention will be higher than with inspection-based attempts at quality control, but total costs will be reduced dramatically. Figure 4.8 shows the sort of relationship that generally applies, with total cost reductions in the range 10–15%. The initial extra attention to appraisal and prevention results in an increase in internal failure costs, as more faults are found in-house. These higher internal failure costs are more than offset by lower external failure costs, as less faulty product reaches customers. As process control becomes more effective, defects are prevented and internal failure costs also fall dramatically. Such cost reductions go straight to the bottom line – that is, the profit of the organization – or can be used to reduce prices to gain a competitive advantage without reducing profit margins.

Cause-and-effect analysis can be used for process improvement as well as problem solving, until a point where the process, the equipment or the design has to be changed to make the next advance in quality performance. Such changes are extremely costly once a product or service is in use. The later in the product cycle that a change is made, the higher the cost will be.

For example, the design of an aircraft's doors was changed for safety reasons some years after the plane entered commercial use. The cost of the design change included lost revenue, as all the planes of that design were grounded until modifications were made, and future sales of planes and flights may have suffered because of damage to the plane builder's and airlines' quality reputations.

To avoid such changes late in the product cycle, the concept of defect prevention therefore has to be taken further back – to the design phase.

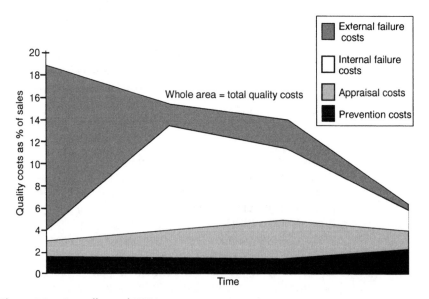

Figure 4.8   Cost effects of TQM

4.4   QUALITY BY DESIGN

The shift from quality control, using inspection to detect defects, to quality assurance by defect prevention was a gradual one. So was the movement to more effective defect prevention through improved design management. There were two forms of delay in the spread of both these changes. One delay was from assemblers and end producers, the so-called original equipment manufacturers (OEMs), to their suppliers; the other was from East to West. Figure 4.9, derived from a Ford engineer's trip report of his visit to Japan, illustrates how the new concepts spread through the Japanese automotive industry. Introduction and development of the ideas in the western automotive industry followed between five and ten years later, so that in some cases it was still going on in the early 1990s.

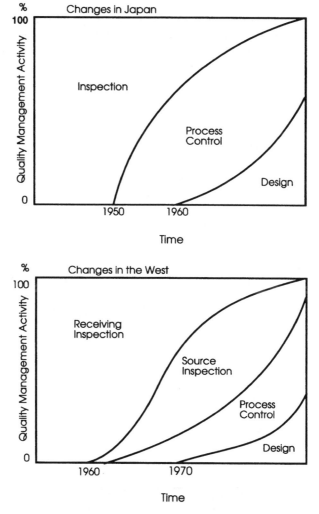

Figure 4.9   A comparison of changes in quality management between Japan and the West

The cost incentive to 'design it right' is enormous. For example, in the early 1980s, Ford of Europe introduced a new engine (known as the DOHC, because it was its first high-volume double overhead camshaft engine), which suffered so many design changes in the run-up to production that over $40 m had to be spent to modify the three major pieces of production equipment. These were obvious costs, but even greater penalties were incurred through delays in market introduction and lost sales. This is not a unique experience – the pattern is repeated in other companies and other industries. Figure 4.10 shows an assessment by British Aerospace of the way opportunities to make changes decline, and the cost of making changes increases at each stage of its product cycle [4]. This aspect of design management is covered more fully in Chapter 6, but the significance of design change and subsequent process change for quality management should be noted – each change creates a risk to quality. The later in the product cycle that the change is made, the greater the risk, since the effects of the change may not be fully evaluated under pressure to get the product to market, or to keep it there.

To minimize late changes, techniques have been developed to assess risks to quality while the product is still being designed. One of the most effective is potential failure mode and effect analysis (FMEA), which can be applied to the design itself or to production or operating processes. Examples of design FMEA are shown in Appendix 2, and a book on the subject is published by the SMMT [5]. The essential steps of FMEA are:

- Identify every conceivable failure mode.
- Consider the consequences of failure for each failure mode and assign a severity index on a scale 1 to 10. (Lethal would be 10, insignificant would be 1.)
- Assess the likelihood of occurrence, again on a scale of 1 to 10. (Almost certain would be 10, just possible 1.)

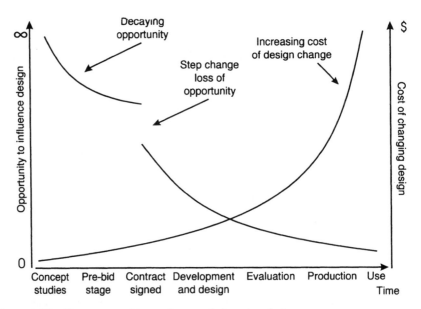

Figure 4.10 The opportunities and costs of changing design

- Assess the likelihood that the failure would be detected before the product is in use, using the same scale. (Almost sure to be spotted would be 1, very hard to detect 10.)
- Calculate a risk priority number (RPN) as the product of the three indexes.

(The RPN is sometimes shown as Severity × Occurrence × Detection. Maybe this is the origin of 'Sod's Law' – what can go wrong, will go wrong.)

The RPN can range from 1 to 1000, and is used to prioritize the designer's actions to reduce risk. If the significance of failure is rated high, say 7, 8, 9 or 10, the design or process would be changed to lower the occurrence level and to increase the likelihood of detection. Reducing occurrence (prevention) is always preferred, since detection processes (inspections) are always fallible.

Other techniques to 'design in' quality, such as design of experiments (DOE) and quality function deployment (QFD), are covered in Chapter 6.

By the time that the importance of design for quality was recognized, the definition of quality had changed from 'meeting specification' to something much wider. The three 'gurus' (Crosby, Juran and Deming) changed their own definitions from time to time, but they can be roughly summarized as:

- conformance to requirements – Crosby;
- fitness for purpose – Juran;
- predictable degree of uniformity and dependability, at low cost and suited to the market – Deming.

These differing views reflect the variety of pressures on the designer, from the external customer and from a range of internal customers. This set of sometimes conflicting demands has been labelled 'the design dilemma' and is covered in Chapter 6, but some idea of the range of demands can be obtained from Figure 4.11.

The designers have to first consider quality in use, where customer experiences that determine new product features can be fed back directly or through dealers and sales organizations. They also have to ensure that the design assists producers and distributors to develop reliable processes; simplifies service and maintenance actions; and does not give rise to concerns about disposal and recycling. By minimizing complexity, the designer can even help the sales force to do a better job – simplifying the task of finding the right product to meet a customer's needs in a brief phone call or showroom visit.

So far, this section on quality by design has dealt with defect prevention through consideration of the needs of internal and external customers. There is, however, the possibility of an even greater contribution to quality at the design stage. This is through the anticipation of customer needs, and the designing-in of features that will surprise and delight the customer. Oddly, one or two pleasant surprises can lead to a customer forgiving and overlooking some shortcomings. This was dramatically illustrated when Ford introduced the Taurus car range in the USA. As with all new car introductions, great care had been taken to identify the 'things gone wrong' (TGW) with the outgoing model, and to aim to put them right by improved design and manufacturing processes. (TGWs per 1000 vehicles during the first three months and the first 12 months in service are one of the principal ways of measuring vehicle quality. The score reflects not only how

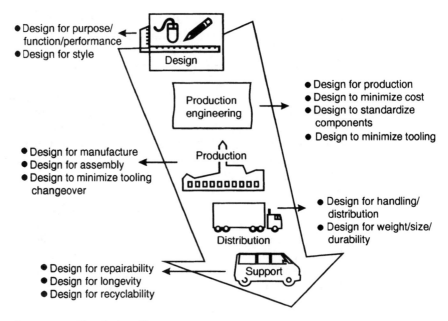

Figure 4.11 The design dilemma

well the vehicle was designed, but also how well it was put together. Producers benchmark themselves against other makers' TGW scores through syndicated market research.) Although delayed by over two years to modify design features that gave rise to manufacturing concerns, the new car disappointed the producers by scoring more TGWs than targeted in the design brief. Yet it was a huge success. The 'European' styling and other features so delighted customers that they reported very favourably on their ownership experience. The 'things gone right' greatly outweighed the 'things gone wrong', and the cooperation between designers and producers in the two years' delay became something of a legend, under the name Team Taurus. The car might have been even more successful had it been developed in the five years originally targeted, and designed 'right first time'.

4.5   QUALITY MANAGEMENT SYSTEMS

To ensure that good quality management practices are applied consistently, most organizations have a quality system. In the defence, health, aerospace and automotive industries, many developed their own systems designed to comply with extensive legislation and regulation, as well as driving to meet their own quality objectives. The existence of so many private systems can be very demanding on suppliers, who may have to deal with several different sets of requirements. There has been significant progress towards resolving this issue in the past decade.

Much of the progress is due to acceptance of the messages of W. Edwards Deming [6], particularly his advice to view production as a system, to acquire some knowledge of variation, and to use the Plan–Do–Study–Act cycle to manage the transformation of an organization. (Plan what to do; Do it; Study the effects of what is being done, using numerical measures and statistical analysis; Act when the measurements show deviations from plan.)

ISO 9000. The international standard ISO 9000 has been transformed by the introduction of ISO 9000:2000 [7], replacing ISO 9000:1994. The 1994 version was a close copy of BS 5750, a standard developed from quality assurance practices of the US and UK munitions industries during World War II (1939–45). It was largely concerned with compliance and consistency, and paid little attention to customer satisfaction or product and process improvement. This is rectified by ISO 9000:2000.

Version 2000 uses a process approach, defined as 'The application of a system of processes within an organization, together with the identification and interactions of these processes, and their management.' This approach incorporates the Shewhart/Deming 'Plan–Do–Check–Act' methodology that is briefly explained in the Introduction to ISO 9000:2000, where it is claimed that the process approach, when used within a quality management system, emphasizes the importance of:

(a) understanding and meeting requirements;
(b) the need to consider processes in terms of added value;
(c) obtaining results of process performance and effectiveness; and
(d) continual improvement of processes based on objective measurement.

This makes the standard a more effective management tool, and the usefulness of version 2000 is further enhanced by its division into two sections. One section, ISO 9001, 'specifies requirements for a quality management system that can be used for internal application by organizations, or for certification, or for contractual purposes. It focuses on the effectiveness of the quality management system in meeting customer requirements.' The other, ISO 9004, 'gives guidance on a wider range of objectives of a quality management system, particularly for the continual improvement of an organization's overall performance and efficiency, as well as its effectiveness . . . However it is not intended for certification or contractual purposes.'

Chapter 23 is a case study where ISO 9000:2000 was incorporated in a successful business start-up.

ISO 9000 with TickIT. There is an extension of ISO 9000 specifically designed to assist and qualify organizations working in information technology – as software or systems providers, and as consultants. The extension, called TickIT, mirrors ISO 9000, with special emphasis on project management. It requires that every systems project has a quality plan, with timed key events such as design verification, design validation, customer sign-off and so on. It is this version of ISO 9000:2000 that features in Chapter 23.

QS 9000. The continual improvement of ISO 9000 (Physician heal thyself!) has enabled the automotive industry to reduce the QMS compliance task facing its suppliers. In 1995, after extensive consultation with suppliers, the 'Big Three' of the US automotive industry (General Motors, Ford and Chrysler) introduced QS 9000. This replaced the companies' three overlapping QMS and used ISO 9000:1994 as a basis, supplemented by industry- and company-specific requirements. Industry-specific requirements were those that all three companies called for. Customer-specific requirements are unique to one auto company, or are shared by two of the three. QS 9000 was applied by the three founder companies worldwide, and other auto companies joined the programme.

The original QS 9000 was designed for component suppliers, and a modified version was developed for suppliers of nonproduction materials, goods and services. Ford integrated a quality operating system (QOS) into its version, which is itself an effective QMS. This required suppliers to:

- identify at least seven parameters that reflect customer (i.e. Ford) expectations;
- establish internal key processes to meet those expectations;
- identify measurables for the key processes;
- monitor and report trends of the measurables monthly;
- predict external performance.

Where a shortfall against target performance is predicted, the 8D Team Oriented Problem Solving tools described in Appendix 2 are used to avert the predicted failure.

Recognition of ISO 9000 certification as a valid demonstration of an effective QMS, and improvements to the standard itself, have helped to make it a valuable management tool rather than a bureaucratic burden. An article in *Logistics Information Management* [8] gives an example of how ISO 9000 and QS 9000 were used to improve business performance and provide a basis for total quality management (TQM). The article also describes Ford's QOS in more detail.

Six Sigma. If a process can be controlled so that outcomes within +/− six standard deviations ('sigmas') of the mean (target) value comply with specification, it can be expected to generate just over three faults in a million. Distribution (c) in Figure 4.7 illustrates such a process. Motorola achieved this level of control in its manufacturing processes in the late 1980s, and the resultant product quality enabled it to improve market share and reduce costs dramatically. Motorola publicized its accomplishments in a global advertising campaign that helped to encourage widespread adoption of Six Sigma programmes. These programmes are project based, with each project led by a Champion who is assisted by specialists trained in Six Sigma techniques. Titles from the martial arts, such as Master Black Belt, Black Belt and Green Belt, are used to recognize the specialists' level of training. Service-sector companies such as American Express have used Six Sigma, and Ford Motor Company has made it the basis of its QMS throughout all its operations – sales, product development, finance and purchasing, as well as manufacturing, and outside to its dealers and suppliers. To support its programme, Ford publishes *Consumer Driven 6-Sigma News* monthly in several languages, and provides more information online [9].

There is debate among statisticians about the validity of Six Sigma techniques [10], but there is no disputing the effectiveness of the approach. It has proved to be a great way to involve and motivate a workforce, it produces quick results and it is claimed to have saved billions of dollars in a wide range of companies. In short, mathematically sound or not, it works.

ISO 14000. This international environmental standard is not a QMS, but some organizations regard it as complementary to ISO 9000 and manage the two standards in parallel. The global power technology group ABB has made 'sustainability' a key component of business strategy, and compliance with ISO 14000 a key element of its corporate image and hence an aspect of quality – see

Chapter 8, section 8.4.2. There is a complete section of the ABB website devoted to sustainability [11].

4.6   TOTAL QUALITY MANAGEMENT

The consideration of the customers' needs from concept, through design, production, distribution, use, service and disposal, can be labelled 'total customer care'. Figure 4.12 illustrates the scope of this quality concept.

The way in which a company deals with its customers or potential customers, with its suppliers and with the business and wider community is all part of its quality performance. The involvement, motivation and training of employees in all of these areas require the participation of all managers, not just those in quality control.

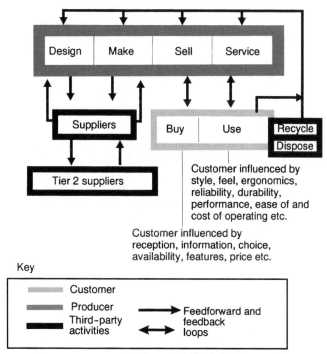

Information about the customer's experience during buying and using the product is fed back to the sales and service activities and to the manufacturing organizations, and on to suppliers. This information is used to improve the 'buy, own and use' experience on current products, and to build up a list of customer needs for new products. New products and process ideas are fed forward by suppliers to the producer, and are implemented or logged for future use. Feedback is also obtained from recycling and disposal agencies, which may also influence current and future products.

The producer takes care of customers' concerns from the time they think about buying, through purchase and delivery, use and service to disposal — and to their next purchase.

Figure 4.12   Total customer care

The way in which people behave as members of a team, or how they care for their customers – the designer for the 'customer' process engineer; the process engineer for the 'customer' production manager and so on – is also part of quality. This is the result of the treatment of staff, the provision of skills, the company's style and its shared values, as well as its systems, structure and strategy – all the 7Ss defined by McKinsey.

The actions of all the people in every function and in every customer contact contribute to quality performance, and managing them is 'total quality management' (TQM).

Some organizations have dropped the 'management' from the name of their programmes to introduce the TQM concept, because it can be interpreted to mean that the programme is management's task alone. This is just one way to stress the significance of 'total', which implies that everyone is involved and responsible for their own contribution to quality performance.

TQM does start with management, but it is a philosophy or style of behaviour that has to permeate the whole organization to be effective. When Xerox Corporation determined that it had to become a total quality company, it set itself a five-year programme, summarized in Figure 4.13. This covered every aspect of its operations, every employee and every supplier. The programme involved training

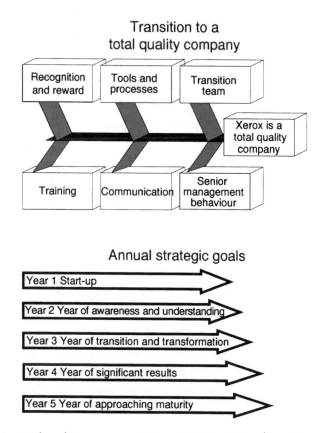

Figure 4.13   Total quality management: Xerox's attempts to achieve it

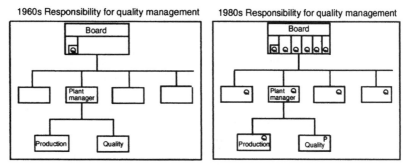

Production performance measured by
adherence to schedule.

Quality performance measured by
adherence to specification.

Figure 4.14   Quality management responsibility

each employee in the Xerox quality 'system' – both training and system are
essential elements of TQM.

Xerox set out to change every aspect of corporate behaviour, in order to
meet its business objectives. This made quality management everyone's responsibility, and helping staff to exercise that responsibility was part of every
manager's task.

Organization charts showing management's quality responsibility 50 years
ago and currently might look something like Figure 4.14. In the 1960s chart,
quality appears as part of the assignment of one board member, and the whole of
the assignment of one department in the production plants. In the plants there
was conflict between the production volume objectives of the plant manager and
the quality objectives of the quality control manager. In the later chart, everyone
shares the quality goals, and the quality manager in the plant – if there still is
one – may even have an interest in supporting the production goals.

4.7   SUMMARY

The most important points in this chapter were:

- the changes in the meaning of 'quality' from compliance with a specification
  (most of the time) to meeting or exceeding customers' changing expectations
  (on time, first time, all the time);
- the shift of responsibility for quality management from specialists to every
  employee;
- quality improvement by defect prevention instead of defect detection.

The influence of design on quality in all of the later stages of the product cycle
was stressed, and some of the tools and techniques of quality management,
such as statistical process control and potential failure mode and effect analysis,
were described.

REFERENCES

1. Kooger, J. (1993) Operations Director, DuPont Europe. From the promotion material for PERA International *Briefing Programme – Making Quality Pay.* PERA publishes an illuminating summary of case studies in TQM.
2. Crosby, P. (1979) *Quality Is Free*, McGraw Hill, New York.
3. Hewitt, B. (1990) 'Total quality at Port Talbot works', ISATA Conference Proceedings, Wiesbaden.
4. From a DTI publication *The Case for Costing Quality.* DTI publications are available through www.dti.gov.uk/publications.
5. *Failure Mode and Effect Analysis*, Society of Motor Manufacturers and Traders, Forbes House, Halkin Street, London.
6. Neave, H. R. (1995) 'W. Edwards Deming (1900–1993): The man and his message', *Proceedings of the 1ˢᵗ World Congress on TQM*, Chapman and Hall, London.
7. *ISO 9001*: 2000, published by British Standards Institution and available from BSI Customer Services, phone 020 8996 9001, fax 020 8996 7001, or online at www.bsi-global.com.
8. Chelsom, J. V. (1997) 'Performance-driven quality', *Logistics Information Management*, Volume 10, Number 6, MCB University Press. Online from www.mcb.co.uk/lim.htm.
9. Details of the Ford programme are at www.6-Sigma.ford.com.
10. An article by Ronald D. Snee, 'Six Sigma and the statistician', in *RSS News*, June 2001 and subsequent correspondence provided some of the background to this section. The publication can be viewed at The Royal Statistical Society, 12 Errol Street, London EC1Y 8LX. Dr Henry Neave shows that effective process control does not depend on the normal (i.e. Gaussian) distribution of results in 'There's nothing normal about SPC!' in *Proceedings of the 2ⁿᵈ International Conference on ISO9000 & TQM*, University of Luton, 1997.
11. There are position papers, product declarations and question and answer documents about sustainability on www.abb.com/sustainability.

5

Materials Management

5.1 INTRODUCTION

A knowledge and understanding of the characteristics, uses and business impact of materials at every stage of the product life cycle is essential to every engineering discipline. Scientists are deeply involved in the development of materials to meet functional, performance and business objectives. Engineers, often working in cooperation with scientists, are concerned with the selection of materials during the design phase, with the processing of materials into components and subassemblies, with the assembly of processed materials, with materials handling and storage, with servicing and maintenance of products, and possibly with recycling or disposal of products and their constituent materials.

Data about the availability and performance of materials and components are critical inputs at the design stage. What the designer does with this information can determine up to 80% of costs throughout the rest of the product cycle. So the designers' outputs are of interest to process engineers, construction engineers, layout engineers, safety engineers, quality engineers, materials handling engineers and service engineers – and others who may not be engineers. Availability, on time, of perfect quality parts, assemblies and materials is vital to successful production or construction, and is greatly influenced by the decisions of engineers, scientists and technologists. Materials in one form or another make up a high percentage of operating costs, and in the form of stock can be a major element on the balance sheet.

Their management therefore has a significant impact on a company's financial performance. Component quality, combined with the quality of assembly processes, determines total product quality to a large extent. Supplier performance is therefore one of the major influences on user company performance in both quality and financial terms. The sourcing process is increasingly a team activity, and engineers and scientists thus have an opportunity to share in materials management from the earliest stage. The engineers' interaction with suppliers in manufacturing plants and on building and construction sites continues this involvement, and influences the user/supplier relationship just as much as the work of the purchasing department.

This chapter indicates the scope of materials management, its influence on operating costs and the balance sheet, and hence its dual effect on profitability.

Some of the interactions of engineers with other functions also involved in materials management are described, showing how engineers and scientists have become part of the decision-making team for materials and supplier selection. The growing significance of new materials and associated technologies is covered in Chapter 6, and inventory management, purchasing and logistics are dealt with in more detail in Chapters 20 and 21.

5.2 MATERIALS SELECTION

The choice of material by the designer has an influence right through the product cycle and deep into the supply chain. This is illustrated in the next three sections, which give examples from three engineering industries.

5.2.1 An example from the electronics industry

The following paragraph is an extract from a Siemens publication [1].

'At the design engineering stage, materials are normally selected according to functional, production-specific, aesthetic and financial criteria . . . Electrical engineering and electronics owe their technical and economic success in no small measure to the use of custom-made and design-specific materials . . . . Until recently, the development and application of materials has been dictated by technical and economic considerations, with little attention being paid to the possibility of retrieving and re-utilizing waste materials . . . This situation is, however, changing rapidly in leading industrial nations as the "throw away society" makes way for an economy based on avoiding, minimizing and recycling waste, as well as environmentally compatible waste disposal.'

5.2.2 An example from the automotive industry

In the automotive industry, if a body design engineer decides that to improve corrosion resistance a door inner panel should be made from steel coated on one side with zinc/nickel (Zn/Ni), it has a series of effects:

- The body stamping engineer has to review the stamping process, die design and die life.
- The body construction engineer has to review welding processes.
- The paint engineer may have to review pre-paint and paint processes.
- The component engineer has to inform purchasing that new supply contracts will be needed, with revisions to scheduling and materials handling.
- The materials handling engineer needs to note that special scrap has to be segregated.
- Service engineers have to check whether dealer body repair processes need to be changed.

If the panel is on a vehicle produced in large quantities, say a few hundreds or even thousands a day, there may be an impact on sourcing decisions – for example, coating capacity may be limited at the present supplier. If the panel is on a product to be made in a number of different locations, say in Europe and North America, the availability of Zn/Ni coated steel might have even more serious implications.

The American industry favours Zn/Zn and discussions with several suppliers on both continents may be necessary to secure supplies.

Life-cycle costing would have to take account of all these factors, as well as the forecast trends of zinc, nickel and other commodity prices.

5.2.3 An example from the construction industry

In Germany, environment protection laws have led to a ban on window frames, doors and other building products made from uPVC, which has created surplus capacity in some parts of the plastics processing industry. Elsewhere in Europe uPVC products may still be used, and German extruders have vigorously marketed their products in these countries, where the selection of uPVC window frames, doors, fencing or fittings by designers would set up a transaction train crossing national borders and stretching from petrochemicals to landfill.

5.2.4 Advanced materials

Developments in materials sciences have importance for engineers and technologists of all disciplines. Their significance is indicated in the following extract from the introduction to a report for the *Financial Times* by Lakis Kaounides, published in 1995 [2]:

'Advanced materials have now emerged as a major new technology upon which further progress, innovation and competitive advantage in high technology and the rest of manufacturing industry increasingly depend. The materials revolution is irreversible, is gathering momentum and its reshaping of the map of world industry will accelerate from the late 1990s onwards.'

The implications for new product development are described in Chapter 6, but other implications are summarized in a note by P. Dubarle of the OECD [3]:

'Advanced materials are increasingly the result of conventional materials blending, associated with various types of mechanical and chemical treatment. Such a materials mix is more and more designed at the molecular level. It therefore reflects the growing integrated and horizontal nature of materials innovations. This new pattern of materials design and fabrication technologies calls for new skills and especially multidisciplinary knowledge among firms and laboratories, engineers and R&D staff, and even among technicians and supervisory personnel involved.'

These quotations emphasize the need for engineers, technologists and scientists to learn to work with other professionals, and to take a holistic view of the implications of materials research and development and materials selection to meet end-product needs, both physical and commercial.

5.3 ENGINEERS, TECHNOLOGISTS AND THE SUPPLY SYSTEM

In the early 1970s 'the supply concept' was thought to cover everything to do with materials, components and bought-out subassemblies 'from womb to tomb' – from release of individual parts by the design engineer to delivery of the end product to the customer. In the 1990s, many industries extended the scope of materials management to coordination of the research and development phases prior to 'conception' of parts or subassemblies, and beyond the 'tomb' of sale and use, to disposal or recycling. Nevertheless, a description of the earlier supply concept can be used to identify some of the downstream involvement of engineers and technologists following their selection of materials and development of designs and specifications.

Figure 5.1 shows the 'supply' activities within the heavily outlined box. It is these activities that have the 'womb to tomb' materials management responsibility. There are three closely linked areas. *Procurement* organizes the purchasing and issues delivery schedules to suppliers, and follows up to make sure that suppliers deliver on time. *Materials planning* sets up and maintains the records of each part used in each plant and determines target inventory levels, and hence delivery frequency. *Traffic and customs* is responsible for movement of materials at every stage – from suppliers to the production plants (sometimes called 'inbound logistics'), between plants, and from plants to the outside customer (sometimes called 'outbound logistics'). Linking these three functions together in a single organization, and so eliminating duplication of activities and conflicting goals in materials management, became fairly widespread in the late 1960s/early 1970s. In the 1990s, increasing application of IT brought the internal functions and external supplier and customer members of the system even closer together.

This is an early example of fitting an organization to a horizontal process or series of related processes, rather than having the separate elements of supply report

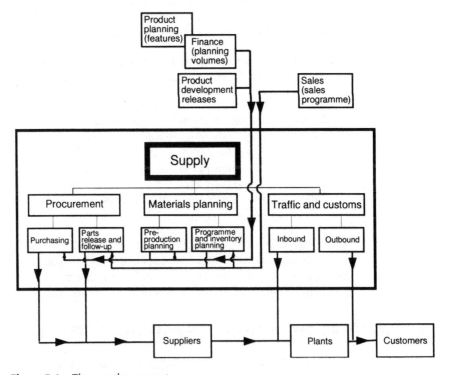

Figure 5.1 The supply concept

upwards within separate organizational chimneys. This approach later became fashionable under the 'business process reengineering' banner. ('Business process reengineering' (BPR) is a term used to describe radical change of organizations and processes – 'breaking the mould' or 'starting with a clean sheet' – as opposed to incremental change or improvement. See Chapter 8 for details of BPR at Chrysler Corporation.) Although this organizational grouping of supply activities improved liaison between each element, and between them and the company's suppliers, it sometimes led to supply being another chimney, separated from the manufacturing and product development activities (other chimneys) in a way that was an obstacle to constructive dialogue between R&D, designers, buyers, producers and suppliers. Further developments of supply systems, and their integration with product planning, design, production, distribution and sales and marketing, are dealt with in Chapters 6 and 21.

Information was fed into the system shown in Figure 5.1 in two phases. First the system was primed (shown by the arrows from finance and engineering into the supply box) by taking each engineering release as an input to establish:

- bills of materials (BOMS) for each end product and each producer location;
- lead times for each commodity group in the BOM (greatly influenced by the designers' specification);
- sources (suppliers) for each item (which often involved the design engineers, who may have designated 'approved' sources);
- shipping modes, inbound and outbound (involving materials handling engineers);
- target inventory levels for each item (involving production engineers);
- databases for scheduling, material control and accounting.

All these data would have to be changed at every point in the system where they were recorded if a component specification were changed. Changes could be made as a reaction to concerns raised:

- by customers, fed back through sales or service; or
- from production or construction difficulties in plants or on sites; or
- from materials handling and transportation.

They could also be made by:

- designers seeking to improve performance or reduce cost.

Most of these sources of change are problems or obstacles to the achievement of separate functional goals, and so are prone to bring user engineers and suppliers' engineers or sales people into contact in situations likely to produce conflict, which could adversely affect user/supplier relationships.

In the second phase, once the system was primed, information was fed in to make it 'run' (shown by the arrows from sales). This information was based on customer orders and was input to establish:

- plant build schedules;
- supplier delivery schedules;
- end-product shipping schedules.

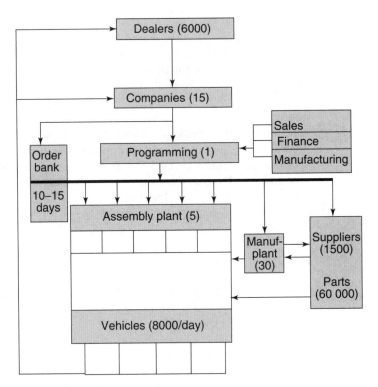

Figure 5.2   Vehicle producer operating system

This second phase is illustrated in another representation of a supply system in Figure 5.2, where the scope is extended to show dealers and suppliers. The numbers of the various groups of players (for example, 6000 dealers) have been shown in brackets for a typical European vehicle producer.

The overall process is similar for most manufacturing industries. The materials management task for a typical vehicle manufacturer starts with the gathering of orders from the 6000 dealers, through the 15 national sales companies, and arranging these orders to form a production programme for each of the five assembly plants. It continues with 'exploding' the orders into production and delivery schedules for 60 000 parts from 500 outside suppliers and the company's own 30 manufacturing plants, and planning the delivery of the 8000 finished vehicles each day to the 6000 dealers. Deriving schedules for the manufacturing plants from the overall production programme, and preparing the plants to meet them, is often called manufacturing requirements planning (MRP II), and calculation to derive schedules for suppliers is called materials requirements planning (MRP I). There are several standard software packages designed to perform MRP I and II, but most companies find it necessary to modify them for their own circumstances or to develop bespoke systems.

Using such a supply system, any one of the six major European car and truck producers processes about 100 m materials-related transactions a day. Turnover (the value of materials moving through the system) for one producer is in the order of £35m a day, and production material inventories held within the system have historically been in the range of £200m to £400m. These figures suggest a high level of activity and vehicle production plants are in fact very busy places. Despite this, the time lapse from the dealer placing an order and delivery of the vehicle to the customer has been in the region of two to three months for the major European producers.

Reduction of inventory and of the time to fill a customer order are two tasks of materials management, and the historical values quoted above do not represent 'world-class' performance. Greater attention to these tasks resulted in major improvements in the 1990s. For example, in 1994 Ford announced a trial scheme at its Dagenham plant that would reduce delivery times for UK customers from two months to two weeks. Customer orders that could not be met from dealer stocks would go straight from the dealer to the plant, where more flexible production and scheduling systems had been introduced. This was a brave attempt to use materials management to develop a competitive advantage, but to no avail. The overall economics of vehicle production in the UK led to Ford's discontinuing car assembly at Dagenham in 2002.

On top of the daily sales and production activities, more materials management tasks arise from the implementation of design changes. Stocks of a superseded part or material have to be exhausted if the designer gave a 'stocks use' disposition, or removed from the system immediately if there is a quality or safety implication. The exact timing of the change has to be recorded for each plant, each process and each product that is affected. It has been estimated that each change involves administrative costs in the region of £10 000, and a vehicle production system as shown in Figure 5.2 historically suffered about 20 000 changes a year. Simple arithmetic: £200m total cost.

With the volumes and values of material movements, the numbers and costs of transactions involved and their impact on product quality and customer satisfaction, materials management systems such as those described above were among the first candidates for change in response to increasing competitive pressures in the 1970s.

To simplify the process by reducing input transaction volumes, as a precondition for reducing inventories and developing a more responsive system, the first step would be to reduce the number of suppliers and the number of parts being scheduled and handled. The most effective way of doing this is through designing and specifying at a higher level of assembly – for example complete seats from one supplier, instead of 28 seat parts from eight, nine or ten suppliers. To make this change, a materials management strategy is required. This would have to be developed and implemented jointly by designers, producers and buyers.

The level of assembly specified by the designer also has an impact on materials management in industries other than vehicle production. For example, a building can be designed with bathrooms that can be put in place as a single unit, which creates a materials management task quite different from that for a building where tiles, taps, toilets and toothbrush holders have to be bought, delivered, stored and handled separately. Similarly, the construction or production process also affects the 'supply' task. Fabrication on site of roof sections, structural steelwork or bridge spans creates quite different materials management tasks compared with off-site prefabrication and sequenced delivery. The Lloyd's building in London provides an example of higher-level assembly design and new production methods – for instance, the toilet blocks are attached to the outside of the structure as complete 'pods' and can be removed (with suitable warning) as a complete unit. Engineers' actions at the design stage also influence another important aspect of materials management – timing. In setting up bills of material (BOMs) for each unit of production and for each location where the product is to be made, long lead-time items have to be identified so that action is taken early enough to ensure that the items are there at the appropriate time in the production or construction process. In fact, the lead time for all items has to be reviewed with the purchasing and scheduling activities in order to set up the project control system. Ideally, this review should take place before design starts, so that the choice of some exotic material or unusual fabrication can either be avoided or built into the timing plan.

The BOM multiplied by the number of units to be built determines quantities to be made or bought in total. The production plan determines the rate of delivery and capacity required – capacity to make, store, handle, assemble, inspect, deliver and service. Planning these capacities involves engineers of many kinds.

Even with the use of computers – or especially with the use of computers – the task of controlling data and physical product becomes complex and error prone when every component part is designed, specified and released by the producer's own design activity. Quality management and materials management tasks can be greatly simplified if a higher level of assembly is specified, and there are fewer sources with whom to discuss quality, delivery and cost issues.

Improved materials management therefore starts at the design stage, with cooperation between the designers, producers, buyers and schedulers to identify materials and components and suitable sources of supply, which support the design intent without compromising cost and timing objectives.

5.4   COST INFLUENCE OF MATERIALS MANAGEMENT

In Chapter 3 it was shown that 'material' constitutes about 60% of total cost in manufacturing. Material is also an important element in construction costs, ranging from 30 to 35% of total costs for dams and civil works, through 40 to 45% for housing, to over 50% for industrial, commercial and public buildings [4].

It follows that a 5% cut in materials costs can have a greater effect on profit than a 5% increase in sales or a 5% improvement in productivity. It may also be less difficult to achieve, for, as shown in Chapter 4, total quality management can cut costs by between 15 and 30%. So a 5% materials cost reduction is well within the range of savings that can be achieved through early cooperation between user and supplier to improve quality.

Figure 5.3 shows, in the first column, Model A, a simplified corporate cost model for a typical manufacturing company, with total sales of 100 units and profits of 5 units after deducting costs of 95 units. The effect of various management actions to improve profit is shown in the other three columns.

As shown in the second column, Model B, a 5% improvement in labour productivity does not affect the costs of materials, services, selling or distribution, but may lead to a small increase in overhead if capital investment in new equipment is required. Profit increases from 5 to 5.5% of sales, a welcome improvement, but not a transformation in business performance.

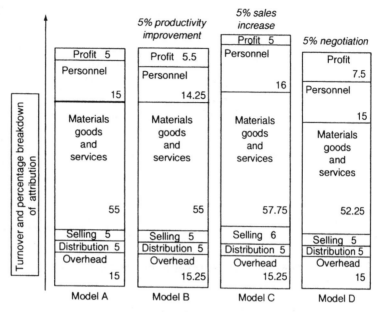

Figure 5.3 Corporate cost models

The third column, Model C, shows the likely effects of a 5% sales increase. This would lead to a 5% rise in materials and services costs (ignoring volume discounts) and would probably entail higher selling costs, including special discounts, increased advertising and promotional campaigns, as well as higher personnel costs for extra hours worked – possibly at premium rates. Distribution costs might be unaffected if this part of the system were less than 100% loaded at previous sales levels. Costs rise by the same amount as sales, and profit remains at about five units (or to be unreasonably precise, increases to 5.25 units), which is poor reward for the extra effort.

However, a 5% reduction in the cost of materials, goods and services, illustrated in the right-hand column, Model D, does not increase any of the other costs and goes directly into profits, which rise to seven and a half units – a 50% increase.

The cost and profit effects described above would appear in the company's profit and loss (P&L) accounts, and hence in the numerator of the return on capital employed (ROCE) formula. (See Chapter 14 for further information on the construction of P&L accounts, a balance sheet and the ROCE formula.) Materials management also has a major impact on the denominator in the ROCE formula. In the balance sheet, materials and components appear as inventory or stocks under the 'current assets' heading. Inventory can be as much as 20% of total assets, and the absolute sums involved mean that a 20% reduction in inventory could fund a whole new product programme or a new manufacturing complex! Their 2002 worldwide accounts, for example, showed Ford's stocks valued at $7.0 bn, DaimlerChrysler's at $16.4 bn and ABB's at $2.4 bn (down from $7 bn ten years earlier).

Inventory also affects the balance sheet through the capital cost of buildings in which to store it. The importance of this was brought home forcibly to teams

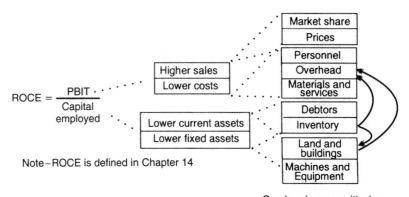

$$ROCE = \frac{PBIT}{Capital\ employed}$$

Higher sales
Lower costs

Lower current assets
Lower fixed assets

Market share
Prices
Personnel
Overhead
Materials and services
Debtors
Inventory
Land and buildings
Machines and Equipment

Note – ROCE is defined in Chapter 14

Overheads are positively affected by reductions in inventory and land and buildings. Inventory levels also positively affect land and buildings.

Figure 5.4   Effects of materials management in return on capital employed

from the West who visited Japan in the 1970s. They found that only about 10% of factory floor space was used for storage, compared with 40% back home. With building costs at that time in the region of $60 per square foot, that had serious cost implications – and still does.

Inventory carrying costs, such as depreciation of the costs of floor space, heat and light for the storage area, and interest on the money tied up, are P&L items. Therefore, inventory reductions improve financial performance both by reducing assets and by cutting operating costs. An elaboration of the ROCE formula illustrating these linked effects is shown in Figure 5.4.

Table 5.1   The profitability effects of reductions in materials costs and inventory

Company	Sales	Cost of sales (COS)	COS as % of sales	Net capital employed (NCE)	Stocks	Stocks as % of NCE	Operating profit	Return on capital employed	Return on capital employed (2)
BBA	1322.6	1028.6	78%	372.8	245.7	66%	77.6	21%	40%
T&N	1390.1	1017.5	73%	535.0	281.3	53%	90.7	17%	30%
VICKERS	718.5	589.4	82%	206.6	153.1	74%	12.6	6%	24%
600 GP	98.5	73.9	75%	58.3	39.2	67%	−1.7	−3%	4%
LUCAS	2252.7	2168.1	96%	743.2	437.1	59%	58.3	8%	25%
GKN	1993.5	1857.8	93%	814.9	306.9	38%	125.9	15%	29%
APV	947.5	732.6	77%	149.8	142.4	95%	16.8	11%	44%
TI	1149.3	1038.0	90%	279.3	250.7	90%	111.3	40%	71%
SMITHS	635.3	428.6	67%	330.0	114.6	35%	93.5	28%	37%
CMB (FF)*	24830	21286.0	86%	25375.0	3534.0	14%	2454.0	10%	14%
IMI (1991)	968	889.8	92%	364.4	256.3	70%	78.2	21%	39%

*Figures for CarnaudMetalBox shown in French Francs, otherwise all figures shown in Sterling.
ROCE (2) assumes 5% COS reduction and 20% reduction in inventory.
All figures calculated from 1992 company reports, except where shown.

Combined with the sort of materials cost reductions described in section 5.2, inventory reductions can change ROCE performance figures significantly. Some examples of theoretical material cost reduction and inventory reduction are shown in Table 5.1 for a selection of engineering companies.

5.5 INVENTORY MANAGEMENT

Inventory is likely to build up in four parts of a production system, and in similar ways in construction:

- on receipt;
- in process;
- on completion;
- in repair, maintenance and nonproduction material stores.

In manufacturing there may also be a fifth type of inventory in transit from suppliers to the producer/assembler, and possibly in transit or storage from the producer to the customer. Material belongs to the producer, and hence appears in the producer's inventory, from the moment it is bought (which may be ex suppliers' works) until the moment it is sold as part of the end product (which may be after it has passed through a distribution system and has been held by the retailer waiting for a customer). This is illustrated in Figure 5.5.

Inventory is held on receipt to allow administrative processes to be completed, which may include inspection, and to provide protection of the production process

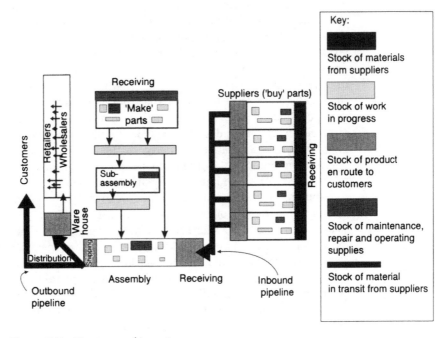

Figure 5.5 Five types of inventory

from interruption due to late delivery, or the delivery of the wrong or faulty items. It may also be held to allow a short-term increase in the production rate. In other words, stock is held to guard against uncertainty in usage or availability or while information is processed. Reduction of stocks at this stage depends on several factors, which can involve engineers and technologists, such as supplier quality performance and inbound logistics, and on the design of information and materials handling systems.

Work-in-process (WIP) inventories held during processing are unavoidable, but can be reduced by faster processing. WIP is held between processes for the same reasons as inventories are held on receipt – uncertainty and poor information flow – and for other reasons. If output from one stage of production is unreliable, stocks may be held after that stage to protect later stages. They may build up before a process if it is interrupted or slows down, or simply because it is not known that the material is there ready to be used – or perhaps it needs to be inspected. Additionally, WIP may be held because of mismatches in process capacity or competing demands for the same equipment – which could be handling or process equipment – or because of poor layout that requires material to be moved around the site between processes. In some businesses such as machine tools, civil engineering, shipbuilding and aerospace, WIP can amount to almost 100% of total project value, so completion on time and prompt handover are vital to profitability, even if the supplier's burden is lightened by stage payments. If interest charges are 1% a month and other inventory carrying costs are a further 1% monthly, a few months' delay can wipe out the profit margin. Thus the true cost of a missing component or a quality concern in the last stages of completion could be thousands of times its nominal cost.

Reduction in WIP depends on machinery and equipment reliability, capacity planning, plant layout, process capability (i.e. quality) and information and materials handling systems, which again are the concerns of engineers.

Completed, finished goods, or end-of-line inventory, may be held because the goods are not in fact finished – there may be more inspection or packaging operations. They may also be held for consolidation with other material prior to delivery, or for collection (poor information processing, again, is a possible stock generator). On the other hand, they may have been completed sooner than required, or they may be on time and demand has reduced. This form of inventory can be reduced by responsive (flexible) production processes, by improved process capability (quality), better outbound logistics, and efficient information and materials handling systems, which means that engineers and technologists of various kinds may again be involved.

Maintenance, repair and operating supplies (MRO) or nonproduction material inventory has a habit of growing through neglect. It consists of such things as spare parts or wear parts for equipment, tooling, cleaning materials, lubricants, fuels, protective clothing or workwear, paint, timber, metals, building materials for small works or maintenance, and packaging materials. Spare parts and tooling are often bought before production starts and are sometimes kept after production ceases or a change makes them obsolete. This is one class of inventory that really is the direct responsibility of engineers, and can be reduced by arrangements made for them, or by them, with suppliers – for example to hold emergency spares, fuels, industrial gases and other stocks on consignment at the producer's site. The time to make such arrangements is before orders are placed for facilities or services. This requires a team effort between engineers, buyers and suppliers at the product or project planning stage.

Ways of reducing or eliminating (i.e. managing) these four types of inventory are described in Chapter 20. Even at this stage, however, it should be apparent that inventory reduction, like other aspects of materials management, involves cooperation between engineers, scientists and technologists of many kinds at an early stage of planning.

5.6 SUMMARY

This chapter indicated the importance of materials management to business success. It showed how engineers' selection of materials and their designation of levels of assembly can affect the scope and scale of the materials management task, with ramifications throughout the production, construction and supply system. The effectiveness of materials cost reduction and inventory reduction in raising return on capital employed was shown to be greater than that of other actions to improve profitability. Four categories of inventory were described, and for each category it was shown that engineers can contribute to inventory reduction. It was emphasized that simplification of the materials management task, materials cost reduction, inventory reduction and lead-time reduction all start with teamwork in the conceptual and planning phase of a project or product programme.

REFERENCES

1. 'Towards a comprehensive materials balance', *Siemens Review R&D Special*, Fall, 1993.
2. Kaounides, L. (1995) *Advanced Materials: Management Strategies for Competitive Advantage*, Management Report for the Financial Times, London. See also Kaounides, L. (1994) *Advanced Materials in High Technology and World Class Manufacturing*, UNIDO, Vienna.
3. From correspondence with Lakis Kaounides in preparation of the Financial Times Management Report.
4. Cassimatis, P. J. (1969) *Economics of the Construction Industry*, National Industry Conference Board, USA.

6

Managing Design and New Product Development

6.1 INTRODUCTION

Earlier chapters have shown how, throughout the world, all engineering and technology-based companies face similar challenges – ever more demanding customers, rapid technological change, environmental issues, competitive pressures on quality and cost, and shorter time to market with new product features. They operate against a background of common external factors – slow growth, excess capacity, increasing legislative control, demographic changes, market complexity and increasing globalization of industries.

Responding to these challenges against this background has led to a focus on the product development process:

> 'Three familiar forces explain why product development has become so important. In the last two decades, intense international competition, rapid technological advances and sophisticated, demanding customers have made "good enough" unsatisfactory in more and more consumer and industrial markets' [1].

The chapters on materials management and quality described how efforts to achieve step-change improvements in performance in these areas have led to multifunctional teamwork in the design phase as the key to success. It is also the key to shorter development times and greater customer satisfaction. However, multifunctional teamwork was not the style of product development practised in western industries until competitive pressures forced them to change, starting in the mid-1980s.

This chapter describes the transition to collaborative new product development. It starts with the design dilemma, first mentioned in Chapter 4, and proceeds through a summary of the pressures for change to conclude with an outline of simultaneous (or concurrent) engineering, which provides a resolution to the design dilemma.

6.2 THE DESIGN DILEMMA

The design dilemma was introduced in Chapter 4, Management of quality, since many of the 'design for' objectives are aspects of quality. The 'design to' objectives

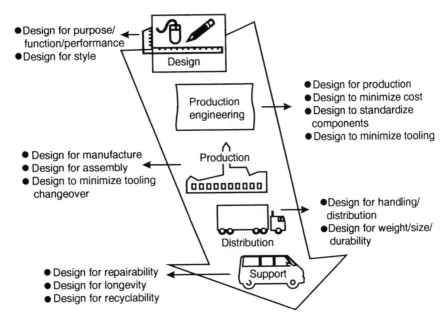

Figure 6.1 The design dilemma

are among the corporate objectives shown in Chapter 2. The designers' dilemma is the reconciliation of their own objectives with those of the production engineering, production, distribution and support activities. The illustration of the design dilemma is reproduced in Figure 6.1.

Not shown in the figure are some other important business issues that need to be considered as part of the management of design. These include time to market, picking winning products, risk in new technology and the cost of new technology – which should all have a familiar ring from reading the Preface.

Many attempts to resolve the design dilemma, and to handle the major business issues, founder when new product development is handled in a sequential way. If design for manufacture and design for assembly, or design for 'buildability', are tested by sequential trial and error, time is wasted and design work has to be repeated. This approach is illustrated in Figure 6.2. FMEA (failure mode and effect analysis) was mentioned in Chapter 4 and worked examples are shown in Appendix 2. 'DFA' is design for assembly.

The diagram shows how some infeasibilities can be identified by analysis at the process planning stage, but that others only show up as a result of physical trials in manufacturing or assembly. (The word 'alternation' in the figure was coined by a German engineer. It is fit for its purpose here, combining 'alteration' and 'alternate'.)

The trial-and-error approach to testing the effectiveness of designs can be extended beyond manufacturing to the customer – the ultimate inspector. Performance in the marketplace and in use will show whether a product really has been designed for performance, function, repairability and longevity.

This expensive, unsatisfactory way to evaluate designs has been labelled 'over-the-walls' engineering (OTW), illustrated in Figure 6.3. In over-the-walls

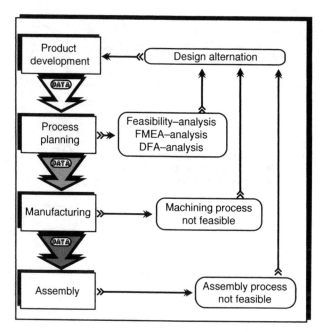

Figure 6.2 Design alternation

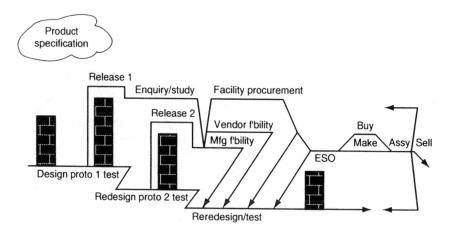

Figure 6.3 Over-the-walls engineering

engineering, product planning information (features and timing) is drip fed to the designer who, in the privacy of his/her own cell, prepares a design and starts to test it. Before tests are complete, timing pressures lead to design information being released (over the wall) to production or construction engineers and buyers so that they can make their own studies and enquiries. Some of these feasibility studies by suppliers and manufacturers will reveal potential problems and lead to requests for design change (back over or under the wall). The designer may already be making changes as a result of his/her own testing of the first design. Meanwhile, timing pressures dictate that procurement of facilities and long-lead items must proceed. The concerns about feasibility and quality lead to design changes, which in turn lead to process and facility changes. Evaluation of these changes may not be fully completed when the time comes for engineering sign-off (ESO). ESO is given in the hope that the product will be 'good enough'. The customer will probably decide that it is not, and this information comes back to confirm the internal final test results. Sales fall below objectives.

Living through one programme managed in this fashion is enough to persuade any intelligent engineer that there must be a better way. There is, and it is called 'simultaneous engineering' (SE). It is also known as 'concurrent engineering'.

Poor business performance as a result of OTW engineering is one powerful reason for adopting SE, but there are others, which may be the root causes of the poor business performance. The most important of these are:

- quality management;
- cost management;
- time management;
- customer requirements;
- the impact of new technologies.

These reasons for adopting SE are described in sections 6.3 and 6.4. Simultaneous engineering is described in section 6.5.

6.3 PRESSURES TO CHANGE

6.3.1 Pressure from quality management

In Chapter 4, cause-and-effect analysis, with its fishbone diagrams, was described as a tool for identifying and removing causes of variation and so improving process control. This produces diminishing returns – smaller and smaller improvements from more and more analysis. A point will be reached where a step change in quality performance requires a change to the process itself, possibly involving a change to equipment and a change in design [2]. It was also shown in Chapter 4 that the cost of making such changes increases exponentially when production or construction is underway. Clearly, there would be advantage in identifying these process, equipment or design changes at an earlier stage, but this requires some way of envisaging or simulating production or construction conditions. Ways exist, but they require input from the production or construction experts at the design stage – that is, the formation of a design team with the knowledge to define alternatives, the skills to evaluate them and, ideally, the authority to implement the jointly agreed optimum solution. Establishing such teams, and vesting them with authority, will normally require a corporate culture change.

6.3.2 Pressure from cost management

It is not just changes made to improve quality that cost more to implement later in the product cycle. All changes cost more the later they are made, and the costs impinge on all parties. It is a mistake to believe that contractors or suppliers profit from changes – they make more profit from doing things right first time, on time.

For manufacturers, some idea of the waste resulting from specification changes can be derived from Chapter 5, section 5.3. There it was stated that a typical major European vehicle producer would process about 20 000 component changes a year, at a cost of about £10 000 for each change. That means there is a £200m prize for the elimination of these changes.

6.3.3   Pressure from time management

Time to market for consumer goods and time to build for construction projects are important competitive factors. In the Preface it was shown that UK business leaders considered time to market the top issue of the 1990s. The principal reason for this is the impact of reduction in development time on profitability over the total product life. This was demonstrated by the Arthur D. Little Company's research published in 1990 (described in [3] and summarized in Figure 6.4).

The chart shows that the $1.95 bn net present value (NPV) of a project requiring an investment of $1.5 bn would be increased by $0.06 bn if R&D expenditure were reduced by 25%. It would also be increased by $0.12 bn if facilities costs were cut by 20% and so on. But the biggest impact on NPV (an increase of more than $0.3 bn) would result from a reduction in the new product development time from five years to four years. See Chapter 15 for an explanation of NPV.

The gap between US and Japanese performance in consumer goods product development timing in the mid-1980s is illustrated in Figure 6.5. The example is for car programmes, but the relationship was similar for home entertainment

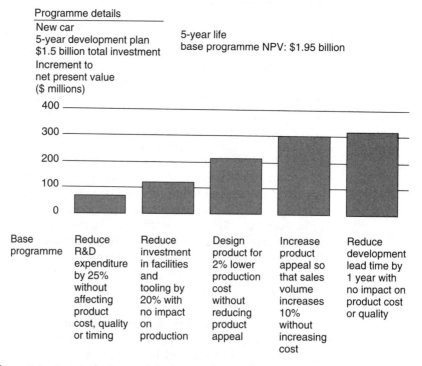

Figure 6.4   Fast cycle time and the bottom line (Arthur D. Little)

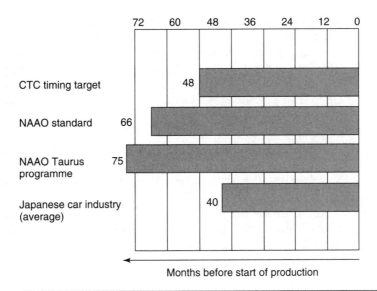

Figure 6.5   Development lead time – competitive comparison

products, cameras, copiers, communications equipment and others. In many industries, US and European performance was similar – they lagged behind their Japanese counterparts.

Compression of product introduction or project completion timing spans can be achieved by doing things faster, possibly by the use of advanced information technology. However, improvements by this method are small compared with the time savings from doing things right first time, and doing several things in parallel rather than sequentially. Chapter 17 describes ways of achieving parallelism in implementation, and such methods have been widely used in the West as well as in Japan for more than 50 years. The difference in performance between these groups in the 1980s was found to be in the planning, design and sourcing phases prior to implementation. SE introduces parallelism to these phases.

6.3.4   Pressures from customers

'Picking winning products' was another top issue identified by UK business leaders, and would certainly feature in similar lists elsewhere in the world. The history of personal and domestic entertainment equipment provides several examples of 'nearly right' products – audio and video tape systems that came second in the competition to be the industry standard; alternative compact disc formats and satellite TV receivers. Whatever happened to digital audio tape (DAT) and high definition television (HDTV)?

In the European motor industry, some producers pursued 'lean-burn' engine technology as their way of meeting expected emission control legislation, but their efforts were wasted when three-way catalytic converters became the mandatory specification. They had picked the wrong product option.

Making the right choice, and accurately forecasting consumers' perceived or latent requirements, is obviously difficult, but the choice has to be made at the start of design. Masatoshi Naito, head of design at Matsushita, the Japanese electronics group whose brand names include National Panasonic, JVC and Pioneer, stated the task in the following way:

'Design is not just a shape or a form but realising what a consumer needs and making a product that meets those needs. Consumer needs have to be the starting point rather than just seeking to differentiate a product superficially' [4].

Consumer needs can only be incorporated in designs if designers are well informed about those needs, and have techniques to reflect them in designs that recognize material and process capabilities. Capturing and using the necessary information is part of SE.

6.4   THE IMPACT OF NEW TECHNOLOGIES

The continually increasing rate of technological change is both a threat and an opportunity. At the consumer level, rapid changes in products such as personal computers, mobile phones and audio equipment render obsolete, within months, equipment that has been touted as 'state of the art'. Buyers who see the value of their purchase halved when scarcely out of its packaging become resistant to the next wonder product. At the producer level, the pressure to be first in the marketplace, or first in the factory, with the latest material, machine or method calls for new ways of evaluating the alternatives in order to avoid costly mistakes. The French politician François Mitterand has been credited with the assertion that:

'There are three ways of losing money – women, gambling and technology. Women is the most pleasurable way, gambling is the fastest, and technology the most certain.'

There is obviously no published research to support this statement, but examples of unsuccessful investment in new technologies feature regularly in the press. This does not make them representative. On the other hand, since most organizations do not willingly publicize failure, the reported cases may only be the tip of an iceberg.

For example, the opening of Amsterdam's Schiphol international airport was delayed by several months by problems with the automated baggage-handling systems. The interim solution was allegedly to hire circus acrobats to scale the racking and retrieve luggage. In 1991, the new high-speed train service from Munich to Hamburg had an inauspicious launch. The train suffered 'a spate of breakdowns, including the electric engines, the super flush toilets and the microwave ovens and beer cooler in the restaurant car' [5]. More recently, another attempt to improve a rail service failed through lack of 'system thinking':

throughout 2003 and into 2004, 150 new, state-of-the-art carriages built for the Connex South East service between Kent and London stood idle in sidings. The new carriages have electrically operated sliding doors, whereas the old rolling stock had manually operated 'slam' doors. The 'third rail' electricity supply system in the region does not have sufficient power to operate the new carriage doors, and it will take further investment over three to five years to upgrade the power supply and bring the new carriages into use.

These examples indicate the need for thorough testing of the product prior to public availability. With new materials and processes this means that producers have to reach down the supply chain to involve the scientists and technologists who are the developers of the new materials and methods in the end-product design and test programmes. It also means that the producers have to evaluate the new technologies in partnership with their materials, component and equipment suppliers in conditions representative of volume production and customer use. Failure to do this can delay and diminish the benefits of innovation, at great cost in lost market opportunity and idle facilities.

In some industries, such as air transportation and pharmaceuticals, the evaluation requirements are determined by government authorities, but even here simultaneous engineering can help. Boeing's development of the 777 aircraft was an example of 'total system' SE within such regulatory constraints [6]. A research programme to reduce development lead times in the UK pharmaceutical industry was launched in 1994 as part of the UK government's Innovative Manufacturing Initiative. It is understandable that this should be a national priority, since evaluation of a new medicine within the regulatory framework can take 12 years of the 20-year patent protection period. The phases of development are shown in Figure 6.6, which is reproduced from a Confederation of British Industries publication [7].

Something not shown in Figure 6.6 is the task of the industry's process engineers. Traditionally, starting late in the clinical development phase, when the likelihood of success is emerging, the process engineers have had to devise production methods that retain the integrity of the laboratory processes used in the discovery and small-scale production phases. The engineers' proposals

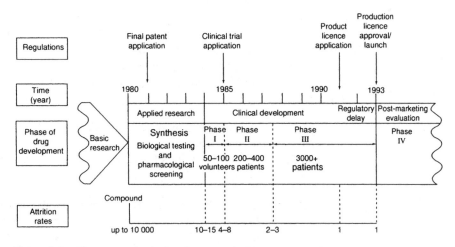

Figure 6.6 Pharmaceuticals development timing

for scaling up production also require approval by the regulatory authorities, which can result in delays to marketing the new product. The engineers' earlier introduction to the development team, to work alongside the research scientists, will be an important part of any plan to reduce overall new product development lead times.

The benefits of saving time in new product development and the faster introduction of new technologies have been demonstrated by success stories in the automotive industry, computer peripherals, aircraft and electronics. In most instances, the time saving is between 20 and 30%, which equates to a year on a typical car project – worth over $300m, according to A. D. Little in 1990 (see Figure 6.4 above). Several case studies from a variety of industries are given in a special report *Concurrent Engineering*, by Rosenblatt and Watson [8].

6.5   SIMULTANEOUS ENGINEERING: RESOLVING THE DESIGN DILEMMA

6.5.1   The aims of simultaneous engineering

In 'over-the-walls' engineering, what the external customer really wanted, and what the internal customers would have liked to have, is eventually provided through application of downstream expertise to resolve problems. This results in late changes that endanger quality, cost and timing. The aim of SE is to avoid the changes and remove the dangers.

This can be done by bringing the downstream expertise of process engineers, service engineers, machine operators, materials and component suppliers, equipment suppliers, sales people and even financial analysts to bear at the same time, and early enough to resolve design and manufacturing concerns before production requirements of components and equipment are ordered.

6.5.2   Simultaneous engineering teams and processes

Although the SE process was developed in manufacturing industry, its general applicability has been increasingly recognized. For example, by the early 1990s there were pleas for its introduction to civil engineering. In the construction industry, SE is achieved by what is known as 'fast-track' construction, in which construction operations are in parallel with, but slightly behind, design. The Latham Report 1994 [9] identifies two factors that are significant in the success of this approach:

(1)   The degree of involvement of the client/owner: clients should be more involved than in the past.
(2)   The degree of coordination between design and construction: design processes all too often exclude any construction expertise.

The Latham Report challenged the UK construction industry to reduce its costs by up to 30% by improving relationships between clients, designers, contractors and subcontractors. Chapter 17 includes further discussion of SE in construction and civil engineering.

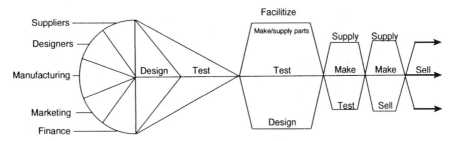

Figure 6.7   The team process of simultaneous engineering

The following steps reflect SE's origins, or revival, in a programme to introduce a new car engine, but can be adapted and adopted for use elsewhere. The stages of SE are:

- form a team;
- agree objectives;
- agree working methods;
- establish control and information systems;
- work together on the design;
- stay together through the implementation.

The process is illustrated in Figure 6.7.

The team consists of representatives from *marketing*, to provide input of customer needs; from *design*, to translate those needs into product features; from *manufacturing and suppliers* (of materials, components and equipment), to realize those features; and from *finance*, to provide assistance in the evaluation of alternatives – not to control costs, which is part of the team responsibility. The inputs from the team members are not spontaneous thoughts, but the documented results of learning from experience and from internal and external customer feedback. See Chapter 4, Figure 4.12.

Selection of individual team members has to be on the basis of their technical skills. In theory, team performance would be improved by incorporating the right mix of personalities to perform the allegedly different team roles. In practice, teams have to be formed from the talent that is available. If the team members' companies have paid due attention to the soft Ss of management (see Chapter 3), all the members will have the interpersonal skills and shared values needed to contribute willingly to the team effort. If some of the companies have neglected the soft Ss, the team leader has an additional task: to create a 'microclimate' of style and shared values that will extract the full potential of the shy, lazy, selfish or obdurate members.

In the early phase, when the team focuses on developing and testing designs, the design engineer is the team leader – the chosen alternative must reflect the design intent. As the project progresses towards production, it may be appropriate for leadership to move to the manufacturing engineer or the manager responsible for production, and, towards market launch, marketing may take the lead in setting priorities for the product mix and production sequence.

At every phase, the task of the team leader is to harness the expertise required to dissolve problems and concerns. Only a small part of the expertise will be available within the producer company – which is why there are supplier members of the team. Since more than 60% of materials and about 90% of production equipment comes from outside the company, in manufacturing industries it follows that this is where most of the expertise lies. Similar logic applies to other industries.

An important early step is recognition of this outside capability by the internal experts, and to move from the stage where SE is seen merely as cooperation between design and manufacturing to stages where inputs are contributed by the whole external supplier base and by customers – simultaneously and in a spirit of partnership. These stages are shown in Figure 6.8.

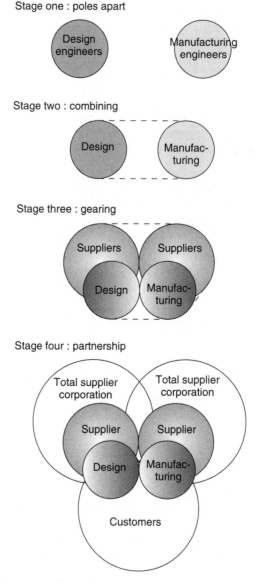

Figure 6.8   The four stages of simultaneous engineering

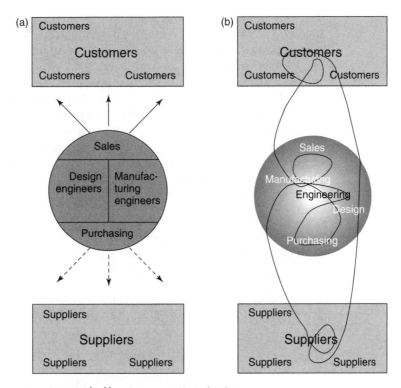

Figure 6.9   Removal of barriers to communication

The spirit of partnership has to be created, both at corporate level and at the personal level. This entails breaking down barriers between functions and between companies, which will not happen until sales people share their access to customers, purchasing staff share their access to suppliers, and physical and mental barriers between groups of engineers, technologists and scientists are removed. For many companies this requires a complete change of corporate culture or style, which may take years to achieve. Figures 6.9 (a) and (b) illustrate the change.

Robert Lutz, chief operating executive of the Chrysler Corporation, describes the situation in Figure 6.9(b) as 'the extended enterprise', where there are no boundaries between the company and its supply base, nor between the company and its dealers [10]. In such a company, the organizational location of the SE team would not be a major concern. However, in the more common hierarchical bureaucracies of western industry there has been extensive debate about the choice between:

- lightweight programme management – where the programme manager has use of the services of team members who remain in their organizational location;
- heavyweight programme management – where members are withdrawn from their organizational location and work for the programme manager;
- independent business units – where a self-contained unit is set up with its own support services and structure.

There are examples of success and failure for each arrangement, and there is no universal 'right' choice. There is, however, a universally wrong choice, which is to superimpose a project management structure on top of the operating functions to coordinate, to gather progress reports and to second-guess decisions made by the line managers. For effective SE, the team members have to be the people who do the design and process engineering, the buying and the supplying, and they should be given authority to make decisions within clear guidelines.

When the team has formed and has been briefed on the corporate objectives, targets can be developed and expressed in the team's own terms. Figure 6.10 shows (verbatim) the targets set by the team that rediscovered SE for Ford of Europe in 1986 on the Zeta engine programme.

The abbreviations in Figure 6.10 are part of the car industry's jargon. Their translation is:

- DQR – durability, quality and reliability;
- NVH – noise, vibration and harshness;
- P&E – performance and economy.

For each of these characteristics, evaluations of competitors' engines provided benchmarks of 'best-in-class' values. 'Performance feel' is harder to translate, but

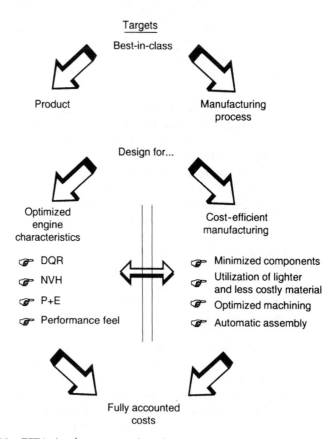

Figure 6.10   ZETA simultaneous engineering team targets

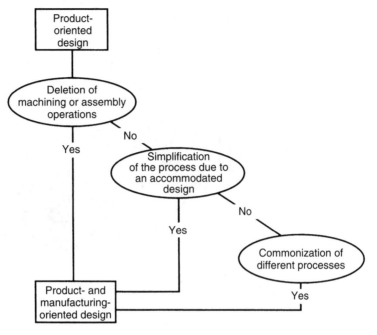

Figure 6.11   Design optimization strategy

there is something about the responsiveness of an engine that can be detected in the first moments of driving, and this term aims to describe it.

Having agreed their targets in terms they all understood, the ZETA team members then developed a simple algorithm to express their design optimization strategy, which is shown in Figure 6.11. It may appear from this abbreviated account that the ZETA team went straight to the 'norming' stage of team behaviour described in Chapter 10. In fact they had their share of 'storming' – at the first meeting of what was then a task force rather than a team, half the design engineers wanted to disband because they had been called together too soon. The other half thought they had come together too late. The task force leader and some members had the sense and skills to keep the group together. Other teams have developed other ways to proceed, but all the successful teams have had an agreed method and style of working.

The algorithm indicates that the team's starting point was a design that met the product objectives (best in class for DQR, NVH etc.). They then sought ways of making the design a 'design for manufacture' by eliminating machining or assembly operations – or, if that were not possible, simplifying operations – until they had a design that was both 'product and manufacturing oriented' – it would meet the product objectives and would be easier to produce with consistent process quality, and both piece cost and capital cost would be within the cost objectives.

Having agreed their targets and strategy, the team's next task is idea generation. By pooling their individual knowledge and experience, and by tapping the knowledge base of their 'home' organizations, the team members can compile a list of opportunities – ways in which the targets may be met. The core team will not be able to handle the evaluation of all the opportunities themselves, and

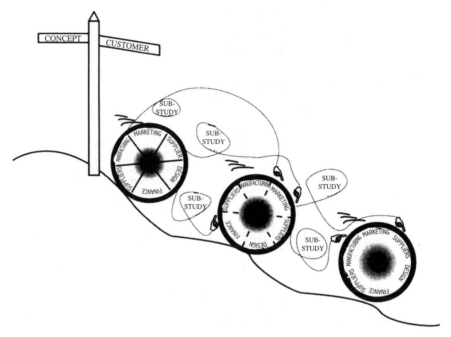

Figure 6.12   The team rolls on, helped by substudies by satellite teams

substudies will have to be spun off to other ad hoc satellite teams, coordinated by core team members.

Figure 6.12 is a representation of the way the emphasis of the team's activity changes as the project progresses (illustrated by the changes in the function at the top of the wheel) and the way in which substudies are spun off and their results fed back.

Barriers that may have existed when the team was formed are removed by shared experiences and team leadership (shown by gradual removal of the spokes in the wheels).

The work of the core team and the satellite substudy teams has to be managed. Tasks and timing have to be set, results evaluated and decisions made. A control format for this purpose is shown in Figure 6.13, which is another of the ZETA team's ideas.

The table in Figure 6.13 is headed 'design for competition', which itself shows a change in focus of the team's thinking – they have a common overriding goal instead of a series of 'design for' statements. The first column in the table, which is headed 'design feature', lists all the opportunities that came out of the idea-generation or brainstorming sessions. There were about 20 ideas, some of which were mutually exclusive – hence the asterisk identifying investment costs that could only be deleted once. The advantage, disadvantage and cost columns are almost self-explanatory. The key control information is shown in the action/responsibility/timing column, which names the core team member (by organization) who will coordinate the feasibility and evaluation study, and summarizes the agreed scope of the study. The last column shows whether the idea has been accepted by the whole team. The technical terms and processes

1992 ZETA Engine Programme Design for Competition Cylinder Head

Number	Design feature	Advantage	Disadvantage	Piece cost effect	Investment effects ($000 max)	Action Responsibility Timing	Concurred as prime programme
1 i	Oil drain holes cast finished	-Machining operations reduced -Tool cost reduction	-NAAO comment- increased fingers on oil jacket core which may result in increased core breakage -Montupet state no change in cost	TBE	(400) x2	Oil drain holes on exhaust cast or machined may result in water circulation problems around exh. port -PDG to review	YES TBA
ii	Self-contained hydraulic tappets	-Machining operations reduced -Minute cost reduction -Product improvements claimed by tappet supplier	-Tappet design not fully developed (DQR concern) -Piece cost increase -Longer assy leakdown required -May require additional squirt holes between tappet and valve stem. -No fallback route in the case of failure if oil hole machining is not protected	TBE	(4000)* x2 (600) x2	Action plan established -Resource and facilities to be established -EAO/NAAO joint testing	TBA TBA
iii	Delete camshaft lubrication holes in half bore	-Machining operations reduced -Tool cost reductions	-External system for CAM bearing lube required DQR			-Design study to be established -Build prototypes for test	
iv	Delete machined oil gallery system by casting the oil gallery and	-Machining and assembly operations deleted	-Piece cost increase	TBE	(4000)* x2		TBA
2	Delete machining	Machining and		TBE	(1200) x2	Series I heads will be made to new level	Montupet YES NAAO

Figure 6.13   Sample of ZETA control format

mentioned are not important to understanding the role of the document, and readers not familiar with them should not be distracted by them or worry about them.

The control format used by the ZETA team, like the other examples of their documents, was something developed by the team as they went along. In later programmes these simple hand-drawn sketches or typed tables were replaced by more sophisticated media, and more formalized management structures were put in place. Figure 6.14 shows a project control structure developed by British Aerospace plc, which reflects the same principles of delegation and monitoring as the ZETA team's simple charts.

British Aerospace (BAe) used the term 'concurrent engineering project' (CEP) when introducing the SE concept. Its 'fishbone' chart, shown in Figure 6.14(a), identifies six areas of opportunity, and Figure 6.14(b) is an example of how issues and opportunities were listed under one of the six headings.

Control of information can be greatly assisted by setting up a shared computer database, enabling all participants to have access to current data. All the powers of information technology (IT) can be harnessed for this purpose, including computer-aided design (CAD), computer-aided manufacture (CAM) and computerized data exchange (CDX). Rosenblatt and Watson [8] describe a range of research programmes in the USA aimed at greater use of IT for communication between geographically separated groups within SE teams. However, IT has its limitations. Some are technical, such as the very real difficulties of

(a) • Six areas of opportunity identified:

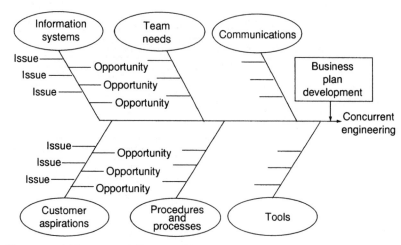

(b) 'Team needs' Issues and Opportunities

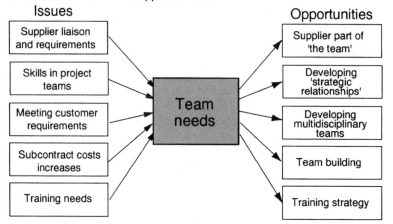

Figure 6.14 BAe's control format (a) the areas of opportunity control and (b) example of issues and opportunities control

CDX between different CAD systems. Some are geographical, such as the limited overlaps in working hours between countries and continents. Some are psychological – screen-to-screen communication is no substitute for face-to-face communication, does not allow full debate and discussion, and is slow to create a sense of shared purpose and teamwork. All SE teams therefore need to meet regularly face to face in the same room, even if it means some of them flying half way round the world. *Business Week* [11] reported that Honda, the Japanese car company, brought nearly 60 production engineers and their families to Japan to make sure that the car designers paid as much attention to the manufacturing needs of the US plants as they did to those of the Japanese plants.

When the team does get together it is important that they follow some house rules for communication. One team built up the following list of 'Dos' and 'Don'ts' over the first few months of working together (shown verbatim):

DO	DON'T
Let one talk, all listen	Talk forever
All contribute	Dominate
Open your mind to all options	Interrupt
Begin and end on time	Tell lengthy anecdotes
Be prepared	Use jargon
Be brave	Confine comment to your own discipline
Be a team player	Patronize or be sarcastic
Share facilitization	Make personal attacks
Set an agenda	Conceal vested interests or constraints of home organization
Reach consensus (try!)	Be a star
Check the process	Create unnecessary paper
Record agreements	
Gather facts; analyse; reach decision; let the facts speak	

A parallel group used the illustration in Figure 6.15 to show what they were trying to achieve. In retrospect, they wished they had illustrated their new approach as shown in Figure 6.16, where there is no 'we' and 'they' and there is only one speaker at a time.

For all SE teams, in addition to their own house rules, there are generally applicable tools and techniques that they can employ. Some of these are described in the next section.

6.5.3   SE tools and techniques

The most powerful SE technique is quality function deployment (QFD), developed originally in Mitsubishi's Kobe shipyard in the early 1970s. It is a systematic way of considering customer requirements, turning them into product features, and developing specifications, manufacturing processes and production plans. Three stages of QFD are shown in Figure 6.17. It is possible to continue the process to a fourth stage covering production requirements, and a fifth stage covering service requirements, but three stages are enough to illustrate the technique.

The whole SE team is involved throughout QFD to provide expert input and to reach balanced decisions. At the first stage, customer requirements are listed in everyday terms, such as 'better economy' or 'more occupant protection', and translated into design requirements. One of the design measures to improve economy might be weight reduction, and a measure to protect the occupants might be improved resistance to side impacts. At the second stage, design requirements are turned into part characteristics. A part characteristic to reduce weight could be an engine part made from aluminium instead of cast iron, and a characteristic to resist side impacts could be additional spars of high-strength alloy steel between the inner and outer door panels. In the third stage, the manufacturing requirements to implement the part characteristics are developed.

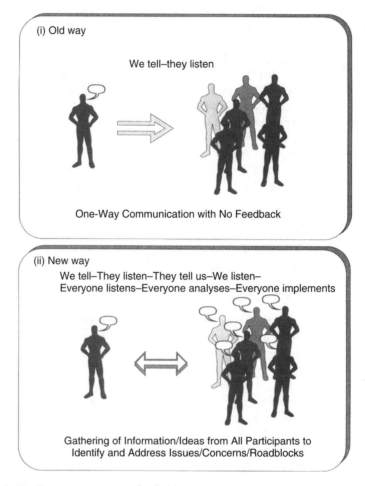

Figure 6.15   Communication methods (a)

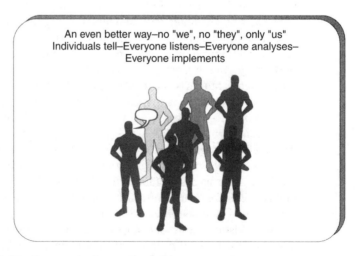

Figure 6.16   Communication methods (b)

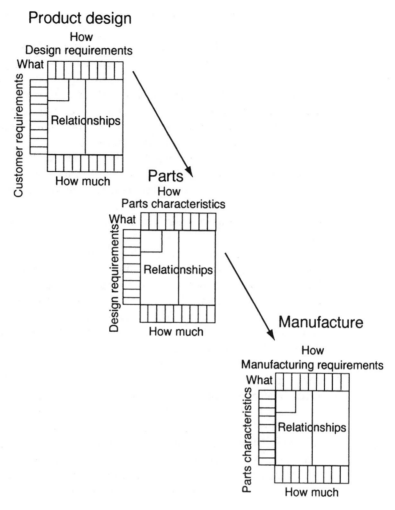

Figure 6.17 Three stages of QFD

At each stage the interaction of the various inputs is noted and identified as a positive or a negative relationship. The added door spars to improve safety would have a negative relationship with measures to reduce weight aimed at improved economy. The team has to consider such conflicts and make trade-offs, which are all recorded on the QFD charts. Rankings are assigned to each input to determine priorities and likely return for engineering effort. Comparisons with competitor products can be built in as another factor in determining priorities. The end result is that the most important customer requirements are incorporated in the product in the most cost-effective, quality-assured way.

QFD was first applied in the Japanese shipbuilding industry, but is now most widely used in the automotive industry. For those who wish to pursue the subject further, a worked example can be found in the *Harvard Business Review* [12].

Potential failure mode and effect analysis (FMEA) is another powerful tool that helps to prioritize engineering effort. The two forms – design FMEA and process FMEA – were introduced in Chapter 4 and a worked example is shown in Appendix 2.

Team effort can be focused by identifying for each part or system the *critical characteristics* – which affect safety or compliance with legislation – and *significant characteristics* – those which affect customer requirements. Using FMEA, designs and processes can be made sufficiently 'robust' to ensure that these characteristics are always within specification. Advanced planning of robust designs and processes can start by review of existing production methods and process capability for 'surrogate' parts, similar to the new parts.

In some organizations, including the US Defense Procurement Executive, SE teams use a suite of decision support tools known as interactive management (IM) [13]. The three phases of IM are idea generation, nominal group technique (which synthesizes the team's ideas) and interpretive structure modelling (which establishes a hierarchy of relationships between factors leading to the group objective).

The workload and time required to evaluate alternative designs and processes can be dramatically reduced by another tool, design of experiments (DOE). Factorial design of experiments to measure the effects of varying each of several factors dates from the work of A. L. Fisher in the 1930s. More recent developments by the Japanese statistician Genichi Taguchi allow variation of several parameters simultaneously to provide reliable results with fewer iterations. Lipson and Sheth [14] provide more details.

Pre-sourcing, the selection of suppliers by ongoing assessment rather than competitive bidding, is another time-saving technique that can reduce new product development times by many months, for example six to nine months in a 24-month lead time. Pre-sourcing is necessary in any case, since SE requires the involvement of key suppliers during product design. This means that the old ways of buying, which entail issuing detailed specifications and soliciting competitive bids, cannot be used. Pre-sourcing cuts out the enquiry/quote/evaluate/order cycle, and uses team assessment of suppliers to select partners to join the SE team. How this is done is described in Chapter 21, section 21.5.

6.5.4 Benefits of SE

In general terms, it can be said that SE:

- reduces time from design concept to market launch by 25% or more;
- reduces capital investment by 20% or more;
- supports total quality with zero defects from the start of production, and with earlier opportunities for continuous improvement;
- supports just-in-time production with total quality supplies, and advanced planning of inbound logistics;
- simplifies aftersales service;
- increases life-cycle profitability throughout the supply system.

In short, it resolves the design dilemma and helps to achieve all the fundamental business objectives.

Rosenblatt and Watson [8] describe extensive research into the application of information technology to simultaneous engineering (or concurrent engineering, as they prefer to call it), but they also quote Roy Wheeler of Hewlett-Packard:

'What tools does an engineer need to get started in CE? Pencil, paper, some intelligence, and a willingness to work with peers in other functional areas to get the job done. Computer based tools can be added as the budget permits.'

Wheeler makes a good point – open minds are more important to SE than open systems.

6.6   NEW DIMENSIONS TO SIMULTANEOUS ENGINEERING

Rapid technological change and advanced materials offer new opportunities to the designer. This brings a new challenge to SE – having reduced the time from design to market launch, the new task is to close the gap between basic research and its application. The reach of the SE team has to be extended to participation of the science base, and the range of possibilities evaluated has to be extended in two dimensions. One is to broaden the coverage of the team's studies to include revolutionary new materials and new processes, not just incremental change. The other is to lengthen the life-cycle time horizon so that the implications of using new materials are considered through to disposal and recycling.

SE teams have been doing design FMEAs and process FMEAs in many industries and many parts of the world since the early 1980s. In considering the use of advanced materials, they now have to perform something akin to life-cycle FMEAs, and will need new analytical and control tools. An example has been provided by Siemens AG, the German electrical and electronics company, where scientists and engineers developed a decision support system for their work on printed circuit board design [15]. Their decision support wheel for selection of plastic materials is shown in Figure 6.18.

The designer starts in the material production subdivision of the manufacturing sector, to identify raw materials and their polymerization, and proceeds in a clockwise direction to cover formulation and supply, before moving to processing, utilization, recycling and disposal. This is a comprehensive system!

Getting R&D closer to the market can be assisted by actions within companies, but is even more dependent on relationships between companies.

An example of action within companies is Nestlé's restructuring in 1991–92, which established an independent business unit (IBU) for each product group and assigned an R&D person to each IBU [16]. This type of action is likely to be more effective if the whole company is alert to the need to encourage the passage of new developments horizontally through the organization, without having to refer up and down functional chimneys (see Chapter 7). Pralahad and Hamel emphasize this in a way that links technology management with two of the soft Ss of management, style and skills:

'Technology can be narrowly held by a group of experts, but the competence to utilize it as a way of improving the performance of . . . products . . . requires many people across the company to understand the potential and the skills to integrate it in new product development' [17].

Outside the company, simultaneous engineering of new designs, new processes and new materials requires new alliances. The science base for new materials is

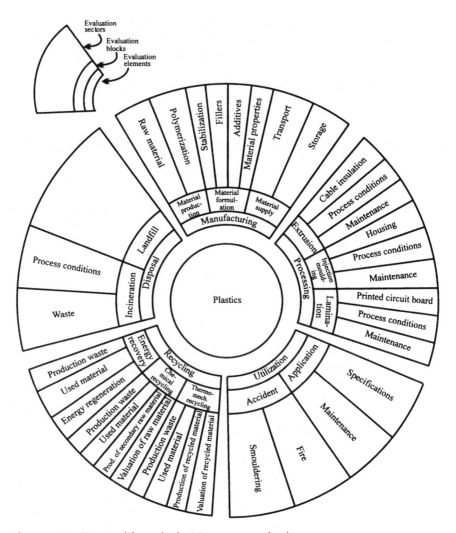

Figure 6.18   Siemens life-cycle decision support wheel

down in the supply chain among the basic materials suppliers and in academic institutions, which are outside the normally accepted scope of the supply chain. Application of the science base output may be at intermediate stages of the chain, at the levels of component and subassembly suppliers. End-product manufacturers, constructors or assemblers, therefore, have to set up complex, market-driven alliances between themselves, their suppliers, their suppliers' suppliers and the science base. If these organizations are to work concurrently, in the style of SE, the concept of the supply chain has to be set aside and replaced by the supply system. Managing this system, with its multiple interfaces within an SE framework, is a complex task. This is illustrated in Figure 6.19.

In the 'normal' linear process of new product development, there is some feedback from production to the production system designers, and from them to the product designers. This is 'over-the-walls' engineering – ineffective and inefficient,

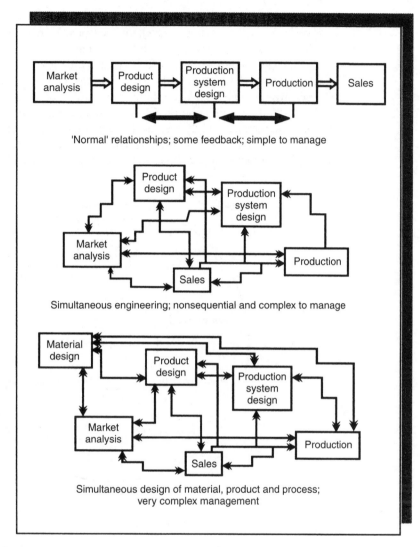

Figure 6.19   Increasing management complexity

but fairly simple to manage. In SE the multifunctional team is generally trying to make incremental improvements to known products and processes. Because design is by a team, relationships between individuals and between organizations become more complex and are more difficult to manage. When revolutionary new materials and technologies also have to be considered, the design team have to think in terms of radical change and possibly the complete replacement of processes rather than their improvement. Members of the team may have to be drawn from different industries, and from academic or government institutions not familiar with the industrial and business worlds. Management of these new partners in a new environment may become very complex.

The importance of such collaborative alliances for new product development was recognized at national level in the USA and in Japan, where 181 state-sponsored projects were launched in 1993, to hit world markets early in the next century [18]. At corporate level, too, new product development programmes of this complexity are being successfully managed.

The European Fighter Aircraft (EFA) is an example of a complex project, both in terms of the product and in the extent of the collaborative alliances involved. The EFA was being designed and built in the late 1980s to early 1990s by teams in Germany, Italy, Spain and the UK. The UK partner, British Aerospace, used SE (or CE in its terms), through a series of design build teams (DBT) coordinated into zone teams, which were coordinated further by integration teams. Figure 6.20(a) shows the composition of a DBT, with a core product database. Figure 6.20(b)

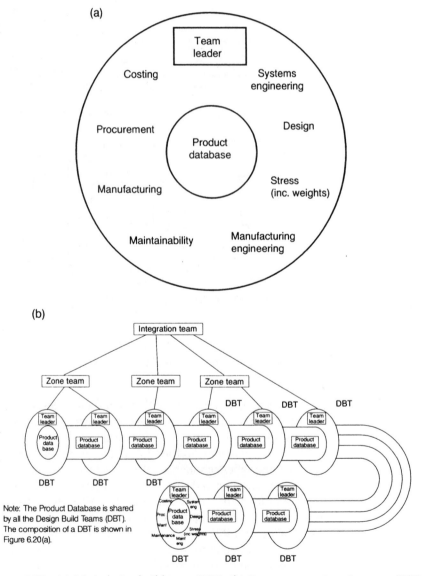

Figure 6.20   (a) BAe's design build team (DBT); (b) Concurrent engineering team (CET)

shows how the DBTs were coordinated, linked by an organizational structure and by the shared database.

Ford's CDW27 car programme, which produced the Mondeo in Europe and the Mystique and Contour in the USA, is another example of a complex programme. The key to its success was identified at the first briefing of the 300 or so pre-sourced suppliers from all over the world: a huge majority (about 70%) voted 'communication' as the most important issue.

The implication for engineers, scientists and technologists is that they too must be able to communicate if they are to contribute effectively to collaborative new product development, and to work in teams with other teams.

6.7   SUMMARY

This chapter described how the new product development process has become the key to achieving the universal business objectives of customer satisfaction, continuous improvement of quality, operating cost reduction, capital cost reduction and reduced 'time to market'.

It was shown that to meet all these objectives, 'over-the-walls' sequential development has to be substituted by simultaneous engineering, performed by teams that include materials suppliers, component suppliers, equipment suppliers, internal experts and, possibly, experts from the science base. Pre-sourcing, rather than competitive bidding to a detailed specification, is necessary to identify the supplier members of the SE team.

Corporate cultures and individual training must be conducive to cross-functional and intercompany partnerships and teamworking, and the teams need tools and techniques that enable them to contribute their specialist expertise.

Some of the tools and techniques were described. The need to expand SE to facilitate early application of new materials and processes was outlined, with its implications for collaborative alliances embracing the science base. This led to substitution of the 'supply system' concept for the 'supply chain' if all participants are to work concurrently. Management of the resultant complex relationships was shown to require engineers, scientists and technologists with interpersonal and communication skills, as well as technical expertise.

REFERENCES

1.   *Harvard Business Review* (1990), 4th Quarter
2.   Box, G. (1994) 'Statistics and quality improvement', *Journal of the Royal Statistical Society*, 157, part 2. This article describes the removal of process disturbances as 'statistical process control' and the step change to compensate for disturbances as 'engineering process control'. This is rather kind to engineers. The same article gives some interesting views on the education of engineers and statisticians.
3.   Dussauge, P., Hart, S. and Ramanautsun, B. (1993) *Strategic Technology Management*, John Wiley & Sons, Chichester.
4.   Leadbeater, C. (1991) 'Design-led change: From quantity to quality', *Financial Times*, 14 August.

5. Murray, I. (1991) 'Bonn's rail jewel is crown of thorns', *The Times*, 11 April.
6. An article entitled 'Triple seven' in the May 1993 issue of British Airways' *BA Engineering* house journal describes the scope and benefits of SE on this project. A shorter article, 'Boeing, Boeing' in BA's *Business Life* December/January 1993/4 in-flight magazine, provides similar information. A more accessible article, 'Genesis of a giant' by Guy Norris, appeared in *Flight International*, 31 August–6 September 1994.
7. 'Pharmaceutical successes', *CBI Manufacturing Bulletin*, No. 5, September 1993. The article gives a brief profile of the UK pharmaceutical industry.
8. Rosenblatt, A. and Watson, G. (eds) (1991) 'Concurrent engineering', *Spectrum*, magazine of the IEEE, New York, July. The report also gives a list of terms and techniques and a bibliography.
9. Latham, Sir Michael (1994) *Constructing the Team*, HMSO, London.
10. Lutz, R. A. (1994) 'The re-engineering of Chrysler', Hinton Memorial Lecture, Royal Academy of Engineering, London, October.
11. 'A car is born', *Business Week*, 13 September 1993, describes Honda's development of the 1994 Accord. The article gives some insights into Honda's new product development and corporate strategies. It mentions that 33 Japanese and 28 American suppliers were involved, which between them provided 60 to 70% of the car's value.
12. Hauser, J. R. and Clausing, D. (1988) 'The house of quality', *Harvard Business Review*, May–June.
13. IM is described in Hammer, K. and Janes, F. R. (1990) 'Interactive management', *Journal of Operational Research Society OR Insight*, 3, 1, Jan–March.
14. Lipson, C. and Sheth, N. J. (1973) *Statistical Design and Analysis of Engineering Experiments*, McGraw-Hill, New York.
15. 'Towards a comprehensive materials balance', *Siemens R&D Review*, Fall 1993.
16. 'Research comes back to the nest' (1992) *Financial Times*, 14 July 1992.
17. Pralahad, R. and Hamel, G. (1990) 'Core competence and the concept of the corporation', *Harvard Business Review*, May–June.
18. Kaounides, L. (1995) *Advanced Materials Technologies: Management and Government Strategies in the 90s*, Financial Times Management Reports, London.

BIBLIOGRAPHY

Concurrent Engineering: Concepts, Implementation and Practice, edited by C. S. Syan and U. Menon, Chapman & Hall, London, 1994, gives an overview of the history, tools and techniques of concurrent engineering. Chapter 2, 'The Ford Experience' by J. V. Chelsom, describes two examples of concurrent engineering from the European automotive industry and compares management of these new product programmes with similar programmes without CE.

Simultaneous Engineering: the Executive Guide, by J. Hartley and J. Mortimer, Industrial Newsletters, Dunstable, 1991 (2nd edn), with contributions by J. V. Chelsom, provides many examples of automotive industry practice and explanations of some of the SE/CE tools.

7

Organizations

7.1 INTRODUCTION

The development of organizational structures has proceeded along lines similar to early attempts to improve labour productivity. Activities became more and more specialized, and at the same time there were increasing efforts to measure and control the performance of the specialists. The main debate was about centralization or decentralization of management control, with general acceptance of a broad division between 'staff' and 'line' functions. These divisions of labour, combined with personal reward systems and status linked to numbers of personnel controlled, led to the growth of bureaucratic hierarchies ill-equipped to deal with changing customer requirements and social developments.

Various alternatives to staff and line were attempted, with much attention being paid to organizational change in the hope of achieving faster development of new products. Where western manufacturers have modelled their organizations on successful Japanese companies the results have often been disappointing – because the changes have not been accompanied by fundamental changes of attitude and corporate culture. To be competitive in quality, cost and innovation, it is necessary to remove internal barriers to communication and teamwork, and to give the new product design teams direct access to external customers, who may be the best source of new product ideas, as well as to team members from suppliers (who may be the best source of new process ideas).

A major reason for the internal barriers and lack of teamwork in western engineering organizations is the widespread adoption of the structures and control systems introduced at General Motors Corporation of the USA in the early 1920s. Centralized control of these western companies has often been dominated by the finance function, which, as a result of consistently recruiting and training the best graduates, was staffed by very talented people who were able to out-debate engineers, even in engineering matters. Manufacturing and research and engineering people have not only been out-debated, they have been outstripped in the promotion stakes as a result of their lack of business and communication skills.

This chapter examines the development of organizations from simple structures designed to assist management, through a stage where centralization provided effective management control, and a further stage where organizational complexity became an obstacle, until the present day when efforts are being made to recreate simplicity and responsiveness.

7.2   GROWTH OF THE HIERARCHIES

The original divisions between staff and line were simple enough: those on the line were the doers – product designers, producers, and sellers and service people – and staff were enablers – they kept the accounts, handled personnel matters, interpreted the law and so on. Organizations were correspondingly simple, as shown in Figure 7.1.

In this first stage of development, the boss employed an office manager to take care of everything in the offices, and a works manager to take care of everything in the works.

As the business grew, the division of labour principle was applied to the offices as well as the works, and the organization became more complicated. The boss acquired a grander title, such as chief executive officer (CEO). This is shown in Figure 7.2.

Then the organization became more complicated still, with sales becoming sales and marketing, accounts becoming accounting and financial analysis, personnel becoming labour relations, organization planning, training and recruitment, management development and so on.

Complexity bred complexity, and financial analysis developed into project analysis, budget analysis, manufacturing cost analysis, purchase cost analysis, analysis cost analysis (!) and many more. Then, with the setting up of separate product divisions or regional divisions, the whole structure was replicated at lower levels and the original staff activities were elevated to a new 'corporate staff' status – see for example the chart for DuPont in Figure 7.3.

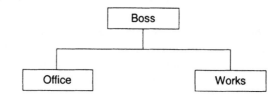

Figure 7.1   The first stage of organizational development

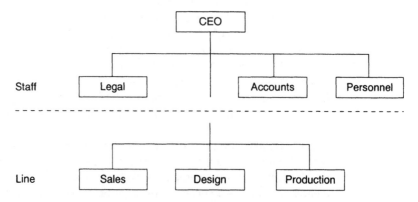

Figure 7.2   The second stage of organizational development

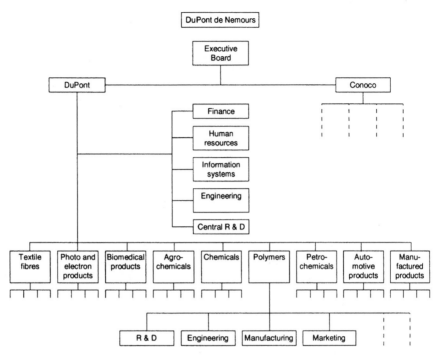

Figure 7.3   The organization chart of DuPont de Nemours

The chart in Figure 7.3 shows some of the structure for the polymers division, with some duplication of the central staff functions – engineering and R&D, for example. Similar structures exist in the other product divisions. What the chart does *not* show is the extent to which the polymers division is allowed to act independently, or to what extent its management is controlled by the executive board or central staff. The difference is a matter of corporate *culture* or *style*, rather than one of *organizational structure*, a very important distinction. If the divisions are fully empowered, with delegated authority to manage their own affairs, they can be regarded as separate or strategic business units (SBUs), which are discussed later in this chapter. If the divisions are centrally controlled, they are just part of the total DuPont bureaucratic hierarchy, and will behave in a completely different way from an SBU.

The growth of such hierarchies in many western engineering companies provided wonderful career opportunities, particularly if combined with a job evaluation or grading system that gave credit for the number of employees supervised. Three or more analysts doing similar work would be coordinated by a senior analyst, and a section of nine or ten analysts with three senior analysts would acquire a supervisor. The need to coordinate three or four supervisors would lead to the appointment of a manager to manage them, and four or five managers would be headed by a director. Directors might be exalted to the status of executive director, and enough of these would generate their very own vice-president, or even an executive vice-president.

These organizational developments are part of the problems of growth (doing more of the same thing) and development (doing something different), which

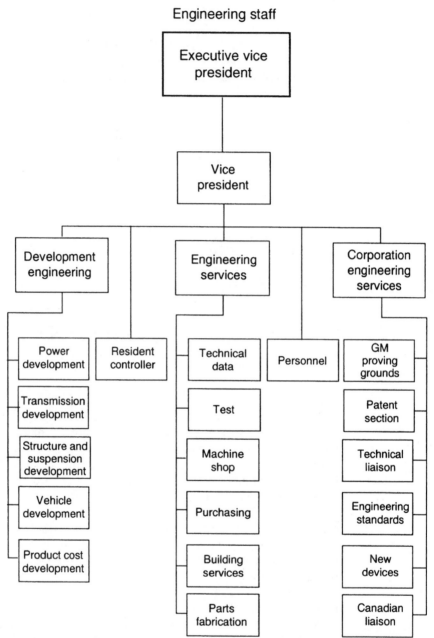

Engineering staff

sooner or later lead to the business becoming too big for the owner/manager or small generalist team. Hence the 'division of labour' in the offices as well as on the shop floor.

The dangers of monolithic structures – slow reactions, stifled initiative, poor communication etc. – were recognized by many companies, and for a long time the most effective response was some form of decentralization. One of the most successful examples was General Motors (GM), which grew by acquisition over a period of less than ten years from one car producer to a raft of vehicle and

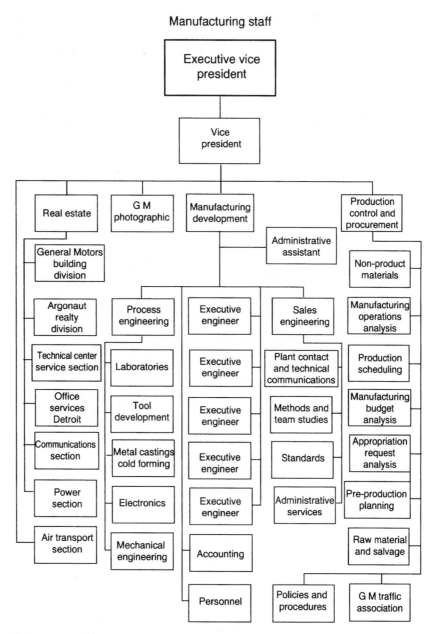

Figure 7.4 Engineering and manufacturing organizational structure at General Motors
Source: © 1963 by Alfred P. Sloan. Reprinted by permission of the Harold Matson Co.,
Inc.

component companies, too complex to be managed by the original entrepreneur,
William C. Durant. It took the systems of DuPont, who bought a major share-
holding in GM, and the genius of Alfred P. Sloan, who came to GM with one of its
acquisitions, to establish the pattern of central financial control and decentralized
operations that provided the model for many other successful organizations. From

the early 1920s, when Sloan became head of GM, the combination of his skills and DuPont's financial control systems enabled GM to become the world's largest industrial company by 1962. But by the 1970s, Sloan's creation was beginning to prove inadequate for the changed customer requirements of that decade. Some of the GM organization charts from 1963 are shown in Figure 7.4. Though dated, these charts still carry important messages:

- *'Size matters'* – true. To operate globally, engineering and science-based companies need huge resources. Witness the continual mergers and alliances in pharmaceuticals, aerospace, automotive and petrochemical industries.
- *'Big is beautiful'* – false. Organizations with hierarchies like those in Figure 7.4 found it difficult to cope with the pace of changes in technology and in customer requirements.

Later in this chapter, section 7.4 describes how one company, ABB, continually reorganizes (that is, *manages*) on a global scale, and uses its human resources to deliver local responsiveness from its massive technological resources.

It is possible to get some idea from the charts in Figure 7.4 of the bureaucracy that weighed GM down, but there were even more barriers than the charts can display. The chart 'engineering division divisions' in Figure 7.5 hints at these barriers.

The product development division is shown divided according to the length of time separating the subdivision's work from its application to the product. The manufacturing division is divided according to the type of engineering function performed. In an automotive company there would be further vertical divisions by vehicle type or by manufacturing activity, as shown in Figure 7.6.

Product engineering	Manufacturing engineering
Vehicle engineering System engineering Component engineering Advanced engineering Development engineering Research engineering	Process engineering Industrial engineering Tooling engineering Safety engineering Quality engineering Plant engineering

Figure 7.5   Engineering division divisions

Product engineering				Manufacturing engineering		
Small car	Medium car	Large car	Van	Power train	Body and assembly	Component plants
Vehicle engineering System engineering Component engineering Advanced engineering Development engineering Research engineering				Process engineering Industrial engineering Tooling engineering Safety engineering Quality engineering Plant engineering		

Figure 7.6   More engineering division divisions

There is no way to illustrate the barriers caused by personal attitudes within and between these divisions, but the most frequent analogy is 'organizational chimneys', as illustrated in Figure 7.7.

Information or an idea had to go all the way up one chimney, say the product development chimney (with wider and wider gaps to bridge between levels of organization within the chimney) before it could cross to, say, the manufacturing chimney, down which it had to make its way to the expert who could provide a sensible response. The response had to follow the same tortuous route in reverse. In many companies, this crossover at the top of the chimneys was complicated by conflict between the two functions.

The finance activity, which had responsibility for providing management with the means to exercise central control, was often blamed for fostering this conflict. They had become disablers rather than enablers, allocators of goals and blame, rather than providers of advice and assistance.

Companies with strong finance activities and with 'divide-and-rule' policies were common in North America, and their management style was widely copied in the West until the 1980s. The strength of the finance departments stemmed from three factors:

(1)   Their location at the top centre of organizations.
(2)   The role of 'controllers' assigned to them by 'Sloanism'.
(3)   Their policy of recruiting the best graduates.

Hayes, in a chapter entitled 'The Making of a Management', describes Ford's recruiting policies in the USA as follows:

'The top three graduates from Carnegie regularly went to Ford year after year. By 1969 the company's remarkable hiring offensive had recruited

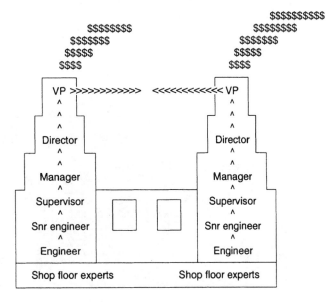

Figure 7.7   Organizational chimneys

twelve hundred executives with MBA degrees, and twenty two hundred with undergraduate degrees in business administration. Half of the MBAs had engineering degrees as well. Most of them were in finance ... Neither Manufacturing nor Engineering placed the same emphasis on the calibre of its recruits ... It was not surprising that they were so often out-debated in their own specialty by finance men' [1].

Not only were the manufacturing and engineering people out-debated, they were also out-stripped in the promotion stakes as a result of their inability to shine in discussions and their lack of business skills.

Placed in corporate staff at the top of organizations divided into functional chimneys, talented people like those recruited by Ford finance became one of the major inhibitors to teamwork and rapid new product development. Halberstam quotes Hal Sperlich, who was vice-president of product development at Ford, and later at Chrysler to illustrate this:

'One of the worst things about being in a finance-driven company was that it took the men from manufacturing and product, men who should have been natural allies, and made them into constant antagonists' [2].

The organization, recruitment and personal development policies of Japanese engineering companies enable them to avoid this divisiveness. Promising engineers and scientists are given the opportunity of education in management and are placed in a variety of functions within the company. As a result, there are people in sales, finance, personnel and corporate planning with both technical and business skills.

Two charts in Figures 7.8 and 7.9 illustrate (for fictitious companies) the differences in organizational location of engineers and technologists in Japanese and western engineering companies. The shaded areas in the charts represent engineers and technologists. In the Japanese company, these people are dispersed throughout the organization and they form the majority in the upper echelons. In the western company, the engineers, scientists and technologists are shown to

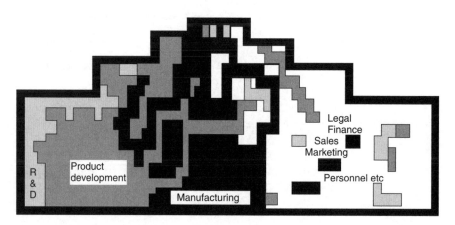

Figure 7.8   Typical Japanese engineering company

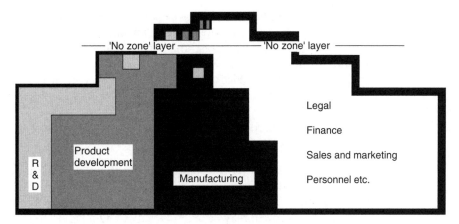

Figure 7.9   Typical western engineering company

stay in R&D, product development and manufacturing – the functions to which they were recruited – and to be outnumbered in the upper echelons.

7.3   TEAM TIME

It was the early 1980s before most of the megalithic organizations recognized the need to change, and in particular to change their new product development process and their attitude to employees. Teamwork and employee involvement became the order of the decade.

In the late 1970s and early 1980s (as described in Chapter 3), two separate sets of studies came to similar conclusions. One set was conducted by the wave of westerners going to Japan to try to identify reasons for the success of so many Japanese companies in export markets. The other – in some ways a deliberate counter to the first – was an examination of US and European companies by the management consultants McKinsey and Co., which came up with its 7Ss of management and subsequently the 'search for excellence' in American companies, using the 7S structure as an analytical tool. (The work was popularized by Peters and Waterman in their book *In Search of Excellence* [3] and in Peters' roadshow and videos.)

The conclusions included the view that teamwork was essential for rapid new product development and for continuous improvement of quality and, hence, productivity. Since then, several organizational schemes designed to encourage cross-functional cooperation have been developed, and many programmes to change corporate cultures have been started.

General Motors, for example, in 1982 began a restructuring of its product development and manufacturing staff [4]. Instead of being in two separate chimneys, product design engineers and manufacturing process engineers were brought together in two huge teams. One was concerned with existing products, called 'current engineering and manufacturing services'; the other dealt with longer-term projects and was named 'advanced product and manufacturing engineering'. At the same time, the whole of the North American GM operations embarked on programmes to promote employee participation, under the heading 'quality of worklife'. Ten years later they were still working on it.

The main reason for this, and other changes mentioned below, not being pursued to completion was a dramatic recovery in the US car market in the mid-1980s. This enabled GM and Ford to make record profits by doing what they had done in the past – sell big cars and small trucks – and to benefit briefly from their partly implemented programmes.

Ironically, Alfred P. Sloan, the long-time head of GM, had warned of this 'danger of success'. In 1963 he wrote:

> 'Success may bring self-satisfaction. In that event the urge for competitive survival, the strongest of all economic incentives, is dulled. The spirit of venture is lost in the inertia of the mind against change. When such influences develop, growth may be arrested, caused by a failure to recognize advancing technology or altered consumer needs, or perhaps by competition that is more virile and aggressive' [5].

GM also initiated a change to 'programme management', so that each new car development programme was directed by one senior manager, with their own teams of product engineers, production engineers, stylists, market researchers, finance people and materials management and so on, and responsibility from product concept through to the start of production. This type of structure is often called 'heavy programme management' (see for example Dussauge *et al.* [6]). A similar style, where the specialists participate in the team effort, but remain part of their parent function, is known as 'lightweight programme management'.

Both of these styles are western versions of something seen in successful Japanese companies, but they have rarely been successfully copied. The teams have instead become another part of the bureaucracy, watchers and reporters rather than doers.

However, the Chrysler Corporation developed a successful form of programme management that was *not* copied from the Japanese. In 1990, faced with disastrous business performance and the prospect of bankruptcy, Chrysler embarked on a complete 'reengineering' of the company, which included the introduction of what it calls the 'platform team organization'. Robert Lutz, President of Chrysler, has said:

> 'We did indeed study, and learn from, many companies, including – yes – some Japanese companies. And, tough as it was to do, we also said "Rest in peace" to the "old Chrysler" and left it to die' [7].

Lutz describes how Chrysler moved from development of new products by 'traditional, vertically-oriented functional departments' (shown in Figure 7.10), which had become '... little bureaucratic empires ... just chock-full of "re-do loops" – mis-communications, false starts, doubling back to do again what should have been done right the first time ... because nobody really worked together as a team'.

He also describes why Chrysler changed its methods and organization:

> '... this system did indeed serve Detroit fairly well for decades. But in this new era of intense global international competition, products simply weren't

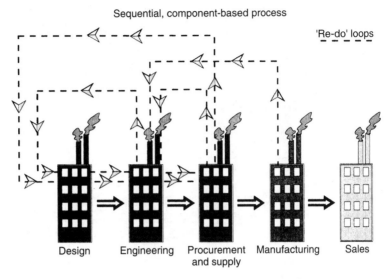

Figure 7.10   Traditional vehicle development process at Chrysler

being developed fast enough, inexpensively enough, or – truth be told – *good enough*' [7].

Chrysler introduced four 'platform teams', named after the basic underpinnings or 'platform' of any given vehicle type, as shown in Figure 7.11.

Each of Chrysler's platform teams was headed by a truly empowered leader. The success of these teams was due in large part to the fact that the leaders were not empowered by top management – the leaders *were* top management. Each leader was also an operating vice-president, as shown in Figure 7.12. According to Lutz:

'What this does is create a natural interdependency among and between our key people and their organizations. For instance, our team leader for large

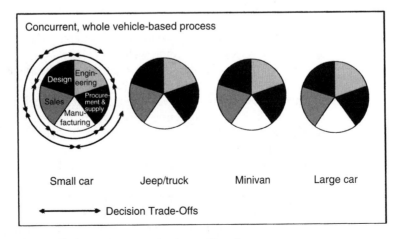

Figure 7.11   Platform team organization at Chrysler

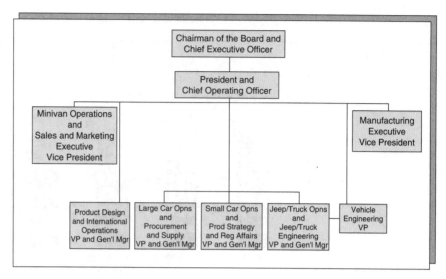

Figure 7.12   The Chrysler corporation

cars – who's also our Vice-President for Procurement and Supply – is motivated to help our team leader for minivans in any way he can. That's because he, in turn, needs the help of the minivan leader – who's also our Executive Vice-President for Sales and Marketing – when it comes to marketing his large cars' [7].

Chrysler's reengineering of its entire organization produced a turnaround in its business performance within three years. A description of the transformation of Chrysler's business performance is given in Chapter 8.

Another American automotive industry example of organizational change has been slower to benefit total corporate performance, but, nonetheless, shows recognition of the need to change from functional chimneys to a team-based organization.

The example comes from GM, which embarked on another dramatic effort to shake off the burden of bureaucracy: it formed a completely new company, the Saturn Motor Co., entirely separate from the rest of the corporation – separate geographically, financially and psychologically.

This was the first all-new GM company since 1918, and provided a sort of test bed for new ways of managing just about everything – from production to procurement, and design to distribution. It was an expensive test – the cost of the new facilities in Nashville, Tennessee was about $17 bn – but it may point the way to GM's salvation. The Saturn Car Company has already achieved a great deal, though the experience has not yet been used to change GM's overall organization or practices. Saturn's successful just-in-time or 'lean' manufacturing is used as an example in Chapter 20, at the end of section 20.4.2.

The Saturn Car Company is a separate (or strategic) business unit (SBU). Dividing a company into a series of SBUs is another of the organizational alternatives for companies seeking the same results as heavy or light programme management: that is, closer contact with customer needs, closer links between

R&D and the production process, better communication and greater employee participation.

Pralahad and Hamel [8] argue that SBUs have a number of disadvantages:

- No SBU feels responsible for maintaining a viable position in core products, nor able to justify the investment required to build world leadership in a core competence.
- Resources can become 'imprisoned' in an SBU.
- World-class research can take place in laboratories without it impacting any business of the firm.

The Nestlé food and confectionery company restructured in 1990 in an attempt to overcome this gap between research and the rest of the business, while securing the benefits of SBUs. Each SBU had a representative of R&D assigned to it as a sort of research sales representative, since research was expected to treat the SBUs as clients [9]. Nestlé's recognition of the need to exert 'market pull' on its R&D activities is an example that engineering managers should follow, even though a Swiss chocolate company may seem to be rather different from their own. However, a fine example of engineering organizational innovation from Switzerland is described in the next section.

7.4 THINK GLOBAL, ACT LOCAL

The phrase 'Think global, act local' summarizes the strategy of ASEA Brown Boveri (ABB), probably the most successful example of another organizational style, known as 'matrix' management.

The usual matrix organization has products on one axis and functions on the other, as illustrated in Figure 7.13. In the matrix organization each product has a product manager, who draws on the specialist functional resources to handle implementation of their plans. This is not far distant from the lightweight programme manager concept, or the brand manager approach of companies such as Procter & Gamble or Unilever, where the brand manager has responsibility for design, manufacture, marketing and distribution of a washing powder, a food product and so on.

Matrix organizations can be effective on a national scale, to help multiproduct companies coordinate the use of resources that are required for the design and manufacture of more than one of the products. The most common, and most effective, application of matrix organizations, however, is when they are used to manage resources on a global scale.

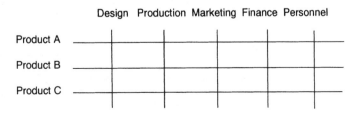

Figure 7.13   The matrix organization

Electronics and electrical companies such as Korea's Samsung and Japan's Matsushita and Sony, as well as the Japanese automotive companies Toyota, Nissan and Honda, have global strategies of designing a product in one place only and manufacturing it in several places geographically 'close to the customer', but generally their organizations are not in the matrix form. The Dutch electrical company Philips, as well as IBM and Ford, has an organization that is somewhere between the Korean/Japanese pattern and the matrix form. The variation involves setting up 'centres of excellence' or 'centres of responsibility', which have global or regional responsibility for designing or making one product line for which the centre has special skills. Without firm direction from corporate management, however, such centres will find their work being redone or duplicated on the grounds of 'not invented here'.

Where ABB differed from most companies using the matrix structure, and from those using the centres of excellence approach, was in the use of local, national heads, who had shared responsibility for all ABB products in that country – shared with product line managers who had global responsibilities. Up to 1992 the ABB product lines were organized into eight groups, as shown in Figure 7.14. The national CEOs were coordinated in five regional groups: Europe EC; Europe EFTA; North America; Latin America/Africa; Asia/Australia.

Having started a customer focus programme in 1990, ABB shifted management efforts away from the reshaping of organizational entities to concentrate on

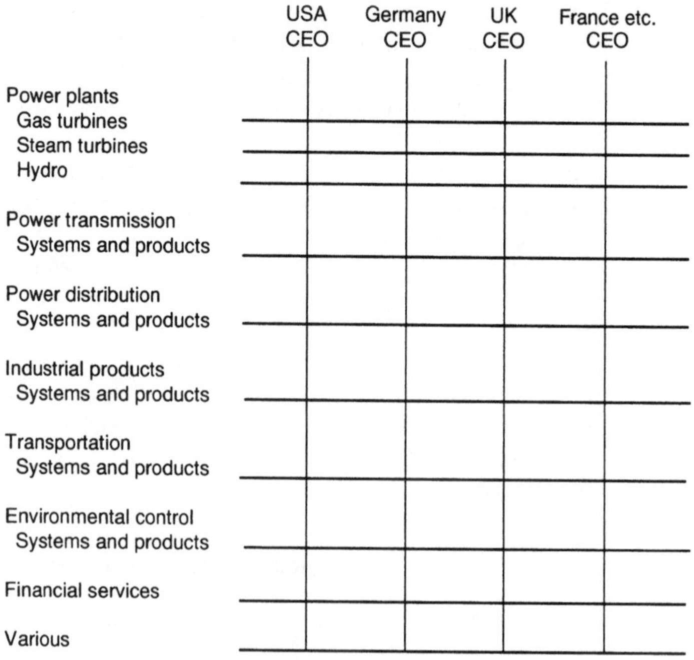

Each of the product divisions – power plants, power transmission etc. – had its own design, production, marketing and other activities shown for Product A, Product B etc. in Figure 7.13.

Figure 7.14 ABB's product line/national matrix (1992)

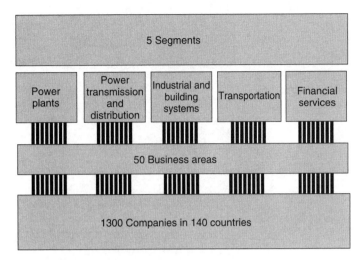

Figure 7.15   ABB global structure (1993)

improving business processes as the route to improved quality, faster completion of orders and all the other business objectives that lead to customer satisfaction (listed in Chapter 2). As a result, in 1993 ABB simplified its organizational superstructure by consolidating into four product segments, plus financial services, and three regions: Europe, the Americas and Asia/Pacific. Its central staff in Switzerland was reduced to fewer than 100 employees. The revised structure is shown in Figure 7.15.

At the same time, ABB strengthened local capabilities by establishing about 5000 independent profit centres around the world. Each profit centre had about 50 employees and complete delegated responsibility for running its part of the business. (This kind of delegation is called 'empowerment'. It should be preceded by training in decision making, so that the employees are able to use their newly delegated power.) Some of the coordinating levels of management were removed, so that there were only two levels of management between the employees in the profit centres and the executive committee in Zurich. These two levels were the 5000 profit centre heads and the 250 business area/country managers. (This removal of supervising and coordinating management has been widely practised in the West and has been described as 'delayering'.) The flat, decentralized organization is shown in Figure 7.16.

However, differential economic regional growth rates (illustrated in Chapter 2, Table 2.2), combined with declines in some product areas and potential growth in others, created strains, stresses, pressures and resistance in the 'system' that is the ABB organization. Like any mechanical or electrical system, the organization had to be modified to retain its functionality and efficiency. Similarly, the patterns of growth and decline meant that ABB had to restructure its product and services portfolio. Like a biological system, the portfolio needed removal, remedy or reinforcement of various elements to remain healthy. There were commercial pressures, too, described in Chapter 8. Combined, these led to more organizational change – still motivated by striving to provide better customer satisfaction, but with the added imperative of cost reduction.

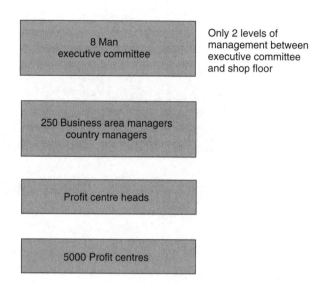

Figure 7.16 ABB – a flat, decentralized organization (1993)

In January 1997 there were major top management changes, but the basic product/regional global organization remained until August 1998. Then a new structure was announced, aimed at 'boosting business growth areas where ABB has technology advantages and unique capabilities' and at 'greater speed and efficiency by further focusing and flattening the organization'. Regional organizations in Europe, the Americas and Asia were 'dissolved after having fulfilled their mission to coordinate ABB's expansion in their respective areas' [10]. In an increasingly difficult business environment, these changes were not enough. A one-year review through 2000 led by a 'Group Transformation division', headed by an executive vice-president, resulted in the structure shown in Figure 7.17, 'the first to organize around customers rather than technologies', effective 1 January 2001, under a new president and chief executive officer [11].

These changes also were not enough to restore ABB's business performance. In 2001 the company posted its first annual loss, and in 2002 again 'streamlined its divisional structure', as shown in Figure 7.18.

The Group Transformation and Group Processes divisions, having been combined at the end of January 2002, had been dismantled by the year end. From October 2002, cost reduction, improved business practices and cultural change were pursued under a programme called 'Step Change'. In a significant shift from traditional responsibilities, this programme was led by the head of human resources – putting him in an operational rather than a staff role.

As ABB has demonstrated, organizational structures are really secondary to business processes and strategies, and may have to be adapted quite frequently in a rapidly changing business environment. It can be argued that with the right attitudes, any organization can be made to work. Equally, with the wrong attitudes, no form of organization will work. To get the right attitudes, the right processes and the right organization requires a holistic approach to business management. Chapter 8 gives some examples of companies that have managed to succeed in this respect.

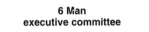

**10 Man
executive committee**

**Group
transformation**

Group processes

Two product divisions

Power Products

Automation Products

Four end-user divisions

Utilities

Process industries

Manufacturing and
Consumer Industries

Oil, Gas and Petro
chemical industries

**Financial
services**

Global customer base

(Active local companies reduced to 400)

Figure 7.17   ABB global structure (2001)

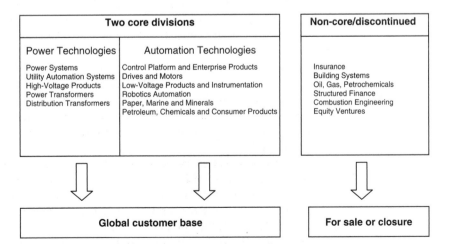

**6 Man
executive committee**

Two core divisions

Power Technologies   Automation Technologies

Power Systems   Control Platform and Enterprise Products
Utility Automation Systems   Drives and Motors
High-Voltage Products   Low-Voltage Products and Instrumentation
Power Transformers   Robotics Automation
Distribution Transformers   Paper, Marine and Minerals
                              Petroleum, Chemicals and Consumer Products

Non-core/discontinued

Insurance
Building Systems
Oil, Gas, Petrochemicals
Structured Finance
Combustion Engineering
Equity Ventures

Global customer base

For sale or closure

Figure 7.18   ABB global structure (2002/3)

7.5 SUMMARY

This chapter has traced the development of organizational structures from a simple means of sharing management responsibilities and tasks into a series of boundaries that divide companies:

- vertically – into functional 'chimneys'; and
- horizontally – into 'staff' and 'line' activities.

It was noted that organizations with hierarchical structures resulting from the management style of A. P. Sloan Jr, with their centralized financial control of separate operating divisions, were unable to respond to the challenges of the 1970s. Despite having been successful at General Motors for 50 years and serving as a model for many other engineering companies, these structures had become obstacles to communication and inhibitors to rapid response to customer needs. Some alternative organizational arrangements were described, such as heavy and light programme management, separate or strategic business units and matrix management.

The special form of matrix management organization developed by ABB was given as an example of the 'delayering' of hierarchies and the benefits of focusing first on process improvement, then on organization, rather than the other way round. ABB's experience also showed that organizations may need frequent change in today's fast-changing business environment.

REFERENCES

1. Hayes, W. (1990) *Henry: A Life of Henry Ford II*, Weidenfeld and Nicholson, London.
2. Halberstam, D. (1986) *The Reckoning*, William Morrow, New York. This 750-page book compares and contrasts the development and management of Nissan and Ford. Chapter 17, 'Deming finds an audience', reveals the basic differences in corporate culture between American and Japanese car companies.
3. Peters, T. and Waterman, R. (1982) *In Search of Excellence: Lessons from America's Best-Run Companies*, Harper & Row, New York. Tom Peters has changed his mind about what the lessons were, and some of the 'excellent' companies have suffered in the marketplace since the book was written. Nevertheless, they were successful for many years and the lessons are there to be considered.
4. 'Will success spoil General Motors?' *Fortune*, 22 August 1983.
5. Sloan, A. P. Jr (1963) *My Years with General Motors*, Pan, London.
6. Dussauge, P., Hart, S. and Ramanautsen, B. (1993) *Strategic Technology Management*, Chapter 8, John Wiley & Sons, Chichester.
7. Lutz, R. A. (1994) 'Re-engineering the corporation for the '90s', The 1994 Christopher Hinton Lecture, delivered to The Royal Academy of Engineering, 4 October.
8. Pralahad, R. and Hamel, G. (1990) 'Core competence of the corporation', *Harvard Business Review*, May–June.

9.   'Research comes back to the nest', *Financial Times*, 14 July 1991 describes Nestlé's reorganization of its R&D activities to bring them closer to the marketplace.
10. 'ABB realigns business segments, names new Group Executive Committee', Press release, Zurich, August 12 1998.
11. ABB Group 2002 Annual Report.

BIBLIOGRAPHY

'The logic of global business' (an interview with ABB Chief Executive, Percy Barnevik), *Harvard Business Review*, March/April 1991, describes Barnevik's vision of ABB's development.

'Resetting the clock' and 'ABB managers strip for action' also give information about ABB's restructuring. These articles appeared in the *Financial Times* on 10 February 1993 and 25 August 1993 respectively.

The ABB website www.abb.com provides 10 years of press releases, financial information and 'some fun elements like online learning tools, electronic technology games and an interesting timeline covering ABB over the past 120 years'. It can also be used to 'find the configuration data for a substation or get monitoring statistics for preventive maintenance'.

8

Managing to Succeed

8.1 INTRODUCTION

This chapter draws together the separate themes of Chapters 1–7. It starts with a general diagnosis of the shortcomings of many western companies, but shows that there are companies that have managed to succeed against a background of changing customer requirements and a changing business environment.

Earlier chapters have shown that technological change and globalization are the most demanding management challenges of the 1990s and the early years of the twenty-first century. Zero defects quality and just-in-time (JIT), or 'lean manufacturing', are the performance base from which excellent companies build their lead in customer care through technological advances in materials, products and processes, applied on a global scale.

Engineers and technologists will be the key players in creating that performance base and managing the new technological developments, through *teams* containing in-house and supplier and customer members. In these teams, engineers, scientists and technologists not only have to work with other engineers, scientists and technologists in new cooperative ways, they also have to work with economists, sociologists, accountants and lawyers.

This chapter looks for models of the organizations, methods, skills and style that will typify the excellent companies of the twenty-first century. By learning from these examples of companies that have 'managed to succeed', potential managers can prepare for a role in the development or creation of other successful companies. There are also lessons to be drawn from companies that have not kept up with competition and change, and the chapter includes some of these.

8.2 SLOWLY SINKING IN THE WEST

There are some general explanations for the widespread failures of western management in the past two decades. The inability of many companies to respond to the changed needs of their customers can be attributed to five major weaknesses.

8.2.1 Western weakness number 1

Application of 'Taylorist' specialization and 'Sloanist' divide-and-rule policies created bureaucratic, top-heavy organizations that have:

* obstructed internal communication;
* created warring internal 'empires';
* destroyed cooperation and teamwork;
* focused attention on internal conflicts rather than the outside customer.

The narcissistic style of large corporations has been copied by some small and medium-sized enterprises too, which, far too soon, turned themselves from 'Stepfoot and Son' into 'The Stepfoot Group of Companies', complete with executive directors or vice-presidents.

Consequently, in large and small companies, employees have been placed in organizational boxes with restricted responsibilities. They have been invited to leave their brains and initiative behind when they come to work.

To overcome this weakness, engineers, and other specialists, have to rise above the rim of their boxes and look in one direction to their customers – internal and external – and in the other direction to their suppliers, also internal and external. They have to ensure that they understand their customers' needs, and that their own needs are understood by their suppliers.

8.2.2 Western weakness number 2

Performance measurement and reward systems in many companies have encouraged employees to concentrate on demonstrating their individual abilities or, at best, the abilities and strengths of their own department, rather than working in cooperation with other company activities and with customers and suppliers. For design engineers or manufacturing engineers to seek or accept advice (particularly from suppliers) has been regarded as a sign of weakness, or even a sign of incompetence. Dictatorial management styles, combined with divisive organizational structures and individual-focused performance measurement, have led many managers to adopt confrontational attitudes and to have a self-regard verging on arrogance.

To rectify this, when they do come out of their boxes, managers need to listen and learn, to communicate and cooperate.

8.2.3 Western weakness number 3

The third weakness is success: something that Sloan [1] warned could bring:

* self-satisfaction;
* dulling of the urge for competitive survival;
* loss of the spirit of venture in the inertia of the mind against change;
* arrested growth or decline, caused by failure to recognize advancing technology, or altered customer needs, or competition that is more virile and aggressive.

General Motors itself, as well as IBM, Xerox and Pan American Airways, were blinded by their own success and suffered in the way that Sloan predicted.

'Benchmarking', the practice of comparing a company's performance with that of successful companies in other industries or with the company's most successful competitors, has similar risks. If the company making the comparisons finds that it performs better than the industry average, or that some parts of its organization are 'world class', there is the same danger of self-satisfaction and a reduced spirit of venture.

The dangers of success can be avoided by embracing the policy of never-ending improvement. As soon as one success is looming, managers should be working towards the next success and the next after that – constantly reinventing their companies, as described in Chapter 3.

8.2.4   Western weakness number 4

The fourth weakness is imitation. Too many western companies identified what they believed to be the practices, policies, strategies or structures that generate superior performance by their competitors, particularly Japanese competitors, and sought to copy them. When this was done without complete understanding of why and how these practices, policies and so on work, the attempts at imitation were detrimental rather than beneficial.

Robert Lutz, when president of the Chrysler Corporation, said that Chrysler learned from studying the Japanese, but warned against superficial attempts to copy them. He used an analogy to make his point. This is taken from a discussion following the 1994 Christopher Hinton Lecture:

'A four year old child wants to learn how to swim. The child's father takes the child next door where they have a pool, and encourages his child to watch the neighbour's four year old swimming up and down the length of the pool. After ten minutes father asks "Have you got the idea?" The child says "Yes", so the father throws the child in and it drowns.'

Attempts to introduce total quality management, just-in-time materials management, employee empowerment or simultaneous engineering without first providing employee and supplier training in the new philosophies, or without management belief in and understanding of the philosophies, can be almost as disastrous as that swimming lesson.

Managers should make sure they fully understand what they observe, add their own expertise to adapt what is beneficial to their own circumstances, and practise the new methods on a limited scale until their local applicability is proven.

8.2.5   Western weakness number 5

The fifth weakness is the inability to recognize, foresee and manage change. This is particularly damaging when combined with the dangers of success. Sloan [1] actually lists it as one of the dangers of success. He explained GM's replacement of Ford as leader of the US car market by saying of Henry Ford I: 'The old master has failed to master change.'

It was change in market requirements that Henry Ford I failed to master. The same can be said of IBM, Xerox and most of the European automotive component producers.

It is not only marketplace changes that need to be managed. All the external factors listed in Chapter 2 have to be carefully monitored. These include technology, legislation, exchange rates, demography, political, educational and sociological trends, as well as internal factors such as employee aspirations and abilities.

Most companies can react successfully to dramatic change, such as a factory burning down or being flooded. Such disasters receive focused attention and often generate a strong team spirit among the whole workforce in their efforts to recover the situation.

It is the slower, more fundamental changes that are not well managed. Senge [2] borrows an analogy from Tichy to illustrate this management weakness. With apologies to animal lovers and environmentalists, it is as follows:

> 'Attempts to boil a frog by dropping it into boiling water fail because the frog reacts to the dramatic change by jumping out of the pot. However, if the frog can be lured into the pot with a few flies and a lily pad, and the water is gradually warmed, the result is different. The frog may feel more comfortable as the water warms, and not notice until it is too late that the water is boiling, and end up dead.'

Managers should be constantly alert to changes in their own working environment and in their company's environment to avoid a comfortable, but premature, curtailment of their careers.

8.2.6   Forward to BASICS

The five weaknesses can be overcome by BASIC behaviour. The new BASICS for managers are as follows:

- *Boxes* – get out of them to listen and learn with your customers and suppliers.
- *Attitudes* – be team players, not stars; seek cooperation not confrontation; consider the other point of view; accept that there are experts other than yourself.
- *Success* – don't be blinded by it. Remember Sloan's advice.
- *Imitation* – what works for others may not work for you. 'It is no use simply to take on board the latest flavour of the month such as quality circles … only to find that these concepts cannot be successfully transported to an environment which remains alien to it' [3]. Listen, learn, but add your own magic.
- *Change* – don't be a boiled frog.

If an 'S' is required, be 'Systemic' – take a holistic view and work on management of people, quality, materials, new products and new technologies, and customer and supplier relationships concurrently. That is what the successful companies do. Who some of them are and how they do it are covered in the balance of this chapter.

8.3   DOES ANYBODY KNOW THE WAY? SIGNS FROM THE 1990S

What should be examined and who should be followed to find the way to succeed?

There is no point in looking at organizational structures, they tell us nothing. It is the way people behave within, between and outside their organizational boxes that makes the difference. So are there models of successful behaviour from which general rules can be derived?

The two long-time models and one-time largest industrial corporations in the world, General Motors and Ford, are both struggling with the same issues of inadequate quality, bureaucracy, long lead times and uncompetitive costs. Ford launched a plan in 1990 labelled 'Ford 2000', designed to create a truly global corporation for the twenty-first century. Harvard Business School called this 'the mother of all Business Process Reengineering' and it was expected to be a model for managing on a global scale. The plan failed, mainly because it took the company ten years to recognize that there is not always a global solution to manufacturing and marketing problems. GM has continued with 'business as usual'. Where else can we look?

From the top European companies, Royal Dutch/Shell may give one of the best leads towards the way to succeed. (They also give one of the worst examples of the dangers of success – see section 8.5.3 below.) In the closing decades of the twentieth century, Shell moved from being one of the marginal members of the 'Seven Sisters' (as the world's leading oil companies were known in the 1970s) to being the largest company in Europe, ahead of BP, and consistently one of the most profitable oil companies. One of the management actions contributing to this improvement was a process that Shell called 'planning as learning' [4]. This involved all the senior managers in 'What if?' business games – such as 'What would Shell do if the price of crude oil increased fourfold?' The resultant plans, which had been developed jointly by the top managers, were filed, and when one of the apparently unlikely events actually happened (the world price of crude oil *did* increase fourfold in the 1970s), Shell was much faster to respond than its competitors. The planning-as-learning process also helped Shell's performance through the 1980s by changing managers' perceptions of themselves, their jobs, their markets, the company and their competitors.

This gives three interesting characteristics of a successful company and its management:

- a dramatic performance shift;
- a built-in learning process;
- a change in management outlook.

There are other companies that have done something similar. Two that had a dramatic performance shift forced on them are Xerox and IBM, and their responses are described below.

Having lost its world dominance of the plain-paper copier market to the Japanese, Xerox used to be the case-study example of American business downfall, but it recovered sufficiently to be used as a benchmark company that has turned itself around from loser to winner.

It embarked on a five-year programme to become a total quality company, after a six-month study by a 25-strong task force. The key elements of the Xerox

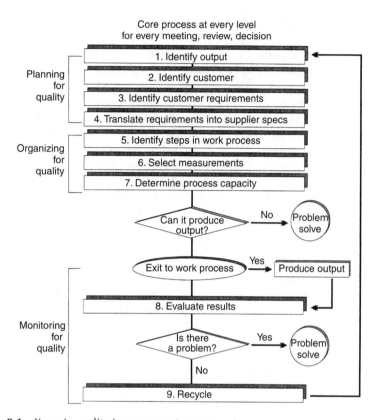

Core process at every level
for every meeting, review, decision

Planning for quality
1. Identify output
2. Identify customer
3. Identify customer requirements
4. Translate requirements into supplier specs

Organizing for quality
5. Identify steps in work process
6. Select measurements
7. Determine process capacity

Can it produce output? — No → Problem solve

Exit to work process — Yes → Produce output

Monitoring for quality
8. Evaluate results

Is there a problem? — Yes → Problem solve

No

9. Recycle

Figure 8.1   Xerox's quality improvement process

transition process and the five stages of its 'leadership through quality' plan are shown in Chapter 4. Aspects of the Xerox recovery strategy that are generally applicable in managing to succeed, and therefore relevant to this chapter, are:

- Quality was identified as 'the basic business principle'.
- Xerox developed a universal quality process (see chart in Figure 8.1).
- Xerox went beyond customer care or customer focus to 'customer obsession'.
- Cooperation was encouraged throughout the supply chain by what was called 'team Xerox' – members were *all* customers, *all* employees, *all* suppliers (but the plan involved reducing the Xerox supply base from 5000 to 400 suppliers worldwide).
- Benchmarking, looking out, not in at Xerox.
- Business unit teams, not functional structures.
- New attitudes to employees and suppliers.

The Xerox quality improvement process illustrated in Figure 8.1 includes two features that are particularly important for improving business performance: customer focus (steps two and three) and continuous improvement (step 9). However, the heading is even more important for young managers: use of the process at every level, for every meeting, every review and every decision is not confined to Xerox. Every manager can use it.

Xerox did not fully recover its market dominance by the end of its five-year plan, but it managed to reverse its decline, which is a major step towards managing to succeed. From its improved base, Xerox repositioned itself as 'the document management company', offering complete electronic information management solutions rather than document reproduction. This brought it into more direct competition with two other model US companies – IBM and Hewlett-Packard.

IBM was seen to be in deep trouble in the early 1990s, but was already planning corrective action before the full extent of its difficulties was made visible in the press. Its decline was rapid: in 1979 IBM was one of Peters and Waterman's 'excellent' companies, and from 1982 to 1985 *Fortune* magazine's 'most admired' American company. By 1989 it was 45th in the *Fortune* list! Like Xerox and GM, it was a victim of its own success (remember Sloan's warning) and of the habit of introspection. As a result of looking inwards and backwards at itself, IBM believed that it was a leader in quality and felt some pride in its achievement of almost halving product defects every two years, but when it looked outwards it found that competitors had passed it to set new quality standards. It also thought it was a cost leader, but found its costs were 20% above those of its main competitors. Through the 1980s, IBM had attempted various piecemeal improvement programmes, but by 1990 the company had realized that it had to change its whole view of competitiveness, and set out to achieve total customer satisfaction through 'market-driven quality'. This programme was described in a lecture by Heinz K. Fridrich of IBM, from which the following material has been drawn [5].

The IBM research in preparation for its recovery programme identified five characteristics of world-class companies, which are shown in Figure 8.2. The

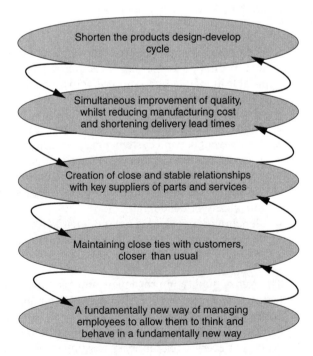

Figure 8.2   Fridrich's five 'component technologies' of market-driven quality at IBM

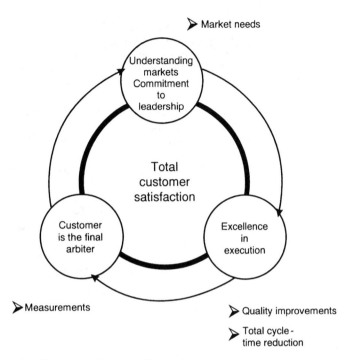

Figure 8.3   IBM's market-driven quality circle

five 'component technologies', as Fridrich called them, have to be practised simultaneously as a single strategy. In terms of the McKinsey 7S structure (see Chapter 3), the last three are 'soft' S requirements: person-to-person interactions with customers, suppliers and employees.

The starting point of IBM's programme was an understanding of the market and the end point was the customer, as shown in Figure 8.3.

To implement its programme, IBM simultaneously pursued the twin goals of quality improvement and total cycle-time reduction. This is illustrated in Figure 8.4. (It was pursuit of the goals that was expected to generate the improvements shown – for example, cycle-time reduction does not itself reduce defects, but striving for reduction may have this effect.)

Like Xerox, IBM did not recover its dominance of world markets and had to recognize that the structure of the industry had changed. However, it has shown that it, too, can change. It *has* introduced new products faster, it *has* improved quality and reduced cost, and it *has* returned to profitability (having briefly held the world record for making the largest one-year operating loss). Also like Xerox, IBM has repositioned itself for the twenty-first century as a provider of complete business management solutions, rather than a supplier of data-processing equipment.

There are elements common to the Xerox and IBM programmes. They were both striving to:

- consider the customer (to the point of obsession);
- change the way they manage people (so that all are involved);

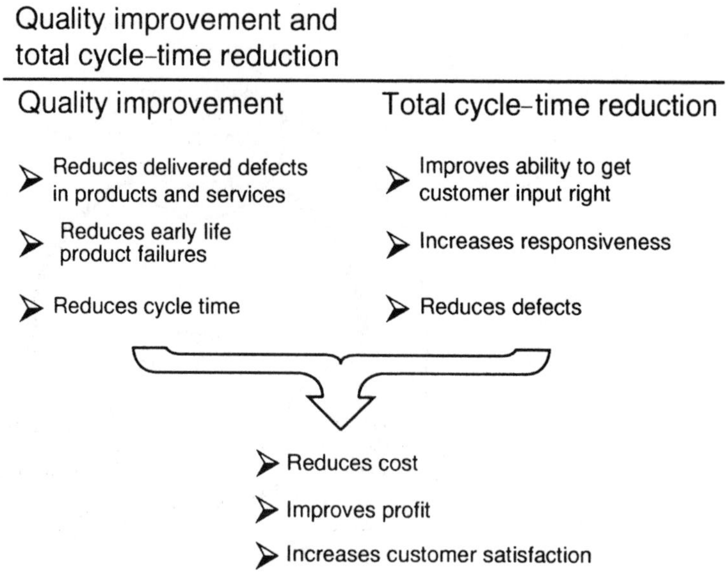

Quality improvement and total cycle–time reduction

Quality improvement	Total cycle–time reduction
➤ Reduces delivered defects in products and services	➤ Improves ability to get customer input right
➤ Reduces early life product failures	➤ Increases responsiveness
➤ Reduces cycle time	➤ Reduces defects

➤ Reduces cost

➤ Improves profit

➤ Increases customer satisfaction

Figure 8.4   IBM's twin goals

- simultaneously improve quality, cost and delivery time (by orders of magnitude, not by the 'old' target of 5% a year);
- include customers and suppliers in close and lasting relationships ('cooperative contracting', Xerox called it);
- concurrent design of product and process by teams that include suppliers;
- improve execution by shared learning from past mistakes.

There was another company that did all these things, was outstanding in managing technological change, and was not forced to do so by near disaster. This consistently successful company was Hewlett-Packard (H-P). Though now, like IBM and Xerox, repositioned as an IT solutions provider rather than an innovative IT product company (and merged with Compaq as the two companies strive to compete on a global scale), H-P's history provides lessons that are still useful. Phillips [6] described some of the successful H-P practices, which are summarized below.

Communication was informal – H-P invented the term and practice of 'management by wandering around'. Information was exchanged freely between people at different levels and in different functions, helping them to behave 'more as a team than as a hierarchy'.

Goals were achievable – with effort. This gave everyone the chance to be a winner and people to share in one another's success: 'you don't have to embarrass a loser to be a winner'. Stretch goals were regarded as '... bad motivationally, and they also lead to organizational self-deception, as people begin misrepresenting what is really going on, in an ultimately futile effort to hide the infeasibility of stretch targets.'

Learning was built in: H-P gave overwhelming attention to training and development, not just to enhance the value of human assets or make employees happy, but for two other reasons:

(1) H-P believed it was dependent on the initiative and judgement of the individual employee; it therefore followed that it should do everything possible to instil in the employee the values and expectations – and the capabilities – that enabled him or her to do what was best for the company. (If that can be done there is no need for the cost-generating controls that typify most cost-reduction programmes.)
(2) H-P believed that it had continually to adapt to a changing environment and therefore had to create a climate of continual learning within the company.

The H-P annual planning reviews illustrate how learning was part of the regular management process:

> 'These are full day affairs, attended by corporate management and top divisional managers. They are structured conversations . . . with few charts or numbers . . . used to seek a common understanding about how to develop the division's business. . .
>
> About 20% of the time is devoted to reviewing the history of past product development programmes and major investments – seeking to learn. . .
>
> The next third is devoted to new product plans . . . shaped on the basis of careful study of past lessons. . .
>
> About 15% is spent discussing a special theme. . .
>
> 20% on the management by objectives programme – not more than three or four critical things that just have to be done. . .
>
> Only about 10% is devoted to next year's financials – if they get the rest of it right, the financials will follow.'

Reviews are shared – when divisional managers review performance with their boss, the other managers are there, to stimulate and learn from each other.

Development is on the job – learning from assignments, membership of task forces, and maybe from a course to provide a special skill.

This section has described how some outstanding companies survived, avoided decline or recovered in the latter decades of the twentieth century. They are now seeking to lead the way into the twenty-first, albeit in a new guise in H-P's case.

There is no guarantee that the solutions used by these companies will apply elsewhere, but there does seem to be a general indication that centralized, heavily structured organizations, with command-and-control styles of management, cannot respond to customer needs changing at the rate they did through the 1980s and 1990s. The final section of this chapter looks at two companies that managed to succeed in the 1990s but have since had mixed fortunes.

8.4   LEADING IN THE 1990S: TWO WAYS TO SUCCESS

Two engineering companies, one American and one European, demonstrated leadership and examples of excellent management in the early 1990s. They are the Chrysler Corporation of the USA and Asea Brown Boveri (ABB). Chrysler, like Xerox and IBM, had greatness thrust on it by the prospect of extinction. ABB, like Hewlett-Packard, steadily acquired greatness, or maybe it was born great. Chrysler's achievement is described first.

8.4.1   Reengineering Chrysler for the 1990s

The achievements of Chrysler in the early 1990s can be measured by extracts from two articles in *The Economist*.

An article in the 14 April 1990 issue asked 'Are America's carmakers headed for the junkyard?' and advised 'One wise move would be for the Big Three to embrace Japanese partners still more closely . . . And if Chrysler were taken over, nobody should weep.' The article also stated: 'The only way to save America's car industry is to leave the present one to die. Detroit, rest in peace.'

An article on 17 April 1993 headed 'Japan spins off' told a different tale. It stated that 'America has changed its methods of making cars' and 'Chrysler is emerging as a star after narrowly escaping its second brush with bankruptcy. The firm is winning sales from the baby-boomers, a group that had been a keen buyer of imported cars.' It also speculated '. . . sales look so good that Chrysler could be the most profitable car company in the world this year'.

Chrysler was not taken over and it was the most profitable car company in 1993, measured by profit per vehicle. How did it make such big changes so fast?

There were five major steps:

(1)   A dramatic change from vertically oriented functional departments to 'vehicle platform' teams.
(2)   A complete change in the way Chrysler dealt with suppliers.
(3)   A lot of progress in bringing teamwork and empowerment to the shop floor.
(4)   After achieving one level of success, Chrysler determined to 'shoot for more' by adopting a mindset of continuous improvement.
(5)   A new employee evaluation system that relies 50% on the attainment of performance goals, and 50% on behaviour.

The re-organization into 'platform teams' was described in Chapter 7. The result of the reorganization was 'lots of teamwork and innovative thinking'. In turn, this led to a flood of new products that were affordable, different and well received by the market. One model, the Vision, was named by the press 'Automobile of the Year' in the USA, and number one among family saloons by *Consumer Reports* magazine. That was the first time in 12 years that the magazine had named a domestic US car as number one. The new products were brought to market faster and with fewer resources than before the reorganization. The new levels of performance are shown in Figure 8.5.

According to Lutz, Figure 8.5 'shows that, with true teamwork, success is not only repeatable, but that it can even be improved upon' [7].

	Viper	LH Sedans	Ram Pickup	Neon	Cirrius/ Stratus
Development time (mo's)	36	39	32	31	35
Development costs (US$bn)	0.08	1.6	1.2	1.3	0.9
Team size	85	850	700	740*	592

*Includes other small car platform products

Figure 8.5   Chrysler 'getting better all the time'

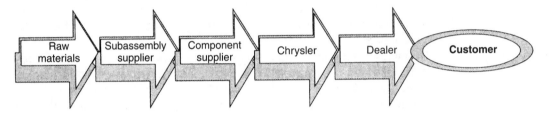

Figure 8.6   The 'extended enterprise' concept

Chrysler's new relations with suppliers involved treating them as 'true extensions of our company – as an integral, creative link in the value-added chain, just as we ourselves are merely a link'. Chrysler called this the 'extended enterprise' concept, which is shown in Figure 8.6.

Chrysler stuck to its new style of relationships with suppliers even when the competitive cost pressures of the early 1990s drove some other automotive companies to revert to type, and to seek short-term savings by transferring business to less-qualified suppliers or by threats to do so. By its consistency Chrysler retained the trust of its supply base and the latter's continued willingness to participate in the early supplier involvement (ESI) programme. Chrysler regarded ESI as an essential part of simultaneous engineering (SE), and SE as an essential part of its faster new product development. Moreover, it became clear that Chrysler's approach to cost reduction was more effective than the old '5% off everything' approach. Chrysler reckoned that its SCORE (supplier cost reduction effort) programme generated nearly 9000 cost-reducing ideas, worth more than $1 bn annually. The rate of saving increased from $250 m in 1993 to $500 m in 1994, and $750 m in 1995. This was not achieved at the expense of suppliers' profit margins, but by the elimination of waste, including waste in the Chrysler link of the extended enterprise. (More information about Chrysler's purchasing policies and performance is given in Chapter 21, section 21.4.2.) Suppliers reacted favourably to Chrysler's new methods: Chrysler was rated a close third to Nissan and Toyota in terms of 'partnership', and ahead of Honda, Saturn, Ford, Mazda and GM, in that order.

Figure 8.7   Empowerment on the factory floor

Empowerment on the factory floor also involved radical change. One change was the reduction in numbers of supervisors, as shown in Figure 8.7.

Reduced supervision was possible because 'self-directed work teams' were introduced. These teams developed their own work assignments, scheduled their own vacations, solved productivity and process problems, and trained other team members.

The mindset of continuous improvement demonstrated itself in the new product introductions and cost reductions already described. Another example was the reduction in 'order-to-delivery' time, as shown in Figure 8.8.

The new employee evaluation scheme was designed to institutionalize the principles behind the other changes. The scheme gave equal weight to the achievement of goals and the behaviour of employees. Behaviour included teamwork, empowerment and innovation, and was judged by several people in the work group as well as the employee's boss. The aim was to reward 'the champions and the foot soldiers of constructive change' rather than 'the person who meets his own goals only at the expense of others'.

While these changes concentrated on Chrysler's North American activities, the company did not neglect the long-term potential of the European market, which it

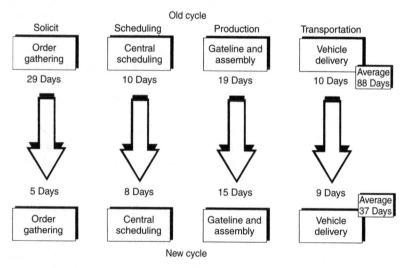

Figure 8.8   Continuous improvement in the Chrysler order-to-delivery process

re-entered in 1988. At that time, all Chrysler's European sales came from vehicles built in the USA. From 1992, however, Chrysler has been assembling minivans in Austria for sale in Continental Europe, and in 1994 assembly in Austria was expanded by the addition of 'luxury sports-utility' Jeeps. Austria's accession to the European Union in 1995 and the introduction of right-hand-drive vehicles to the Austrian-built range continued this expansion. Chrysler's low manufacturing costs in the USA allowed them to export saloon cars to sell profitably in Europe.

Summarizing the reengineering of Chrysler, Lutz [7] said:

> 'If there's one overriding lesson from our experience, I think it's that there are indeed great rewards for organizations that pay as much attention to the engineering going on in the so-called "soft" side of their business as the "hard" side.'

Chrysler's balance between hard and soft aspects of management and its use of development and delivery time compression as competitive weapons are mirrored in the next example in this chapter, which relates to ABB.

8.4.2   ABB at the threshold: Ready to 'go for growth'

ABB was formed in 1987 through the merger of ASEA AB of Sweden and BBC Brown Boveri Ltd of Switzerland. The new company began operations in January 1988 and set out to be a global leader in a range of electrical and mechanical engineering products.

Its strategies closely followed Ansoff's matrix (see Chapter 22, Figure 22.8). Ansoff's strategy 3 (new markets/existing products) was pursued by selling the group's 'star' and 'cash cow' products (see Chapter 22, Figure 22.6) into central and eastern Europe and into Asia (see Figure 8.9).

ABB in China

➤ 11 joint venture companies–more in the pipeline

➤ ABB China Business School established

➤ 1000 managers a year to be trained in business administration

➤ Employees will increase to 10 000 by end of decade

ABB in central and eastern Europe

➤ 61 majority-owned joint venture companies

➤ 25 000 employees in 16 countries

➤ Enormous production and design potential– will become a major competitor to western Europe

Figure 8.9   ABB's moves into emerging markets

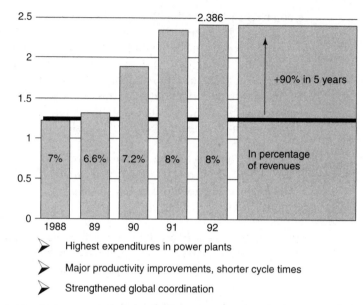

Figure 8.10   ABB research and development

Strategy 2 (existing markets/new products) and strategy 4 (new products/new markets) were strongly supported by investment in R&D. ABB almost doubled its R&D investment over a five-year period while eliminating overlapping work in different R&D centres – see Figure 8.10. This R&D was customer driven and enabled the product divisions to make more than 50% of their 1993 sales from products developed in the previous four years.

Strategies 2 and 4 were also pursued through acquisitions. ABB bought 15 companies in its first year of operation and 40 in 1989. Acquisitions in 1990 included Combustion Engineering Group (CE) of Stamford, Connecticut, USA, which turned out to contain an environmental 'time bomb' that almost ruined ABB 10 years later, through compensation claims for industrial injuries related to use of asbestos. (See section 8.5 below.)

Through the 1990s ABB, like Chrysler, pursued the business goals listed in Chapter 2. ABB's way of achieving these goals differed from Chrysler's in the extent of its response to the external factors influencing the business environment, which were also listed in Chapter 2. ABB gave the same sort of attention as Chrysler to customer satisfaction through new technologies, faster delivery and higher quality at competitive prices. It went somewhat further in reducing management structure and empowering employees. It went much further than Chrysler in extending its operations on a global basis to position itself to use external factors to its own advantage.

The difference of approach between ABB and Chrysler derived from Chrysler's concentration on only one of what Eric Drewery, chief executive of ABB UK, called 'the two fundamental changes in the international environment for multinational corporations' [8]. The two changes are:

(1)   'An unprecedented wave of restructuring measures in the business world [which] has been called the "third industrial revolution" [and] has to do

with all the processes of management aiming at producing the best, most innovative products and services, in the shortest amount of time, at the lowest price...'

(2)   '...[the realization that] today the "world economy" comprises the whole world...a new world with new growth poles in China, Central and Eastern Europe, India, the Middle East, Brazil and Mexico.'

ABB did not merely observe the second change – it participated in it. Figure 8.9 summarizes the group's actions in two of the 'growth poles' at a time when it was reducing its total, worldwide employment levels (by 47 000, or about 25% in the western world).[1]

ABB's activities in the developing regions of the world also helped it to benefit from what other companies saw mainly as a threat: long-term trends in currency exchange rates (as opposed to daily or monthly changes). The 1993 ABB Annual Report stated:

'About one third of ABB's total value-added is in countries where currencies have significantly depreciated against those of their major trading partners. The increased overall competitiveness of the products, parts and projects from those countries will support growth as well as defend our domestic market positions.'

Another thing that ABB did more extensively than Chrysler was to use its financial services division (subject of a forced sale in 2003) as an integral part of its marketing strategies. This enabled ABB to supply a 'turnkey' paint plant for a US automotive company, a combined heat and power plant in Scandinavia, and rolling stock to London Underground in the UK. (Chrysler has a customer finance arm, but its offerings are mainly like other car purchase finance schemes, while ABB's schemes were innovative and distinctive in its markets.)

ABB did more than Chrysler to confirm its 'good corporate citizen' role in its management of environmental issues. This went beyond the development of 'cleaner' power plants: ABB made environmental protection management a top corporate priority. It signed the International Chamber of Commerce Business Charter for Sustainable Development in 1991, and set out a five-year programme of internal and external actions to establish environmental management systems. The programme included the adoption of the international environmental standard ISO 14001 for implementation at all ABB factory and service sites worldwide. By the end of 2002 98% of sites in more than 50 countries complied with the standard. This commitment to good stewardship brought international awards in the shape of number one ranking in the Dow Jones' Sustainability Index for 1999, 2000 and 2001.

---

[1] Under the heading 'Corporate Unity', ABB's Missions and Values booklet explained such variations in regional employment trends as follows: 'Optimizing our worldwide resources to serve our local and global customers will require strategic re-orientation and major structural changes in local companies. Overall improvement of the Group's competitiveness is the most important criterion for making such changes. We will strive for the best possible and equitable solution from a Group perspective, even if this should result in uneven distribution of hardship among the countries or locations involved.' Also, the company's 1993 accounts showed that expansion into new markets in China and eastern Europe had a mitigating effect on employment reductions by increasing exports from western Europe and North America.

ABB started a customer focus programme in 1990, aimed at responsiveness and continuous improvement worldwide. The faster throughput times and materials management actions resulting from this programme enabled ABB to halve its inventories between December 1990 and December 1993, from $7.3 bn to $3.6 bn. ABB also contributed to its customers' inventory reductions by providing 'fast, comprehensive and reliable equipment maintenance and service' and by introducing services such as 'short time no stock' (STNS) deliveries. An STNS system in Scandinavia enabled delivery of cable in any length directly from the factory to the customer's construction site, in many cases within three days.

ABB's 'delayering' and 'employee empowerment' actions have been described in Chapter 7. They were comprehensive and sustained, like Chrysler's.

Suppliers do not feature strongly in ABB's public statements. They are explicitly involved in ABB's environmental management plans and implicitly must be involved in ABB's quality and JIT programmes, since purchased materials account for over 40% of ABB sales and other purchases push the total 'buy' beyond 50% of sales. ABB's lesser emphasis on suppliers is a major difference from Chrysler, IBM, Xerox and the other models of successful engineering management.

In total, ABB's internal restructuring and outward customer focus enabled it to sustain modest growth in sales and profits during the 1991–93 period, which saw deep recession in the OECD countries. The group's long 'reach' in geographical terms and in planning horizons put it in a strong position to take advantage of the expected recovery in the OECD countries and the forecast strong growth in emerging countries in the decade from 1995. Percy Barnevik, ABB's president, described this period as 'the third phase in ABB's short history, the expansion phase' and said that the company was well-positioned to 'go for growth'.

8.4.3   Why compare and contrast?

ABB is primarily an electrical engineering company that is also a world leader in process engineering and is based in Europe. Chrysler is an automotive industry company, based in the USA. The reason for comparing them is that they were both successful in the early 1990s and it is possible to identify factors common to their success. These factors are:

- removal of organizational barriers to fast communication and decision making;
- use of the simplified organization to motivate all employees to focus on their customers and to pursue never-ending improvement;
- use of speed as a competitive weapon – speed of new product development and speed of order execution;
- balanced management of 'hard' and 'soft' issues.

The reason for contrasting them is that '. . . mere rationalization and restructuring are not enough to safeguard the future. Much more is needed today to differentiate a company, because most companies are restructuring successfully' [8].

Chrysler differentiated itself in its industry by its concept of the 'extended enterprise' and its cooperation with its supply base. It differentiated itself in global terms by the speed of its restructuring and the extent of its turnround in profitability.

ABB differentiated itself by the extent of its planning horizons, the extent of its delegation and empowerment, and the extension of its offering to its customers.

In its first 12 years ABB had not needed the sort of turnround that Chrysler achieved, because it never sank to the low levels of performance that Chrysler suffered. This can be construed as 'better' management, but there is little point in trying to determine who is 'better'. Both companies provide examples of excellent management, which was recognized by their peers:

- In June 1994, ABB was rated equal first with the UK retailer Marks & Spencer in terms of commitment to their customers, shareholders and employees, in a survey by the *Financial Times* and Price Waterhouse that questioned 2000 executives of the largest organizations in nine countries. In September 1995 the survey put ABB first on its own, while Marks & Spencer fell back to fifth place [9].
- In December 1994, Chrysler was named by Alex Trotman, chairman of Ford Motor Company, as Ford's most dangerous competitor in North America. For over 80 years previously that accolade had gone to General Motors.

From these peaks in the mid-1990s both ABB and Chrysler suffered 'the slings and arrows of outrageous fortune', with ABB suffering most, as described in the next section.

8.5 TWISTS IN THE TALE

8.5.1 Chrysler's wrong turn?

By the mid-1990s Chrysler held a sound, profitable position in the US automotive market, had an alliance with Mitsubishi that assisted new product development and gave some presence in the Far East, and was following a strategy for building its participation in western Europe – the world's largest car market. However, it lacked the benefits of scale achieved by the largest automotive companies, General Motors, Ford and Toyota. To share these benefits, and to boost performance outside North America, in 1998 Chrysler agreed a merger with Daimler, to form DaimlerChrysler AG. The merger suited Daimler's 'targeted occupation of the automotive industry's Number 1 spot', as the chairman claimed in the 2001 annual report.

When detail of the new group's top management structure emerged it became clear that most of the *leaders* of the 'reengineering of Chrysler' had left. The sole survivor of the top team was Thomas T. Stallkamp, who had been Chrysler's vice-president of purchasing and supply, and *leader* of the Large Cars platform team (see Figure 7.11). Stallkamp became president of Chrysler Corporation and vice-chairman of DaimlerChrysler. By the end of 1999 he too was gone, and most of the senior positions in Chrysler were filled with ex-Daimler executives.

One of the pillars of Chrysler's restructuring was retained. DaimlerChrysler built on the platform team concept to form 'product innovation teams', 'to link the entire organization'. This was reinforced in early 2001 by formation of an

Executive Automotive Committee (EAC), described as the 'key steering instrument for the world-wide automotive business' and working in four areas:

- product portfolio;
- technology;
- production capacity/purchasing and supply;
- sales and marketing.

The aims of the EAC were similar to those of the old Chrysler management and other industry management teams – reduced vehicle development times; improved quality; sustained profitability.

The first products jointly developed under this regime were scheduled for the market in 2004, but earlier cost-reduction benefits were claimed from the global coordination of purchasing. This was more like playing 'catch-up' with GM, Ford and Toyota than industry leadership.

Chrysler division's performance after the merger was patchy. The US car market was already stagnant or declining from 2000, and plunged with the rest of the US economy after the 11 September 2001 attack on the World Trade Center. The new management declined to join the discount war waged by GM and Ford, electing to 'improve profits rather than protect market share at any price'. Chrysler's US car market share did decline (from just over 14% to just under 13%) and profits did improve from $1.9 bn negative in 2001 to $1.4 bn positive in 2002, but fell back again in 2003 to a $50 m loss, excluding losses of $594 m on disposals and restructuring. From 1998 to the New York stock market's low point in early 2003, the price of shares in DaimlerChrysler fell by almost 90%, compared with about 50% for General Motors and Ford.

This patchy performance led to public criticism from shareholders, many of whom were unhappy with the style of management introduced by Daimler. Some held that they had been misled regarding the 1998 merger, contending that it was in fact a Daimler takeover and calling for compensation. In November 2003 one group action by former Chrysler shareholders was settled by DaimlerChrysler for $300 m, but they vowed to fight other similar claims. One fight was with Kirk Kerkorian and his investment firm Tracinda, which had owned 14% of Chrysler. Seeking $9 bn in compensation, on 2 December 2003 Mr Kerkorian told a Delaware court that the 1998 deal was not a 'merger of equals' but a takeover, based on 'deception' and 'fraudulence'.

So, having provided a stand-alone model of successful corporate reengineering and turnround in the early 1990s, with a different top management team, Chrysler was struggling to be any sort of success as a division of DaimlerChrysler AG 10 years later.

8.5.2   ABB at another threshold

The growth from the mid-1990s for which ABB prepared did not materialize, except in the newer, smaller markets it had entered – western Europe, Japan and the USA stagnated or declined.

Nevertheless, ABB continued its product portfolio development by acquisitions and divestments, by strategic alliances and by focused R&D. Major acquisitions included Elsag Bailey Process Automation NV, bought in 1998 for $2.1 bn, 'making ABB a world leader in the global automation market'. Among divestments were its railway transportation interests, sold in 1999 for $472 m, and its nuclear power-generation activities, sold in 2000 for $485 m. Alliances were formed with the French ALSTOM power-generation company in 1999, and with SKYVA, an American software developer, in 2000.

The joint venture with SKYVA supported ABB's decision to create a *unique selling point* by embedding advanced information management capability in all its products. ABB patented its industrial information technology as 'Industrial IT' and from late 2001 focused its major R&D efforts on software and Industrial IT. This IT capability not only helped ABB to secure individual contracts, it also provided a new revenue stream in its own right, such as the 10-year global agreement with Dow Chemical to 'infuse Dow's plants with a new generation of ABB's industrial IT'.

Earlier R&D contributions to ABB's product portfolio included:

- In 1998, 'Powerformer', the world's first high-voltage generator to supply power direct to the grid without transformers, saving up to 30% in overall plant *life-cycle costs* (see section 15.5, Cost–benefit analysis and life-cycle costing).
- In 2000, 'Windformer', a new wind-power technology that makes wind farms competitive with conventional large power plants by providing 20% more output and cutting *maintenance costs* by half (see Chapter 16, Maintenance management).

These three innovations – Industrial IT, Powerformer and Windformer – show how ABB combines advanced technology and comprehensive financial analysis to support its corporate image and marketing platform. (See section 22.12, Integrated marketing.)

Changes in the balance of ABB's product and service offering were accompanied by organizational changes as described in Chapter 7, and by changes in senior personnel. From January 1997 Percy Barnevik, who, as leader of the group since its formation, was widely regarded as the world's foremost proponent of globalization, became nonexecutive chairman. He was succeeded as president and CEO by Goran Lindahl, who steered the company through a reorganization in 1998 that consolidated product groupings 'to tap market trends' and did away with regional coordinating management layers. Lindahl stepped down in October 2000 in favour of Jorgen Centerman, 'a younger leader with a true IT profile', as part of ABB's preparation for 'the next phase in its transformation, driven by the far-reaching IT revolution affecting the global technology group and its customers'. Barnevik resigned in November 2001, saying that he had to 'take my share of responsibility for the less good performance of ABB in recent years'. Centerman departed 'unexpectedly' in September 2002.

ABB's 'less good performance' was partly due to simultaneous stagnation in the major geographical and product areas where it competes, and perhaps largely due to bad luck in the shape of claims in the USA for compensation for injuries related to asbestos. The asbestos injury claims were against Combustion Engineering (CE), a company ABB bought in 1990, and related to CE's use of asbestos in the 1970s, well before the acquisition. ABB contested the claims, but had to take charges against earnings of $470 m in 2001 and a further $590 m in 2002, when there were 94 000 claims pending. These provisions completely undermined investor confidence, and by late 2002 ABB shares were trading at about 5% of their peak price of February 2000. ABB's credit ratings were cut by leading agencies, Moody's and Standard & Poor, 'to just above junk status' (*The Times*, 24 October 2002). To fund ongoing operations ABB had to borrow over $4 bn in short-term debt, which it then had difficulty servicing. To repay debt ABB had to restructure yet again, to separate those operations that it would continue to develop from those that it would have to close down or sell. The forced sales included ABB's financial arm, Structured Finance, which had been an integral part of its marketing platform. This was sold in September 2002 to GE Commercial Finance for $2.3 bn 'to reduce debt'.

Besides being a financial millstone, the claims against CE required a huge diversion of management attention. In addition to the cost of compensation, ABB spent $52 m on administration and defence of claims in the three years 2000 to 2002. Eventually a potential solution was developed, which involved 'ringfencing' CE and putting it into Chapter 11 bankruptcy. In January 2003, ABB agreed a pre-packaged bankruptcy plan with representatives of asbestos plaintiffs, involving a provision of $1.1 bn in addition to the $550 m settlements paid in 2001 and 2002. To secure preliminary court approval of this scheme, in July 2003 the provision was increased to $1.2 bn. The full court hearing was scheduled for mid-2004. With the cost of asbestos claims capped, ABB was able to secure stockholder approval in November 2003 for a $2.6 bn share issue, 'signalling an end to the worst of its financial troubles' (*The Times*, 21 November 2003).[2] This outcome could be viewed as failure, but compared with some it is another example of superior ABB management. For example, Federal Mogul, a US engineering group with global presence similar to ABB, failed to limit asbestos-related claims against one of its subsidiaries and the whole group was forced into administration.

Claims for asbestos-related illness began in the 1970s, mainly against asbestos mining companies, so maybe the claims against CE were not all due to bad luck. Perhaps ABB could have performed more thorough due diligence audits when acquiring CE in 1989/1990. Looking for other lessons from ABB's problems in the late 1990s and early 2000s suggests that 'loss of focus' might be added to Sloan's list of the *dangers of success* (see section 7.3).

Success seems to have distracted ABB's chairman, Percy Barnevik. In November 1996 he was the first non-American to receive the National Electrical Manufacturers' Award, in recognition of 'inspirational *leadership* in the electroindustry and his role as chief architect of a global company'. Barnevik took positions on the boards of General Motors, Sandvik, AstraZeneca and Investor Corporation.

Goran Lindahl also received recognition. In 1999 he was the first non-American to be given *Industry Week*'s 'Chief Executive Officer of the Year' award. He also took positions outside ABB, on the boards of DuPont and LM Ericsson AB, and as vice-president of the Prince of Wales' Business Leaders' Forum and the only industrialist member of the World Commission on Dams.

---

[2] November 2003 may have seen the worst of ABB's financial troubles, but they continued. In February 2004 it announced that the group's loss for 2003 was $767m, little improved from the 2002 level of $783m.

Barnevik and Lindahl also devoted too much and inappropriate attention to their own futures. They set up a $150 m pension fund for themselves that was investigated in 2002 and found not to have had the board's necessary authorization.

These distractions for the group's two most senior executives may have contributed to ABB's slide from one of *Industry Week*'s 'World's 100 best-run companies' in 1999 and 2000 to the threshold of disaster in 2002.

8.5.3   No longer sure of Shell [10]

Section 8.3 included a description of how Shell's 'planning as learning' process helped to change management attitudes and sharpen the organization's responsiveness, enabling it to overtake BP as Europe's largest oil company. That was the 1980s, and the executives who had been the drivers of change moved on. During the 1990s, BP recovered from Shell its position of number one oil company in Europe, and number two in the world, and it turned out that Shell had again changed its style and had lost its dynamism as well as its number one spot.

This all surfaced in January 2004, when Shell stunned the business world with the 'devastating news' that 20% had been wiped off its proven reserves – 'one of the core measures of an oil company's performance'. The news, and the style of the announcement 'under the arcane heading "proved reserve recategorisation following internal review"', brought a stream of criticism of management and its style, from shareholders, analysts and commentators. The chairman and finance director 'could not be bothered to turn up to a conference to explain what had happened'. The chairman, who had been CEO of exploration and production for five years before his appointment, was 'labelled one of the poorest communicators to head a FTSE 100 company' who 'reeks of all the worst elements of the company – patronising, bureaucratic, secretive, lumbering and underperforming. And let's not forget the arrogance.' At Shell's annual results meeting in February 2004, the chairman apologized for his nonappearance at the January press conference and promised that he would oversee a review of the group's structure. On 3 March 2004 he and the head of exploration 'resigned'. Worse followed: on 17 March the US Department of Justice revealed that it was conducting a criminal investigation into the actions of the two retirees and Shell's finance director. On 18 March the new chairman of Shell's management committee announced a further reserves writedown and postponement by two months of the Shell annual report, due on 19 March. Such a postponement by a 'Footsie 100' company was unprecedented.

In a decade, Shell had gone from one of the most admired industrial companies to one of the most reviled, demonstrating the importance of leadership and the profound effect of individuals in even the largest organizations. The January 2004 incident also emphasizes the importance of *communication* (see Chapter 11).

8.6   SUMMARY

The chapter started with a list of BASIC weaknesses of western engineering management: organizational boxes, antagonistic attitudes, satisfaction with a single success, ineffective attempts at imitating the Japanese and inability to manage change. Companies that had overcome these weaknesses in the 1980s have been examined, and the common factors in their formulae for success have been identified.

Two companies, ABB and Chrysler, which were outstandingly successful in the early 1990s, were considered in more detail. The internal restructuring or reengineering of these companies was found to have much in common, but it was noted that, on its own, restructuring would not be sufficient to secure success in the future. To continue to be successful companies have to match or beat their competitors in quality, innovation and customer satisfaction, but additionally they have to find ways to differentiate themselves. ABB did this by its intense forward look down the value-added chain and the harnessing of IT. Chrysler did it by an equally intense look in the other direction: backwards, up the value-added chain. ABB was seen to use the financial and marketing management tools that are described in other chapters of this book.

However, it was noted that both companies failed to extend their successes into the early twenty-first century – Chrysler largely as a result of the merger with Daimler and the departure of the leaders of its reengineering, which left it less able to deal with difficult conditions in the key US vehicle markets; ABB largely as a result of asbestos injury claims, and possibly too much reengineering and leadership change.

Another company, Shell, was shown to have declined through changes in leadership and management style, and to suffer from poor communication.

It was suggested that Chrysler and ABB faltered because, though they continued with their successful policies, products and practices, they also suffered from a change of style following the departure or distraction of successful people, the very people who had been the *leaders* of the corporations' creation and reengineering.

This prompts the idea of another set of 4 Ps to be added to the two marketing 4 Ps (see The marketing mix, section 22.11). The 4 Ps of corporate management would be: Policies, Products, Practices and People.

REFERENCES

1. Sloan, A. P. Jr (1963) *My Years with General Motors*, Pan, London.
2. Senge, P. (1990) *The Fifth Discipline*, Century Business Books, Boston, MA.
3. Robinson, P. (1990) 'Innovation, investment and survival', Institution of Mechanical Engineers Conference, 4 July.
4. De Geus, A. P. (1988) 'Planning as learning', *Harvard Business Review*, March–April.
5. Fridrich, H. K. (1992) 'World class manufacturing', The 1992 Lord Austin Lecture at the Institution of Electrical Engineers, London.
6. Phillips, J. R. (1982) 'Managing people', *McKinsey Quarterly*, Autumn.
7. Lutz, R. A. (1994) The 1994 Christopher Hinton Lecture, to The Royal Academy of Engineering, 4 October, and private correspondence with T. T. Stallkamp, Vice-President – Procurement and Supply and General Manager – Large Car Operations provided the basis of section 8.4.1.
8. Drewery, E. (1994) Address of the Department of Systems Science Research Colloquium, City University, London.

9. 'Europe's most respected companies', *Financial Times*, 27 June 1994 and 19 September 1995.
10. The quotations in this section are from the business sections of *The Times*, 10 January 2004, and *The Sunday Times*, 11 January 2004.

BIBLIOGRAPHY

'Resetting the clock – Asea Brown Boveri is transforming its factories by slashing lead times', *Financial Times*, 10 February 1993, gives some background to ABB's Customer Focus programme.

'ABB managers strip for action – The Swiss engineering group's streamlining', *Financial Times*, 25 August 1993, reports ABB's 'delayering' of its management structure.

ABB and DaimlerChrysler annual reports and press releases provided much of the material for section 8.5 and contain material for further reading. The latest reports and most past press releases can be seen on the companies' websites: www.abb.com and www.daimlerchrysler.com.

Robinson, G. (2004) *I'll Show Them Who's Boss: The Six Secrets of Successful Management*, BBC Books, London. Gerry Robinson's six secrets of success, drawn from his experience in a range of service industries, have a lot in common with the lessons presented in Chapter 8 – and with those to be found in Chapter 23.

Part II

Managing Engineering Resources

Human Resource Management – The Individual

9.1 INTRODUCTION

This chapter and Chapter 10 provide an introduction to the management of people in organizations, a subject known generally as human resource management. The human workforce is arguably the single most important resource of any organization. The human being is also the most complex and perhaps the least predictable of all organizational resources. All technical activity involves people, and no scientist, technologist or engineer can be successful in his or her career without an understanding of human behaviour. These two chapters give an introduction to some basic aspects of human behaviour, and the management of people as one of the resources of a business. The more recent approach to human resource management, which emphasizes the resource aspect rather than the human aspect, is also considered. The logical place to start is with the individual, the basic building block of groups and organizations. This is followed by a consideration of group behaviour. A study of organizations has already been provided in Chapter 7.

In this chapter the behavioural sciences are briefly described and explained, the issues of motivation explored and the theories of Maslow and Herzberg considered, and other theories of motivation outlined. The latter part of this chapter is concerned with individuals in pairs, and the relationship between individuals is examined using transactional analysis. However, some general aspects of the behavioural sciences, and their relationship to engineering, are considered first.

Adopting the differentiation frequently used in systems science between 'hard' and 'soft' complexity, science and technology is found in the 'hard complexity' area. Technology is a complex subject, but most of its problems are quantifiable. There is a basis of science from which the laws of technology are developed. These laws may be derived from a mathematical basis, or may be established empirically by observation and experiment. A mathematically derived theory may be shown by experimentation or by empirical observation to be true, and the linkage will give understanding to the phenomenon under examination.

Consider two examples. In engineering, the study of the behaviour of fluids or materials starts from a mathematical base. It is possible to derive a mathematical model for the flow of fluids from the Navier-Stokes equations. The model can be verified by experimental observation. Similarly, a mathematical representation can be derived for the bending of a beam by consideration of the physical characteristics of the system such as the beam length, the force applied and the modulus of elasticity. This can also be checked by experiment. Conversely, in any area of applied science an experiment can be performed and a hypothesis proposed to explain the results obtained. This is the empirical approach. By reiteration between such empiricism and theory, a complete understanding of the phenomenon under consideration can eventually be obtained.

Students who train as engineers and technologists generally do so because their aptitudes and inclinations are compatible with the scientific basis of their profession and its practical aspects. The further training of engineers and technologists tends to reinforce the understanding of the relationship between theory and practice, and the dependability of the results that can be obtained by the application of established technical knowledge. Engineers and technologists are comfortable with the way in which the practical aspects can be represented in a 'hard' – that is, quantifiable – way. The performance of a machine or a structure can be assessed, hard data generated, and the whole condensed in the form of definitive equations that are known to have a sound basis in theory as well as to work in practice. This is comforting for the technologist, who can predict with reasonable confidence the outcome of the application of technical skills.

Unfortunately, there is no fundamental theory derived from a scientific basis for the behaviour of the human being. People are individual, complex and sometimes unpredictable. Professional engineers and technologists frequently work in teams with other technical personnel and with other professions. Many technical organizations, and organizations that employ technologists, are large and have many people with whom the technologists have to relate. Little if anything is achieved in the world without human effort, so it is essential that the technologist should have a basic understanding of the way people behave. We need to know what makes people tick.

Studies in this area are known as behavioural studies or behavioural sciences, though students of sciences with a mathematical basis might argue about the use of the term 'science' for investigations of human behaviour. Even so, much effort has been put into the study of human behaviour and a mass of evidence collected, giving a body of information that constitutes what is known about the behaviour of human beings, individually and in groups. Some technical students may find their first encounter with behavioural science a little strange, because it does not have the same scientific basis that underpins engineering and other technologies and uses an unfamiliar terminology. The individuality and complexity of the human being do not allow a hard theoretical basis for human behaviour. As people from the North of England might comment, 'there's now't so strange as folk'!

Though the subject may appear woolly and intangible when contrasted with the certainties of science and engineering, behavioural scientists have brought order and structure to their study of the human race. Their method has similarities with the empirical approach to technology mentioned earlier. Human behaviour is observed, individually and in groups, and common characteristics noted. Theories are advanced to explain the behaviour of individuals or groups of individuals in various situations. These theories are better described as hypotheses or models, in that they are constructs based on observation, which suggest a basis for the behaviour or a common thread in the observations. The hypothesis or model may help towards an understanding of human behaviour, and therefore help in the management of human resources.

Publication of these hypotheses allows discussion and further evaluation on a wider basis. The system is similar to the publication of papers on scientific subjects. As work progresses, some models of human behaviour are developed and are accepted as a helpful indication of behaviour. Such models, once they gain acceptance, constitute the theory of the behavioural sciences. Some of the more generally accepted theories, hypotheses or models will be discussed in the next chapters, but it must be remembered that these constructs are a guide to, and frequently a simplification of, the very complex reality of human behaviour.

Human resource management contains many elements that could be considered as 'soft'; that is, not easily quantified. Soft issues are those that involve value judgements, have no right or wrong solution, are difficult to define precisely, and may be interpreted differently by many people according to the values and priorities that they hold. Such issues are clearly difficult to quantify, and readers more familiar with 'hard' data may experience some discomfort, as the phenomena do not slot neatly into boxes or generate equations that will produce 'right answers'. There are, quite often, no right answers to human resource management problems, just better or worse solutions.

Just as good engineering relies on a sound basis of theory and practice, so good human resource management relies on a sound knowledge of the theory and practice of human behaviour. No matter how technically brilliant an engineer or technologist may be, success and career progress will depend on an ability to work with, motivate, influence and manage other people. Even the most sophisticated computer systems and robotic production facilities are designed and controlled by people. Professional engineers are likely to spend the majority of their lives working in organizations with a variety of people: individuals, groups of individuals and the amalgamation of groups that form the organization. Ability to interact effectively with these individuals and groups will be of primary importance to the career of the professional engineer or technologist, so it is of very great value to the student to acquire knowledge and skills in the area of human resource management.

9.2   THE BEHAVIOURAL SCIENCES

Behavioural science is an interdisciplinary collective term for the three elements of social science interested in the study of the human species. It involves the study of psychology, sociology and anthropology, but is often taken to apply specifically to aspects of human behaviour related to organization and management in the working environment. It is therefore very relevant to the circumstances in which engineers and technologists perform their duties.

Psychologists are concerned with human behaviour, individual characteristics and the interaction of individuals in small social groups. Attention is focused on the 'personality system', the characteristics of the individual, such as attitudes, motives and feelings.

Sociologists are concerned with social behaviour and the relationships among social groups and societies, and the maintenance of order. Attention is focused on the 'social system', the interrelationship of individuals in social structures.

Anthropologists are concerned with the study of mankind and the human race as a whole. Their attention is focused on the 'cultural system', the values, customs and beliefs, that influence the emphasis people put on various aspects of behaviour.

The interrelationship of psychology, sociology and anthropology forms the interdisciplinary study area of behavioural science.

9.3   THE INDIVIDUAL AND THE ORGANIZATION

The approach taken in this introduction to human resource management is to consider initially the individual, pairs of individuals, then individuals in groups, groups in organizations, people considered as a major resource of business, and finally organizations within the environment. Leavitt [1] adopted a similar approach in his book *Managerial Psychology*, considering people as individuals, in pairs, in groups of three to twenty, and in organizations of hundreds and thousands. He also considered the ways in which people are similar and the ways in which they are different. It is perhaps from an understanding of the similarities and differences, the way people interact with one another and the way they work in groups, that an appropriate start can be made to human resource management. Certainly, the professional engineer or technologist at the start of a career will be more concerned with interaction with people singly and in small groups, at the operational level. It is only at a later stage of a career when a senior appointment has been won that the professional engineer or technologist will need to consider human resources at the strategic level.

For the *individual*, consideration will be given to personality, attributes and skills, attitudes and values, and expectations and needs. For *individuals* (in pairs and in relation to groups), personal interaction, influence, authority, leadership and manipulation will be considered, but communication will be discussed in Chapter 11. For the *group*, functions and structure, informal organization, role relationships and group influences and pressures will be discussed in Chapter 10.

These aspects will be related to the achievement not only of organizational performance and effectiveness, but also to satisfying the needs of the individuals in the organization. By the use of good human resources management, the goals of the organization are achieved and the needs of the employees are satisfied. At the same time, a climate is created in which the employees will work willingly and effectively.

9.4   THE INDIVIDUAL

9.4.1   The concept of self

The way in which a person behaves in the working environment depends on the 'self', the internal psychology of the individual. Part of this the individual understands, but much of it is unconscious. The 'self-concept' is that part of the self of which the individual is aware. The major part of the self is the 'unconscious self'. An individual also has a 'self-ideal', the self that person would like to be. The self-ideal is quite separate from the self, and the individual continually compares the self-concept with the self-ideal. Another important element of the self is self-esteem, how good an individual feels about himself or herself. The individual tries to increase self-esteem whenever possible.

The self is formed as a result of the complex interaction of expectations, inhibitions and preferences that individuals experience as children and adolescents. Personal growth involves the enlargement of self-awareness – that is, becoming consciously aware of more of the self, and thus enlarging the self-concept. The self develops and changes with time.

Here are some definitions:

- *Self:* the total set of beliefs, values and abilities within the individual, including those not yet realized.
- *Self-concept:* the pattern of beliefs, values and abilities of which an individual is aware. This excludes the unconscious component.
- *Unconscious self:* the pattern of beliefs, values and abilities of which an individual is unaware. This excludes the conscious component.
- *Self-ideal:* the set of beliefs, values and abilities towards which an individual aspires.
- *Self-esteem:* the value placed by individuals on their beliefs, values and abilities.

9.4.2   Psychological energy

Psychological energy is present in all individuals and is variable but not limited. The quantity of psychological energy varies with the psychological state of the individual. Its individual expression can never be permanently blocked. If its expression is frustrated, some alternative form of expression will be found. Psychological energy is largely beyond an individual's control, and finding how to inspire, enthuse, 'turn on' or 'psych up' oneself is very important. In a leadership or managerial role, it is just as important to be able to inspire and motivate the team members or the employees under managerial control. Competitive sports players are now well aware of the need for psychological energy. International sports players, particularly in one-to-one competitive sports such as tennis, may have a psychological coach as well as a physiological one.

An individual will be happy to contribute energy to an activity if it is anticipated that, by so doing, an increase in self-esteem will result. Self-esteem will be increased if a measure of 'psychological success' is achieved in the activity. The judgement of the measure of psychological success will depend on the self-concept and the conditions associated with the activity. There are two broad criteria that will determine whether self-esteem is enhanced by an activity. First, the individual should feel 'comfortable' with the activity. It should feel right or fit the subconscious self. If this is not the case, the individual will not feel happy, however successfully the activity has been performed. The activity should seem important to the individual. Second, the individual should perceive success in the activity, the success being measured according to the individual's own criteria, not those of any other person.

Self-esteem is enhanced in two major ways. The first way is by growth. The individual may try something new. As a result, the individual's self-concept will be enlarged, and the individual will value himself or herself more. With increased confidence, the individual will be encouraged to try more new activities, which will develop that person yet further. The second most important process for increasing a person's self-esteem is dealing competently with the problems of his or her environment. If the person can attribute the resolution of personal, domestic or work problems to his or her own efforts, skills or abilities, self-esteem will increase. It is essential that the individual can see a direct connection between his/her own efforts and the resolution of the problem.

To summarize, the self-esteem of an individual will increase as the individual is better able to define the goals; these goals relate to his/her basic values and needs; and he/she is able to select the means to achieve those goals. Thus it is important when dealing with people in a working environment to be aware of their need for self-esteem. People will be more responsive and more willing to help or to carry out a manager's wishes if they are allowed some autonomy. The manager may not be able to allow them to select the goal – this is most likely to have been decided already by higher authority – but the manager can be sensitive to the needs and values of subordinates, and allow them as much freedom as possible in the way they do their jobs.

9.5   BEHAVIOUR, MOTIVATION AND WORK

Why should engineers and technologists be interested in the psychological basis of human behaviour? Why should they need to understand what motivates people?

All organizations consist of complex interrelationships of individuals and work functions. In most cases, individuals can be subjected to hard (i.e. quantifiable) analysis of their intelligence, skills or work output. However, it is their level of motivation (their enthusiasm for the job) that will determine the performance, efficiency and ultimate success of the organization. Thus a manager needs to understand the behaviour and motivation of those who work in the organization. The individual also needs to know how to self-motivate to give a better performance, for the good of self-esteem or perhaps to enhance career progress. The manager needs to motivate subordinates, convince peers and persuade superiors.

There are many views on the principal factors that motivate people. An individual may be motivated by personal gain or by concern for the good of others; by a desire for security or a need for excitement; by physical gratification or intellectual stimulus. A person may experience all of these motivations at different times. Studies of the psychology of the individual have identified three elements of behaviour that can be applied to all [1]:

(1)   Behaviour is caused.
(2)   Behaviour is motivated.
(3)   Behaviour is goal directed.

These elements can be related by a simple model, Figure 9.1.

The model has a feedback loop, and engineers will recognize it as a control model. A simple example might be the feeling of hunger. This need would motivate an appropriate behaviour or action, which could be the opening of a can of soup or attending a restaurant. Either might achieve the desired goal of a satisfied stomach, and the choice, which is by no means restricted to the examples given, would depend on many things, the most significant being the time and money

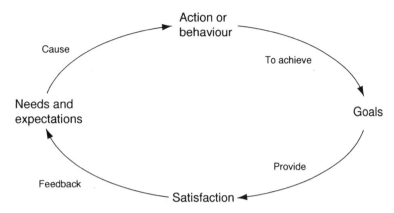

Figure 9.1 Simple motivation model

available. Having consumed the soup or the restaurant meal, the individual might or might not have resolved the need for sustenance or achieved the goal of satisfaction. The feedback loop will allow for further action if required.

This physiological example is a very easy one. Satisfaction of one's psychological needs may be a little more complex and a little more difficult to define. They will certainly be more variable from individual to individual than the hunger example, which will apply to all people at some time. Such psychological needs as safety, friendship, recognition and perhaps achievement come to mind, and apply to most people. These will be examined in more detail later.

Before moving to consider human psychological needs, the simple motivational model needs to be examined a little more. All is well if the desired goal is achieved. The feedback loop will identify that the need or expectation has been fulfilled and therefore no further action is required. It could be that the goal is only partially fulfilled and further action is required, but in this case one can assume perhaps that the progress made will encourage further effort to achieve the required goal fully.

However, what if the person is completely obstructed in achieving the goal? There are a number of possible results. The barrier or blockage to achieving the goal could be considered as a problem that requires resolution and various means could be sought to resolve the problem. The problem-solving method might be successful and the goal would then be achieved. By a somewhat longer and more complex route we arrive at the same situation given in the earlier model. The action taken would be regarded as constructive behaviour. Seeking an alternative acceptable goal would be another example of constructive behaviour. This might involve some element of compromise. The original goal might now be perceived as unrealistic in the circumstances and a less difficult goal substituted. If the alternative goal is acceptable and can be achieved, the result will be satisfactory, though some residual minor frustration might remain.

If the required goal, or an acceptable substitute, is not achievable, the needs and expectations are unfulfilled and frustration is the result. This can be very damaging psychologically, and a number of possible nonconstructive, negative and defensive behaviour patterns can emerge. These include aggression, regression, fixation and withdrawal, and the response to the obstruction may involve combinations of these reactions.

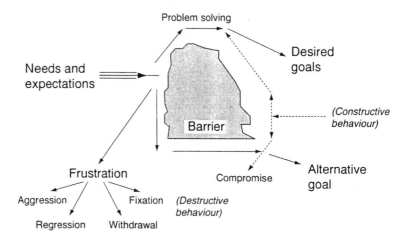

Figure 9.2   Simple frustration model
Source: adapted from Mullins [2], reproduced with the kind permission of Pitman Publishing

Aggression could be in the form of a physical or verbal attack on the cause of the obstruction. If the cause happens to be the boss, and career preservation suggests that violence in that direction is unwise, the aggression may be displaced in other directions, such as abusing subordinates or kicking the cat. *Regression* is reversion to a childish form of behaviour, such as sulking or throwing a temper tantrum. *Fixation* is persisting with the action or behaviour and disregarding the obvious failure and likelihood of continuing failure. *Withdrawal* is giving up, not just with the action or behaviour that has been the basis of the frustration, but with all aspects of the situation in which the problem occurred. The frustrated individual can become generally apathetic and may lose totally the will to continue with the activity. A model of frustration is given in Figure 9.2.

Frustration will occur in some form and to some degree in most jobs, and in engineering as in other professions. It is therefore the function of the manager or team leader to minimize the extent to which these naturally occurring problems demotivate the team. The manager needs to be aware of the frustration potential within any job and to assess how important this element is to the team member. This will depend on such factors as the personality of the individual, the nature of the obstruction and how intensely the individual desires the goal. The manager can help the situation generally by sound guidance and leadership, particularly by good communication, good work organization, and perhaps by adopting a participative style of management.

9.6   PERSONALITY

9.6.1   Individuality

Even within the areas considered so far in the search for commonality in human behaviour, it is evident that only partial success has been achieved. Some general similarities in human behaviour have been identified, but they all tend to be hedged about by implied deviation. Examples of individuals can always be found to provide an exception to the generalizations put forward. Leavitt [1] moves quickly from a short chapter headed 'People are alike: some basic ideas' to a much longer evaluation commencing in his second chapter headed 'People are different: the growth of individuality'. This could imply that the differences are greater than the similarities, or that they should be given more consideration. It does not matter which, provided the significance of individuality is understood.

We are all different, and most of us are conscious of and proud of our individuality. No one can get to the stage of commencing a university course without being aware of differences in personality. Friends at home and school display personality differences. Even close relatives near in age such as brothers and sisters can be quite different in personality, and sons and daughters differ from their parents in many ways. In the last two examples, common inherited characteristics would suggest some similarities. Similar experiences of brothers and sisters, and closeness of age, might make for greater likeness than between children and parents, where age and experience disparity might cause greater differentiation.

When encountering another person for the first time, perhaps the characteristics noticed initially are physical ones such as sex, age and physique. The other person, for example, is a tall, slim teenage girl or a short, stocky middle-aged man. Greater acquaintance will reveal other aspects of the person, such as abilities, motivation and perhaps attitude. An even greater knowledge of the individual might be needed before the perceptions and the cultural and social aspects of that person are established. If the initial meeting was in a social setting rather than a work environment, the cultural and social aspects of personality might become apparent earlier.

With the exception of sex, most characteristics will change during a person's lifetime. Of course, most people have physical characteristics that remain with them throughout life, such as a propensity towards slimness or plumpness. Even these characteristics are known to change in the long term. Many people who were slim in their adolescence and early adult period have encountered the phenomenon known colloquially as 'middle-age spread'.

So, what aspects influence an individual's personality? There will be hereditary characteristics that will be derived from the person's parents. These will probably influence such elements as intelligence, and specific abilities such as those concerning, for example, music or mathematics. Early experience in the family will have a profound effect and will bring social and cultural influences. Later, experience in the working environment will influence the personality that has been formed in childhood and adolescence. The roles taken, and the successes achieved, will influence the personality and develop further attitudes and perceptions. However, the debate continues as to which of the two main components is the most significant, the inherited characteristics or those acquired from experience.

9.6.2 Personality studies

As early as the fourth century BC, Hippocrates identified four types of personality: sanguine, melancholic, choleric and phlegmatic. In modern psychology, there are two different approaches to the study of personality, the nomothetic and the idiographic. The *nomothetic* approach assumes that personality is largely inherited and consistent; social and environmental influences are minimal and the personality is resistant to change. The *idiographic* approach regards personality as a developing entity, subject to the influence of interactions with other people and the environment. Though an individual is unique, the self-concept (see section 9.4.1) of every person will develop with time.

It is not appropriate in an introductory text on management for engineers to consider in any depth studies of human personality. However, some studies, notably those of Cattell [3] and Myers-Briggs [4], have enabled tests to be developed that give an indication of the characteristics of individuals, which are generally referred to as personality tests. Some companies use these in their assessment of potential recruits and for internal promotions, and it is in the first of these examples that the young engineer is most likely to encounter.

The *Cattell 16PF questionnaire* assesses personality on the following factors:

Outgoing	_____	Reserved
Abstract Thinker	_____	Concrete Thinker
Emotionally Stable	_____	Emotionally Unstable
Dominant	_____	Submissive
Optimistic	_____	Pessimistic
Conscientious	_____	Expedient
Adventurous	_____	Timid
Tender-minded	_____	Tough-minded
Suspicious	_____	Trusting
Imaginative	_____	Practical
Shrewd	_____	Ingenuous
Insecure	_____	Self-assured
Radical	_____	Conservative
Self-sufficient	_____	Group Dependent
Controlled	_____	Casual
Tense	_____	Relaxed

The *Myers-Briggs (MBTI) test* assesses personality on the following four scales:

Introvert (I)	– (E) Extrovert	How we relate to others and the world
Sensing (S)	– (N) Intuitive	How we obtain information
Thinking (T)	– (F) Feeling	How we make decisions
Judging (J)	– (P) Perceiving	How we assess priorities

This Myers-Briggs classification provides a 4 × 4 matrix of personality types, with 16 possible combinations designated by the letters of the four characteristics that predominate. It is interesting to note that there is a preponderance of ISTJ and ESTJ types among managers in UK organizations. However, in organizations in the Far East, a much wider spectrum of Myers-Briggs personality types is found in management positions.

Students will recognize many of the characteristics mentioned in these tests, and may be able to attribute some of them to themselves or their friends. Experience will also tell you that personalities with combinations of these characteristics occur, and may result in individuals having personalities that are compatible or incompatible with others, or complementary or noncomplementary to others. When people work in teams this can be most important, but that is a subject that will be discussed in Chapter 10, which considers people in groups.

Another popular model of personality is the *big five model*. This identifies five personality traits:

- *Extraversion:* The degree to which an individual is talkative, sociable and assertive.
- *Agreeableness:* The degree to which an individual is good-natured, trusting and cooperative.
- *Conscientiousness:* The degree to which an individual is responsible, dependable and achievement oriented.
- *Emotional stability:* The degree to which an individual is calm, enthusiastic and secure, or conversely tense, nervous and insecure.
- *Openness to experience:* The degree to which an individual is imaginative, intellectual and artistically sensitive.

So what should the student learn from this very short section on personality, or the differences between people? Perhaps no more than that human beings have many characteristics in common, but everyone is different. Each individual has strengths and weaknesses in a wide spectrum of capabilities. The better the manager knows these characteristics and capabilities, the better he or she will be able to manage and motivate the team.

9.7 GENERAL THEORIES OF HUMAN MOTIVATION

9.7.1 Introduction

What motivates the human being is of great interest to the behavioural scientist, and should also be a major concern of the manager. In the latter case, the interest is caused by a desire to obtain more productive performance from members of the workforce at all levels. A highly motivated and enthusiastic workforce can be a major asset to a company and contribute substantially to its progress and profitability. A demoralized and uninterested workforce can have a debilitating effect on an organization. It is a major objective of managers, supervisors and group leaders to achieve high levels of motivation from the subordinates under their control.

The study of human motivation has occupied the time of behavioural science research workers for decades, and many theories and models have been proposed, criticized, modified, accepted or rejected, according to the help they provide in understanding the subject. A totally acceptable model has yet to be found, but two of the longer-established ones that have gained a measure of acceptance will be discussed, together with the criticisms made of them. Some more recent models will then be considered briefly.

The long-established theories or models are those of:

- Maslow – Hierarchy of human needs [5];
- Herzberg – Two factor theory [6].

9.7.2   Maslow – hierarchy of human needs

Maslow's model identifies five levels of human need, which are, in ascending order:

- physiological;
- safety;
- social;
- esteem;
- self-actualization.

Figure 9.3 illustrates these five levels.

The principal elements of Maslow's model are as follows:

- *Physiological needs:* The most important is homeostasis, the body's automatic system for maintaining its normal functions. These include the need to breathe, eat, drink and sleep, and to maintain a reasonable temperature. Also included are activity, sensory pleasures and sexual desire.
- *Safety needs:* These include safety and security, protection from danger and physical attack, protection from pain or deprivation, and the need for stability.
- *Social needs:* These are also defined as love needs, and include friendship, affection, love, social activities and interaction with other people.
- *Esteem needs:* These are also defined as ego needs, and include self-respect and the respect of others.
- *Self-actualization needs:* This is the fulfilment of one's potential, of achieving all one is capable of achieving, and becoming whatever one is capable of

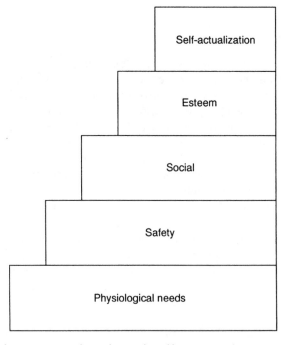

Figure 9.3   Maslow's hierarchy of human needs

becoming. This involves strength, independence and achievement, appreciation, status and prestige.

Maslow maintained that people generally fulfilled these needs in the order given, but that there were many exceptions and there might in some special circumstances be a reversal of the order. Reversal of the social and esteem elements in the hierarchy is the one most frequently encountered. To some people, esteem assumes a greater importance than the social elements, such as love. Some managers may accord greater value to the respect of their associates than to their affection.

Individuals who have experienced continual deprivation may never aspire to the higher-level needs. The peoples of many famine-stricken and disaster-prone areas of the world may regard attainment of modest elements of the safety and social levels as a happy achievement. The permanently unemployed may be willing to settle for any job and its attendant security.

People with high levels of motivation, such as idealists and creative artists, may be little concerned with the lower-level needs such as food, shelter and security, provided they can achieve the objective to which they are dedicated. The artist starving in a garret is willing to accept the absolute minimum of lower-order needs, providing he or she can continue to paint. However, those minima must be available so that he or she can remain dedicated to art. Martyrs, in the most extreme example, are prepared to give up life itself for the sake of their cause, showing that, for a few, self-actualization even transcends homeostasis.

Maslow claims that the hierarchy is universal across different cultures, but that there are some differences between the individual's motivational content in a particular culture. He adds that it is not necessary for an individual to be fully satisfied at one level before having needs at a higher level. Requirement for satisfaction at a particular level will vary from individual to individual.

Maslow's model has been much evaluated and some criticisms have been made of it. The extent to which individuals vary in their requirements for satisfaction at each level may be unclear, but so too is the time element. How long does it take for a higher-level need to emerge, or are they all present to some degree, latent in our personalities? We may all have needs for self-actualization, but for many, working to pay the mortgage and provide food for the family will take precedence. The differences in personality briefly mentioned earlier in this chapter may have some influence. Those with timid, insecure and pessimistic characteristics are likely to value security highly and be willing to work in a safe but tedious job. An optimistic, self-assured and adventurous individual might feel more suited to an exciting high-risk, high-prestige post. An aspect that the manager should bear in mind is that employees do not fulfil all their aspirations in the workplace. The home, social activities and activity in organizations outside the workplace all provide means for individuals to fulfil their needs. It is advantageous for a manager to have some knowledge of employees' lives outside the workplace to understand what needs are fulfilled in Maslow's terms.

9.7.3   Herzberg – two factor theory

Herzberg's original investigation involved interviews with some 200 engineers and accountants. They were asked to identify those aspects of their jobs they enjoyed and aspects they disliked. Consistent responses were obtained, indicating two sets of factors that affected motivation, and this led Herzberg to propose his two factor theory.

One set of factors was identified that, if absent, caused dissatisfaction. These relate to the context of the job and were called hygiene or maintenance factors. The other set of factors were those that motivate the individual to superior performance. These relate to the content of the job and were called 'motivators' or 'growth' factors.

Herzberg extended his analysis to other groups of workers and found that the two factors were still present. Components of jobs that fall into each category are indicated below:

Factor name:	*Motivators*	*Hygiene factors*
Alternative name:	Satisfiers	Dissatisfiers
Concerning:	Growth	Maintenance
	Job content	Job context
Factors:	Challenging work	Status
	Achievement	Interpersonal relations
	Growth in the job	Supervision
	Responsibility	Company policy and administration
	Advancement	Working conditions
	Recognition	Job security
		Salary

There are two major criticisms of Herzberg's theory. One is that it has limited application to people in jobs that are monotonous, repetitive and limited in scope. Some studies undertaken subsequently showed that some groups of manual workers were primarily concerned with pay and security, and regarded work as a means of earning money to satisfy outside needs and interests. However, studies of other groups of manual workers tended to support Herzberg's view.

The other major criticism is that the results obtained were influenced by the methodology used. Studies conducted without asking the employee to identify aspects of the job that caused good or bad feelings tended to give different results and there was a possibility that Herzberg's results might be subject to interviewer bias.

It should be noted that Herzberg's hygiene factors correspond roughly to the lower-level needs in Maslow's hierarchy, as shown in Figure 9.4.

9.7.4   Other theories of motivation

There are many other theories of human motivation, a detailed account of which may be found in standard HRM textbooks.

McClelland [7] and others have proposed a *three needs theory*, which identifies three major motives in work: affiliation, power and achievement.

The *goal setting theory* incorporates the proposition that performance is related to the acceptance of goals. More difficult goals ('stretch goals'), when accepted, generate better performance than do easy goals. Goal commitment, national culture and self-efficacy (an individual's belief in his or her capability) all contribute to goal achievement.

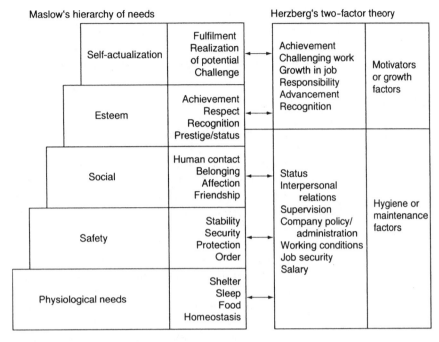

Figure 9.4   Maslow/Herzberg comparison

Reinforcement theory, in contrast to goal-setting theory, suggests that behaviour is a function of its consequences. Behaviour is controlled by reinforcers, consequences that occur immediately after a behaviour and increase the probability that the behaviour will be repeated.

Equity theory proposes that an employee will consider what is put into a job and what is obtained as benefit from it (the inputs and outputs). After comparison with the lot of other employees, the employee will perform according to the perceived equity of the employment, or lack of equity.

Expectancy theory assumes that an individual will act in a certain way in the expectation that this act will be followed by an outcome that is attractive to the individual.

This collection of different theories may suggest some element of confusion as to the basis of human motivation, but many of the theories have evidence in support. With an issue as complex as human motivation, it is very likely that all the theories represent facets of the issue and identify elements that can contribute to employee motivation.

The practical application of some elements of the theories can be found in work to improve the design of jobs so that they are more motivating for the employee. Job design (the way tasks are linked to form jobs) can result in work being done in a way that may be more or less motivating for the employee. The *job scope* (the number of different tasks incorporated in the job) may be subjected to *job enlargement*, thereby including a greater variety of work. *Job enrichment* is the addition of allied components to the job, such as planning or assessment functions. It can increase the degree of control the employee has over the job, thereby increasing autonomy. Other aspects contribute to job enrichment, such as the variety of skills, talents or specializations that the job requires; the extent to which the individual is responsible for the totality of an identified piece of work; and the significance of the activity to the objectives of the organization.

9.8   INDIVIDUALS IN PAIRS

9.8.1   The importance of personal relationships

Some of the most enjoyable and enriching experiences and also some of the most annoying, frustrating and distressing experiences for people in all walks of life occur on a one-to-one basis. In the workplace, relationships are usually more formal, less close, yet can have a powerful influence on the lives of the people involved and on others. A good relationship between a young engineer or scientist and his or her supervisor might lead to rapid promotion within the company. The synergy of a good relationship between two engineers working on the same project, or two scientists jointly conducting research, might lead to a greater interest in their work, more creativity and a more successful outcome to the project. A relationship of shared confidence and trust between the engineer and the client might result in more commissions for that engineer from the client, as well as recommendations to other clients.

A key to successful relationships is good communication. Even in the most formal situations, and even where the personal feelings of the parties towards each other need to be suppressed in order to achieve an effective working relationship, good communication between the parties is essential. (This chapter should be read in conjunction with Chapter 11, which begins with an examination of the processes of communication and then considers ways in which technologists and managers can improve their ability to communicate with others by the most appropriate means.) The remainder of this chapter deals with the *psychology of personal relationships*.

9.8.2   Development of relationships

Close relationships start very early in life. The human baby's first and most critical relationship is with its mother. The bond that is made during the nine months of pregnancy is the precursor to the maternal instincts that encourage a mother to feed and protect her offspring. This is fundamental behaviour that is found throughout the animal species, with few exceptions.

The father is perhaps a shadowy figure in the early stages of the life of a new baby. His role in procreation, short but enjoyable, is well removed in time from the birth of the baby. In most cases, however, 'daddy' is the second person with whom the infant has a unique relationship. Often, 'mummy' and 'daddy' are the first words a baby learns. This brings us to another fundamental in human behaviour, the desire and need to communicate with other individuals. The very important activity of *one-to-one communication* provides a major part of the later material on communication.

As the child develops, so do other relationships. There are interactions with family members: brothers and sisters; grandparents; the wider members of the family. At school, relationships develop with teachers and particularly with other children. These relationships may be formal, as with the teacher, or friendly (most of the time) with other pupils.

With the onset of puberty, a need arises for closer relationships. The human being is one species with a tendency to 'bond for life'. There are others, swans for example. With some species, the bonding is only for one breeding cycle. The increasing incidence of divorce in the more highly developed countries might suggest that life bonding is decreasing, or that the ability to make an accurate choice is deteriorating. However, most human adults have a need for a special relationship with another person, and although the phrase 'a meaningful relationship' has become a cliché, the phenomenon it describes is still important.

Special relationships sometimes occur in the working environment, and when they do they can cause problems as well as benefits. Relationships at work are basically formal. There is work to be done; each person has a job function and a role to perform. Ideally, friendly relationships exist in the workplace, but this does not always occur and in some instances the reverse is true. Technically it is not necessary for there to be friendship between co-workers. Some bosses may opt for a relationship with subordinates where there is an element of fear. 'You don't have to like me, you just have to do what I say,' observed the boss of one of the authors, when as a young engineer he unwisely expressed discontent with a decision.

At work, the technologist tends to interact with a wide spectrum of colleagues and needs to establish a viable working relationship with each and every one of them. If this can be friendly, so much the better, and you may even socialize after work with some of your colleagues. This can be a helpful element but it is not essential; you need only to establish a practical relationship for the duration of working hours. The key to this is communication.

9.9 INTERPERSONAL PSYCHOLOGY

9.9.1 Transactional analysis

In Part I of this book, teamwork was identified as a positive factor helping to enhance business performance, and confrontational attitudes were cited as an obstacle. Effective teams depend on transactions between individuals, and a person's manner or the words chosen in the early stages of a dialogue can strongly influence the transaction. It is therefore important for engineers and technologists to be aware of the ways in which transactions between individuals can be 'managed', to improve communication and elicit cooperation.

Personality dynamics

The following is the introduction to an article by John Martin [8] on personality dynamics:

'Interpersonal work relationships display the complete spectrum of possibilities; cooperative and competitive, loving and hateful, trusting and suspicious and so on. Most individuals would like to improve some of their work relationships so as to make their job easier or more rewarding or more effective. The problem is that although we can all recognize relationships that aren't right, only a few know how to improve them. Often attempts to improve relationships are based on analyses of what the other person is doing wrong. Such an approach never works, and usually makes things worse.

There are three requirements to improving relationships. The first is to have a well-tried framework for thinking about and understanding what goes on in relationships – a theory. The second is to apply the theory to oneself in order to discover how one elicits the responses one gets and why these are unsatisfactory. The third is to use the theory and one's self-understanding to set about doing something different, i.e. relating to others in a different way. Experience in all sorts of management training, relating workshops, therapy groups, training groups and so on has demonstrated that such an approach can work; individuals who follow the above steps do find their relationships improving. One of the theoretical frameworks used in this sort of exercise is the Transactional and Structural Analysis developed by Eric Berne.'

A better understanding of personality dynamics can be obtained from transactional analysis (TA), and the basic principles of TA will be considered in the rest of this chapter. For an excellent analysis of the main features of TA, refer to the article by John Martin [8]. For further reading, the original works of Eric Berne [9] should be consulted.

The simple theoretical model

TA is all about interactions between people. The interaction is normally limited to two people, and concerns the *transactions* that pass between them. These are frequently in the form of a conversation, but can be by other forms of communication, such as body language, gesture and physical contact. Eric Berne calls these interchanges *strokes*. A remark by one person that elicits a response from another will be a two-stroke transaction.

The theoretical framework assesses the psychological state in which the participants find themselves during the transaction, and relates this to the psychological development of the individual from infancy. Three major *ego states* or subpersonalities are identified, which coexist in every individual over the age of five years. They are parent, adult and child, as shown in Figure 9.5.

The child is the biologically driven subsystem that provides emotions, feelings and energy (pleasure, sadness, excitement, curiosity etc.). Almost all the basic components are there at birth, and a stable configuration is achieved by four or five years of age and remains unchanged until the emergence of adult sexuality at puberty.

The parent subsystem is established in the child during the first few years of life by copying and relating to its parents and immediate environment. Cultural background is assimilated by this process, and desirable and undesirable behavioural patterns are passed on from generation to generation.

The adult subsystem is the computer-like part of behaviour, which collects data, makes estimates and deductions, attempts rational judgements, and tries to

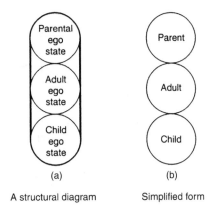

(a)   (b)

A structural diagram   Simplified form

Figure 9.5   Major ego states.
Source: adapted from *Games People Play* by Eric Berne © 1964; published by Penguin Books 1968. With permission

make the individual behave in a logical manner. The adult ego state is present in the small child, but it is the part of the individual that 'grows up' as that person gets older, whereas the other two states are formed very early in life and remain substantially unchanged. Even when the child and the parent ego states appear to be in control, the adult state can regain control if necessary.

John Martin suggests a simple method of distinguishing the major substates by the phrases used while in those states. The key adult phrase is 'I think I could do ...'; the key parent phrase is 'I ought to do ...'; and the key child phrase is 'I want ...'. Other words frequently used that are indicative of the three states are listed in Figure 9.7.

The extended model

There are subdivisions of the three main states. The parent state can be subdivided into the two parts of the parental role: the nurturing parent, who comforts and supports, and the critical parent, who makes demands and sets standards.

Although it is not usual to subdivide the adult ego state in transactional analysis, it is possible to make a division into the primitive adult for the intuitive problem-solving abilities possessed even by young children, and the rational adult for the ego state that develops as the child's linguistic and conceptual skills increase.

The child starts as the free child, expressing simple and immediate needs and becoming angry if these are not obtained. The free child soon has to compromise and the adapted child, and its converse the rebellious child, develop.

The major substates and their subdivisions are shown in Figure 9.6. The ego states, their characteristics and the words often associated with them are summarized in Figure 9.7, from John Martin's article [8].

Complementary transactions

In order to maintain a satisfying transaction, it is important that the initial transaction and response are in the same ego state, or a compatible one. Clearly,

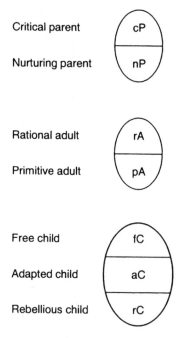

Figure 9.6   Subdivision of substates

if one person talks to another in an adult ego state, that person will not be kindly disposed to a critical parent response, still less to a rebellious child response. Also, if an individual wishes to complain about current difficulties (on a rebellious child basis), a rational adult response precisely analysing the problems may not really be what the individual is looking for.

Berne analyses such transactions as follows: the comment or action that initiates the sequence of transactions is termed the *stimulus*, and the reply (again a comment or action) is termed the *response*. The simplest transactions are those between individuals in the same ego state, adult to adult for example, or child to child. Also satisfactory are transactions in compatible states such as parent to child. Berne designates these as *complementary transactions* and differentiates them as Type 1 and Type 2 (see Figure 9.8). Each response is in turn a stimulus, so the interaction can continue almost indefinitely, or as long as it is worthwhile for the participants, provided the ego states remain compatible and the transactions therefore complementary.

Crossed transactions

Disruptive to useful interaction and communication is the *crossed transaction* and it is this which causes social and communication problems, in both personal and working relationships. Berne comments that the problems caused by crossed transactions provide work for psychotherapists. He defines two types; see Figure 9.9. The stimulus could be adult to adult: 'Do you know where my pocket calculator is?', expecting to elicit an adult to adult response such as 'It is under

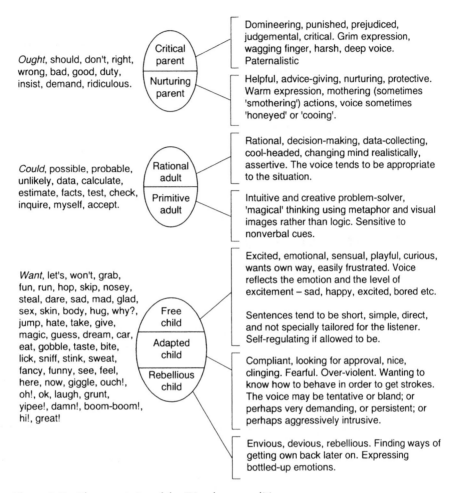

Ought, should, don't, right, wrong, bad, good, duty, insist, demand, ridiculous.

Critical parent — Domineering, punished, prejudiced, judgemental, critical. Grim expression, wagging finger, harsh, deep voice. Paternalistic

Nurturing parent — Helpful, advice-giving, nurturing, protective. Warm expression, mothering (sometimes 'smothering') actions, voice sometimes 'honeyed' or 'cooing'.

Could, possible, probable, unlikely, data, calculate, estimate, facts, test, check, inquire, myself, accept.

Rational adult — Rational, decision-making, data-collecting, cool-headed, changing mind realistically, assertive. The voice tends to be appropriate to the situation.

Primitive adult — Intuitive and creative problem-solver, 'magical' thinking using metaphor and visual images rather than logic. Sensitive to nonverbal cues.

Want, let's, won't, grab, fun, run, hop, skip, nosey, steal, dare, sad, mad, glad, sex, skin, body, hug, why?, jump, hate, take, give, magic, guess, dream, car, eat, gobble, taste, bite, lick, sniff, stink, sweat, fancy, funny, see, feel, here, now, giggle, ouch!, oh!, ok, laugh, grunt, yipee!, damn!, boom-boom!, hi!, great!

Free child — Excited, emotional, sensual, playful, curious, wants own way, easily frustrated. Voice reflects the emotion and the level of excitement – sad, happy, excited, bored etc.

Adapted child — Sentences tend to be short, simple, direct, and not specially tailored for the listener. Self-regulating if allowed to be.

Compliant, looking for approval, nice, clinging. Fearful. Over-violent. Wanting to know how to behave in order to get strokes. The voice may be tentative or bland; or perhaps very demanding, or persistent; or perhaps aggressively intrusive.

Rebellious child — Envious, devious, rebellious. Finding ways of getting own back later on. Expressing bottled-up emotions.

Figure 9.7   Characteristics of the TA subpersonalities

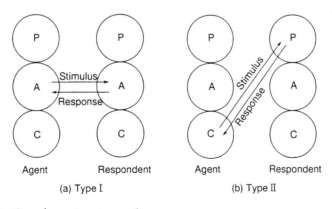

Figure 9.8   Complementary transactions.
Source: adapted from *Games People Play* by Eric Berne © 1964; published by Penguin Books 1968. With permission

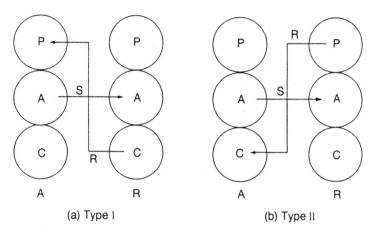

Figure 9.9   Crossed transactions.
Source: adapted from *Games People Play* by Eric Berne © 1964; published by Penguin Books 1968. With permission

that pile of drawings.' If the response is 'Don't blame me, I haven't borrowed it!' (a child to parent response), this is a crossed transaction. Another response could be 'Why don't you look after your things properly?', which is also crossed (parent to child) and just as unhelpful.

If someone encounters a situation involving a crossed transaction, the first aspect to realize is that the transaction will have no useful outcome if it remains crossed. Therefore, the first reaction must be to find a way of uncrossing it, and there are two possible ways. One is to try to reactivate in the respondent the ego state wanted originally; that is, adult in the example given. The other is to find an ego state that is compatible with that of the respondent. If the respondent is in the child state, then a parent response (preferably nurturing parent) would be appropriate. If the respondent is in the parent state, a child response would be suitable, preferably adapted child.

John Martin [8] gives some examples using the extended and subdivided model, and identifies that in most cases the person initiating the transaction assumes a preferred ego state in the respondent, and the best result is a reply in that ego state. The response may be from another acceptable ego state (complementary) or it may not (crossed). An acceptable response to the adult enquiry concerning the whereabouts of the calculator could be: 'I have not seen it, you can borrow mine' (nurturing parent to adult). A crossed response would be: 'If you were more careful and put it away in your desk drawer, you would not waste time looking for it!' (critical parent to child).

Berne has identified other, more complex interactions, such as *ulterior transactions*, those in which the transaction occurs simultaneously at two levels. A transaction, often a spoken one, occurs at the social level. However, another transaction also occurs at the same time, and unspoken, at the psychological level. *Rituals* are transactions such as standard greetings that are largely formalities and the precursor to more significant interactions. *Games* are transactions in which the ego states are deliberately manipulated by one party to the disadvantage of the other. These levels of the model are perhaps too sophisticated for an introductory text such as this, but the reader is referred to *Games People Play* by Eric Berne [9].

Conclusion

An understanding of transactional analysis allows individuals to be aware of the ego state they find themselves inhabiting and to recognize it in others. It is possible thereby to respond more appropriately to the needs of the other person, and so perhaps achieve better understanding and cooperation. Recognition of a crossed transaction will allow the interaction to be modified so that a fruitful outcome can be achieved. It is best to aim for an adult/adult interaction in most situations that the young engineer or scientist might encounter, though there are some instances when other complementary relationships may be appropriate. For example, the nurturing parent and adapted child mode will suit many superior and subordinate relationships.

9.10 SOME GENERAL CONCLUSIONS ON INTERPERSONAL RELATIONSHIPS

Many of the problems of life are concerned with achieving good relations with other people. This can be on a personal, domestic, social or work basis. The key to this is an ability to communicate, and an ability to understand something of the workings of the other person's mind. In one-to-one relationships the objective is often simple: to obtain understanding of one's views and requirements, and to solicit the cooperation of the other person. The method is also simple, but not always easy to achieve. It involves communicating at all times in a clear and appropriate way (see Chapter 11) and being aware, as much as possible, of the perspective of the other individual.

9.11 SUMMARY

This chapter has indicated the difference between the hard systems base of engineering and the soft systems base of behavioural studies. The methods used by behavioural scientists have been discussed, and similarities noted with the empirical approach used by scientists and engineers. The roots of behavioural studies in psychology, sociology and anthropology were described. The basis of the psychology of the individual was briefly discussed, and the way individuals seek to achieve personal goals considered. Individuality, or the difference between people, was discussed, and two methods of assessing this were summarized. The discussion of individuals concluded with an analysis of human motivation, and two examples of models of human motivation that are regarded as helpful to the understanding of individual behaviour were discussed.

The psychological state of individuals during interaction was then considered. Personality dynamics was discussed on the basis of transactional analysis. A simple and an extended theoretical model was introduced, together with the concepts of complementary and crossed transactions. The importance of awareness of the other person in any one-to-one interaction was stressed in the conclusions.

REFERENCES

1. Leavitt, H. J. (1964) *Managerial Psychology*, 2nd edn, University of Chicago Press, Chicago, IL.
2. Mullins, L. J. (1993) *Management and Organisational Behaviour*, 3rd edn, Pitman, London.
3. Cattell, R. B. and Kline, P. (1977) *The Scientific Analysis of Personality and Motivation*, Academic Press, London.
4. Myers-Briggs, I. (1987) *Introduction to Type*, Oxford Psychologists Press, Oxford.
5. Maslow, A. H. (1987) *Motivation and Personality*, 3rd edn, Harper and Row, London.
6. Herzberg F., Mausner, B. and Snyderman, B. (1959) *The Motivation to Work*, 2nd edn, Chapman and Hall, London.
7. McClelland, D. C. (1988) *Human Motivation*, Cambridge University Press, Cambridge.
8. Martin, J. (1984) 'Personality Dynamics and Transitional Analysis (TA)' in R. Paton, S. Brown, R. Spear, J. Chapman, M. Floyd and J. Hamwee (eds), *Organisations: Cases, Issues, Concepts*, Harper and Row, London.
9. Berne, E. (1964) *Games People Play*, Grove Press, New York. Also published in 1966 by Andre Deutsch, London.

10

Groups of People

10.1 INTRODUCTION

In the old joke, a psychologist is defined as a man who goes to the Folies Bergère and looks at the audience. Those who attend entertainments such as the theatre or a sports match, and who observe the crowd as well as the spectacle, will testify that the behaviour of a crowd differs in many respects from the behaviour of individuals. People will take actions as members of groups that they would not consider taking as individuals, and are sometimes swayed by the influence of others to perform actions that are inconsistent with their normal behaviour. It is therefore not surprising that psychologists have devoted great interest and much study to the behaviour of groups.

Since, in the world of work, much activity is undertaken by people in groups, and most organizations are set up with teams, committees, sections, departments and other agglomerates that could be broadly classified as 'groups', the industrial psychologists' use of their time in the study of group behaviour is worthwhile. Engineering and technical work is performed in the main by people of various disciplines and experience working in teams. Managers are believed to spend about 50% of their time working in groups, and senior managers may spend more than 80% of their time working in groups. Thus engineers and technologists who aspire to management roles have a particular need to understand the purposes for which groups are established, how they function and how the characteristics of the group members affect the group.

There is another old saw, which defines a camel as a horse designed by a committee. This implies that something put together by a group of people may be an uncoordinated product, containing a hotchpotch of features acceptable to the majority, and other features included to placate particular individuals. It might also imply adverse criticism of the way committees operate. If this is so, why do people spend so much of their lives in groups? It may be because performing a task with a group of people can often be a very efficient way of achieving the objective. It is difficult to see how a task of substantial size could be achieved without using large numbers of people. The completion of the Channel Tunnel or getting a man on the moon are examples of tasks that needed vast resources of money and people, and the coordination of such activity was a major management activity.

For individualists, groups and committees tend to be inhibiting. More gregarious 'team' people enjoy working in groups and want the involvement and participation of everyone. The manner of operation of project teams with a well-defined task is usually easier to comprehend than the operation of committees, where more complex personal interactions tend to occur. In this chapter, the workings of some of the most frequently encountered groups will be studied, together with the way the individual member relates to the group.

10.2   TYPES AND PURPOSES OF GROUPS

10.2.1   Introduction

There are many types and purposes of groups. Some may be established in organizations in the form of semi-permanent formal work groups, such as sections, departments, divisions and so on. Another system of group organization is the project team, which has a specific task and exists only for the period necessary to complete that task. The status, importance and power of the group can vary from the board of directors of an organization, which will have statutory responsibilities, to an informal (*ad hoc*) discussion group, which has no power and is significant only to its members. The purpose of the group is usually a major factor influencing its type, structure and characteristics. The basic purposes for which groups are formed will now be considered.

10.2.2   Organizational purposes

Work groups may be assembled within an organization for any of the following purposes:

- distribution of work;
- to bring together skills, talents and experience appropriate to a particular task;
- to control and manage work;
- problem solving and decision making;
- information and idea collection;
- to test and ratify ideas;
- coordination and liaison;
- negotiation and conflict resolution;
- inquest and enquiry into past events.

10.2.3   Individuals' purposes

An individual may join a group for one or more of the following purposes:

- to satisfy social and affiliation needs, to belong or to share;
- to establish a self-concept, in relation to others in a role set;
- to gain help and support;
- to share in common activities;
- to gain power, promotion or political advantage.

10.3   GROUP EFFECTIVENESS

10.3.1   General

Groups utilize the abilities, knowledge and experience of the individual members, to a greater or lesser degree of efficiency. Groups tend to produce a smaller number of ideas than the sum of the ideas that the individuals might produce working separately. However, the ideas produced are likely to be better because they are more thoroughly evaluated and have the benefit of the greater collective knowledge of the group during this evaluation. Groups tend to take greater risks in decision making than the individual members might take on their own because the responsibility is shared. People behave in a more adventurous way in a group, because there is 'safety in numbers'.

How effective a group might be depends on many things, some of which are inherent in the purpose, structure, membership and so on of the group (the 'givens'); another is the maturity of the group, which changes during its life. The performance of a group can be enhanced by education in group behaviour and training in teamwork skills.

10.3.2 Givens

Group size

The larger the group, the greater the diversity of talent, skills and experience on which it can draw. However, the chance of individuals participating to their full capability and making a useful contribution reduces with increase in group size. The optimum group size is usually considered to be about five, six or seven.

Member characteristics

People with similar characteristics tend to form stable groups. The groups have homogeneity, but tend to be predictable and not particularly innovative. Groups with members having dissimilar characteristics exhibit higher levels of conflict, but are often more innovative and more productive.

Individual objectives and roles

If all group members have similar objectives, the group's effectiveness is much enhanced. However, some group members may have specific objectives of which the rest of the group are unaware ('hidden agendas'). These can include such activities as covering up mistakes, scoring points, making alliances, paying off old scores and so on. When this occurs, group effectiveness can be much reduced.

Nature of the task

Effectiveness will vary with the nature of the task; whether it is urgent; whether the result can be measured in terms of time and quantity; how important the particular task is in terms of the objectives of the individual; and how clearly the task is defined.

10.3.3 The maturity of the group

The way a group performs will change radically during its existence. When the individuals who are to constitute the group first come together they may have no knowledge of one another. They may not know at the outset the skills, abilities and experience that the other members possess. They may not know how to cooperate and coordinate with each other. They are at the bottom of the learning curve.

As time progresses, so the group will develop and with effort (and perhaps a little luck) will become a coordinated working unit. How this happens has been the subject of much study and a number of theoretical models have been proposed. Here are two examples.

Tuckman's model of group development

Tuckman [1] suggested that there were four stages in the development of a working group: forming, storming, norming and performing.

Forming: The group members meet for the first time. They are a set of individuals. They may not know each other or much about each other. They need to establish what the group purpose is, who is to lead the group and who is to be influential in the group. Each individual wishes to establish a personal identity and to become known to the other members of the group.

Storming: This is the conflict stage. More than one individual may feel a desire to lead the group. Leadership may be challenged. Interpersonal hostility may develop. Hidden agendas may become apparent. The successful handling of these problems by the group will lead to the formulation of realistic objectives.

Norming: Norms of group behaviour are established. Greater knowledge by group members of the skill resources available within the group allows appropriate work practices to be established. Greater trust and understanding by group members of other group members allows realistic decision-making processes to be set up.

Performing: The group is fully mature. It works in an efficient and coordinated way. There is open behaviour by group members, with mutual respect and trust.

Bass and Ryterband model

Bass and Ryterband [2] also suggest a four-stage model that is very similar to that of Tuckman. The four stages are:

- Developing mutual acceptance and membership.
- Communication and decision making.
- Motivation and productivity.
- Control and organization.

Although there are some minor differences of definition and emphasis between these two models, there is a great deal of similarity and a large measure of agreement on the key elements of the learning process.

10.3.4   Behaviour of mature groups

A mature group, one which is long established and which perhaps has an ongoing task, will exhibit a different pattern of behaviour. One model of this is the 'creative cycle' [3]. This also has four phases.

Nurturing phase: The members meet. There may be some social discussion. Perhaps coffee or tea is served. Late arrivals may be sought. Papers are circulated, minutes of previous meetings may be discussed.

Energizing phase: The meeting proper starts. Discussions relevant to the group's purpose commence. Input is made and the required decisions are identified.

Peak activity phase: Interaction between members is high. Important matters are evaluated. Conflict may occur and be resolved. Decisions are made.

Relaxation phase: All important business is complete. Participants begin to 'wind down' and consider their next activity or appointment. This is not the time to introduce significant new matters.

An example of the application of the creative cycle model would be the regular meetings of a management team or a technical project design team. These meetings might be monthly in the first case, or weekly in the second example. The purpose in both cases would be to review progress against schedule and to resolve any problems that had occurred since the previous meeting, but could include other activities. Since the meeting is employed frequently as a means of communication and decision making, it will be considered in some detail.

10.4   MEETINGS

10.4.1   Introduction

The meeting format is much used in the working environment. Many would argue that it is too much used, for it can be a very inefficient means of conducting business. Many managers spend a major part of their working time in meetings, of greater or lesser productivity, either as the chairperson or leader of the meeting or as a participant. The efficiency and effectiveness of the meeting are critically dependent on the way it is run by the chairperson and on the behaviour of the meeting members. Engineers and technologists will encounter meetings with increasing frequency during their careers and therefore would be wise to learn about their characteristics. Section 11.6 gives more details about meetings.

10.4.2   Types of meeting

Meetings can be generally classified according to their size, frequency, composition and motivation.

Size

Meetings can be placed in three broad categories according to size:

(1) Large – the 'mass meeting', with perhaps 100 or more people, divided into speakers and listeners.
(2) Medium – the 'council', with speakers, listeners, questions and comments.
(3) Small – the 'committee', with a maximum of about 12 people and a free range of discussion and interaction.

Frequency

Four groupings for meetings according to frequency are suggested:

(1) Daily – people liaising as part of their work.
(2) Weekly or monthly – meetings such as formal review meetings, or the coordination and decision meetings of functional heads in organizations.
(3) Annual – the type of meeting that is required by statute or some other governing ordinance, such as the annual general meeting of a public limited company involving the directors and the shareholders. These meetings tend to be very formal.
(4) *Ad hoc* – meetings of an irregular nature, occasional and usually called to deal with a special problem or issue.

Composition

In this case we can categorize meetings according to the commonality of the participants' activities:

(1) Same activity – scientists, technologists, engineers, nurses, doctors, accountants.
(2) Parallel activities – managers of sections of a production plant, leaders of sections of a project, regional sales managers of a marketing company.
(3) Diverse – a meeting of strangers of diverse interests, united only by one interest, which is the reason for the meeting.

Motivation

This category relates to the purpose of the gathering, which could be business or social, active or recreational:

(1) Common objective – process improvement task force, product design group, senior citizens' club, football team, pressure group.
(2) Competitive – subsidiary company managers within a conglomerate discussing resource allocation.

10.4.3 Functions of a meeting

The most obvious functions of a meeting are communication and decision making, but the meeting also fulfils a rather broader spectrum of functions for the group. The meeting *defines* the team, the group, the section or the unit. An individual's presence at the meeting signifies full membership of the group, and allows identification with the other members of the group. Group members are able to share *information*, authority and responsibility, and to draw on the knowledge, experience and judgement of the other members. The meeting enables each individual to understand the collective *objectives* of the group and the way each member can contribute to the attainment of those objectives. The activities of the meeting create a *commitment* to the objectives and decisions of the group.

The concept of 'collective responsibility' is generated. The meeting allows the group leader (whether so designated or titled chairperson or manager) to *lead and coordinate* the whole team. The meeting allows each individual to establish a *status* within the group, and the group membership can confer a status on the individual within the organization.

10.4.4   Managing a meeting

There are a number of training videos that give guidance on the correct procedure for managing a meeting. Two also demonstrate what can go wrong if the established guidelines are not followed [4]. In these, a demonstration is given of inept chairmanship that renders the meeting totally ineffective (group members confused, essential decisions not taken) and inefficient (group members' time wasted). It is suggested that the five essentials for getting the meeting to operate well are (with alliteration to assist easy memorizing): planning; pre-notification; preparation; processing; and putting it on record.

Planning

It is worthwhile considering the objectives of the meeting at the outset. Is a meeting the best way to achieve these objectives? The chairman should be clear precisely what the meeting is intended to achieve.

Pre-notification

All the other members of the group need to be informed about the meeting and its purpose. The time, place and duration need to be conveyed to the members, and these should be realistic to accommodate the known or anticipated commitments of the group members. Sufficient notice should be given to allow members to attend without having to cancel other engagements, and to acquire whatever information is needed to make a contribution to the meeting. Occasionally, a manipulative chairperson or secretary may arrange the meeting at a time and place that ensure that certain members of the group cannot be present!

Preparation

To some extent, preparation runs in parallel with pre-notification. The matters to be considered need to be arranged in a logical sequence. The amount of time required to consider each item should be assessed. The relative importance of the items should also be considered. Care should be taken that items that seem urgent should not take too much time. Time should be conserved for the consideration of longer-term matters of greater importance. There is merit in an agenda that allows some quick decisions to be made on urgent but less important items at the start of the meeting, providing more time for more detailed consideration of weightier matters for the greater part of the meeting. Group members need to know the agenda of the meeting in advance so that they can come fully prepared and able to contribute to the discussion and the decision-making process.

Processing

The management of the meeting itself requires considerable skill, and it is surprising how often this skill is lacking in the people who chair meetings. The discussion of each item needs to be structured. Members need to be brought back to the point when they wander into subjects outside those scheduled for consideration (deviation) or attempt to revisit subjects already discussed (reversion). Loquacious members need to be gently but firmly restrained (repetition), so that the contributions of the more reticent members can be obtained. Attempts to have private discussions, or to have a mini-meeting within the meeting (diversion), should be firmly resisted. A skilled chairperson can deal positively with any disagreements that arise during the meeting and steer the group to a consensus. However, the chairperson should not impose his or her view arbitrarily, a fault that often occurs when the chairperson is significantly more senior than the other members of the group. To do so makes the meeting pointless; the same result could be achieved more efficiently by the chairperson sending written instructions to each member. Members should be aware that the chairperson may need their compliance to endorse his or her actions, and should not commit themselves to a course of action that they believe to be wrong.

Putting it on record

There is a need to summarize as the meeting progresses and to record the decisions made, their principal justification, what actions are to be taken and by whom. This ensures that at a later date, perhaps the next meeting if there is one, the 'story so far' is clear and progress can be reviewed without wasting time on establishing what was intended to be done. The notes may be taken by the chairperson, by one of the members designated to do so permanently or on rotation, or by a minute taker or secretary brought into the meeting for that purpose. The latter option has advantages, as a member of the group can participate less if he or she also has to concentrate on keeping an accurate record of the proceedings. It is often helpful if the conclusions, decisions and action points are recapitulated at the end of the meeting to ensure that all members are clear about the outcome and committed to the decisions taken, before the meeting disbands.

10.5   LEADERSHIP

10.5.1   Introduction

Any group, even if it is set up with absolute equality of membership, will tend towards a system in which it has a leader. Of course, many groups are deliberately set up with an officially appointed leader. The England cricket team is selected after the captain has been appointed and the captain has a role in the selection of the team. In other sports, the team may be selected and a suitable person appointed captain from the team members by the team manager or controlling committee. There may even be a nonplaying captain, or a more democratic approach may be taken, in which team members elect one of their number as captain. The situation in sport is comparatively simple, as the objective is clear (to win the game) and the philosophy is established that all members of the team will contribute to the full to achieve this objective. Those who do not are unlikely to be selected for the next game. The democratic election of a leader occurs in other groups, particularly where the group members themselves are elected representatives such as parish, borough and county councils in England, or trade unions. In the former case the elected representatives may elect a leader from among their number, and in the latter case this may also occur. Another option in the case of a trade union is that its rules may require the election of the leader by the total membership.

In the world of work, the leader of a group is usually appointed, by an individual or by a body of individuals. That person then becomes the manager of the department, the director of the division, the leader of the section or the project leader of the project team. Regardless of the title, the role is leadership of that group, be it large or small, and the implication is that an individual can be a leader of a smaller group, or subsystem, and at the same time a member of a larger group or system of which that subsystem is part. There is much commonality between the roles of leader and manager, though they are not the same. Managers without the ability to lead may become effective administrators, but leaders without management skills may fail.

Even an *ad hoc* group, one that forms casually for a particular purpose such as a discussion, a gossip or a need to resolve a minor problem, will after a short while show evidence of one or more people attempting to 'lead' the group.

Leadership is a process whereby one person influences other people towards a goal. Leadership capability may be a function of personality; it is related to motivation and to ability to communicate. Leadership cannot be separated from the activity of the group. Leaders are people who do the right thing to accommodate their team's visions. The leader/follower relationship is a reciprocal one, and to be effective it must be a two-way process.

10.5.2 Management and leadership

Leadership and management are closely related. Many definitions of the components of management include leadership, as shown in Chapter 1, where Mintzberg's description of the manager's role included leadership. Even so, a good leader may not be a good manager and vice versa.

Management may be defined as the achievement of stated organizational objectives using other people, and is essentially within the organizational structure, or at the interfaces with other organizations. Leadership is based more on interpersonal relationships and does not need to take place within the structure of the organization.

The components of management were analysed in Chapter 3, but this section concentrates on leadership, which can be regarded as the bridge between the group processes and management. According to Hellreigel *et al.* [5], the principal roles of a *leader* are to set vision and direction, to align employees and to motivate and inspire. The principal roles of a *manager* are to plan and budget, to organize and staff and to control. Managers who are also leaders will perform all of these functions.

Mullins [6] suggests that a leader may be elected, formally appointed, imposed, chosen informally, or may just emerge naturally. Mullins also differentiates between attempted, successful and effective leadership:

- Attempted: an individual attempts to exert influence over the group.
- Successful: the behaviour and results are those intended by the leader.
- Effective: the leadership helps the achievement of group objectives.

10.5.3   Sources of a leader's power

There are three basic sources of a leader's power. They are:

- Organizational power: legitimate – reward – coercive.
- Personal power: expert – referent.
- Other power: connection – information.

Organizational sources of power

With *legitimate power*, the subordinate perceives that the leader has the right to exercise power. Authority is conferred by position within the organizational hierarchy. With *reward power*, the subordinate perceives that the leader has the ability to reward him or her with, for example, pay, privileges, praise, promotion and so on. With *coercive power*, the subordinate fears that the leader has the power to punish him or her, for example by withholding the rewards mentioned previously and/or giving reprimands, fines, dismissal and so on.

Personal sources of power

With *expert power*, the subordinate perceives that the leader has special knowledge, ability or experience that makes the leader especially able in a particular area. With *referent power*, the subordinate identifies with the leader who has particular characteristics that the subordinate admires. This could be respect for the leader's values or admiration for his or her charisma.

Other sources of power

Two additional sources of the power of managers have recently been identified. *Connection power* is drawn from the manager's relationships with influential people, both within and outside the organization. *Information power* comes from the manager's access to information that is valuable to other people and in which they would like to share. While much of the power associated with both these sources may come from within the organization, it may not all be internally related and the sources may be both positional and personal.

Power in a scientific or technical environment

There are no great problems in relating these generalized concepts of power to activity in a scientific, technical or engineering environment. The director of a technical division in an organization, the manager of an engineering department in a company and the leader of a scientific project are all appointed by the appropriate authority of the organization or company concerned. This authority can replace the individual if it so chooses. While the individual holds the job, he or she has legitimate power and the direct subordinates are required by their terms of employment to perform any reasonable instruction given by the leader, within the terms of the job. This is a formal, even a contractual relationship, and the leader is justified in exercising legitimate power. Since the same terms of contract will include arrangements for salary, bonus and other elements of reward, and also disciplinary procedures, the employment contract will also define some elements of the reward and coercive powers of the legitimate leader. However, the other elements such as praise, blame, the allowance of privileges and the use of criticism will be more directly related to the specific task in hand. A junior technologist may be praised by his or her leader for a well-thought-out design study, but criticized for the imprecise phrasing of the accompanying report.

The young scientist may respect another colleague if this individual has evident talent, skills, qualifications and experience, and will defer to him or her in areas where these abilities are significant. The colleague therefore has expert power. If a young engineer respects the ethics, values and principles of a colleague, then any influence exerted by the colleague is derived from referent power. The most successful managers and leaders in technical and scientific activity are very likely to be able to draw to some extent on all the sources of power mentioned previously as they perform their leadership and management function.

10.5.4   Analyses of leadership

Various attempts have been made to examine the basis of leadership. Three aspects are regarded as important by many analysts: *qualities or traits*; *situation*; and *functional* or *group*.

Qualities or traits

This model assumes that leaders are born, not made, and that leadership ability is within the person. Some investigators who have studied acknowledged great leaders of the past have found little correlation of characteristics. Others note the commonality of the traits of intelligence, supervisory ability, initiative, self-assurance, individuality, dependability and so on. The subjectivity of this analysis is a major limitation, as is the inability to assess the relative value of the characteristics. It could be suggested that these would vary according to the circumstances in which the leadership was required. This leads naturally to the situation model.

Situation

People with different personalities and from different backgrounds have become successful leaders in a variety of circumstances. Professional knowledge or specific expertise is important, as is the way the leader performs his or her task. The implication is that training for leadership is possible. However, it seems that knowledge and training do not completely fulfil the requirements. Though professional knowledge, skill and experience would be necessary for the leadership of a high-profile, time-limited and critical engineering project, and for the management of engineering standards within a large, high-technology organization, the personal characteristics required for these leadership functions would probably be very different.

Functional or group

This model considers the functions of the leader and how his or her behaviour interacts with the group of followers. Adair [7] identifies three major activities: *task* needs; *team maintenance* needs; and *individual* group member needs.

Task functions involve achieving group objectives; defining tasks; work planning; resource allocation; quality control; performance monitoring; reviewing progress.

Team functions involve maintaining team morale; maintaining team cohesiveness; setting standards; maintaining discipline; communication; group training; delegation.

Individual functions involve meeting individual members' needs; sorting personal problems; giving praise; awarding status; reconciling individual/group conflicts; individual training.

Another rather simpler version of the same model was devised many years ago by Blake and Mouton [8] and has been a major element of management thought to the present time. This model differentiates the task elements of leadership and the people elements, in the form of a two-dimensional matrix. The model does not differentiate, as does the Adair model, between the team people needs and the individual people needs.

Blake and Mouton apply their model to leadership in an organizational situation and therefore term it a management grid. Various styles of management are defined on the two-dimensional matrix according to the extent to which there is concern for people (vertical axis) or concern for production (horizontal axis).

Minimal attention to either element is termed 1,1 or *impoverished management*.

Maximum attention to the needs of people for satisfying relationships, and a comfortable friendly working atmosphere, is termed 1,9 or *country club management*.

Maximum attention to production, achieved by arranging matters so that human elements interfere to a minimum extent, is termed 9,1 or *authority-compliance management*.

An adequate performance achieved by a satisfactory balance between production and people is called 5,5 or *middle-of-the-road management*.

Maximizing concern for both people and production, with a committed workforce with a common stake in organizational purpose, is termed 9,9 or *team management*.

10.5.5 Functions of leadership

Adair [7] suggests three major functions:

(1) Awareness – group process; underlying behaviour.
(2) Understanding – knowing which function is required.
(3) Skill – to lead effectively, which can be assessed by group response.

Krech [9], in an analysis similar to that of Mintzberg mentioned in Chapter 3, identifies 14 functions:

(1)   Executive – coordinator of activities and policy execution.
(2)   Planner – ways and means, short and long term.
(3)   Policy maker – establishes group goals and policies.
(4)   Expert – source of skills and knowledge.
(5)   External representative – spokesperson for the group.
(6)   Internal relations controller – establishes group structure.
(7)   Rewarder/punisher.
(8)   Arbitrator/mediator – controls interpersonal conflict.
(9)   Exemplar – behaviour model for the group.
(10)   Symbol – provides cognitive focus and entity.
(11)   Substitute for individual responsibility – relieves individuals of responsibility.
(12)   Ideologist – source of beliefs, values and standards.
(13)   Father figure – focus for positive emotional feeling and identification.
(14)   Scapegoat – 'can carrier' in the event of failure.

It would take too long to quote examples of the relevance of all these activities in a scientific or technological context, and the detail would not be helpful. Sufficient to say that any manager experienced in leadership in a technical function would recognize the existence of all of them and be able to quote many specific examples. The extent to which an individual will be required to exercise leadership will depend on the size and nature of the technical organization. In a small and highly specialized engineering consultancy, the senior partner may well fulfil most or all of these functions, even causing some feelings of frustration by other members of the consultancy. In a large, technology-based organization, the various aspects of leadership may be fulfilled by any number of individuals in various functions within the organization.

10.5.6   Styles of leadership

Here are some styles that have been identified in the literature, with some elements of explanation:

- Authoritarian/autocratic/dictatorial – the leader alone makes all the key decisions.
- Democratic – the leadership function is shared with the members of the group.
- Bureaucratic – the leader takes decisions according to the established organizational mechanisms.
- Benevolent – the leader is motivated by the best interests of the group and acts as a 'father figure'.
- Charismatic – the leader's personality is so strong that the group members follow his or her bidding.
- Consultative – the leader takes the views of the team members into consideration but makes his or her own decision.
- Participative – the leader accommodates the views of the group members when coming to a decision.

- Laissez-faire – the leader observes the group and allows them to get on with the job as they see fit, but acts as an adviser and is available when help is needed.
- Abdicatorial – the leader lets the group do the job as they see fit and provides no advice or help. It is arguable that this is the absolute opposite to dictatorial/autocratic leadership, is zero leadership and therefore no longer justifies the name.

These tend to lie on a spectrum from authoritarian and dictatorial at one extreme via consultative and democratic to laissez-faire and abdicatorial on the other. Another way of considering this spectrum of leadership styles is to assign four positions on the continuum, where the leader tells, sells, consults or joins:

- Tells – the leader makes a decision, informs the subordinates and expects them to expedite the decision.
- Sells – the leader makes the decision and explains and justifies it to the team.
- Consults – the leader discusses the problem with the team before making the decision.
- Joins – the leader defines the constraints of the problem and jointly makes the decision with the team.

It is clear that the spectrum of leadership styles is very wide, from the absolute 100% leadership of the autocrat or dictator, to the 0% leadership (even nonleadership) of the abdicatorial leader. It is difficult to say with certainty where the most appropriate leadership style might be, for the obvious reason that technical activities vary and the most appropriate leadership style for each activity will be different. It is possible, however, to narrow the spectrum somewhat, to eliminate some extreme options and to suggest some criteria for style selection.

As the activity is peopled by trained professionals, it is unlikely that the subordinate members of the team would find a dictatorial or autocratic style of leadership acceptable for very long. Also, such a style would not be rational since it would require the leader to be totally expert in all aspects of the subject. If this were so, the expertise of the other members of the team would not be needed and they could be replaced by less-experienced and less well-qualified personnel, with the attendant saving of cost. Though readers may be aware, as are the authors, of instances where leaders have attempted to follow an autocratic style within a technical activity, it is rarely successful and usually proceeds via seething discontent to confrontation.

At the laissez-faire and abdicatorial end of the spectrum, there is little compatibility with the physical requirements of technology in the real world, which involves the delivery of a product or service to meet cost objectives, on time and to specification. Neither abdicatorial nor laissez-faire leadership is likely to provide the necessary impetus and coordination to achieve this, though the latter might just cope in the area of scientific research. The most likely area for an effective leadership style is the 'sells' or 'consults' area. The greater or lesser elements of democracy will depend on the personality of the leader and the nature of the task. Since the leader is likely to be established formally in the role by the organization that commissions the task, and will carry the responsibility for the success of the task, it is likely that the style will veer towards the 'tells' rather than the 'sells' when difficulties arise.

10.5.7   Theories of leadership

The contingency theory of leadership

Three key components are identified in this model proposed by Fiedler [10], which considers the influence of the leader/member relationship, and such contingency variables as the task structure, group atmosphere and position power.

The business maturity theory of leadership

Clarke and Pratt [11] believe that the most appropriate style of leadership may vary according to the maturity of a business:

- Champion – needed for the 'seedling' business.
- Tank commander – growth stage of the business, capture of markets.
- Housekeeper – mature business, keep it running smoothly and efficiently.
- Lemon squeezer – for the business beyond its sell-by date, the need is to squeeze as much out of it as possible.

Leadership theories in a technical context

The Fiedler model is an integration of components that have been discussed in earlier sections of this chapter, and the comments on the applicability in a technical context made for the components apply to the integrated version.

The Clarke and Pratt model was devised for businesses and can be related to technology-based businesses. It can also be applied to the embryo, developing, established and waning elements of the technical functions within a business. An example of an engineering-based business that has followed the Clarke and Pratt model is the UK shipbuilder Swan Hunter. An embryo business in 1860, it grew rapidly in the 1870s and was an established major contributor to Britain's shipbuilding from 1880–1960. Thereafter it waned from 1960–1990.

An example of the changes that may befall the technical function in an organization can be found in those companies manufacturing automatic calculators and cash registers. The early models of both were essentially mechanical devices. The current equipment is mainly electronic. The mechanical engineering function in the companies making such equipment has declined with the change of the basic technology, while the requirements for electronic technology have increased.

10.6   TEAM MEMBERSHIP

10.6.1   Introduction

The team, and being a member of a team, is a concept to which most people are introduced very early in life. The domestic family unit may, with luck, have some of the elements of a team. In pre-school and early school activities, children become members of many teams, usually organized by the teacher. The self-organized group or 'gang' may not be formal, but it will still have some characteristics of a team. At school, most sport is arranged in the form of team games. Some simple management justification can be found for this. If it is necessary to organize some exercise for a large number of children, what more convenient and economical way of doing this than by a team game of football, netball, cricket or hockey? If rugby football is the chosen sport, 30 boys can be exercised simultaneously!

So playing the game, playing for the team and showing team spirit are part of the culture and education of most countries, almost from the start. The broader concept of team spirit is one of those aspects of education that is believed to be character building. It is expected that the collective good of the team and achieving the objectives of the team will be given a priority higher than those of the individual. It is soon learned that the individual who wishes to maximize his or her own achievements, by scoring all the goals or making all the runs, is not the person most desired as a team member. Like the Three Musketeers, there must be an element of 'One for all, and all for one'. It is sometimes possible to create a team of 'star' players, in football for example. Even then, a team with less overall talent but whose players coordinate better together may be victorious. In another field, that of music, it is often found that the best string quartets are formed from musicians who are very good but not of international soloist standard. The individualism and bravura of the soloist may not produce as good and as integrated a performance as slightly less brilliant musicians who have practised long together and therefore work better as a team.

This applies to the selection of teams for technical activity. The recruitment for a design project of a team consisting of 'star' designers of international reputation might not be as successful as its cost would require.

There are jobs to be done in the team and roles to be taken. These can be specific: an individual cannot simultaneously be both striker and goalkeeper, bowler and wicket keeper. They can also be more general: not every member of the team can be the leader; there are other roles or functions that need to be done in the team. In the domestic and social world, people can to a great extent choose the groups or teams they wish to join, and the choice will depend on how the objectives and values of that team or group coincide with their own. At work, the choice of colleagues is rarely available, so it is essential to develop techniques of working with other people to achieve the objectives of the group, team or organization. Engineers, technologists and scientists in particular have to learn to do this to a high standard of professionalism.

However, it is not only in science and engineering that the team approach is becoming more prominent. In other technical and even nontechnical areas, the use of problem-solving teams is increasing. Greater use is being made of project teams for activities within organizations. The simplification of organizational structures and the introduction of greater flexibility in such structures, which was one of the outcomes of the 'restructuring' of many companies in response to the recession of the early 1990s, made it possible for companies to use a project team approach more widely. The need to accelerate the speed of the development of a new concept to a marketable product has encouraged the use of new concepts of multiple interlocking teams such as concurrent engineering. A flexible team approach suited to the operation of long-term high-technology projects will be mentioned later in section 10.7.

10.6.2   Motivation

The way in which individuals are motivated was discussed in Chapter 9, and the extent to which group members will be motivated by the activities of the group will depend on a variety of aspects of the group activity. Satisfaction is one of the possible outcomes of group activity, and although a satisfied group is not necessarily a productive group, it generally helps it to be so. Lack of satisfaction will lead to absenteeism by members of the group, but satisfaction is not all that is required for motivation.

An individual will be satisfied with a group if:

- the individual likes and is liked by other group members, i.e. friendship;
- the individual approves of the purpose and work of the group, i.e. task;
- the individual wishes to be associated with the standing of the group, i.e. status.

One or more of these can lead to satisfaction, depending on the *psychological contract*. In the work situation, there is a *formal contract* between employer and employee, which states, for example, the task, the hours and the pay rate. There is also an *informal contract*, which both parties understand and which covers the reciprocal needs such as the employee's need for time off for personal reasons and the employer's for extra work when the firm is busy. The psychological contract is the often unconscious expectation that the employer and employee, or the group and the group member, have of one another. It often surfaces only at moments of stress or crisis.

It is important that the group members know what is expected of them and the standards that they are required to achieve. High standards will give a sense of achievement when attained, and this will contribute to the motivation of the group members. It is also important that the group members receive feedback on the results of their efforts. For example, if a team of engineers has designed and built a road, a bridge, a piece of equipment or a process or production plant, then three major indicators of team performance would be that the new unit was completed to the target cost, on time and to specification. The team need to know the extent of their success against these criteria, and to be rewarded in some way, or at least praised, if they have fulfilled or exceeded the requirements. They also need to know of any shortfall, so that they can learn from the experience and do a better job next time.

Motivation by involvement will only result if the individual perceives that the group and its task are important to that individual. Sometimes it is possible to generate motivation within a group by looking outside the group into the greater organization, or into the environment beyond the organization, and identifying a 'common enemy'. Unfortunately, this can often generate rivalry and conflict between groups within the organization, which may be counterproductive to the objectives of the organization.

10.6.3   Characteristics of team members

The seminal work in this area is that of Belbin [12], who studied the performance of teams of management students undertaking a business game. The students were mature people who held senior posts in commerce and industry and the studies continued with many teams over a period of nearly 10 years. The objective was to assess what characteristics of the teams and their members made the difference between relative success and relative failure in the management game, which was competitive and designed to simulate real conditions.

The eight-role model

Belbin found that there were a number of characteristics that contributed to the success or failure of the team. The ability of the individuals and their general characteristics were measured by various psychometric (personality) tests, but teams with apparently greater ability and apparently compatible personality characteristics did not perform significantly better than less well-endowed teams. Belbin found that the *roles* individuals undertook in a team were important, and defined eight roles that were significant in team performance: company worker (CW); chairman (CH); shaper (SH); plant (PL); resource investigator (RI); monitor/evaluator (ME); team worker (TW); and completer/finisher (CF).

The functions of the team roles are described as follows:

- *Company worker* (CW): Turns concepts and plans into practical working procedures and carries out agreed plans systematically and efficiently.
- *Chairman* (CH): Controls the way in which a team moves towards the group objectives by making the best use of team resources; recognizes where the team's strengths and weaknesses lie; and ensures that the best use is made of each team member's potential.
- *Shaper* (SH): Organizes the way team effort is applied; directs attention generally to the setting of objectives and priorities; and seeks to impose some shape or pattern on group discussion and on the outcome of group activities.
- *Plant* (PL): Advances new ideas and strategies with special attention to major issues and looks for possible breaks in the approach to the problem with which the group is confronted.
- *Resources investigator* (RI): Explores and reports back on ideas, developments and resources outside the group; creates external contacts that may be useful to the team and conducts any subsequent negotiations.
- *Monitor/evaluator* (ME): Analyses problems and evaluates ideas and suggestions so that the team is better placed to take decisions.
- *Team worker* (TW): Supports members in their strengths (e.g. building on their suggestions); underpins members in their shortcomings; improves communications between members and fosters team spirit generally.
- *Completer/finisher* (CF): Ensures that the team is protected as far as possible from mistakes of both omission and commission; actively searches for aspects of work that need a more than usual degree of attention; and maintains a sense of urgency within the team.

Belbin also provides a self-perception inventory. The questionnaire allows an individual to determine his or her compatibility with the various team roles and his or her relative preferences; that is, the most preferred team role, next most preferred, down to those that are least attractive. Thus it is possible to add to the psychometric tests an assessment of most-preferred and least-preferred team roles. Many years of application of these tests has allowed the prediction of likely success or failure of a particular set of individuals who form a team, though the prediction of failure is easier than the prediction of success. Specifically, Belbin identifies particular combinations of individuals with preferred team roles that are known to cause problems. He indicates ways by which the composition of the team should be adjusted to give a better balance, and suggests that certain combinations can be linked in reporting relationships in ways likely to be successful or unsuccessful.

Two other general points need to be made. The team may not have as many as eight members, and it would therefore be impossible for each role to be assumed by one team member. For smaller teams, a particular team member may need to assume more than one role. In larger teams, there will be some role duplication. Even in smaller teams, in practice a team member may have to assume a second-preference or 'back-up' role if there is more than one individual capable of the first-preference role. This is termed 'making a team-role sacrifice'. To give a cricket analogy, if there are two competent wicketkeepers in the team, one may have to field at first slip, or better still at second slip, to avoid role confusion and conflict. In engineering terms, an individual who is familiar with doing the stress analysis of a mechanical design and is happy and comfortable with this function may have to move to a second-string activity, such as pressure vessel design, if the team happens to contain a stress analyst of national standing.

The nine-role model

In 1993, Belbin published an update on his continuing investigations on the subject of team performance [13]. Two of the team roles were renamed and a ninth role was added to the model:

> 'Two of the roles have been renamed, largely for reasons of acceptability. "Chairman" has become "Coordinator", and "Company worker" has become "Implementer". The former term was originally chosen on the grounds of factually referring to the role of the person in the Chair. In the end it had to be dropped for three reasons: its status implications were judged too high for younger executives; in the eyes of some it was "sexist"; and it was liable to be confused with the title that could signify the head of a firm. "Company worker", by contrast, proved too low in status, being especially resented by managing directors who were so described, and the word "Implementer" was eventually substituted.'

The most significant change was the addition of a ninth role, 'specialist'. This role was added after the eight-role theory had been applied in a substantial number of industrial examples. It was found that in project work (a major user of the team system), relevant professional expertise was significant and could not be ignored without endangering the objectives of the project. The inclusion of the specialist in the team was also an important element in career development, by widening experience or by encouraging personal growth. This observation correlates with those of one of the authors [14] in balancing the expertise and experience of team members in a long and complex project, while at the same time maintaining continuity.

However, the model needs to be extended further to accommodate the increasing use in technical projects of multidiscipline and multispecialist teams, and the requirement for interlinking or even interlocking teams in some complex projects. Though most technologists will have to adopt one or more of the roles defined in the earlier model on many occasions during their careers, the role of 'specialist' will probably be one of the most frequent. With the increasing use of multidiscipline teams, particularly in the larger, more sophisticated and more 'high-tech' projects, the majority of the team will be specialists. The specialists, under these circumstances, will need the capability of adopting other roles to facilitate the team activity. Referring back to the Tuckman model, role adoption could take place during the 'storming' phase of the formation of the group, with some role sacrifices being made before the 'norming' phase can commence. Both the Tuckman model and the Belbin model can be recognized in the description of simultaneous engineering teams in Chapter 6.

There is growing use within technology-based organizations of *ad hoc* groups; that is, groups set up with a specific short- or medium-term objective.

These can be problem-solving working groups (PSWGs), quality circles (QCs), quality implementation teams (QITs), departmental improvement teams (DITs), process improvement teams (PITs), team-oriented problem-solving groups (TOPS) and concurrent engineering groups. The highly interactive nature of such groups, combined with their intermittent meetings and impermanent nature, makes understanding of roles and group behaviour even more important if a useful outcome is to be achieved. The contribution of Tuckman and Belbin is very helpful to such understanding. There is growing use of facilitation as a means of helping the group to a successful result. If the facilitator is familiar with the objectives for which the *ad hoc* group has been formed, understands the techniques being used and is experienced in group behaviour patterns, then he or she can be of major assistance to the group. In particular, the facilitator can recognize and resolve problem situations, both incipient and actual, that the group members may not be able to recognize due to their heavy involvement in the work of the group.

Since the *ad hoc* group tends to meet from time to time, a need becomes quickly apparent for those good practices identified in section 10.4.4 for meetings, such as advance notice of the gathering, a well-prepared agenda, notes on progress of activity at the last meeting and so on. With an *ad hoc* group, there is perhaps even more need for skilful leadership from the leader or chairperson of the group.

10.6.6   Team role behaviour

Belbin identifies six major factors that contribute to behaviour and performance in a team role. These are:

(1) *Personality* – psycho-physiological factors, especially extroversion/introversion and high anxiety/low anxiety. These factors can be assessed by psychometric tests.

(2)   *Mental abilities* – high-level thought can override personality to generate exceptional behaviour.

(3)   *Current values and motivations* – cherished values can provide a particular set of behaviours.

(4)   *Field constraints* – behaviour can depend on constraining factors in the immediate environment.

(5)   *Experience* – personal experience and cultural factors may serve to conventionalize behaviour, or behaviour is often adapted to take account of experience and conventions.

(6)   *Role learning* – learning to play a required role improves versatility.

In practice, it is seldom the case that there will be a psychologist on hand to assess the personalities of potential team members and to give advice on appropriate selection. However, those who select people to form teams for technical projects could well consider the factors that Belbin believes underpin role behaviour when considering potential team members. Young engineers, technologists and scientists who wish to achieve selection might also give them some thought.

10.6.7   Teams in technical activity

Belbin tends to use the word 'behaviour' frequently in his analysis of team performance, which is to be expected from a psychologist and behavioural scientist. Teams in technical activity tend to be judged more on performance, and those who make this judgement are generally those who commission the engineers to perform the task. To the judges, delivery is what matters; the behaviour that assists or impedes this is not their immediate concern.

There are two very significant elements that are relevant to teams of technologists: the importance of experience, and the possibility of role learning. This means that experience of working in teams is helpful. Furthermore, people can be trained in teamworking skills. In principle, the available potential team members can be assessed and the Belbin systems of analysis used to ensure that the requisite role-playing skills are represented, and no incompatible combinations are present.

However, a number of factors are significant for the management of teams to undertake technical activity. The appropriate expertise must be available within the team. Work of one of the authors [14] suggests that there is a need for a variety of specialist expertise to be available throughout a long project, but that continuity needs to be maintained. This continuity is not only necessary in leadership and management, but is also essential for technical reasons. Another practical difficulty is that the 'pool' from which the team is selected is likely to have a limited population, and with the downsizing of many technical organizations in the early 1990s it could be very limited. The 'specialist' role of Belbin's revised model is critical in technical and engineering projects, and the specialism is likely to have two components: technical or scientific knowledge or understanding, and practical experience of the application of that knowledge. In some instances, this specialization will be generally available. For example, most major engineering contracting companies can provide experienced personnel over the spectrum of established engineering techniques. However, in the areas of innovation, leading-edge technology and new product and process development, the availability of the essential specialists may be very limited. In any case, the organization may be constrained to use the staff it has, and may be unwilling or unable to recruit extra personnel. In that case it will be necessary to prioritize, and it is recommended that the scarce specializations are given first priority, the other specialisms second and the nonspecialist roles third. Whatever flexibility in team selection is then available can be subjected to review to establish any Belbin incompatibilities, and if possible to adjust to give a team as balanced as possible in Belbin's terms.

10.7 DIVISION OF WORK

10.7.1 Introduction

Groups come together to form multiple groups, which in turn come together to form organizations. Organizations have already been considered in this book (Chapter 7). Examination of an established organization will show that it is composed of groups and subgroups, and this system forms the organizational structure. Likert gives a model that shows how groups of four employees, one leader and three subordinates, can be structured into an organization (Figure 10.1). The leaders in the lower levels of the organization become the subordinates in the next higher level of the hierarchy. The top leader could be the chief executive. Of course the numbers of employees in each group will vary, and structures much more complex than this simple theoretical model will be found in practice. However, the principle it illustrates will be maintained.

There are many ways in which structures similar to Likert's 'linking pin' can be formed, and organizations frequently revise their structures in a search for the most efficient way of achieving their objectives. (See Chapter 8 for the example of ABB's revisions.) Work is divided between work groups, which are then arranged in a way that is logical for the business process. Here are some of the more common methods of division of work.

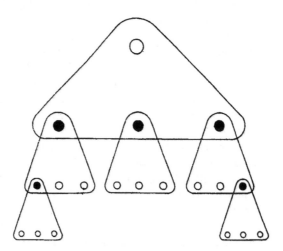

Figure 10.1   'Linking pin' diagram
Source: Rensis Likert (1961) *New Patterns of Management*, McGraw-Hill, New York

10.7.2   Major purpose or function

This is the most common basis. The group is formed according to its specialization, professional expertise, primary activity and so on. Engineering would be an example, as would accounting, safety or medical services. In a large engineering organization, there may be subdivisions of more detailed engineering expertise, differentiating between civil, mechanical or electrical engineering. In very large engineering organizations, there may be further differentiation, for example civil, structural and so on.

10.7.3   Product or service

This is common in large, diverse organizations. The grouping is on the basis of activities associated with a large production line. An example would be a car production facility or a chemical plant complex for product-based industry, or particular services provided by an insurance company or a technical consultancy.

10.7.4   Location

It may be more appropriate to group the activities on a locational basis. This would apply to country-wide businesses such as multiple retailing, which would be grouped on a geographical basis, or such organizations as Railtrack, whose stations are grouped under the control of area managers.

10.7.5   Time basis

Processes that have very high installation and depreciation costs, or cannot easily or economically be started and stopped, may be operated on a shift basis. This applies not only to production plants such as steel manufacture, but also to organizations such as the police, hospitals and the ambulance and fire services. The group may even be named 'B shift' or 'Red watch'. Those industries that operate complex machinery on a continuous basis normally require a maintenance service, also operated on a shift basis.

10.7.6   Common processes

This covers such services as may be provided within an organization that are generally needed by other groups within the organization. Examples would be a typing pool, personnel department or computer services.

10.7.7   Staff employed

The grouping is on the basis of the level and type of qualifications and expertise. Examples would include scientists, technologists, engineers, technical draughts-people and laboratory technicians.

10.7.8   Customers

In organizations devoted to selling a variety of goods to the general public, the grouping may be perceived by market sector or purchasing power. Examples from the UK retailing business could include Woolworths and Marks & Spencer, Harrods and Fortnum & Mason.

The former pair would provide good-quality standard goods at competitive prices, while the latter pair would provide luxury goods at very high prices. The engineering equivalent might be the good-quality motor products of Ford, General Motors or Rover compared with the very high-standard and performance products of Rolls-Royce, Lotus or Lamborghini.

10.8   ORGANIZATIONS AND HUMAN RESOURCE MANAGEMENT

10.8.1   Introduction

The modern view of HRM emphasizes the 'resource' aspect and is essentially strategic. Therefore many organizations develop a human resources strategy to support their overall business or organizational strategy. This complements other support strategies such as financial, operations, technology and marketing strategies. Poole [15] defines HRM as:

> 'a strategic and coherent approach to the management of an organization's most valuable assets – the people working there who individually and collectively contribute to the achievement of its goals.'

Organizations may have an explicitly stated HRM policy, which will contain a commitment to societal wellbeing and well-regarded principles such as 'equal opportunities' or 'investment in people'.

Many organizations believe that their progress and profitability can be enhanced by a staff and workforce that have high-quality, well-trained and committed employees. HRM policies should be designed to attract, retain, motivate and develop competent employees.

10.8.2   HRM models

There have been a number of attempts by academics to model the HRM process. One developed at Harvard University by Beer *et al.* [16] is generally known as the '4 Cs' or 'Harvard model' and seeks to incorporate the expectations of the stakeholders in a business (shareholders, employees, government) and various situational factors (labour market, management styles, skills and motivation of the workforce). The HRM strategy and the effectiveness of its outcome should be formulated and evaluated by appraising attainment under four headings: *commitment* (employees' loyalty); *competence* (employees' skills); *congruence* (shared vision of employees and management); and *cost efficiencies* (operational efficiency).

Commitment

This concerns employees' loyalty to the organization, personal motivation and liking for their work. It can be assessed by attitude surveys, labour turnover and absenteeism statistics, and by interviews with departing employees.

Competence

This considers employees' abilities and skills, training needs and potential for promotion. It can be assessed by employee appraisal systems and preparation of skills inventories.

Congruence

Managers and workers share the same vision of the organization's objectives and work together to achieve them. Ideally, employees at all levels share a common perspective on how the organization operates. This is an organizational culture that can be facilitated by good management, with appropriate communication, leadership style and organizational system, but staff and workers should feel involved. Congruence can be assessed by lack of grievances and conflicts, and by good industrial relations.

Cost-effectiveness

This concerns operational efficiency. Human resources must be used efficiently, productively and at the lowest cost. The organization must be able to respond to changes in the labour market.

The Harvard model has some problems. These include measurement of the variables, possible conflicts between congruence and cost-effectiveness, the wide variety of factors influencing any HR situation, and the limit to which it is possible to improve some of the Cs, particularly in menial jobs.

An expanded model by Hendry [17] incorporates a broader range of external societal influences. These include socioeconomic, technical, political, legal and competitive issues, which vary considerably in different international situations.

10.8.3   Human resource planning and forecasting

This is the key part of the preparation of a human resources strategy. It starts with an assessment of the organization's current personnel resources. This may include a *skills inventory*, which should include skills available but not necessarily required for current activities. Comparison is then made with expected future needs, related to the organization's future plans. Factors to be considered should include potential labour shortages or surpluses; changes in working practices; skills and potential of current employees; and succession planning. Plans should be reasonably detailed but not extended too far into the future, and should be regularly updated.

10.8.4   Recruitment and selection

Organizations usually recruit against a detailed job specification and a clear profile of the qualifications, experience and general characteristics of the required employee. This could be based on:

- physical requirements;
- communication needs;

- formal qualifications;
- experience required;
- specific skills;
- personal ambition.

Most organizations will request the completion of an application form, which will check these key requirements. Those who fulfil the requirements will be entered on a *short list* and interviewed. The recommended approach to interviewing will be discussed in Chapter 11.

Some large organizations, particularly large multinationals, may recruit graduates of high calibre and potential to supply a general need for young executives with the possibility for development into the organization's future managers. A graduate training scheme may be provided that will give experience in various roles and in various parts of the organization's activities.

In the section on interviews (11.4.1), *recruitment* will be used to signify the acquisition of personnel from outside the organization and *selection* to signify the appointment of individuals from within the organization to new jobs. This use of the terms is not followed by all texts. The recruitment/selection process should be evaluated for validity and cost-effectiveness; that is, were the most suitable candidates attracted and recruited, and was this done in the most economical way?

Internal appointment has the advantages of being cheap and more reliable, and improves staff goodwill. Some organizations have a policy requiring the advertisement of posts both inside and outside the organization, either on principle or to check the availability of talent. External recruitment is expensive, complicated, lengthy and unreliable. Advertising is a major cost in the UK if national coverage is required, but can be reduced if a more focused approach is adopted.

Recruitment agencies also provide a means of supplying personnel and human resource departments may keep a file of unsolicited enquiries. 'Word of mouth' and personal recommendation are also significant recruitment mechanisms. Headhunting is practised by many organizations in the search for individuals suitable for senior management posts. Agencies established for this purpose will search the ranks of competitor organizations, professional associations and trade association literature. 'Old boy' and other networks may be used, and the approach to a possible candidate will be discreet. There are advantages and disadvantages to the use of this method.

10.8.5   Training and development

Training is done, either internally or externally, to enable the employee to perform the required tasks, to improve the performance of those tasks, or to prepare the employee for different or more challenging work. The usefulness of the training needs to be assessed prior to the training being given. If no significant improvement in organizational performance can be expected, then the justification of the training should be questioned. The effectiveness of the training also needs to be assessed in terms of improved skills and capabilities.

Personnel development may also be valuable, particularly for management and potential managers. This may involve specialist training perhaps in management techniques, but can also include secondment to learn more about the organization's broader activities, or a series of progressive appointments to a variety of posts. It can even include leadership training of the 'outward bound' type.

10.8.6 Performance appraisal

This is a standard procedure in most companies and many do an appraisal exercise annually. An interview with a superior, often but not always the interviewee's direct superior, takes place. This type of interview will be discussed in more detail in Chapter 11, section 11.4.1.

Past performance should be evaluated, realistic targets set for the future, and appropriate training provided where necessary. The targets should be SMART (Specific, Measurable, Agreed, Realistic, Time related). Some organizations use self-appraisal, peer-group appraisal or even '360 degree appraisal' (superior, peer and subordinate), or arrange for appraisal by a more senior employee who is not the direct superior of the person being appraised.

Appraisal is often a flawed process, mainly due to poor manager training, bias, stereotyping, inadequate information and overemphasis on major incidents.

10.8.7 Personnel movement

The four main ways of changing job are:

(1) *Promotion*: A move to a more senior job, one with more power, scope or a higher position in the hierarchy of the organization. This would occur with the need to fill a vacancy at a more senior level. It would be based on superior performance at the current level, and anticipation of competence at the higher level.

(2) *Transfer*: A move to a job with similar power, scope and position in the hierarchy, but in a different part of the organization or in a different function. This would often occur to fill a vacancy and/or to give an employee wider experience of the organization's activities as part of career development.

(3) *Demotion*: A move to a less senior job, one with less power, scope or a lower position in the hierarchy of the organization. This would occur only if the employee was demonstrably unable to perform adequately at the current level.

(4) *Dismissal*: Removal from the job and the organization. Normally this would occur only following serious dereliction of duty. Protection of employment legislation would ensure that minor infringements or poor performance would require formal warnings, opportunities to rectify any deficiency, and possibly even remedial training, before dismissal followed if these measures were ineffective.

Whatever the reason for the job change, the employee who has been moved faces the same tasks of forming new relationships – with new colleagues, new superiors and, possibly, new subordinates.

10.8.8 Remuneration

Pay, salary, remuneration, reward system, the compensation package or the emoluments are the most visible indication of the organization's appreciation of the services of an employee. The remuneration may be group or individual related, and money related or non-money related. Also the selection of a reward system by an employee can be driven by security, tradition, employability or contribution.

Total payments can have a component of salary/wages and a component that is incentive based. There may be elements of 'perks', deferred payments or payment equivalents.

In designing a salary structure, it is important that there is internal comparability, that jobs of similar importance to the organization's objectives are similarly rewarded. It is helpful to go for 'broadbanding', the replacement of complex, narrow and detailed salary scales by fewer scales with wider 'bands', which will allow flexibility according to personal performance. Most salary schemes include annual increases on the assumption that the performance of the individual will improve with time. Some organizations have formal schemes that give annual increases almost unrelated to performance. This is falling out of favour with most competitive organizations. A salary scheme must consider the 'market rate' for the particular skills in demand, and the availability of suitable personnel in a particular area. Since the 'resources' approach to HRM assumes that labour is one of the many resources required by an organization, it must accept that it is subject to market forces.

10.9   INTERNATIONAL HRM

International HRM reflects all the elements of HRM discussed so far, though this discussion has been related largely to the UK environment. International HRM has to consider also the management of an organization's human resources across national barriers. There is a need to consider the differences that may exist in the employment environment in a particular country and, if a multinational organization is involved, the extent of the centralization of that organization. The major factors that give rise to differences are the national culture; the state of economic development of the country; the legislative framework of employment; and the political system in place. These factors are reviewed below.

10.9.1   National culture

It has been established that there are significant differences in national culture, and these differences will influence the way employees are managed and therefore employment policy. The established model of national cultures is that of Hofstede [18]. By a global study of the employees of IBM, he established four fundamental dimensions of national culture:

(1) *Power/distance*: How power and influence are distributed; acceptance of unequal distribution of power; access to sources of power.
(2) *Uncertainty avoidance*: The extent to which there is a need for order and certainty; the extent to which uncertainty is considered disquieting.
(3) *Individualism/collectivism*: The extent to which individuals are expected to look after themselves; the extent to which a common purpose is valued and adopted.

(4) *Masculinity/femininity*: The extent to which the values associated with gender stereotypes are followed in a country; 'masculine' values are those such as wealth, possessions and competitiveness; 'feminine' values are those such as sensitivity, care and concern.

The way business is operated, and the manner in which people are managed, will be influenced by these factors. High power/distance will generate a more authoritarian style of management and a more tractable workforce. High uncertainty avoidance will be consistent with lifelong employment by one organization and long-established business partnerships. Individualism will promote a competitive employment environment, while collectivism will stimulate strong group cooperation. Masculinity will promote a desire for financial reward for good performance, while the economies of countries with a feminine bias will be strong on social support.

10.9.2 Economic development

Developed countries have very well-embedded systems and mechanisms for the employment of labour. These frequently include well-established trade unions and professional employee representative organizations. In less-developed and partially industrialized countries, there may be a large pool of untrained cheap labour available, but a shortage of highly skilled, technical and professional recruits. As a country's economy develops, so will its need for more technical and nontechnical skills, and these will be developed to fulfil the need.

10.9.3 Legislative framework

The legislative framework on employment varies from country to country. In particular, the extent of the protection of employees may be low, as in many eastern countries, to high, as it is in some European countries (such as Germany, where employee representatives have a place on the supervisory board of any company).

Major employment-related legislation in the UK has many elements designed to counter discrimination on an unreasonable basis, for example the *Sex Discrimination Act* and the *Race Relations Act*. There is other legislation that is not specifically antidiscriminatory in its objectives, for example the *Protection of Employment Act* and the *National Minimum Wage*. Many companies and noncommercial organizations show their intention to follow the requirements of all the legislation by stating an equal opportunities policy.

The Social Charter

The Social Charter is legislation within the European Union that puts further restraints on employers. It includes requirements for:

- a fair wage;
- workplace health and safety;
- training;
- association and collective bargaining;

- integration of disabled people;
- information, consultation and worker participation;
- improvement in working conditions;
- social protection;
- equal treatment of men and women;
- protection of young people.

With legislation varying, the influence of the Social Charter will also be different from country to country within Europe.

10.9.4   Political and economic system

A country's political system is linked to, and is the generator of, its legislation. The political system will be committed to a particular economic system. The economic system will be at a point on the spectrum from a free-market system (a *demand* system), with little intervention from the government, through to a completely centralized and government-controlled system (a *command* system). The economic system of the USA might be considered an example of the demand system and the closest to an example of completely free-market capitalism. The former Soviet Union might have been considered an example of the command system and the closest to an extreme example of government-control communism.

However, there are many examples of hybrid systems, the *mixed* economy, in which the free market continues with varying levels of government control: light in the USA, greater in the UK and with even more intervention in the EU. In countries such as Japan, Singapore and Korea, government intervention is high. China was previously a command economy, but some elements of free-market activity are now permitted.

10.10   SUMMARY

This chapter was concerned with people in groups. Initially the types of group were considered, and then their purpose, both for the individual and for the organization. The relationship between the effectiveness of groups and their size, member characteristics, individual objectives and roles, task and the maturity of the group was discussed. Some models of group development were presented, together with a model of the behaviour of the mature group. The chapter continued with a discussion of personal interactions within groups. The meeting was considered in some depth, with an analysis of types of meeting, their functions and how they should be managed. The next main topic was leadership. The relationship between leadership and management was discussed, together with the sources of the power of a leader, the styles and functions of leadership, and some theories of leadership. The following topic was teams. Some motivations to become a member of a team were given and the work of Belbin on team member characteristics was outlined. Two models of team roles were described and the factors contributing to team member behaviour summarized. The importance of team building and team roles was considered, particularly that of the specialist, and its importance in engineering projects was emphasized. How groups are formed into organizations was considered using Likert's linking pin model, and the methods used for division of work were assessed. Recent management thinking on the use of human resources from a strategic viewpoint was considered. Finally, the international aspects of human resource management were discussed.

REFERENCES

1. Tuckman, B. W. (1965) 'Development sequence in small groups', *Psychological Bulletin*, 63.
2. Bass, B. M. and Ryterband, E. C. (1979) *Organizational Psychology*, 2nd edn, Allyn and Bacon, London.
3. Open University Course T244, Block II Work Groups, The Open University Press, Walton Hall, Milton Keynes.
4. *Meetings, Bloody Meetings*, Video Arts (London and Birmingham), 1993; *More Bloody Meetings*, Video Arts (London and Birmingham), 1994.
5. Hellreigel, D., Slocum, J. W. and Woodman, R. W. (1976) *Organizational Behaviour*, West Publishing, St Paul, MN.
6. Mullins, L. J. (1985) *Management and Organisational Behaviour*, Pitman, London.
7. Adair, J. (1979) *Action Centred Leadership*, Gower, Aldershot.
8. Blake, R. R. and Mouton, J. S. (1985) *The Managerial Grid III*, Gulf Publishing, Houston, TX.
9. Krech, D., Crutchfield, R. S. and Ballachey, E. L. (1962) *Individual in Society*, McGraw-Hill, Maidenhead.
10. Fiedler, F. E. A. (1967) *Theory of Leadership Effectiveness*, McGraw-Hill, Maidenhead.
11. Clarke, C. and Pratt, S. (1985) 'Leadership's four part progress', *Management Today*, March.
12. Belbin, R. M. (1981) *Management Teams: Why They Succeed or Fail*, Butterworth, London.
13. Belbin, R. M. (1993) *Team Roles at Work*, Butterworth-Heinemann, Oxford.
14. Reavill, L. R. P. (1995) 'Team management for high technology projects', Management of Technology Conference, Aston University, April.
15. Poole, M. (1990) 'HRM in an international perspective', *International Journal of HRM*, 1, 1.
16. Beer, M., Lawrence, P. R., Mills, D. Q. and Walton, R. E. (1984) *Managing Human Assets*, Free Press, New York.
17. Hendry, C. and Pettigrew, A. (1990) 'HRM: An agenda for the 1990s', *International Journal of HRM*, 1, 1.
18. Hofstede, G. (1980) *Cultures Consequences*, Sage, London.

11

Communication

11.1 INTRODUCTION

Communication is critical to the interaction of individuals, to groups of people undertaking a variety of activities, and most particularly to organizations trying to coordinate their work towards a common objective. Engineers and scientists are not exempt from the need to communicate clearly and effectively, both to other engineers and scientists as part of the team activity that business often requires, and also to non-engineers and non-scientists who may be less able to understand the technical concepts on which the engineer's or scientist's input is based. No matter how skilled an engineer, scientist or technologist may be, he or she will not succeed without the ability to inform, persuade, instruct, argue with, and convince other people. The mirror image of this also applies. The engineer, scientist or technologist needs to be able to receive communication, to understand instruction and to interpret information.

Young engineers and scientists may believe correctly that attaining technical competence in engineering and science and their practical applications is the most important objective. To communicate this acquired competence is also important, and it is for this reason that a substantial chapter of this book is assigned to theoretical and practical aspects of communication directly relevant to engineering and scientific activity.

However, before starting the theoretical analysis, it is worth making a point about the context of a communication. We tend to interpret the meaning of a communication by relating it to the context in which it is made. A simple example will illustrate this. An unkempt man with a blanket and a paper cup sitting outside an London Underground station may ask a passing stranger, 'Have you any change?'. This is not likely to be interpreted by the stranger as an enquiry about whether he or she has sufficient small coinage for immediate needs, but as a request for a contribution towards the speaker's living expenses. Further down the street, the stranger might encounter a man in a business suit standing by his car at a parking meter, who also asks, 'Have you any change?'. This request is for an exchange of coins so that the car owner can feed the particular denominations of coins required by the parking meter. It is unlikely that there will be any misinterpretation of the message. If both requests are followed, the passer-by will have fewer coins after the first encounter and different denominations after the second one.

11.2   COMMUNICATION IN THEORY

11.2.1   Theory of one-to-one communication

Communication is a process by which information and data are passed between individuals and/or organizations by means of previously agreed media and channels. Communication can also be defined as any means by which a thought is transferred from one person to another.

Let us consider the simple situation of communication between two individuals. There are four key elements. The person with the thought or information to communicate is designated the *transmitter*. The person to whom the thought or information is to be communicated is designated the *receiver*. The means by which the thought or information is to be transferred is called the *medium*. The method of connecting the two participants is called the *channel*.

11.2.2   The medium

Though the two participants could be individuals, groups or organizations, for the purposes of this chapter and of establishing the basic principles, two individuals will be considered. The medium can be any form of symbols or code understood by both the transmitter and the receiver, and could include words (both spoken and written), figures, graphs, diagrams, pictures, facial expression and so on. In times past, media such as Morse code were used for international communication and flag signals (semaphore) were used by ships at sea.

11.2.3   The channel

The word media (plural of medium) is much used currently as a collective noun for the many means of mass communication such as newspapers, magazines, radio, television and the Internet. These are really channels of communication. Each newspaper or magazine will convey information in the form of written words or pictures (media). The newspaper or magazine itself is the channel of communication, the direct link between the writers of the news reports and articles and the reader. For the radio broadcast, the medium is the spoken word. Other information may be given in the form of 'sound effects'. The channel is the radio station to which you tune your set. In the case of television, the station to which you tune your television set is termed a channel. In the UK currently five 'terrestrial' channels are available in many parts of the country, together with other systems such as the satellite broadcaster Sky and cable television. A cable system can be used for the Internet, but many users of this communication medium use an Internet Service Provider (ISP) and make connection with this Internet link via a modem and their telephone connection. Provided the appropriate equipment is available, the signals can be received, and television programmes watched and information received from a wide range of Internet sources. The media used include the spoken word, pictures and occasionally diagrams or written words. So in these examples of major means of public communication, all the terms so far discussed are present: medium or media, channel, transmitter and receiver.

However, we cannot be sure that true communication has taken place. Has the receiver understood the message that the transmitter was trying to convey? Has the message even arrived at its intended destination?

11.2.4 Feedback

Transmission is not communication. The other person may not be listening. We need some response from the receiver to know that the person has both received the message and understood it. For successful communication, the response of the receiver must be the one intended by the transmitter. To be fully effective the information must flow both ways, and this return message is termed *feedback*.

By the nature of the feedback message, the transmitter can judge how well the original message has been understood. It should be noted that in the feedback process, the roles of the two participants have reversed: the original receiver transmits the feedback message for the original transmitter to receive. The feedback message does not necessarily have to be verbal or written, the information sent may be stored – that is, remembered – by the other person, there may be a change of attitude or behaviour, or the receiver may act as intended. Suppose you see someone about to step into the path of a speeding car and shout 'Stop!'. If the person stops, the action is the feedback and a genuine (and helpful) communication has been achieved. If the person does not hear (channel of communication not established), or does hear but does not understand your instruction (not the response intended), another casualty could be added to the road accident statistics.

A simple conceptual model of communication, relating the elements introduced so far, is shown in Figure 11.1.

11.2.5 Verbal communication

Verbal communication, the conversation of two people together face to face, has a number of advantages. The most significant is the immediacy of the feedback. The receiver of the initiating message can ask questions of the original transmitter to clarify the message, or can supply answers if the original message asked a

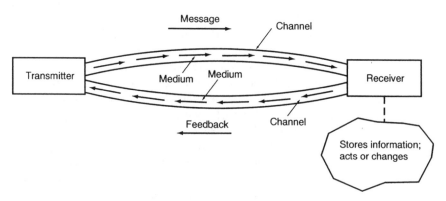

Figure 11.1 The fundamentals of communication

question. As the conversation continues, new subjects may be introduced by either party and information will flow in both directions.

11.2.6 Conversation

Not all the information exchange in a conversation is verbal. The message can be attenuated or emphasized by tone of voice (harsh, friendly, jocular) and people sometimes communicate unintentionally by facial expression or posture ('body language'). This will be discussed further in section 11.7.8.

11.3 COMMUNICATION IN PRACTICE

11.3.1 Introduction

The theory of communication suggests a simple process that should work easily and efficiently. All that is required is to formulate the message clearly, select an appropriate medium that will allow easy feedback, and all will be well. Why is it, then, that our questions receive inadequate answers, our messages are misunderstood and our clear instructions are not carried out properly? Well, there may be many reasons, but most of them can be traced to poor communication. Though simple in theory, achieving good communication is quite difficult in practice. The more complex the message to be communicated and the more people involved, the more the difficulty is compounded. Complex communication in large organizations presents particular problems. For the moment, if the basic problems of one-to-one communication can be identified, they will provide the basis for the more complex variations.

11.3.2 Opening the channel

A person has a message to give to another person. How can the channel be established or opened? If both people are present in the same room, it can be merely a matter of one person attracting the other person's attention, for example by establishing eye-to-eye contact or making an opening remark: 'Excuse me'. This may elicit a response, perhaps 'Yes, what do you want?', and the channel is open. The message has yet to be passed, but already there is evidence of feedback, in that the other person is at least paying attention and demonstrating a willingness to listen. The transmitter must ensure that the channel remains 'open'. The would-be listener might be distracted by some other event. For example, the telephone might ring, or the remarks made might become so boring that interest fades.

Let us suppose that the other person is not present. A means of communication must be selected. Possibilities include the telephone, fax, e-mail and the postal service. If a letter is sent by post, the message can be written clearly in a form that the correspondent will understand, but it may take a day or so to reach her. Even if she replies 'by return of post', it will take some time to learn if the message has been received and understood. She may receive the message and be so convinced that she understands it that she believes it requires no reply. In this situation, information has transferred, but the transmitter does not know that the message has been understood, or even received. Hence the importance of feedback at the most fundamental level. By a system of letter and formal acknowledgement, accurate communication can be achieved, but it can take a long time.

Perhaps it might be quicker to telephone. Once the number is dialled and the telephone receiver is picked up, the channel is established. Well, perhaps it is, if the would-be receiver is at home and answers the telephone. Maybe there will be no reply as she is not there. There may be an answering machine, so a message can be left. Even so, the transmitter will not know when she will receive it. With ill luck, her father may answer and the message may not be one that we want him to intercept!

Fax or e-mail may present a compromise of speed and accuracy, but still lacks the immediacy of one-to-one contact. Even the telephone in its traditional form has disadvantages, in that the extra information available visually from your correspondent will not be available. However, a letter, fax or e-mail message does have the advantage of putting the information on record, which will not be the case with an exchange of information based on a conversation, unless one of the participants takes the trouble to record all conversations, like the late US President Richard Nixon. As another American, the film tycoon Sam Goldwyn, commented, 'A verbal contract isn't worth the paper it's written on!' In law, a verbal contract is worth as much as a written one, but proving the verbal contract can be more difficult.

11.3.3   Choosing the code

In section 11.2.2, the medium was described as a previously agreed system of symbols or code that is understood by the transmitter and the receiver. The examples mentioned were words, graphs, diagrams and pictures. The important factor is that the code or symbols are fully understood by both parties. This might at first sight appear simple, but often it is not.

Consider this example. In the UK, someone might address a remark to another person in English with a reasonable expectation of being understood. By mischance, the person could encounter in London a first-time visitor from Latvia who understands very little English. An English-speaking person, visiting the USA or Australia, could confidently start a conversation in English, as the inhabitants of those countries speak their own version of English. When on holiday in Italy or Greece, an attempt to converse in the language of the country would be recommended. Someone whose Italian or Greek is weak or nonexistent might be forced to resort to other means of communication, such as gesture or drawing a picture or diagram. Indeed, there are some messages that might be more easily expressed in these terms. For example, explaining which particular spare part is required for a car repair may be helped by a drawing of the part, or pointing at the appropriate section of an exploded diagram in the car manual.

Thus, at the start of the attempted communication above, a medium was chosen (words) and a code selected (English). The choice might, or might not, be appropriate, and this will soon become clear. If English is not a code understood by both parties, an attempt might be made to identify a common language. The Latvian visitor could be directed to his destination using a few ill-remembered phrases of German and diagrams drawn on the back of his guidebook! The scenario is that a code has been selected and an attempt has been made to *encode* a message. It has been found that the receiver does not understand the code and therefore cannot *decode* the message. Other codes are tried until a common one is found. To some extent the medium has been changed; that is, from words to diagrams.

11.3.4   Encoding and decoding

Consider the simplest of situations, in which Smith wishes to give Jones a message. Both are present and speak the same language. The stages in the activity are as follows:

(1)   Smith has the content of the message in his mind.
(2)   Smith formulates this message into words (encoding).
(3)   Smith engages the attention of Jones.
(4)   Jones gives Smith his attention (channel open).
(5)   Smith enunciates a series of words (transmission).
(6)   Jones hears the words (reception).
(7)   Jones interprets the meaning of the words (decoding).
(8)   Jones now has the message.

Even in this most simple of situations, without feedback it will not be clear that the message has been accurately received and Smith will have to judge from Jones's response how successful the process has been. Even with a common language, people may interpret the same phrases differently for many reasons.

11.3.5   Barriers to good communication

A surprisingly large number of obstacles stand in the way of good communication, which perhaps accounts for the difficulty in maintaining a high standard in this activity. Here are some of the more significant obstacles.

Language

As mentioned before, a language must be found that is understood well by both parties. The linguistic *style* must not be too complex for both participants. The linguistic style of an academic may not be appropriate for the shop floor of an engineering factory, or vice versa. If there are specialized components, linguistic short cuts such as acronyms, or technical or other forms of *jargon*, these must be fully understood by both parties.

   The *accent* of the speaker may give problems if the listener is unfamiliar with that accent. Most 'Home Counties' English people can cope with the accents of Devon and Cornwall, Wales, Scotland and Ireland both north and south, and have learnt to understand, perhaps with the help of motion pictures and television 'soaps', American and Australian accents. The metropolitan accents of London, Birmingham, Liverpool and Newcastle can easily be identified, and accent can often be used to identify geographical origins and social standing within the UK, though not perhaps in the precise detail suggested by G. B. Shaw's Professor Higgins in *Pygmalion*.

Dialect is a different problem, since this usually involves additional vocabulary and regional turns of phrase, as well as different pronunciation. People who tend to use dialect expressions in their conversation should remember to eliminate them when talking to someone who is not from the immediate locality and may be unfamiliar with the expressions.

Culture and education

Culture tends to influence communication largely in the way that information is understood and interpreted. Also, to some extent the importance and significance attributed to the content of the communication may be influenced by cultural differences. *Education* will determine the extent to which the concepts of the conversation are understood by both parties. A technical education will help the understanding of engineering matters and a classical education will help in the literary area. It is helpful to have some knowledge of the culture and education of your communicant, and a few exploratory exchanges, ostensibly for social purposes, can assist business communication. Some Japanese companies, for example, have a stock list of questions aimed at securing information about new business associates.

Personal

People may be selective when receiving a complex communication or one containing a number of elements. A subordinate in discussion with a superior may give greater emphasis to the more complimentary aspects of the superior's remarks. To an extent, the subordinate may 'hear what he or she wants to hear'.

In a rather similar example of the superimposition of other mental processes on the simple one of decoding the message, more credence will be given by the receiver to information from someone with status or expertise, or for whom the receiver has great respect. This is called the 'halo effect'. The converse is also true. The information from an individual in whom the receiver has little confidence may be disbelieved, regardless of its true value. A metaphor for this is 'the singer, not the song'. Advertisement creators use this effect to influence buyers. The person advertising the product may wear a white coat, the image being that of a 'scientist', thereby giving the information presented a bogus credibility. When the actor concerned adds a stethoscope to the costume, the even greater credibility of the 'doctor' image is created.

The interpretation of the message is very much dependent on the perception of the receiver and on that person's 'frame of reference'. Perception is how the individual 'sees' the information and its context; that is, how it relates to that individual. The frame of reference can be defined as those aspects of our principles, beliefs and experience that go together to make the context against which we judge events. A young, single woman who is thinking of leaving her job to get married and live in another area of the country will be little concerned by a message of possible redundancy circulated by her employers. A middle-aged married man in the same organization, with a sick wife, a large mortgage and five school-age children, and who had been made redundant from two previous jobs, would probably react entirely differently to the information.

Behaviour of the transmitter

There is much that the transmitter can do to help the communication process. The aim should be accuracy, precision and conciseness. This means that the information sent should avoid the woolly, the vague and the ill-defined, the long-winded and the overdetailed. The style is also important. A person who presents the information in an enthusiastic and confident manner is much more likely to convince his or her listener than one who approaches the communication process in a morose and apologetic fashion.

Behaviour of the receiver

The receiver can do as much to help the process as the transmitter. The major problems to avoid are:

- *Inattention* – The channel is not open, so no messages can flow.
- *Distraction* – The receiver allows himself or herself to be distracted during the transmission of the message and the channel closes in mid-transmission.

Process maintenance

Communication is a process and, like all processes, it requires maintenance; that is, attention to ensure that it continues to operate properly. The major problems to avoid are:

- *Attenuation* – The system must ensure that part of the message does not get lost.
- *Noise* – Extraneous signals that prevent the receiver from being able to pick up the message, for example background noise during a conversation, or other messages being transmitted, must be eliminated or reduced in order that the message can be received.
- *Distortion* – The system must avoid the distortion of the message in transit or an inaccurate version will be received.
- *Overload* – If the transmitter is sending more information than the receiver can assimilate, then a situation of overload exists. Rate of transmission should not exceed the receiver's capability to receive, assimilate and understand the message.
- *Redundancy* – Sending the same information more than once does not always help. The duplication wastes time and may fray the receiver's patience. However, it is noticeable that advertisers, particularly on television, tend to deliver multiple repeats of the name of the product. Their assumption may be that the listener is stupid (or uninterested), or possibly that they do not have the listener's undivided attention. Repetition is necessary to increase the chance of getting the message through when the channel is open.

A more detailed conceptual model of the communication process is given in Figure 11.2.

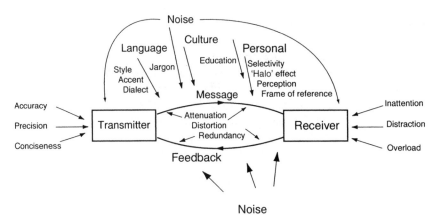

Figure 11.2   Conceptual model of communication

11.3.6   A short exercise

Consider the following sentence, which encapsulates many of the problems of poor communication:

'I know that you believe that you understand what you think I said, but I am not sure that you realize that what you heard is not what I meant!'

Try what is called a reductionist analytical approach to the sentence; that is, break it down into its component phrases and see how many can be related to the processes and problems outlined so far in this chapter.

11.4   FORMAL COMMUNICATION SYSTEMS

11.4.1   Interviews

An activity that is frequently used in the working environment for communication purposes is the interview. Generally, the interview is on a one-to-one basis, though in certain cases, for example the recruitment or selection interview, the format may involve the interview of one person by a selection panel composed of two or more people.

At the workplace, interviews are used for a number of purposes. The most common include the recruitment/selection interview already mentioned, the induction interview, the progress or appraisal interview, the discipline interview, the counselling interview and the termination interview. These will be examined in a little more detail later. Examples of many types of interview can be demonstrated by means of the many excellent training videos currently available [1].

The primary purpose of an interview is the acquisition or exchange of information. There may be other subsidiary purposes such as decision making, but these are usually dependent on what the information that has been obtained indicates to both parties. All the general principles considered so far apply to the interview situation, and every effort should be made to ensure good standards of communication or the exercise will be ineffective.

Selection and recruitment interviews

These terms are almost synonymous. We could perhaps differentiate the *recruitment* interview as the situation where an organization wishes to add to its complement of employees and recruits externally. Candidates may be found by advertisement or from agencies, according to the type of job involved. The *selection* interview might be to fill a post within the company from current employees, either by transfer or by promotion. In this case, the candidates would be identified by the personnel department from company records.

The recruitment interview is something that a student is likely to encounter in the later stages of a university career. If the student has been interviewed for a place at university, the basics of the exercise are similar. The interviewer will be anxious to discover as much as possible about the candidate, for example qualifications, experience, aspirations and potential. The interviewer will have guidelines on the general characteristics required for the organization for its employees and the particular skills required for the specific job. The consequent decision based on this information will be whether the candidate is suitable and, if there are a number of candidates, which one is the most suitable. The candidate will be equally keen to obtain as much information as possible. He or she will want to know the conditions of employment, what the duties will be, whether the company will provide adequate career development and whether he or she will be happy to work for that organization. If the potential employee is lucky enough to obtain more than one offer of employment, this information will facilitate the choice of the post that is most suitable.

As in all interview situations, preparation is essential. The interviewer must have a mental or written specification of the job and clear guidelines on company norms and policies. He or she will have the details supplied by the candidate on an application form (a documentary communication of information) and should use this as a basis for confirmation and expansion. The candidate should also have done his or her homework, and obtained details of the company, its policies and, if possible, the opinions held about it by employees.

The induction interview

This might be considered the second phase of information exchange by the successful candidate. Details of the company's ways of working will be imparted to the new employee. Training plans, company rules and regulations and other information given only to employees will be discussed at such an interview, and introductions made to site, premises and colleagues.

The progress or appraisal interview

Most organizations with a well-developed personnel policy will have a scheme of appraisal interviews. These will normally take place once a year and will involve a representative of the management, usually the employee's supervisor, but sometimes also a representative of the personnel function or a more senior manager. The purpose is to review the employee's performance and progress.

It enables the organization to indicate the areas in which the employee has performed well and those in which performance could be improved. It enables the employee to discuss any problems he or she may have and how his or her career is progressing and might be allowed to progress further.

There are four major activities in an appraisal interview:

(1)   Review the case history in advance.
(2)   Listen to the evidence and agree an analysis.
(3)   Face up to any problem areas.
(4)   Agree and review a plan of action.

The discipline interview

In the working environment, discipline is about having a clear and fair framework in which everyone can work: rules on safety, time keeping, standards of work and so on. It is also about ensuring that these rules are kept and standards are maintained and, if they are not, taking action to close the gap between the actual performance and the required performance.

There are three major activities in a discipline interview:

(1)   Establish the gap between required and actual performance.
(2)   Explore the reasons for the gap.
(3)   Eliminate the gap.

The counselling interview

This covers the situation where an employee, a member of a team or department, has a problem. The employee may come to the supervisor for advice, or the problem may be so evident that the supervisor notices that it is beginning to affect the employee's work. Two possible courses of action that the supervisor may be tempted to take are to disregard the problem (telling the individual 'stop making a fuss, pull yourself together!') or to impose a solution. Both are unwise and doomed to likely failure. Usually the best action is to help the individual find his or her own solution to the problem.

There are four stages in a counselling interview:

(1)   Set up the counselling session.
(2)   Encourage the person to talk.
(3)   Help the person to think things through.
(4)   Help the person to find his or her own solution.

The termination interview

This can incorporate the termination of employment by the employer (dismissal, redundancy) or by the employee (resignation, retirement). In the former case, there may be elements of recrimination. Even so, there still remains information that needs to be transferred, such as terms of compensation, pension entitlement, the address of the Department of Employment and so on. In the case of the resignation of an employee, it provides an opportunity for the organization to establish, if it does not know already, the reason for the departure. Much valuable information may be obtained by the company from the termination interview. It costs substantial sums of money to recruit employees and train them, and a low staff turnover is considered by many organizations to be advantageous. It also gives the employee an opportunity to air any comments on the organization and its practices that he or she feels impelled to make.

Summary

There are four major activities in *any* interview:

(1) The interviewer (and the interviewee) should always be clear as to the *purpose* of the interview and what results are required from it.
(2) The interviewer (and the interviewee) must always *prepare* thoroughly for the interview and obtain all the required data beforehand.
(3) Adequate time must be allowed for *performing* the exercise and privacy assured. Questions must be put clearly and concisely. The interviewee must be encouraged to talk and attention paid to what is said and left unsaid. Both interviewer and interviewee should use, and note the use of, body language. Notes should be taken as unobtrusively as possible. The next action should be agreed before the interview is ended.
(4) The results of the interview should be recorded and progress on any agreed actions monitored.

11.4.2   Written communication

There is very little that is formal in a short note scribbled to a friend or to let a colleague know of a telephone message. However, the fact that the message or information is written constitutes a record for as long as the 'hard copy' is kept. Even if the message is captured in some other form, for example as an e-mail or on a computer file, the element of formality exists. A person's word may be their bond, but a piece of paper with a signature is more acceptable to the lawyers.

Here is a comment on the activity of writing from a professional writer, the novelist Simon Raven [2]:

> 'I enjoyed writing, and it was my pride to render clear and enjoyable what I wrote. A reader, I thought, must pay in time and in money for his reading; both courtesy and equity are therefore required, not indeed that one should defer to his possible opinions, but that one should attempt to entertain him while demonstrating one's own. This is a matter of the simplest common sense, but as relatively few aspirant writers ever seem to grasp the point, those that do set out with a distinct advantage.'

The professional engineer, scientist or technician, unlike the novelist, will rarely be paid for writing as such, except perhaps for an occasional article in a technical journal, for which the fees are likely to be minimal. Also, unlike the novelist, it is not the duty of the engineer or scientist writer to entertain the reader. However, the rest of Raven's comments and his emphasis on clarity, on getting one's point across and on doing this in a way that the reader will find acceptable are germane to our purpose. Focus your attention on the reader! Always have a mental picture of your intended reader when you write a piece. Knowledge of the reader's needs and capabilities are crucial. What is the point of writing anything if it is not read by the intended recipient? Make your material 'user friendly'!

Professional engineers and scientists are frequently called on to communicate their ideas or report on their work to others. Similarly, engineering and science undergraduates are required to write essays and reports during their course of study. These are frequently to explain the results achieved in laboratory experiments, the bases of design studies, and as a general training for future employment. The following paragraphs give some general guidance on written communication.

Syntax

Your syntax (the grammatical arrangement of words in sentences) must be correct. So must the punctuation and the spelling. There is no excuse for bad spelling. Use a dictionary!

If you have a word processor, establish whether the software package includes a spell check. Most advanced software packages may include a grammar-checking system. Use of the spell check will eliminate the majority of spelling errors, but not all of them. The spell check operates by comparing each word with the computer's databank of words. If the system recognizes the word it will pass on to the next. If it does not, it will stop and ask you to check the word. Remember, it only checks that the word exists. The computer does not check that you have used the word properly, or that your spelling is correct in the context in which you have used it. Also, the package may have American spelling rather than English. For example, the element sulphur (English) is the same material as sulfur (American), and aluminium (English) is the same metal as aluminum (American).

Difficulties arise with grammar and with clarity of exposition because sentences are too long. Short sentences are easier to construct correctly and easier to understand. An average sentence length of between 25 and 30 words is desirable. Try not to exceed 40 to 45 words. The tense of verbs constitutes another problem. Reexamine sentences and paragraphs to see whether the tense has shifted between past, present, future, conditional and so on, without reason. Another frequent error is to mismatch the plural form of a verb with a singular form of subject or vice versa.

The most effective way to eliminate punctuation errors is to read the piece aloud, or get a friend to read it to you. Lack of fluency, hesitation or breathlessness will indicate any shortcomings.

Content and structure

The engineer, scientist or technician must ensure that the content of his or her writing is clear, correct, concise and courteous. Every report or essay should have a discernible structure: an introduction, a main body of material and a conclusion.

Do not use five words when one will do. 'Because' is much simpler than 'due to the fact that'. 'Now' means the same as 'at this present moment in time'. The use of simple words makes for more economical, more easily understood and often more elegant writing. Avoid jargon (words the reader might not understand), undefined acronyms and meaningless abbreviations. Resist the temptation to use long words that seem impressive, but of whose meaning you are not certain.

Do not use clichés or hackneyed phrases, such as 'leave no stone unturned', 'free, gratis and for nothing' (five words meaning the same as the first) or 'the acid test'. The latter phrase is a specific term used in assays, and should be avoided in essays. So should weak puns like the last sentence. A cliché should only be used if an original or genuinely witty variant can be found, such as that of the theatre critic who attended a poor variety performance and returned home to write his review, determined to leave 'no turn unstoned'.

Essays

Essays are commonly used for the expression and development of ideas. The purpose may be to inform, analyse, explain or persuade. Success will depend on whether the essay is made interesting and is well presented, whether it contains a clear line of argument and is concise. All essays should have an introduction to tell the reader how the author intends to approach the subject. Where an essay theme is wide ranging or capable of broad interpretation, the writer must define, and possibly justify, the limits of his or her discussion. Essay writing is much used in the UK education system as a training tool for the orderly expression and transmission of ideas. However, in scientific and technical circles it is rarely used, and the more popular form of information transmission is the report, or its baby brother the technical memorandum, which is a very short, simple report.

Reports

These are usually more formal than essays. Their objective is to inform and persuade, and they depend more on measured evidence and less on opinion and passion. A report should have an executive summary, introduction, main body and conclusion.

The *executive summary* should summarize the findings and conclusions, normally in half a page and certainly in less than one page. The *introduction* should indicate the purpose of the report and the context, perhaps the background reasons for undertaking the exercise. The *main body* should describe the procedures and methods adopted and report on the findings. If the findings consist of extensive numerical tabulations or calculations, then these should be summarized in the main body and the bulk placed in appendices. The main body of the text should be divided into sections and subsections to give structure and clarity, with suitable headings and subheadings to act as 'signposts' to guide the reader.

The *conclusions* should state what conclusions the author or authors derived from the findings. Recommendations based on these conclusions may be made, if appropriate. Where the writer is seeking guidance or authority, it is helpful to phrase the recommendations so that the reader only has to sign the document to achieve the required result.

A report that has been commissioned should always begin by stating the terms of reference. If specific facts, figures, conclusions or quotations are made from published material or other sources, this should be acknowledged in the text and a list of references included at the end of the report. If material is used generally (but not specifically) from published material, this should be acknowledged in a bibliography at the end of the report.

An undergraduate project report should be arranged as follows:

Title sheet

Contents

Synopsis

Introduction

Main body

Conclusions (and recommendations)

References and/or bibliography

Appendices

Readability

A well-known, but rather complex, method of assessing readability is the 'fog factor', devised by Gunning [3]. Calculation of the fog factor can be performed using an equation involving the average sentence length, the number of separate ideas in a sentence and the number of words with three or more syllables. The answer is alleged to equate to the number of years of education required to read the passage with ease, but need not delay us. To reduce the fog factor and improve the readability of our prose, we need merely to keep sentences short, consider only one idea per sentence and avoid the use of polysyllabic words (such as polysyllabic!).

As a check on the readability of a piece of writing, consider the following:

(1) What is the objective in this communication, and is a written document the best way of achieving it?

(2) Is the document well oriented towards the reader, both in form and content?

(3) Does the title explain the relevance of the document to the reader, and does the first paragraph explain this relevance in greater detail?

(4) Is the document as clear and simple as rewriting can make it?

(5) Are graphics (pictures, graphs, diagrams and so on) used to maximum advantage?

(6) Are signposting devices such as contents lists, headings and subheadings used as much as possible to make the message easy to follow?

(7) Are the main points repeated to ensure clear understanding, and are the reactions required of the reader clear or the conclusions he or she is to accept precisely stated?

Some ways to improve clarity

(1) Simple, declarative sentences are easier to understand than more complex forms; active constructions are easier to understand than passive; positive constructions are easier to understand than negative.

(2) Many words widely used in industry such as policy, productivity, security, turnover and so on have different meanings for different groups of people. Choose your words with care, be mindful of your intended audience and define every doubtful term.

(3) Abstruse language often masks fuzzy thinking. When the language is starkly simple, the thought stands out more sharply.

(4) Paradoxically, a simple style demands more time and effort than a long-winded style.

(5) Thinking in managerial jargon inhibits new approaches and imaginative solutions to problems. Writing in managerial jargon produces confusion in the minds of both the writer and the reader.

(6) Jargon can speed communication between people within the same discipline or at the same level, but it causes a breakdown of communication between people of different disciplines and levels.

(7) Charts, graphs, diagrams, posters and so on offer less scope for misrepresentation than words. It is easy for the reader to visualize information presented graphically, so this is a useful method to communicate technical information to nontechnical people.

11.4.3 Summaries and abstracts

Frequently in the industrial work situation the young engineer or scientist will be attempting to convince a more senior colleague of his or her views on plans, policies, purchases, technical standards and so on. A more senior colleague is likely to be grateful to you if you are economic with his or her time. Crosby [4] refers to the 'elevator speech'. The individual with the point to make enters the elevator or lift and the senior colleague is already on board. The senior person is trapped for a minute or two, and there is an opportunity for a very short speech in favour of the plan or proposal. The speech is a verbal version of a summary, and summaries or abstracts are very useful forms of written message, particularly if the recipient is known to be very busy.

An *abstract* is an extended title, useful for information storage and for helping a reader decide if the subject matter of a document is of interest to him or her. However, the paragraph at the beginning of a paper or report may be entitled *summary*. This can also be a very short version of a report or paper for a slightly different readership. A *précis* is an exercise in shortening to a specific length, but not reorganizing. As such, it may be useful as a comprehension test, but has no place in informative writing.

Four uses of summaries

A summary is necessary for the busy reader who cannot read the whole document because of lack of time. The summary must present the reader with the main conclusions and recommendations and the major reasons to support them. This reader will not look at the rest of the report, so the summary must stand alone.

For all readers, a summary helps focus attention on the topic of the paper or report and its conclusions. The summary does a similar job to a quick flick through the document itself. It gives readers an overall picture, a sort of map of the document, which will help them to orientate themselves as they read the document subsequently.

For the marginally interested reader it acts as an abstract, an extended title, to help the reader decide whether to read the whole document.

For all readers a summary is a help to memory, a set of notes, to remind them what was in the report itself. This is especially useful when looking back over a report some time after reading it. The summary acts as a hook to fish out of the recesses of the memory the forgotten details of the report.

Remember the reader

Many readers are prepared to allocate a few minutes to getting a rough idea of what a report is about. This time limit is independent of the complexity of the material. Therefore it is the reader's motivation and needs, rather than the complexity of the report, which will condition the length of the summary. The reader who is given only a few sentences is likely to supplement this by leafing through the paper itself. The writer could better employ the reader's attention with his or her own deliberate choice of supporting detail. Conversely, if a summary is too long (more than a page) the reader may skim even the summary, thus defeating its purpose. Thinking about the reader in these terms provides a sense of realism about length.

11.5   GROUP INTERACTIONS

This section will consider how communication occurs within a group. This will significantly affect the performance of the group and the level of satisfaction obtained by group members. In Chapter 9, communication between individuals was discussed and the point was made that there needed to be a channel for the communication process. If more than two people are involved, it is possible for interaction to take place between all those involved. With two people, there are two interactions, the message and the response. With three people there will be six possible interactions, with four there will be 12, and with five there will be 20. A mathematical formula can be derived for this relationship, which is:

$$\text{Numbers of interactions } (N) = p(p - 1)$$

where p is the number of members of the group.

This is the first and the only instance in this book when it has been found possible to reduce an aspect of human behaviour to an equation.

The increasing complexity of the communication net, as it is sometimes called, is evident with increase in numbers within the group. Between each pair of individuals there is a communication channel, and with more than two individuals we have one form of multichannel communication. However, it is possible to have multichannel communication between two individuals, so the general subject of multichannel communication will now be considered.

11.5.1 Multichannel communication

This can occur in three ways.

One-to-one communication

There may be one or more channels of communication during an interaction between two individuals. One individual may speak. If the other is listening, one channel of communication is established. The medium of communication is verbal; that is, spoken words.

Additional channels may be established. The speaker may display a diagram or picture to illustrate his or her remarks. By doing so, a second channel is established if the receiver also pays attention to the visual illustration. The medium of the second channel is graphic. The speaker may make gestures or display some element of body language that will give information to the receiver if that person is looking, and is perceptive enough to interpret the information. A third channel is established, the medium is visual, but the decoding of the message given by the body language or the gesture could require some skill. In more intimate conversations, information may be transferred by touch or physical contact. If all these means are used simultaneously, we have a multichannel communication with four channels between only two people.

The senses being used in communication discussed so far are sight, hearing and touch. Could communication take place using other senses? The two senses not considered so far are smell and taste. While these senses may seldom be used deliberately for communication, they can provide much useful information. A smell of burning from the kitchen can indicate that the dinner, though unseen, unheard and untouched, has passed the cooking stage and moved into the incineration stage. A tasting of the product of such an unfortunate accident could indicate the extent to which the food had moved from the uncooked state to the inedible state. Blenders of tea, and buyers of wine, rely heavily on their ability to judge quality by the taste or aroma of a sample of the liquid concerned. An example more likely to be encountered by an engineer or scientist would be the product of a water company. Though the quality of drinking water can be tested by chemical analysis, the water is regularly tasted. The smell could be unattractive if some of the chlorine used to treat the water remained when it reached the customer.

One estimate of the extent to which the various senses are used in the gathering of information suggests sight = 83%; sound = 12%; smell = 2.5%; touch = 1.5%; taste = 1.0%. This perhaps explains why material on a television news broadcast, with its strong combination of visual and aural information, can provide a most effective means of communication.

One-to-'more-than-one' communication

In 'one-to-many' communication, even if only one medium is used, there will be a channel between the transmitter and each receiver. If the transmitter uses more than one medium, verbal and visual means for example, there will be a pair of channels to each receiver who is paying attention.

Multiple (many-to-many) communication

In a group with a high degree of interaction, information is passing between group members by many channels and means (media) throughout the session. All channels are unlikely to be open all the time, and the communication may be intermittent and confused. This is the situation of the 'meeting', and there is a need for some ordering of the discussion, by providing, for example, a chairperson, an agenda, a summary of the points made, and the establishment of areas of agreement and disagreement. This will be discussed further in section 11.6.2.

11.5.2   Communication patterns

The pattern of communication that occurs within a group can be quite critical to its efficient functioning. Some theoretical patterns for a simple group of five people are considered below, but as indicated earlier, the patterns can become much more complex as the numbers in the group increase. Figure 11.3 represents these patterns of communication in diagrammatic form.

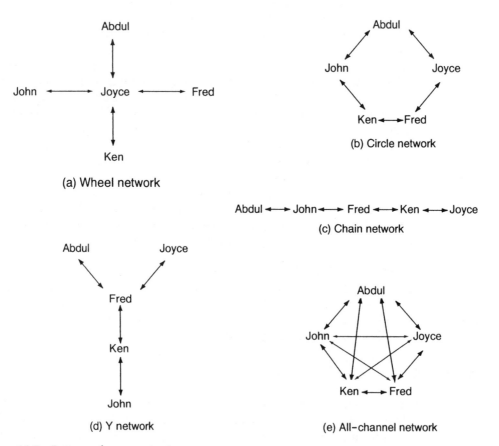

Figure 11.3   Patterns of communication

Wheel

This mode can occur if there is a formal chairperson present who insists on all communication being via the chair, or if one of the group members is very senior or has a dominating personality. The system may be quickest to a solution, but is inflexible to changes of task. The satisfaction of the member at the centre of the wheel is high, but the satisfaction of those at the periphery is low. If the group members are not together and written means of communication are used, the individual at the centre of the wheel is in a very powerful position to influence the decision (see Figure 11.3a).

Circle

This is the least satisfactory method in all respects. The system is slow to a solution. Member satisfaction is low. The system is inflexible to task changes. It is most unlikely to occur if the group members are all together. However, it is the practice in some organizations to circulate documents for comment and each individual will receive the information in the document, plus the comments of those who have previously received the document. It eventually returns to the originator. If all members are to be aware of the views of all the other group members, the document needs to be recirculated a second time. This may still not be completely satisfactory, since the comments of those members who received the document late may stimulate further thoughts by those earlier in the sequence (see Figure 11.3b).

Chain

This is also a poor method. The message will take some time to reach the person at the end of the chain, and may become attenuated and distorted in the process. Those who have played the parlour game 'Chinese Whispers', in which a simple message is passed from person to person, will know how easily this can occur. Member satisfaction is low. There is a possibility of malpractice, in that the intermediate members of the chain may deliberately distort the message to achieve their own goals. Even so, this is a method that is in frequent use in organizations. The views of a director on a particular issue are passed to the divisional manager, then to the departmental manager, and then to the individual responsible for the issue. The director's views may be subject to some interpretation or distortion on the way. Each of the intermediate individuals may put his or her own 'spin' on the content of the message (see Figure 11.3c).

Branched chain or Y

This is very similar to the chain, except that the individual at the branch point is in a powerful position, in that he or she can differentiate the way the message is passed to the group members beyond the branch. Information possibly helpful to the career development of people in a particular section of an organization may be passed by a manager to the section head. If the section head decides to pass this information to one member of the section and not to others, the receiver of the information will have the advantage (see Figure 11.3d).

All channel

This system has the potential for the best solution in complex situations. It is highly participative and therefore has high member satisfaction. The quality of output is high. However, it can disintegrate or revert to the wheel mode under time pressure. At a distance and by written communication, an example would be that all information generated by all members of the group is sent to all other members. When the members are present at a meeting, even with the simple example of five people, if everyone attempted to speak at once there would be pandemonium and no real communication would take place. In practice, and with some discipline, awareness of others' needs and an element of 'give and take', individuals would listen most of the time and speak only a little of the time, perhaps ideally about one fifth in our simple example (see Figure 11.3e).

If the point of discussion were emotive or critical, there might be a greater need for constraint on all members making their contribution at the same time. Should this happen, one individual might attempt to introduce some order to the proceedings, by leading the discussion or by moving into a 'chairperson' role. The discussion would then have aspects of the 'wheel' model. With the advent of modern communication systems such as e-mail or video conferences, the possibility of very cost-efficient long-distance communication becomes a reality, but the requirement for an orderly procedure is enhanced.

The need for good communication not only within groups but also between groups working on a coordinated project has been discussed in Chapter 6, and Figure 6.16 shows the inter-group communication requirements in simultaneous engineering. The roles that individuals play in facilitating communication within working groups was discussed in sections 10.6–10.9.

While the theoretical possibilities of communication patterns can be illuminating, the realities of personal style and behaviour and many other factors will influence the way the communication pattern develops, particularly if the group is together in a room. As the discussion develops, it may change from a pattern that is represented by one of the models to one similar to another of the models.

Consider the situation where five students are meeting to plan a group exercise. Though all are of equal status, there will probably be natural leaders and natural followers in the group. The initial all-channel communication may develop into a 'chaired discussion' if the initial exchange of information is confused and incoherent. One of the students with a stronger personality may feel a need to take a grip of the random discussion and focus it on the task in hand. In the professional engineering world, the discussion held by a small group to consider a design point or a problem on a project is likely to have members with different status, seniority, expertise and responsibility. Other group members will assert their right to comment, or defer to others, according to these factors. If the knowledge, experience and expertise of all members of the group are to be harnessed to progress the group task, and particularly if the time resources of the group are to be used efficiently, there is a need to ensure the best possible ways of communicating within the group. Often, the formal meeting is adopted as a means to achieve this.

11.6 MEETINGS

11.6.1 Introduction

The meeting as an example of a group activity has been discussed in section 10.4, and it is worth repeating that the major objectives of a meeting will include communication and decision making.

11.6.2 Communication at meetings

The comment has been made before that the efficiency and effectiveness of a meeting are critically dependent on the way it is run by the chairperson, and on the behaviour of the meeting members. The ordering of the agenda of the meeting helps good communication, as the items can be addressed in a logical order. A good chairperson will guide the discussion, for example by keeping group members to the point at issue, persuading the more verbose members to be more concise and eliciting comments from those who are more reticent. The group members themselves can assist the communication process by commenting only when they have something useful to contribute, by avoiding all speaking at once, and more subtly by making supportive comments after other speakers have put forward good ideas and offering constructive criticism of those proposals with which they do not agree. The chairperson can assist the communication process by summarizing from time to time in a long discussion and agreeing very carefully with the group the summary and conclusion of each item of discussion. This conclusion can then be committed to paper as one of the 'minutes' of the meeting.

It should be noted that communication in meetings generally follows the all-channel model, in that all members can both comment and receive information. However, if the chairperson is heavy-handed in controlling discussion and monopolizes most of the speaking time, the 'wheel' model will apply, at least to the transmitting aspects of the communication. Even in this situation, all others present will be able to receive information.

The fact that the major points of the meeting are committed to record as 'minutes' indicates the selection of a different communication technique to allow group members to be able to refresh their memories of the outcome of the meeting, to notify absent members of the results of the discussion and to provide a basis for continued discussion at the next meeting if it happens to be one of a series.

11.7 NONVERBAL COMMUNICATION

11.7.1 Introduction

The media of communication discussed so far have used words, both written and spoken. The written word is a form of visual communication, but there are many other methods that use visual means. The engineer and scientist will be familiar with many of these means: graphs, technical drawings and logic diagrams, for example. The principle in visual communication is the same as in verbal communication: be sure that the receiver understands the code. One engineer can communicate easily with another engineer by means of a technical drawing using the standard conventions, but the same drawing would probably not be easily understood by a company employee with a financial or personnel background. A simplified version of the drawing might be useful to explain components that will justify expenditure, or space that will be available to accommodate staff.

11.7.2   General

A verbal communication may present difficulty for the receiver if the vocabulary is unfamiliar, the statements are long and complex, and there are too many interrelated clauses to be held in mind. The direct analogy applies to visual means. Difficulty can be caused if the visual vocabulary is unfamiliar, the visual material is voluminous and complex, and too many interrelated elements are presented simultaneously.

11.7.3   Major elements in visual communications

Numbers

Much useful and precise information can be transmitted by numbers, provided they are tabulated clearly and the interrelationship of the groups of numbers is not too complex. When more complex relationships can be deduced from the numbers, careful tabulation may be required to demonstrate the interesting relationships.

Lines

Lines can be composed into graphs and drawings. Entities can be defined in boxes and circles, and the relationships between such boxes indicated by lines and arrows. The graph indicating the relationship between one variable and another is a simple combination of three lines, but the information conveyed can be very significant.

Words

Tables and diagrams will have no meaning without some words. Many student engineers' and scientists' reports are submitted with interesting graphs, but with the units of one or both of the variables missing, and sometimes even the names of the variables. It may be obvious to the transmitter, but is sometimes not immediately clear to the receiver. All visual communications deserve a title. Tables of numbers need headings, diagrams need captions, graphs require the axes defined and the units of measurement stated, and figures need their components annotated. It should be noted that all observers of a visual image may not interpret it in the same way, so it is advisable to give the receiver an indication of what the illustration means.

Shapes

In symbolic charts and drawings, shapes are frequently significant. Many shapes acquire a conventional meaning, often worldwide, and are known as 'icons'. Their use, for example, on road signs and other useful facilities such as public toilets allows identification of the facilities even by individuals who cannot read. Representational drawings and photographs convey information by means of a multiplicity of shapes. The well-selected picture will certainly save a thousand words, as it conveys multiple information to the eyes of the receiver.

11.7.4   Visual cues

Spatial cues

The two-dimensional space available allows the transmitter to incorporate additional information by the method of alignment of figures in rows or in columns, the balance of the elements, and the degree of prominence with which they are presented.

Typographical cues

Information can be conveyed by the size of type, the design and layout of the typography, the use of special typefaces such as capitals and italics, and the weight of type employed.

Colour and shading cues

In a monochrome diagram, meaning can be conveyed by density of shading. With colour, more detailed information can be conveyed, for example the various lines of the London Underground. Colour of print and colour of the background paper can also be made significant.

Ruling

The ruling of lines between elements or groups of figures can help to differentiate between these elements or groups of figures, or relate them to one another, or indicate common relationships.

11.7.5   Impact

Each visual communication should make a point, usually only one point but occasionally more than one. The greatest impact is generated if the visual communication is *clear, organized* and *uncomplicated.* It is best to give the receiver as much data as is needed to make the point and to avoid the visual equivalent of waffle. It is worthwhile to use a fresh visual item to complement each significant point, and best to resist the temptation to use a diagram or a photograph just because it is impressive or merely available.

11.7.6   Tables

Tables should be arranged in numerical or alphabetical order, should be grouped in a logical sequence, and should accommodate those brought up in the English or European culture to read from left to right. Units should be quoted in the headings where appropriate. Footnotes may be needed to explain figures that do not follow the trend and the dash that might accompany the absence of a particular reading may need some explanation.

11.7.7   Charts

There are a great number of different charts, from the simple graph or line diagram to the semi-symbolic diagram. If some measure of time is a component of the illustration, it normally occupies the horizontal axis. Some brief comments on the most frequently encountered types are given here.

Bar charts

Bar charts can be vertical, horizontal or floating. The vertical type may be termed a histogram, and an example would be the monthly rainfall figures for a particular year and location. Horizontal bar charts are used for time-related activities such as the various jobs in a project, can have floating elements if the times of start and finish of a particular job are variable, and may provide input for more complex charts such as PERT (Project Evaluation and Review Technique) charts (see Chapter 18).

Pie charts

The simple circular image of the pie chart is good for indicating the proportions of a particular entity that relate to different contributors. Examples could be the proportion of sales from the five product lines of the company, or the elements of cost attributable to raw materials, labour, overhead and profit. The information is usually recorded in cyclic fashion, starting from '12 o'clock', with the segments in descending order of size.

Graphs and line charts

When well presented, these have high visual impact and numerical accuracy. Scales should be appropriate and clearly marked, as should the units of any numbers represented. If more than one line or graph is included on the diagram, the significance of each should be clearly labelled. Sometimes space economies encourage communicators to include too many graphs on the same diagram, and it is worthwhile to ponder the inverse relationship between economics and clarity.

Flow charts

The receiver's expectation to read left to right and downwards should not be forgotten, at least when considering communication in the western world. In some parts of the East, the conventions are different. Some types of flow chart in frequent use have developed standard conventions. Genealogical charts such as family trees have a convention of each succeeding generation occupying a new lower level on the chart, which can create practical problems for the compiler of the diagram if the family is prolific. Progeny are recorded in birth sequence from left to right. Some flow charts use algorithms that show a sequence of decisions and actions. They may incorporate conventions for decision points, information generation and actions.

Semi-symbolic charts

These may involve the use of simple symbols as units in a type of bar chart, for example a motor manufacturer illustrating the monthly output of cars by using rows of car symbols, each equivalent to 1000 cars manufactured. A more complex example would be an Ordnance Survey map, where a wide variety of symbols is used. In such cases, a 'key' should be included that explains the symbols to those who are not familiar with them.

Technical drawings

Engineers should have no problem with the type of illustrative diagram that is a major tool of the profession, the technical or engineering drawing. To this can be added the more specialized drawings of the various disciplines of the engineering profession, such as the wiring diagrams of the electrical engineer, the process flow diagram of the chemical engineer, and the reinforcing steel drawings of the civil engineer. The pitfall here is that familiarity and continual use of such diagrams may tempt the young engineer to assume that most people are able to interpret the information contained within them. If the recipient is an engineer, this assumption is justified. To those not so familiar, the document might well be as meaningful as Egyptian hieroglyphics are to most engineers.

11.7.8   Body language

This is a form of visual communication that is often unwitting, and therefore uncontrolled. An individual may convey information about his or her feelings or views by facial expression, deliberate gestures, nervous movements of the hands or feet, stance or posture, the manner of moving and in many other ways. The poker player may note the gleam of appreciation in an opponent's eye as the other player picks up a strong hand. The experienced poker player will attempt to appear passive at all times; the cunning player may attempt to exude elation when holding a weak hand and gloom when dealt an array of aces and kings. Most people in the interview situations discussed in section 11.4.1 would try to display confidence and calm, but might indicate tension by nervous hand movements or rapid eye movements.

It is impractical in an introductory text such as this to go into detail about body language, but there are a number of specialist texts to which the reader can refer. Study of such texts and, better perhaps, the study of ourselves and others when attempting to communicate can help to eliminate the more obvious problems of our body language being at variance with our 'official' message. Viewing of a recorded video of ourselves in the act of communication can be enlightening, though sometimes a confidence shaker. While the real views of the situation might betray individuals when attempting to communicate a message in which they had little faith, the converse fortunately is true. If the belief in the message is strong and the speaker is committed and enthusiastic about the subject, then body language can reinforce the message and make the speaker more convincing.

11.8 MULTIMEDIA COMMUNICATION

11.8.1 General

As we saw in section 11.2.3, the term 'media' is often used to indicate the various means of mass communication that are now available, such as national and international newspapers and magazines, radio and television. The newspaper will use text, pictures and sometimes graphs and diagrams, for example for stock-market prices and the weather forecast. More than one medium of communication is used by a newspaper, but there is only one channel, the visual contact between the reader and the newspaper. Radio also uses only one channel, the auditory one between the broadcaster and the listener. Words, music and 'sound effects' are three media that convey information to the listener, who can perhaps imagine the farming activities of a radio programme such as *The Archers*.

Films, stage plays, musicals and television are multimedia activities. Many people believe that television is the most powerful means of mass communication. It communicates by both the visual and the auditory channels. We see video and film clips of current events and hear the verbal views of commentators as 'voice-overs'. Programmes on economics can use graphs and diagrams, weather forecasters can use a sequence of computer-generated semi-symbolic diagrams and, at election times, political experts can make assessments supported by tabulated figures and charts. The medium of television also has the immediacy of bringing us the 'news', be it the football results or the numbers winning the National Lottery; ministers answering questions (or avoiding answering them) in the House of Commons; or the tragedies of conflict in yet another area of the world. The events are reproduced right in our living rooms.

While the young engineer and scientist is unlikely to be able to mobilize the communication resources available to the makers of television programmes, there are some basic lessons that can be learned. Use media with which you can cope and which are appropriate to the message you are trying to project. Use both visual and audible media in an appropriate and realistic combination and try for immediacy, strong projection and close contact with your audience. The situation in which the young engineer or scientist is likely to encounter such a problem is the presentation, so the following sections give some advice on how to approach this activity.

11.8.2   Presentations

Engineers, scientists and technologists are required to make presentations with increasing frequency. These may be to customers to explain the technical aspects of an engineering product or service, to senior management to report progress on a research project or an area of activity, or to directors to justify budgets or specific items of major expenditure. For the customers and the senior members of the company, the communication method has the major advantage of allowing questions, discussion, feedback and possibly an immediate decision. It may also be preceded by preliminary written information such as a briefing document, and may be followed by a more detailed report that the recipients can study and consider at leisure.

It is therefore very important that young engineers and scientists should become competent at making presentations. Ability in this activity is very helpful to a career, since it allows the presenter to demonstrate ability and make a good impression with influential people. The converse also applies, so the trend towards more frequent presentations has disadvantages for those less able to stand up and communicate well. If one's only exposure to the directors of the company was a totally disastrous presentation, this could be unhelpful for career progress. It is therefore sensible to take every available opportunity to practise making presentations, thereby improving one's technique.

Some basic principles

A presentation tends to be a formal occasion, often with little or no audience participation. You will need to work to sustain a good level of interest. Try to avoid turning it into a dreary lecture. Remember that you are not providing entertainment, you want people's hearts and minds. Of course, if you can be mildly entertaining at the same time, it helps to maintain attention. The normal maximum attention span is said to be 40–50 minutes. Many members of the audience may have a lesser span of attention. As the audience may not have a chance to ask immediately about a confusing or difficult item or a missed point, it can be helpful to recapitulate key points. It is also helpful to indicate the general subject of your talk at the start and to summarize at the end. This basic plan can be summarized as: 'Tell them what you are going to tell them; tell them what you want to tell them; then tell them what you have just told them.'

Preparation

- *Decide* What are your objectives? What should the title of your talk be? What should the content be for the presentation? Establish the context in which you will make your presentation; that is, the circumstance that has made this presentation appropriate or necessary.

- *Discover* How much time is available? What are the characteristics of the audience? What is the environment in which you have to speak (the room, the equipment in it, the lighting and seating)? Adjust the presentation to suit these constraints.
- *Determine* Avoid unnecessary detail. Do not swamp the audience with information and do not overestimate their ability to absorb new information.
- *Design* Design your presentation to support the conclusions you wish to promote. Introduce the key points or facts in a logical order. Demonstrate the linkages between these facts, and use visual aids to emphasize the important points.

Visual aids

Visual aids provide added impact if they are well prepared and correctly used. They are excellent for conveying related information such as diagrams and maps by utilizing our parallel visual communication channel. They need to be a focus for the audience's thinking, to be large and clear, to require a minimum of reading and to be left in view long enough to be fully assimilated. Avoid the 'now you see it, now you don't' trick of removing the viewfoil before the audience has time to read it. However, it should be removed from view when it is not supporting the comments of the speaker, as in this situation it can be a distraction. Visual aids may detract from, rather than support, the speech if they are confusing, hard to read or sloppily prepared. They are unhelpful if they do not fully relate to the speaker's current comments or if reading and assimilating them removes attention from the content of the continuing speech. They must not contain spelling mistakes, which can be a distraction to some members of the audience.

In many cases, visual aids will be limited to an overhead projector, a blackboard or a whiteboard, and perhaps a flip chart. If the latter items are available in the presentation room or lecture theatre, it is worthwhile to take chalk, marker pens and a board rubber, or at least check that these are available. You may need to produce an impromptu illustration in an answer to a question at the end of the presentation. It is therefore worthwhile to have available a couple of blank viewfoils and a viewfoil pen, for the same purpose.

In these days of high technology, the location for your presentation may have additional facilities such as more than one overhead projector, allowing the projection of more than one image at a time, for example for comparison purposes. There may be television screens or a video projector that would allow you to add video clips to your talk, or a slide projector that would allow the use of coloured pictures. More sophisticated auditoria may have a computer facility linked to a 'projection tablet' that sits on the overhead projector and will allow you to use computer-generated data or diagrams. There may even be audio equipment that would permit music or other sound projection to be used. The important thing to remember is that the more aids you use, the more complex the presentation becomes and the more there is to be controlled. There is also more equipment that can fail at the most inappropriate moment. If you do wish to take advantage of these modern aids, it is well worth training an assistant or assistants to operate the equipment, leaving you free to do the talking. It is also wise to test the equipment before the presentation, and to perform additional trial runs to ensure that the material is integrated and the exercise is well rehearsed.

The speaker

The speaker should check voice, physical appearance and bearing, and try to avoid nervous or habitual movements. It is helpful to face the audience directly, repeatedly gaining eye-to-eye contact with its members, but avoid permanently addressing one member of the audience, as that person may bask in the special attention or may become nervous at being the constant focus of the speaker's performance.

The speaker should attempt to speak clearly at a reasonable speed and loudness, with pauses after key points to allow them to sink in. Try not to speak from a script, which often causes the inexperienced presenter to slip into reading the script and thereby losing contact with the audience. More time should be given to key points and key words and phrases should be repeated.

The speaker's personality and mannerisms are significant and the speaker should try to appear confident and enthusiastic. Habits and movements that could distract the audience, such as playing with a pen, chalk or watch, should be avoided, as should tendencies to repetitive speech patterns such as 'um', 'ah' and 'you know?'. Small habitual mannerisms such as scratching an ear lobe or, worse still, some more intimate part of the body are unconscious, so the comments of a frank but kindly friend or colleague can be helpful. The opportunity to see a video recording of the performance is very helpful, if a little damaging to one's self-confidence, as we have said.

Good practices

Write notes well beforehand so that you can review and correct them. An early start will allow you to practise the speech several times at the correct volume and speed, to check the timing, and possibly to practise before at least one critical audience member. Try to reduce the speech to 'trigger notes' or 'bullet points'. This will give the speech more immediacy, will avoid the trap of reading a script, and the 'bullet points' can be put on the viewfoils and on small cards that are convenient to use.

Before you start to speak, check your appearance and that your notes and viewfoils are complete and in the right order. Numbering slides and viewfoils will help to sort them if they are dropped or muddled. Check that the equipment is set up correctly, that the image is the maximum possible for the screen, and that it is in focus. If a microphone is available and necessary, check that it is in working order and the volume is correct. If you are not introduced, give your name and a short self-description. Then give the title of the talk and why it is worthwhile for the audience to give you their attention.

When using an overhead projector, avoid having your shadow on the screen and blocking the view of some members of the audience. Try to ensure that all can see and hear well. To explain or emphasize something, point to the surface of the viewfoil with a small pointer or the tip of a pen, not to the screen, otherwise you will tend to turn your back to the audience and your comments may become inaudible. Remember, a finger is a very blunt instrument when used in conjunction with an overhead projector. Place the pointer on the surface of the viewfoil if you are nervous and your hand is shaking! It is particularly important not to overrun the time allocated, especially if this occurs in conjunction with a slow and rambling early section and a manic speed-up towards the end.

11.9   SUMMARY

The chapter has considered the importance of communication for engineers, scientists and technologists. A theoretical framework was given for one-to-one communication, which introduced the concepts of the medium of communication and the channel of communication. The importance of feedback was stressed. Verbal and written communication were discussed and the use of codes such as a language was introduced, with the procedure of encoding and decoding the message. Some of the problems encountered in achieving good communication were discussed. Formal communication systems such as the various types of interview were defined. Means were suggested of achieving a good standard in such written documents as essays and reports, and uses indicated for summaries and abstracts.

The chapter continued with a discussion of group interactions, particularly the patterns of communication and different forms of multichannel communication. A popular form of multichannel communication, the meeting, was considered. Next nonverbal communication was addressed and the use of tables, graphs, figures, diagrams, charts, technical drawings and pictures. Finally multimedia, the simultaneous use of a number of media, was considered and a detailed guide given of the methods of making a professional presentation.

REFERENCES

1. *Can You Spare a Minute?*, Video Arts (London and Birmingham), 1988; *I Want a Word with You!*, Video Arts (London and Birmingham), 1979.
2. Raven, S. (1961) *The English Gentleman*, Anthony Blond, London.
3. Gunning, R. (1952) *The Technique of Clear Writing*, McGraw-Hill, Maidenhead.
4. Crosby, P. (1979) *Quality Is Free*, McGraw-Hill, Maidenhead.

12

Work Study

12.1 INTRODUCTION

Work study derives from the work of F. W. Taylor, *The Principles of Scientific Management* [1], referred to in Chapter 3. 'Taylorism' has been blamed for much that was wrong in the management of large engineering enterprises: for the way in which it led to a fragmentation of labour, the breakdown of jobs into tasks that were reduced to simple movements at set speeds, depriving the workforce of any need for skill or intelligence. This does not mean that the principles of work study should be rejected. Processes in the best-run factories can still benefit from studying the sequence and timing of operations, and method study can be of great utility in small factories and workshops, especially if it is complemented by teamworking, delegation and empowerment (see Chapters 7 and 8). Work study can also be a useful tool for teams engaged in some of the newer process improvement approaches such as Six Sigma teams (see Chapter 4) and Team Oriented Problem Solving (see Appendix 2).

This chapter explains the purposes of *work study*, describing its two parts, *method study* and *work measurement*. Simple examples to illustrate the principles of work study are given.[1]

12.2 OBJECTIVES OF WORK STUDY

British Standard 3138 [3] defines work study as 'the systematic examination of activities in order to improve the effective use of human and other resources'. The proper objective of work study is to ensure that the time and talents of every individual and the operational time of every machine are used effectively. Ideally, this should result in improved productivity and profitability, maintaining employment at higher rates of pay and with greater job satisfaction.

12.3 METHOD STUDY

BS 3138 defines method study as 'the systematic *recording* (sic) and *critical examination* of ways of doing things in order to make improvements'. Therefore,

---

[1] For those interested in pursuing their studies a little way beyond the principles described in these next few pages, Chapter 11 of Hill, *Production/Operations Management Text and Cases* [2], is recommended.

method study can be applied at all operational levels from the narrowly defined interaction between an operative and a machine to, say, a just-in-time materials supply system. Method study procedures are defined by the steps below:

- *Select* the work to be studied.
- *Record* all the relevant facts.
- *Critically examine* (analyse) the recorded facts.
- *Develop proposals* for improved methods.
- *Choose a solution* from a short list of solutions.
- *Implement* the solution.
- *REPEAT* the procedure as appropriate.

Selecting a process for study should not be difficult. Any process in which there is evidence of machinery or operatives being inactive for substantial periods of time will benefit from such a study. Other reasons might be irregular production rates, excessive movement by people, machines or materials, production bottlenecks, low quality and/or unsafe processes.

It is with the methods of recording and examining the facts that this section will be concerned.

BS 3375 Part 2 [4] provides the following standard symbols for the purposes of recording a task:

O = *operation* (a process which furthers the job or changes the state of the material).

⇨ = *transport* (any form of movement from one position to another).

∇ = *storage* (material in a prescribed storage location).

D = *delay* (some form of interference with the flow).

□ = *inspection*

◎ = main function is inspection but a *process* is also involved.

These symbols are used in the tables, charts or diagrams that have been drawn up to represent the task or process. Sections 12.3.1 and 12.3.2 will examine the use of flow process charts and diagrams and multiple activity charts.

12.3.1   Flow process charts and diagrams

A flow process chart shows the sequence of the individual operations that make up the task or process. It should be noted that preparing a flow process chart is a usual first step in any form of process or quality improvement, for all sorts of processes, not just engineering ones.

Example 12(1)

An aluminium alloy window frame consists of three elements. Assume that the three elements are manufactured in a workshop and that within the same workshop the three elements are assembled and then insulated, after which

Table 12.1   Flow process chart

Line 1	Line 2	Line 3	Description
O	O	O	Make elements in workshop
∇	∇	∇	Store elements
⇨	⇨	⇨	Move to assembly point
	O		Assemble unit
	□		Inspect unit
	⇨		Move to insulation bay
	D		Await insulation
	O		Apply insulation
	⇨		Lift to glazing bay
	O		Fit glass
	⇨		Move to inspection bay
	□		Inspect
	⇨		Lift to storage
	∇		Store

the glazing is fitted to complete the window. These activities are represented in Table 12.1.

For the sake of simplicity this chart does not show the whole process. Immediately after each inspection there is the possibility of nonacceptance, which would lead to an alternative path of rejection or dismantling and some form of reassembly. These paths are not shown. The flow chart gives rise to two immediate observations: Why is there a delay in the insulation bay? Why are there two inspections? Would one suffice, or should there be a process to prevent production of components and assemblies of unsatisfactory quality? (See Chapter 4, Management of Quality and the use of statistical process control.)

Figure 12.1 shows a more complex, nontabulated flow process chart. The actual manufacturing process is not described, but it is clear from the chart that three parts are processed separately (two of these include inspection and correction procedures), after this they are assembled (task 13) then inspected. Inspection leads to acceptance or rejection; the rejected assembly is removed to a place of storage; the accepted assembly is stored then moved and after a delay is combined into a larger assembly (task 21) before being placed in final storage.

A flow process diagram provides similar information to the flow process chart except that the flow lines are superimposed on the layout of the factory or workshop. This is useful where the layout itself is a prime cause of the way/sequence in which the work is carried out. Figure 12.2 shows how the flow process chart in Figure 12.1 can be drawn as a flow process diagram.

After preparing the chart or diagram, the process is analysed using the questioning technique set out in section 12.3.3. Note: The very act of recording what is actually happening will usually yield some useful information. In this case two delays (D) have been observed before any analysis has been conducted.

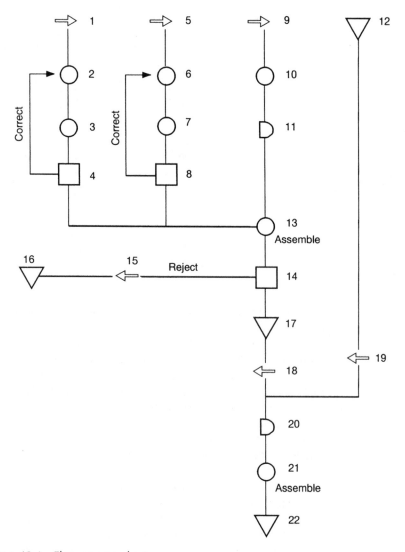

Figure 12.1   Flow process chart

12.3.2   Multiple activity charts

Multiple activity charts examine how processes and movements relate to one another chronologically. The times in which men and machines are operational and nonoperational are recorded.

Example 12(2)

Figure 12.3 represents a materials-processing plant in which a mixer is filled from a hopper at Position 1 with Material A, moves to Position 2 where Material B is added from a second hopper, moves to Position 3 where the contents

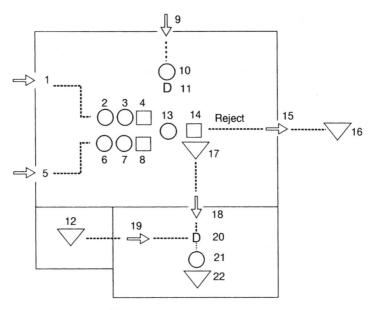

Figure 12.2 Flow process diagram

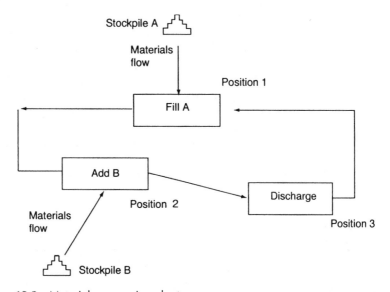

Figure 12.3 Materials processing plant

are discharged, then returns to Position 1 to repeat the cycle. Transportation between stockpile and hopper is by electrically operated barrows.

Time to load material at Stockpile A and transport to Position 1	6 min
Time to load material into empty hopper at Position 1	2 min
Time to return barrow to Stockpile A	4 min
Time to fill mixer from hopper at Position 1	2 min

Time for mixer to travel to Position 2	2 min
Time to load material at Stockpile B and transport to Position 2	12 min
Time to load material into empty hopper at Position 2	2 min
Time to return barrow to Stockpile B	6 min
Time to fill mixer from hopper at Position 2	4 min
Time for mixer to travel to Position 3	2 min
Time to discharge mix at Position 3	2 min
Time to return mixer to Position 1	4 min

(a) Draw a multiple activity chart for the process.

(b) Given that $4 \, m^3$ of Material A is placed in the mixer at Position 1, determine the steady rate of use of Material A.

(c) Management is interested in speeding up the process. What one small improvement in the speed of operation would have considerable benefits?

Solution

(a) Figure 12.4 shows the completed chart. Usually at least three operational cycles must be constructed. In this case it becomes clear when constructing the third cycle that the latter elements of the third cycle are identical to those of the second cycle. These, therefore, are not shown.

In constructing the chart each activity should start at the earliest possible time. So, at the start, both the barrows at A and B are loaded immediately and the materials are transported to the mixer location, then loaded in the hoppers. In the second cycle the barrow at A loads and returns to Position 1 to fill the hopper (completed at *time: 20 minutes*), even though the hopper cannot be emptied until *time: 26 minutes* when the mixer returns to Position 1. The second cycle shows two changes from the first:

(1) **Fill mixer at 1** is delayed 6 minutes, waiting for the mixer to return to Position 1.

(2) **B to Position 2** must await the return of the barrow to Stockpile B.

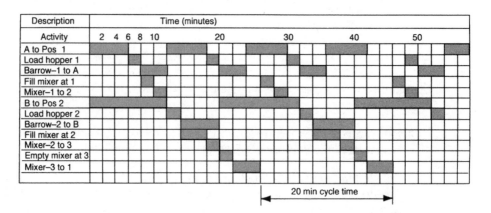

Figure 12.4   Multiple activity chart

The third cycle shows one change from the second:

(3)   The time between the two operations **fill mixer at 1** has increased from 16 to 18 minutes as the return of the mixer from Position 3 is awaited.

The remainder of the third cycle is now identical to that of the second cycle and the regular cycle time of 20 minutes for the operation can be observed from the chart.

The fourth cycle shows the effect of the delayed return of the mixer to Position 1. The hopper at Position 1 cannot be loaded at *time: 42 minutes,* as it has not yet discharged its contents into the mixer. Only between the fifth and sixth cycles does the regular pattern of operations become clear: every operation is repeated at 20-minute intervals.

(b)   A 20-minute cycle time means three operations per hour, giving a steady usage of materials at stockpile A of

$$3 \times 4 \text{ m}^3 = 12 \text{ m}^3 \text{ per hour}$$

(c)   The controlling sequence is

B to Position 2
Load hopper 2
Barrow – Position 2 to B

There are no delays, no waiting for other operations. If this could be reduced by four minutes, the cycle time would be 16 minutes, increasing the average hourly usage at stockpile A to 15 m^3 per hour. This could be achieved, say, by speeding up the operation **B to Position 2** by four minutes.

The chart shows when equipment is operational and when it is not. By the fourth cycle it can be seen that the barrow serving Position 1 works for only 12 minutes out of every 20 and that the mixer is either being filled, moving or being emptied for 16 minutes out of every 20. This shows that the 16-minute cycle advocated in (c) above would be attainable by one change to one operation, namely **B to Position 2** to take 8, not 12, minutes.

Thus, a multiple activity chart is a simple but effective tool in the analysis of cycles of concurrent and dependent operations.

12.3.3 Interrogative method

This method will usually be used in conjunction with the analysis described in the preceding subsections. The principle is that questions are asked of each activity or task. The questions are:

- WHAT is done?
- HOW is it done?
- WHO carries out the task?
- WHEN is it done?
- WHERE is it done?

and for each of these

- WHY is it done at all, at that time and so on?

There is always value in questioning what is being done, in taking nothing at face value. By questioning even the obvious it is possible that some fundamental insight might lead to radical change and significant improvement.

Example 12(3)

A yard has been set up for cutting and bending reinforcing steel for a building site. Standard lengths of different diameter steel bars are delivered to the yard by road, unloaded at the roadside, lifted into storage racks from which they are then drawn as demand dictates, cut to length, bent to shape, transferred to a second storage location and eventually transported to site. Table 12.2 shows a

Table 12.2   Steelyard analysis sheet

No.	WHAT is done? (a)	WHEN is it done? (b)	HOW is it done? (c)	WHERE is it done? (d)	WHO does it? (e)	WHY?
1	Offload steel into racks	Weekly, on delivery	By crane	At the roadside	Gang from building site	(d) Yard entrance is too narrow
2	Storage of steel	Between delivery and use	In labelled racks	In labelled racks near the roadside		(a) Continuous operation needs a stock at the yard
3	Move from racks to cutters	On demand by the cutter	Sliding by hand	Between racks and cutters	Cutter operators	(e) No one else available
4	Cutting to lengths	On arrival at cutter	By machine	On power cutters	Cutter operators	(a) Schedule shows various lengths
5	Move steel to temporary storage	After cutting is complete	By hand	From cutters to temporary stack	Cutter operators	(a) Must be moved out of the way
6	Steel awaits bending	Between cutting and bending	In small heaps	Near the benders		(a) Cutting is faster than bending
7	Move steel to bender	On demand by bender	By hand to slide into bender	From temporary stack to the benders	Bender operators	(e) No one else to do it
8	Bending of steel	On arrival at bender	By hand and machine power	On the bending machine	Bender operators	(a) Schedule requires different shapes
9	Move to storage	After bending is complete	By hand	From benders to storage	Bender operators	(e) No one else to do it
10	Steel awaits transfer to building site	Between bending and delivery	In marked bays	In bays near to site entrance		(a) and (c) Deliveries out are irregular
11	Loading of steel onto site transport	On demand from building site	By self-loading truck	At the storage bays	Site materials gang	(b) Don't want too much stock

typical interrogative analysis sheet. Only illustrative responses are given in the WHY column.

The responses yield some useful observations.

No. 1, offloading the delivery truck from the road may be less than ideal, possibly unsafe and creating a traffic problem. A slightly wider entrance might ease the problem.

In nos 3, 5, 7 and 9, both the cutter and bender operators have their work interrupted moving materials. Is this necessary? Might the addition of one or two unskilled labourers result in a far faster operation?

In no. 6 the benders are slower than the cutters, resulting in unnecessary temporary storage and probably in double handling. The work rates of the two kinds of machine and numbers available need to be investigated.

This method of analysis is always radical in the sense that it gets to the root of the operation. To be successful it needs to be conducted by someone with an enquiring mind who does not accept things at face value.

12.4   WORK MEASUREMENT

Work measurement is concerned with measuring the times needed for specific tasks. This is essential information for planners and for estimators. It will be equally apparent from the foregoing section that measuring the time taken for selected operations will be important in method study. The objectives of work measurement can be summarized as follows:

- To assist in method study.
- To assist in planning and scheduling of labour, plant and materials.
- To assist in cost control.
- To compare different methods of working.
- To provide a basis for incentive schemes.
- To assist in implementing total quality management.

BS 3138 defines work measurement as 'the application of techniques designed to establish the time for a *qualified worker* (sic) to carry out a *task* at a defined rate of working'.

This is a much more precise definition than might at first sight appear to be necessary to meet the listed objectives. But, for a measured time to have any value, certain criteria have to be met. These are discussed briefly below.

The work must be such that, when carried out by different people and/or plant, it is performed in the same way and under similar conditions. Therefore, there are many operations for which work measurement is inappropriate. Herein lies a difficulty. Engineers may be tempted to rely on those things that can be measured and to ignore, when making decisions, those things that cannot be measured. Chapter 9 referred to *hard* and *soft* issues. Engineers commonly use hard data in their analyses; this will not suffice in work study. Once the engineer has completed the numerical analysis, careful thought must be given to the soft issues before a decision is made on how the findings should be used.

What is a *task?* The underlying assumption is that any operation can be broken down into a series of finite tasks. Factories, banks and bureaucracies have had occasion to measure the duration of tasks, from the time taken to machine a pipe end to the number of keystrokes per minute made by someone inputting financial data into a computer. If, however, the measurement and summation of these times do not contain provision for the wider needs of people at work – for time to reflect on what they are doing; time for discourse and for relaxation; time to deal with irregular interruptions – then the process of measurement can make work appear mechanistic. This can create all sorts of problems, for people are not machines. Work study and, in particular, work measurement must not be allowed to separate each worker from his or her colleagues. In many work situations, overall performance is improved by teamwork, where the individual's ability to perform a particular task to a specified standard is less important than the combined ability of the team. In this kind of situation work measurement should not be used to compare the performance of individuals, but still has a place in determining standard times for an operation. Method study techniques will always be useful in improving performance through joint efforts.

The workers who carry out the work must have the necessary skills and be experienced in the particular operations. Measuring the time it takes for an inexperienced worker, albeit a skilled one, to complete a task would be of little value, as once the worker had become adept at the work it would be performed much more quickly. This simple and obvious lesson is sometimes forgotten when incentive schemes are set up for new operations with which management and workers are unfamiliar; too often the incentive scheme is based on the low initial rate of production during the learning phases, and a few months later extraordinarily high bonus payments are being paid to the workers.

Work measurement is used to establish standard times for jobs. These can be used either for planning or for monitoring of performance. Measurement techniques available range from the simple timing of actual jobs (for example, for constructing multiple activity charts), through time study, in which the actual times are corrected by a performance-rating factor, to various forms of activity-sampling synthesis. This chapter will consider briefly the following:

- predetermined motion-time systems;
- time study.

Although few engineers will be involved in these measuring techniques, they may be required to act on the data provided, so they should be aware of the principles.

12.4.1   Predetermined motion-time systems (PMTS)

PMTS come under the field of work study known internationally as methods-time measurement (MTM). MTM examines the time taken for basic human motions such as to grasp something, or press a key, or bend an elbow. The times taken for these movements are very short and MTM's time measurement unit is 0.00001 hour (0.0006 min.). These systems demand a conjunction of film/video recording and high-precision timing methods.[2] Whitmore [5] observes that the time required to analyse a job using the MTM system at its basic level can be as much as 150 times as long as the task takes to perform! Users of MTM systems rely on data tables that provide the times taken for all these 'basic human motions'.[3]

In PMTS a system of synthesis is used. A task under consideration is analysed into its constituent basic human motions and the times for these standard elements are obtained from data tables and aggregated. Underlying the data tables are implicit rates of working for all the classified human movements therein. The principle is that 'the time required for all experts to perform true fundamental motions is a constant' [5]. PMTS data tables have been developed for many areas of employment, from clerical work to maintenance, providing data on higher levels of operation such as inserting a letter into an envelope or changing the wheel of a car. These data tables allow an employer to determine a standard time for any operation selected.

12.4.2   Time study

Time study uses observed times and converts them into standard times. Observed times reflect how well the individual worker applies himself or herself to the task. Consequently, observed times must be adjusted to represent the time the worker would have taken if he or she had been working at a standard rating. Each measurement of time must be accompanied by an assessment of how much above or below a standard rating the employee is working. Clearly this is a subjective measure and one that is most reliable only where the time study engineers have considerable experience. The principle of assessing a rating for an individual is that if he or she works at half the standard rating, the task will take twice as long. The standard rating assumes that the worker is adequately motivated to do the job. A worker provided with an incentive would be expected to operate one third faster than his or her normal rate. In short, the time study engineer, as he or she times the task, must compare the worker's performance with that of a notional rating that must be held in his or her mind for the duration of the task.

There are a number of scales in use:

- The 60/80 scale, in which 60 represents the standard rating and 80 the 'incentive' rating.
- The 75/100 scale, in which 75 represents the standard rating and 100 the 'incentive' rating.
- The 100/133 scale, in which 100 represents the standard rating and 133 the 'incentive' rating.

---

[2]No attempt is made in this chapter to describe these methods which in recent years have become capable of measuring very short durations of motion with a high degree of accuracy. The different approaches to measurement such as consecutive-element timing and selective-element timing are described in Whitmore, Chapters 9 and 17 [5].

[3]These data tables rely on a wide range and large number of measurements. See Whitmore, Chapter 6 for the mathematics of measurement and Chapter 7 for the process of analysis [5].

The latter scale is the one recommended by the British Standards Institution [6]. In each case the 'incentive' rating is one third higher than the standard rating.

Ratings assessed on the 100/133 scale would normally lie in the range 70 to 160 (ratings in excess of 130 should be rare), for most people are unable to properly assess performances that are very slow or very fast. Ratings should normally be assessed to the nearest 5 points, although an inexperienced observer would be wise to keep to the nearest 10 points. Whitmore [5] recommends that an assessment of rating should include pace, effort, consistency, dexterity and, possibly, job difficulty.

The standard time for an operation is found from the sum of the basic mean times of the elements or tasks that make up that operation. Time study engineers will usually subdivide operations into elements of less than one minute's duration. Measurements are usually made in centi-minutes (hundredths of a minute) rather than seconds.

Measurements should be taken over several cycles of the operation and for a sufficient number of workers to provide a statistically representative sample.[4] For each worker completing a task the basic time will be determined (see below) and the mean of all the basic times for the group of workers will be calculated; this is known as the basic mean time.

For the 100/133 scale:

$$\text{Basic Time (BT)} = \frac{\text{observed time} \times \text{rating}}{100}$$

A slow worker will be characterized by a large observed time and a low rating. A fast worker will be characterized by a small observed time and a high rating, so the basic times for each worker should be similar. Where there are major differences in the BTs for a given task there may well be something wrong with either the observed times or, more probably, the ratings. BTs that are exceptional in some way should not be considered in determining a basic mean time. Either the whole measurement process should be repeated or, where only one or two values are distinctly odd, the exceptional values should be excluded from the calculation.

To calculate the basic mean time the BTs are aggregated and divided by the number of observations:

$$\text{Basic Mean Time (BMT)} = \frac{\sum (\text{BT})}{n}$$

Before the standard time for an operation can be determined, allowances must be made for the fact that each cycle of work may have to be repeated many times during the working day. People must be able to relax, sometimes even rest, after physical exertion and attend to calls of nature. In addition there will be occasional interruptions that, although unpredictable, are not unexpected. Allowances for relaxation and so on are referred to as *relaxation allowances*. Allowances for unexpected interruptions are referred to as *contingency allowances*. Allowances are usually made by adding on a small percentage to the sum of the BMTs.

---

[4] See Dilworth [7] for more information and guidance.

Typical figures for relaxation allowances are from 5 to 10% and for contingency allowances certainly not more than 5%. Other provisions such as for unoccupied time may also have to be made.[5]

The standard time for an operation will be found as follows:

$$\text{Standard Time} = \sum (\text{BMT}) + \text{relaxation \%} + \text{contingency \%}$$

Example 12(4)

Manufacture of a product X involves a single operation that time study engineers have subdivided into three tasks A, B and C. One group of six workers carries out the operation. A time study has been carried out on each of these tasks to determine a standard time for the production of X. The observed times and assessed ratings are tabulated in Table 12.3 (note: In reality, measurement of several cycles for each operation for this group of workers would be carried out).

Determine a standard time for the manufacture of Product X.

Solution

Tabulated in Table 12.4 are the basic times for each worker and each task.

Table 12.30

Name	Task A Observed time (min.)	Rating	Task B Observed time (min.)	Rating	Task C Observed time (min.)	Rating
Alyson	0.18	110	0.21	80	0.24	90
Anka	0.17	80	0.22	90	0.19	110
Cosma	0.16	110	0.21	100	0.22	90
Josef	0.17	90	0.23	90	0.24	90
Joyce	0.19	100	0.24	80	0.22	90
Kerr	0.16	110	0.22	100	0.25	80

Table 12.4

Name	Task A BT	Task B BT	Task C BT
Alyson	0.198	0.168*	0.216
Anka	0.136*	0.198	0.209
Cosma	0.176	0.210	0.198
Josef	0.153	0.207	0.216
Joyce	0.190	0.192	0.198
Kerr	0.176	0.220	0.200

[5]Whitmore [5] provides a useful chapter (Chapter 12) on allowances. See also BS 3375 [8].

For each task the mean time is calculated. However, for each of tasks A and B the BT marked by an asterisk has been eliminated from the calculation; the values are considerably different from all the others and are therefore unsafe.[6]

$$BMT_A = \frac{(0.198 + 0.176 + 0.153 + 0.190 + 0.176)}{5} = 0.1786$$

$$BMT_B = \frac{(0.198 + 0.210 + 0.207 + 0.192 + 0.220)}{5} = 0.2054$$

$$BMT_C = \frac{(0.216 + 0.209 + 0.198 + 0.216 + 0.198 + 0.200)}{6} = 0.2062$$

From this the BMT for the whole process is found:

$$BMT_{(A+B+C)} = 0.1786 + 0.2054 + 0.2062 = 0.5902$$

In order to arrive at a standard time, relaxation and contingency allowances should be included. In this case, and in the absence of any further information, the value determined above is multiplied by 1.15. That is, a 15% total additional allowance is provided.[7]

$$\text{Standard time} = 1.15 \times 0.5902 = 0.6787 \text{ minutes}$$

It would be absurd to give this exact figure as the standard time. The selection of 15% was arbitrary; it could equally well have been, say, 14% or 16%. Also, each rating is a subjective assessment and even with an experienced work study engineer working consistently with his or her own notional criterion of a standard rating, only an educated estimate is possible. Therefore it would be entirely inappropriate to give the standard time to four decimal places. A realistic standard time would be 40 seconds (0.6667 min.), which is very close to the calculated figure anyway.

$$\text{Answer : Standard time} = 40 \text{ seconds}$$

12.5 SUMMARY

This chapter has provided a brief outline of the principles of work study and described a number of the techniques used in both method study and work measurement. For method study, flow charts and diagrams, multiple activity charts and the interrogative method have been illustrated. For work measurement, PMTS and time study procedures have been described and illustrated.

REFERENCES

1. Taylor, F. W. (1929) *Principles of Scientific Management*, Harper, London.
2. Hill, T. (1991) *Production/Operations Management Text and Cases*, 2nd edn, Prentice-Hall, London.

---

[6] Consideration of the range of performance times is given in section 7.6 of Chapter 7 of Whitmore [5].
[7] See BS 3375: Part 5 [8].

3. *BS 3138 Glossary of Terms Used in Management Services* (1992) British Standards Institution, London, available via www.bsi-global.com.

4. *BS 3375: Part 2 Guide to Method Study* (1993) British Standards Institution, London, available via www.bsi-global.com.

5. Whitmore, D. A. (1987) *Work Measurement*, 2nd edn, Heinemann, London.

6. *BS 3375: Part 3 Guide to Work Measurement* (1993) British Standards Institution, London, available via www.bsi-global.com.

7. Dilworth, J. B. (1999) *Operations Management: Providing Value in Goods and Services*, Thomson Learning, London.

8. *BS 3375: Part 5 Guide to Determination of Exposure Limits, Recovery Times and Relaxation Times in Work Measurement* (1997) British Standards Institution, London, available via www.bsi-global.com.

BIBLIOGRAPHY

Wild, R. (1998)   *Essentials of Production and Operations Management*, 4th edn, Cassell, London.

Johnston, R., Chambers, S., Horland, C., Harrison, A. and Slack, N. (1997) *Cases in Operations Management*, Pitman, London.

Stevenson, W. (2001) *Operations Management with Student CD-ROM*, McGraw-Hill Education Europe, Maidenhead.

13

Costing and Pricing

13.1 INTRODUCTION

Any organization, whether it is a business, a government department or a charitable foundation, will eventually fail if costs exceed revenue. Therefore, costs are fundamental to all enterprises, including those that employ engineers, scientists and technologists. Businesses need to make a profit. This is the business imperative: revenue must exceed costs. This, however, is not the sum total of an engineer's or scientist's interest in costs. Engineers and scientists will often be required to submit cost plans (budgets) for approval and, once approved, a budget will be used to monitor their production performance. Engineers often tender for work and/or price their services or products for potential customers. These could be either external or internal customers, and engineers and scientists may have to justify their department's continued existence in the face of 'business process outsourcing'. Tendering and pricing cannot be achieved successfully without a clear understanding of the principles of costing.

This chapter will examine the nature of costs (*direct* and *indirect* costs, *fixed* and *variable* costs), cost–volume relationships (*break-even*), *budgets* and *tendering* and *pricing*.

13.2 WHAT IS THE COST OF A PRODUCT?

Consider launching onto the world markets an entirely new product. There are no competitors, nothing like this has been manufactured and sold before. What costs are incurred in the development, manufacture and eventual sale of the product?

Any or all of the following costs might be incurred:

Market research[1]	Will people buy this product and at what price? What is the potential size of the market?
Design and specification	Designers will specify performance, quality, life; they will carry out calculations and produce drawings and schedules.
Prototype manufacture	One or a small number of the products will be manufactured, built or created.
Development	The prototype(s) will be tested, modifications may be made to the design and specification.

---

[1] See Chapter 22, section 22.12, for the role (and costs) of market research.

Tooling and facilities   Setting up the production line, machine tools, process equipment, assembly etc. prior to production.

Production   Materials will be consumed, labour utilized, fuel/power consumed etc.

Marketing   Publicity and advertising.

Distribution   Packaging and transportation etc. to dealers.

Product support   Parts and labour for after-sales maintenance and servicing and even a helpline for user advice and online fault fixing.

If the product is a success, the costs of manufacture and distribution will continue and further expenditure on marketing will be incurred from time to time. The earlier costs must be paid for out of sales revenue. If only 1000 are sold, then the cost of each item sold must include 1/1000 of the total initial costs (market research to tooling); if 1 million are sold, the cost of each item sold could include one millionth of the initial costs. Alternatively, some of these costs may be considered as a company overhead and charged against other products (see section 13.2.3). Whether this is so depends on how soon and by what means the company wants these initial costs to be recovered. Cost, therefore, is a matter of judgement, not a matter of fact.

Even the cost of production is not straightforward. Certainly there will be an identifiable materials cost and, when operatives work on that material, direct labour costs are incurred. But there will be other costs: heating and lighting, depreciation in the value of the buildings and equipment, consumables used in maintenance and servicing, salaries of supervisory staff, administration and so on.

Costs that arise directly from production are known as *direct costs*; those that arise whether production takes place or not are known as *indirect costs*. The allocation of indirect costs is an arbitrary process. In the example given above, the company will be manufacturing other products besides the new one. How should the costs of rent, administration and heating and lighting be assigned to different products? A further question needs to be considered. What would have happened if, during the development stage, a decision had been made to proceed no further? The costs incurred still have to be covered by the revenue earned from selling the company's other products; therefore each product may have to be assigned a proportion of these costs. The subsections that follow will consider the ways in which the costs of *materials* and *components*, of *labour* and of *overhead* (indirect costs) are measured.

13.2.1   Measuring the costs of materials and components

Of all costs, materials costs are the most clearly identifiable as a direct cost. If a production centre uses a material or good then cost is incurred. If production does not take place the material or good is not procured and no cost arises. This suggests that, for a production centre to keep records of its costs, it must 'buy in' the material or good. Most business-conscious organizations today comprise a number of *cost centres*. Although actual money is unlikely to flow from one cost centre to another, each cost centre will record the cost of buying in materials, goods or services from either external suppliers or other cost centres, and will charge subsequent suppliers or cost centres for the materials, goods or services provided to them. Therefore, when a production centre draws materials from stock, a clearly defined cost is incurred.

But what would be the position if the stock has been supplied by different suppliers at different prices over a period of time? Assuming that the stock is nonperishable, that the time it is kept in stock does not affect its quality so that when drawing from stock the goods are selected at random and not according to, say, chronological order of supply, what is the value of the goods that should be charged to the cost centre? Should it be the cost of the oldest goods in stock? Should it be the cost of the most recently purchased goods? Should it be an average of the values in stock? Before the direct costs of materials or goods can be assigned to production the company must decide, as a matter of policy, a method of stock valuation. The decision will affect two measures: the value of stock in the stockpile at any one time and the cost charged to the production centre.

The two most common methods in use for determining the value of stock are first in first out (FIFO) and last in first out (LIFO).

First in first out (FIFO)

In this case the amount charged to the cost centre will reflect the cost of the earliest items delivered to stock.

Example 13(1)

Stock has been delivered to the stockpile on the first of every month in the quantities and at the prices listed below:

1 February	300 no.	@ £20.00
1 March	400 no.	@ £21.00
1 April	350 no.	@ £20.00
1 May	300 no.	@ £22.00

On the last day of May, 800 items are drawn from stock. What should the production centre be charged? Although there may be no means of determining which stock is the oldest, the charges will be based on the costs of the first in. Thus, the production centre will be charged:

300 no.	@ £20.00 =	£6 000.00
400 no.	@ £21.00 =	£8 400.00
100 no.	@ £20.00 =	£2 000.00
	total cost =	£16 400.00

The value of the remaining stock can also be ascertained:

(350 − 100) = 250 no.	@ £20.00 =	£5 000.00
300 no.	@ £22.00 =	£6 600.00
	total value =	£11 600.00

A consequence of this method of costing and valuation is that the costs charged to the production centre are low relative to the current prices of materials. Is this a sensible policy to adopt? A further consequence of this method is that the value of the stockpile is high, reflecting current prices of buying in materials. A conservative approach to stock valuation will favour a low valuation. For example, the price it might command on the open market could be lower than the price at which the company buys it from its suppliers.

Last in first out (LIFO)

The alternative is to charge the cost centre at prices that reflect the current prices of buying in stock.

Example 13(2)

The same data will be used as in Example 13(1).

Stock has been delivered to the stockpile on the first of every month in the quantities and at the prices listed below.

1 February	300 no.	@ £20.00
1 March	400 no.	@ £21.00
1 April	350 no.	@ £20.00
1 May	300 no.	@ £22.00

On the last day of May, 800 items are drawn from stock. What should the production centre be charged?

300 no.	@ £22.00 =	£6 600.00
350 no.	@ £20.00 =	£7 000.00
150 no.	@ £21.00 =	£3 150.00
	total cost =	£16 750.00

This is a higher cost than would have been charged using FIFO. The value of the remaining stock can now be ascertained.

	300 no.	@ £20.00 =	£6 000.00
(400 − 150) = 250 no.		@ £21.00 =	£5 250.00
		total value =	£11 250.00

This is a lower value and therefore might be preferred. However, if this method is adopted a consequence could be that the stock valuation might contain values of small amounts of materials that were purchased many months or even years ago. This would be inevitable if the stockpile were not completely cleared out from time to time.

Other alternatives

Alternatives to FIFO and LIFO include using the average price of the materials bought in, replacement price and standard costing.

If the average price is used, recalculation is needed whenever more stocks are purchased and/or stocks are issued. In an inflationary situation stock valuation will lag behind market value and costs charged to production centres will lag behind current costs, but not as badly as under FIFO.

Another term for replacement price is next in first out (NIFO). A production centre is charged for goods at the costs of buying in new materials. This ensures that the production centre charges its customers at prices that reflect current prices for goods and materials. Actual stocks will tend to be undervalued.

Standard costs are predetermined estimates (see section 13.2.4), in this case, of material costs. Thus the production centre will be charged the standard cost for the materials provided and the stock will be valued at the standard cost. Records of the variances between the actual costs and the standard costs will be kept and reconciled at the end of an accounting period. This works well when costs are reasonably steady.

To summarize, the way in which material costs are assigned to production is not a matter of fact but a matter of policy. It will become clear that the ways in which labour costs and overheads are determined are equally discretionary.

13.2.2   Measuring the costs of labour

There are two elements within most labour costs:

* time related (hours worked × hourly rates);
* output related (piecework, incentive schemes).

Hourly rates will also vary with the number of hours worked in a day and in a week. Higher, overtime rates usually must be paid if the hours worked on a given day exceed a prescribed maximum. Working more than, say, 37 hours in a week or working at weekends may also entitle the individual to overtime payments. In some industries there will be extra payments for working with certain tools and/or equipment or for working under certain conditions. Companies often introduce incentive schemes to improve productivity and the incentive payments are later subsumed into pay norms as productivity rises with capital investment. A possible result of this is that over many years a very complex pay structure develops involving an array of so-called special payments, which no longer represent the nature and extent of the work done. Companies sometimes experience considerable problems when they try to rationalize pay structures.

The wages bill is only one part of the cost of employing labour. In addition there can be provision for holiday pay, sick pay, national insurance, private medical insurance and pension schemes, redundancy funds, industrial training levies, and possibly travel and accommodation expenses.

When a person is assigned to a specific task, a direct cost arises: the cost of the work must include the cost to the company of employing that individual.

However, it should be clear from the foregoing paragraphs that the actual cost may depend on the actual hours worked and on the productivity of the individual and includes a number of indirect costs. Therefore, although these costs will provide a historical record of costs incurred in a production process, they will be in a most unsuitable form for immediate entry as current labour costs. What has to be done in order to allocate labour costs against production is to agree a notional labour cost per hour and to monitor regularly how close notional costs are to the recent past historical costs.

In short, a company, as a matter of policy, will establish a set of labour rates (for different kinds of work or for different skills) chargeable as a direct cost to customers and/or internal cost centres. These rates will be updated from time to time.

13.2.3   Measuring and allocating overhead

Overhead is the term used for indirect costs (in the USA the term 'burden' is sometimes used). The difficulty lies less in the measurement and more in the allocation of overhead. If a company manufactures a single product, then that product must pay for the total overhead of the company. There is no difficulty there! If a company manufactures two products, A and B, how should the overhead (rent, heating and lighting, administration, marketing, even research and development) be allocated? If Product A uses the services of ten workers and Product B uses the services of 40 workers, one method might be to assign overhead to A and B in the ratio 1:4. If, however, Product A benefited from a higher proportion of capital investment, then a system relying on a head count would be inappropriate. Another approach might be to use the floor area allocated to production as a basis for distribution of overhead, but, unless these floor areas could be closed down and no charges incurred by the company if there were no production, there is no rationale for this procedure.

The two methods above are actually used, because they are simple to apply. Another method is to allocate overhead on the basis of turnover, where turnover is defined as:

$$\text{number produced} \times \text{selling price per item}$$

Example 13(3)

A company produces 1500 of Product A per week, which sells for £10 each, and 6000 of Product B per week, which sells for £6 each. Company overheads amount to £8000 per week. Determine the overhead cost that should be assigned to one unit each of Product A and Product B.

$$\text{Turnover for Product A} = 1500 \times 10 = £15\,000$$

$$\text{Turnover for Product B} = 6000 \times 6 = \underline{£36\,000}$$

$$\text{Total turnover} = £51\,000$$

$$\text{Overhead for Product A} = \frac{15\,000 \times 8000}{51\,000} = £2353$$

$$\text{Overhead for Product B} = \frac{36\,000 \times 8000}{51\,000} = £5647$$

$$\text{Unit overhead for Product A} = \frac{2353}{1500} = £1.57$$

$$\text{Unit overhead for Product B} = \frac{5647}{6000} = £0.94$$

This appears to be a clear, rational approach. Overhead could be reassigned, say, annually against production using the previous year's figures for turnover. But what would be the position if Product B was in a highly competitive market and its direct costs amounted to £5.25? By adding a unit overhead cost of £0.94 the total cost would be £6.19 and the product would be losing money. If, at the same time, the unit direct costs for Product A only amounted to £3.00, then each sale of Product A would earn the company a profit of

$$£10.00 - £3.00 - £1.57 = £5.43$$

Both the loss on Product B and the profit on Product A are illusory, a consequence of an arbitrary distribution of overhead. The facts are as set out below:

direct costs of Product A = £3.00 × 1500 =   £4 500

direct costs of Product B = £5.25 × 6000 = £31 500

total overhead =   £8 000

total costs = £44 000

total revenue from Products A and B = £51 000

total profit for Products A and B =   £7 000

Which of A or B is responsible for the greater proportion of that profit is a matter of opinion, not a matter of fact. To pursue the matter a little further, an alternative distribution might be to divide the overhead equally between Products A and B. In this case the total costs of A would amount to £4500 + £4000 = £8500, resulting in a total profit for A per week of £6500. The total costs for B would be £31 500 + £4000 = £35 500, giving a total profit per week for product B of £500. What is evident from this example is that the distribution of overhead can cause a product or cost centre to appear to lose money when in each case the product's sale price exceeds its direct costs. In fact, each product is making money! Perhaps the question that needs to be asked, by the costs centres of their administration, is 'Why are the overheads so high?'. It might be that it is the administration that is 'losing money', not the cost centres!

Example 13(4)

A large agrochemical company has hundreds of distribution centres. Company overhead is £150m p.a. The annual turnover is £1000m. A new small distribution centre has been established to serve what is hoped to be an expanding market. Company policy is to distribute central overhead as a direct proportion of turnover. Using the data below for the first year of operation of the new centre, should it be allowed to continue in business for a second year?

	Costs £(k)	Revenue £(k)
Annual sales of agrochemicals		400
Costs of importing agrochemicals	200	
Costs of distribution to customers	20	
Staff costs	120	
Depot overhead	24	
Company overhead (15% of sales)	60	
Totals	424	400

The new distribution centre is, apparently, losing £24k p.a. Therefore a decision to close it down would not be unexpected. But would that decision make any business sense? If the company overhead is removed from the list of costs, the value of the costs incurred by the new centre is only £364k; so revenue has exceeded costs by £36k. The centre is making money. Since the centre has been located to serve an expanding market, a decision to keep the centre in operation would be far more sensible. Those huge company overheads are so large in comparison with this small operation that they must be absorbed elsewhere, just as they were in the year before opening the new centre.

Overheads must be paid, certainly. If the production centres cannot absorb them then they must be reduced (at head office if necessary!). Managers should be acutely aware that the way in which overheads are assigned to production centres is essentially arbitrary, and therefore should be a matter of discussion, not of imposition.

13.2.4   Standard costs

Managers are required to make decisions on current and future operations on the basis of experience and knowledge of costs. For managers to make the correct decisions, they must have up-to-date data on costs. By now it should be clear that costing is an imperfect science; yet cost data must be held, because the information is needed. In view of this, companies keep records of *standard costs* that are updated at regular intervals. These standard costs are predetermined estimates of the costs of carrying out standard operations under standard conditions. Standard costs can be kept for specific tasks that employ labour, equipment and materials. Standard costs do not usually include overhead; this will be added on at a level decided by management, depending on the purpose of estimating the costs of the work to be done.

Since the actual costs incurred do not usually correspond to the standard costs that make up the cost plan, it will be important to record the differences that arise. These differences are referred to as variances and the cost centre manager will be called on from time to time to examine the findings of variance analyses produced by his or her cost accountants and to explain any large variances. Large variances may arise when:

- the work is not being carried out under standard conditions;
- productivity is significantly different from that assumed by the standard costing (for reasons known or unknown);
- the standard cost has not been updated.

The objectives of variance analysis are twofold: to ensure that the standard costs held on record are suitable for current and future use, and to monitor the performance of the work in hand.

Standard costs will be kept for elements of work on a quantitative or hourly basis depending on the nature of the work. For example, a garage might apply a single standard cost for an annual service for a vehicle, but an hourly cost for body repair work. A builder might apply a quantitative standard cost for laying asphalt (£ per m^2), but a single standard cost for installing a boiler. These standard costs will be kept in the form most suited to the needs of the company.

13.3   MAKING A PROFIT

In order to make a profit, revenue must exceed total costs. There will be direct and indirect costs. Direct costs consist of *fixed* costs and *variable* costs. Direct costs will usually be assigned as a variable cost where the cost varies either with time or with production. A fixed cost is one that arises independently of the quantity of goods or materials produced. The cost of bringing in a machine to carry out a task might have two elements to it. Costs arise in transporting the machine to the site, then setting it up and, after the task is over, removing it and returning it to wherever it came from. The fixed cost for using the machine is the sum of these costs. During production the machine also incurs costs; these could be hire costs, fuel costs or depreciation costs. These are variable costs; that is, they vary with the output, which is determined by time and productivity.

Figure 13.1 shows some of the ways in which costs can vary with production. Figure 13.1(a) shows a fixed cost; the cost does not vary with production. Figure 13.1(b) shows a variable cost in which there is a linear relationship between volume of production and cost, for example a material cost; double the quantity means double the material cost. Figure 13.1(c) shows the most common position, in which fixed and variable costs are incurred.

A linear relationship is not going to arise at all times. There are several possibilities. When production rates rise the company might be able to buy materials at a discount, and the materials' unit costs decline as the volume increases; this could give rise to the curve in Figure 13.2(a). On the other hand, to increase production, overtime might be necessary, increasing the direct labour cost; this could give rise to the curve in Figure 13.2(b).

The simplest approach, and one commonly adopted in practice, is to assume a linear cost function. If one then assumes that, once a unit price has been

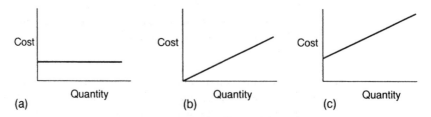

(a)   (b)   (c)

Figure 13.1

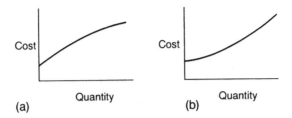

Figure 13.2

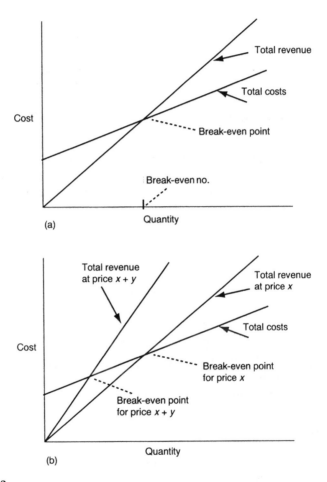

Figure 13.3

established, it will not vary with the volume sold, a *break-even* chart can be drawn (see Figure 13.3(a)). This linear break-even chart has one break-even point, which is the point at which total revenue equals total costs. Knowledge of the break-even point is most useful: it tells the manager the volume of sales needed before a profit can be made. Further analysis of the position could be carried out. An alternative price would give rise to a different break-even point, as shown in Figure 13.3(b).

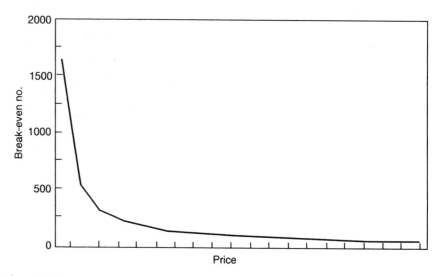

Figure 13.4

Figure 13.4 shows a curve obtained by plotting sales price against break-even number. From this curve can be found either the minimum sales that must be achieved for a whole range of price options or the minimum price the company must charge in order to cover its costs if production and sales are limited.

This curve can be plotted from the equation

$$n = \frac{F}{(S - V)}$$

where n = break-even number
   F = total fixed costs
   S = sales price per unit
   V = direct variable cost per unit

$(S - V)$, the difference between the sales price and the variable cost, is known as the *contribution*; it is the money available for paying off the fixed costs and, eventually, for providing profit.

Example 13(5)

A factory established to manufacture three products, A, B and C, has fixed costs as below:

	£
Salaries per week	7 000
Heating and lighting per week	100
Office supplies per week	100
Marketing per week	600
Building (lease and taxes) per year	12 000
Loan for equipment (repayable over one year)	8 000
Interest rate on loan 10% per year	

Current rates of production are 120 per week for A and 80 per week for B. There is no production of C. The total weekly variable costs are £300 for A and £440 for B. Sales price for A is £100, sales price for B is £110.

(1) Find
  (a) the unit variable cost for each of A and B;
  (b) the unit contribution for each of A and B;
  (c) the current total weekly profit.
(2) Assuming that the fixed costs are distributed to A, B and C in proportion to their turnover (current rate of production × current sales price), determine the break-even number per week for A and B.
(3) If product C is introduced at 100 units per week, to sell at £150 each, and production rates for A and B are halved, find the new weekly profit. Use a unit variable cost of £15 for C. Should product C be introduced in this way?
(4) The directors decide to concentrate solely on product C. Find values of the weekly break-even number for C for a sales price ranging from £50 to £150.

Solution

(1)(a) Unit variable cost for A $= \dfrac{300}{120} = 2.50$

Unit variable cost for B $= \dfrac{440}{80} = 5.50$

(1)(b) Unit contribution for A $= 100 - 2.50 = 97.50$

Unit contribution for B $= 110 - 5.50 = 104.50$

(1)(c) Total weekly profit $= 120 \times 97.5 + 80 \times 104.5 - F$

where $F =$ fixed costs per week

$$= 7000 + 100 + 100 + 600 + \frac{(12\,000 + 8000 + 800)}{52} = 8200$$

therefore total weekly profit $=$ £11 860

(2) Distributed fixed costs: A $= \dfrac{(120 \times 100)}{(120 \times 100 + 80 \times 110)} \times 8200$

$$= 4730.8$$

therefore fixed costs: B $= 8200 - 4730.8 = 3469.2$

and

break-even no. for A $= \dfrac{4730.8}{97.5} = 48.5$

break-even no. for B $= \dfrac{3469.2}{104.5} = 33.2$

Since the company will be unable to sell a fraction of a product, the break-even numbers should be rounded upwards to 49 no. for Product A and 34 no. for Product B.

(3) New weekly profit

$$= 60 \times 97.5 + 40 \times 104.5 + 100 \times 135 - 8200 = 15\,330$$

Since profitability has increased substantially, Product C should be introduced in this manner.

(4)   Using the formula $n = \dfrac{F}{(S - V)}$

where $F = 8200$ and $V = 15$, the values shown in Table 13.1 are obtained. These values have been plotted in Figure 13.5.

Table 13.1   Break-even values for Product C

S	50	60	70	80	90	100	110	120	130	140	150
N	234	182	149	126	109	97	86	78	71	66	61

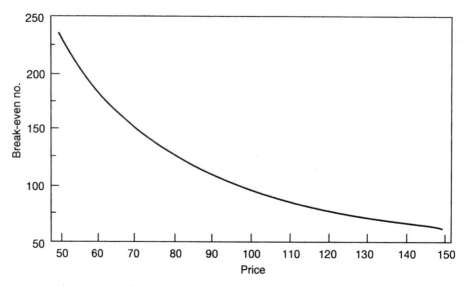

Figure 13.5   Break-even curve for Product C

Example 13(6)

A company is planning to produce and sell widgets. Estimated fixed costs are £9000 per week; variable costs are estimated as follows:

£20 each for the first 400 widgets
£40 each for the next 100 widgets
£80 each for the next 100 widgets

A profit margin of at least 10% is desired.
    For the weekly production rates below, determine an appropriate selling price per widget.

100   200   300   400   500   600

Solution

In this case it is probably easier to apply some common sense rather than resorting to formulae.

For 100 widgets:

Total cost $= 9000 + 100 \times 20 = 11\,000$
Therefore cost per widget $= 11\,000/100 = 110$
Therefore sales price per widget $= 1.10 \times 110 = £121.00$
Similarly, for 200, 300 and 400 widgets, prices of £71.50, £55.00 and £46.75 are obtained.

For 500 widgets:

Total cost $= 9000 + 400 \times 20 + 100 \times 40 = 21\,000$
Therefore cost per widget $= 21\,000/500 = 42$
Therefore sales price per widget $= 1.10 \times 42 = £46.20$

For 600 widgets:

Total cost $= 9000 + 400 \times 20 + 100 \times 40 + 100 \times 80 = 29\,000$
Therefore cost per widget $= 29\,000/600 = £48.33$

From the foregoing, the lowest price coincides with sales and production of 500 widgets per week. A problem similar to this is developed a little further under section 13.6.

13.4   BUDGETS

A budget is a cost plan. Government departments will work to an annual budget. In the private sector annual cost plans may have to be presented to boards of directors for their approval. For example, a research and development (R&D) department of a company will usually have to obtain approval in advance for its operations; a major element of the departmental plan will be its budget. Even where programmes are instituted to run for several years, the cost plan, when submitted for approval, will break the work down into a series of annual budgets. The purpose of presenting a budget is not merely to obtain approval. Once the programme of work is underway it will be used as the basis for monitoring costs and, possibly, for moving funds within the cost plan (if this is permitted).

An example of a budget for a three-year research programme is given in Table 13.2.

13.5   ESTIMATING, TENDERING AND PRICING

There are two determinants of price, *cost* and the *market*.[2] If the price is too high the product will not find a market and either production must cease or costs must be reduced.

---

[2]Economists might argue that there is only one determinant of price: the market. However, costs can be considered a determinant of price for two reasons. (1) In markets with few buyers and/or few suppliers the costs of supply are sometimes used as a basis for negotiating prices. (2) Suppliers can only remain in the market if prices exceed their costs in the long term.

Table 13.2   Budget for research programme

Details	Year 1	Year 2	Year 3	Year 4
Personnel nos.				
Researcher	2	4	4	2
Technician	1	2	2	1
Secretarial	0.5	0.5	0.5	0.5
Salaries				
@ 21 000	42 000	84 000	84 000	42 000
@ 18 000	18 000	36 000	36 000	18 000
@ 14 000	7 000	7 000	7 000	7 000
Consumables	1 000	2 000	2 000	1 000
Travel and subsistence	1 000	2 000	2 000	1 000
Other costs	500	1 000	1 000	500
Equipment	2 000	500	500	500
Total costs	71 500	132 500	132 500	70 000
Grant income	39 300	73 200	73 200	38 700
Net surplus/(deficit)	(32 200)	(59 300)	(59 300)	(31 300)

13.5.1   Tendering

Frequently suppliers and contractors are invited to tender for work. In these circumstances cost will be the major determinant of the tender price. The tender period even for a contract valued at some £100m may be only a few weeks. The suppliers/contractors may not know who else is competing for the work. In these circumstances the tenderer must, in the first instance, rely on its cost records. Therefore it is essential that companies maintain up-to-date cost records at all times. The basis of the tender will be the standard costs kept by the company. When pricing the work the manager will take the standard costs and add what is often referred to as 'mark-up' to cover overhead and profit. This will be at the discretion of the manager or even his or her board of directors. When work is in short supply the mark-up will be low, when the company has plenty of work and/or the risk attached to the job is high the mark-up will be high. When competition for the work is high and the company is anxious to be awarded the contract, the mark-up will be low.

13.5.2   Estimating with inflation

In order to price a product, a service, a specific order or a contract, the costs of that product, service, order or contract must be estimated. The estimator must be able to convert the recorded standard costs into future costs. Cost price indices, giving figures for materials and labour, are published for different industrial sectors and for the national economy. The most commonly quoted index in the UK is the Retail Price Index (RPI)[3] published monthly by the Central Statistical Office. This is based on a range of consumer goods and services, the prices of which are surveyed, averaged across the country and weighted according to a formula. The range of goods and services and the formula weighting are adjusted at intervals

---

[3] See also footnote on page 317.

of about five or ten years to take into account changes in consumer behaviour. The RPI is used by government and economists to measure the rate of inflation in the UK economy (see also Chapter 15, section 15.9). Price indices can be used to estimate current costs from past costs.

$$\text{estimated current costs} = \text{past costs} \times \frac{I_c}{I_p}$$

where I_c = current index value
       I_p = index value at time costs were recorded

The estimator may then have to add a further percentage to account for future cost changes during the period of manufacture, installation or construction. During periods of low inflation fixed-price contracts (contracts that provide for no change in contract prices as costs of labour and materials change) are common for contracts of up to two years in duration. For longer contracts, fluctuation clauses can be introduced that permit price changes in accordance with a formula associated with an accepted price index (see also Chapter 17, section 17.4.4). In these circumstances the estimator need only consider current costs in his or her estimate.

13.5.3 Approximate estimates

Estimating is not only carried out by manufacturers and contractors. The recipients of these goods and services will often produce estimates for their own use. Usually the bases for these estimates will be very different from the bases used by the supplier. The bases used by the customer may include:

(1)   Comparison with previous prices paid for similar goods and services.
(2)   Carrying out a survey of current (published) market prices.
(3)   Using an appropriate industrial formula.

An investor in a new power station, or theatre or office building might adopt method (1). An investor in a new fleet of buses or in re-equipping an organization with word processors could adopt method (2). An investor in a new petrochemical plant might adopt method (3).

Approximate estimates for buildings

Engineers, scientists and technologists may have occasion to commission a new building as their business expands. Approximate building costs can be quickly established either by making comparisons with the costs of constructing similar buildings or, more commonly, by accessing data from commercial providers who give costs per square metre of floor area for that kind of building. Architects and quantity surveyors will have at their fingertips the costs per square metre of floor area for housing, apartments, commercial office space, warehouses, supermarkets and so on. So, for example, if a new office building of 8000 m^2 is desired at a given location, the architects look up their cost data for that area and might quote a construction price for that form of construction at, say, £4000 per m^2, then the approximate building cost will be £4000 × 8000 = £32 000 000. Industrial buildings are more difficult; the usual method is to determine the cost of the shell of the building, which would be done in exactly the way described above, and then to add in the estimated costs of the plant and equipment using suppliers' quotations and/or past records of purchases of similar plant and equipment. For process plant in the UK, approximate costs can be determined by pricing the plant and equipment and multiplying these costs by an appropriate factor; see *Capital Cost Estimating* by Mark Gerrard [1].

13.5.4 Pricing for the market

In order to sell engineering products in the market, the price must be one that the customers are prepared to pay and, of course, one that compares favourably with the competition. Economists speak of the laws of supply and demand.

As the price of a widely available product falls, more of that product is purchased. This is either a result of existing customers being able to afford more, or a result of new customers buying the product, which previously they could not afford: *demand increases as the price falls*.

Further, the higher the price that customers are willing to pay for a product the greater the incentive for companies to provide that product: *supply increases as the selling price rises*.

These two hypotheses are demonstrated in Figure 13.6. The intersection point of the two curves is known as the point of equilibrium. Market prices will tend towards that point. If, for example, the price rises higher than the equilibrium point price, the number of people buying the product will reduce, which will result in the suppliers being unable to sell their stock. Therefore, production will have to be reduced. At the same time prices may have to be reduced to offload surplus stock and prices and quantities will return to equilibrium. There are many reasons why simple economic theory does not always apply in practice, but these forces are present in any market economy. The interested reader will find *Economics* by Parkin *et al.* [2] a useful text for further study of the laws of supply and demand.

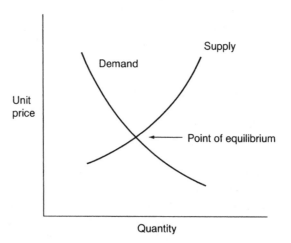

Figure 13.6   Supply and demand curves

An example of how a demand curve can be used in conjunction with break-even is provided on the John Wiley website, at www.wileyeurope.com/go/chelsom.

13.6 EXERCISES

Exercises relating to some of the material in this chapter are provided at www.wileyeurope.com/go/chelsom.

13.7 SUMMARY

This chapter has examined how different costs, direct and indirect, contribute to the total cost of a product and the ways in which these costs can be assessed. The concept of break-even has been examined and the two determinants of price (cost and the market) considered. Methods of estimating have been examined and examples provided. Finally, an example has been provided on the John Wiley website to show how an appropriate market price can be determined using a demand curve and a break-even curve.

REFERENCES

1. Gerrard, M. (2000) *Capital Cost Estimating*, 4th edn, Institute of Chemical Engineers, Rugby.
2. Parkin, M., Powell, M. and Matthews, K. (2002) *Economics: European Edition*, FT Prentice Hall, London.

BIBLIOGRAPHY

Horngren, C. T., Bhimani, A., Datar, S. M. and Foster, G. (2002) *Management and Cost Accounting*, Pearson Education, Harlow.

Curtis, T. (1994) *Business and Marketing for Engineers and Scientists*, McGraw-Hill, Maidenhead.

14

Measuring Financial Performance

14.1 INTRODUCTION

All business managers need to measure the success and performance of their business operations. Engineers, scientists and technologists with managerial responsibilities have exactly the same need. They will want to know how well they have performed relative to their expectations and how well they have performed relative to their competitors.

Some of the most commonly used measures of performance are found in the financial data presented in a set of accounts. It is therefore important for managers to understand the components of a set of accounts, and to be able to analyse and interpret them. In this chapter the basic components are identified and described by examining in detail the *profit and loss account* and the *balance sheet*. The fundamental concepts by which accounts are prepared are briefly outlined. The chapter then examines the *working capital cycle* and shows how data from the primary statements of account can be used by management. Finally, the chapter explains how a number of *financial ratios* can be used to analyse and interpret a business's financial position.

14.2 BUSINESS ENTITIES

Business is conducted in a variety of different forms. The most common types of business operations are sole traders, partnerships and limited companies. Sections 14.2.1 to 14.2.3 will examine how these business entities are constituted.

14.2.1 Sole trader

Sole trader is the term used to describe one person engaged in business. It is the simplest business entity, examples of which include corner shop, dentist, farmer and even consulting engineer or IT consultant. The individual is the owner of the business, although he or she may employ someone to help him or her. The day-to-day affairs of the business will be operated independently of the owner's personal affairs and separate accounts will be drawn up. However, from a legal and taxation point of view, the business is part of the individual's personal assets. His or her liability for business debts is unlimited and all his or her other personal assets (for example, his or her home!) are at stake. On the other hand, he or she is the sole beneficiary of any profits earned by the business.

14.2.2 Partnerships

Partnerships are an extension of the sole trader entity, comprising two or more people engaged in business together. Each partner will have invested capital in the business and will be entitled to a share of the profits. The business therefore has two or more owners. Partnerships are a common form of business entity in the professions. Many small firms of consultants are partnerships.

The share of the profits payable to each partner is normally proportional to the amount of capital each has invested in the business. For example, if three partners contribute capital in the proportions 3:2:1, it is likely that the profits will be shared in the proportions 3:2:1.

In the UK, partnerships are governed by the Partnership Act 1890, which states that partners' liability for business debts is joint and several. This means that the partners are responsible for the debts of the business jointly as a body of partners and also individually (severally). Their liability for the debts of the business is the same as that of a sole trader in that their personal assets may be used to settle the liabilities of the business. In addition, a partner's personal assets may be used to settle a debt incurred by the other partners in their business operations if the debt cannot be settled from the funds of the business.

14.2.3 Limited companies

A limited company is often referred to as a corporation. The term 'incorporation' will be explained later. The owners of a limited company are called shareholders. They subscribe capital to the company and in return are issued with shares in the company. Limited companies can be private or public. The difference between the two is that the shares of a public limited company are traded on a stock exchange. Most large engineering businesses and commercial laboratories are limited companies, although not all are public companies.

The term 'limited' is used because the liability of the shareholders is limited. This means that, in the event of the company being unable to settle its debts, the shareholders only stand to lose what they have contributed by way of share capital. Their personal wealth is not at stake. Contrast this with the position of the sole trader or partnership, where the individual stands to lose everything. This is probably the most significant advantage of incorporation.

Incorporated companies or corporations are so called because they are regarded in law as separate legal entities that are distinct from their owners, the shareholders. A corporation can enter into contracts in its own right; that is, a contract would be between ABC Company Limited and XYZ Company Limited rather than between the individual shareholders.

The day-to-day management of a company is carried out by the directors of the company. In smaller private companies the directors and shareholders may, in fact, be the same people. However, in the large public limited companies (plcs in the UK) management is delegated by the shareholders to a board of directors. This is a further element in the separation of the owners of the company from the company itself. The directors have a duty to act in the best interests of the company at all times and their performance is, of course, reviewed by the shareholders at regular intervals. One of the ways in which the shareholders of a company can appraise the performance of its board of directors is by reviewing the annual financial accounts. The accounts are therefore very important, and this explains why there can be so much pressure on boards of directors to show a healthy financial position.

Businesses that operate as limited companies are subject to much more accounting regulation than partnerships and sole traders. There are obviously costs associated with this degree of regulation and careful consideration must be given by the owners of a business as to which type of entity would be the most suitable.

Sole traders, partnerships and companies are also bound by laws covering, for example:

- health and safety at work;
- public liability;
- employment of individuals;
- taxation;
- ownership of land and property.

14.3   THE FUNDAMENTAL BASES ON WHICH ACCOUNTS ARE PREPARED

Although the different types of business entity are subject to different degrees of regulation, from an accounting point of view the basic requirements for the different entities have much in common. The level of detail required in a set of financial accounts will be different for each entity, but the methods of preparation and the fundamental accounting concepts are the same. Hence, throughout the remainder of this chapter the term business or company is equally applicable to all types of entity.

14.3.1   Fundamental accounting concepts

All accounts should be prepared with the following fundamental concepts in mind.

The matching principle

Income and costs should be recorded in the accounts of a business in the accounting period in which they arise. This may not necessarily be the same period as that in which the cash is received or paid. In addition, the income and costs should be matched with one another so far as a relationship between them can be established. This concept is discussed more fully in section 14.5.1 when the profit and loss account is examined.

The going concern concept

The assumption is made that the business will continue in operational existence for the foreseeable future unless information is provided to the contrary. There is the further assumption that there will be no significant reduction in the scale of the operation of the company and that debts will continue to be paid when they fall due.

Consistency

This provides for a consistent accounting treatment of items that are similar within an accounting period and a consistent treatment of items from one accounting period to the next. For example, there is no reason to treat sales to Company A any differently from sales to Company B, and one would expect the manner in which these sales are entered in the accounts to remain the same from one period to the next.

The prudence concept

This means that accounts should always be prepared exercising a degree of caution. Profits should not appear in the accounts before they are earned, while losses should be included as soon as it is known that they will occur.

For example, if a company is engaged in a long-term contract and realizes that the costs are escalating to a much higher level than the income that will be received from the contract, then the extra costs should be included in the accounts for the current accounting period rather than when the contract is completed.

This operation of the prudence concept appears to conflict with the matching principle. Where there is a conflict, the prudence concept should prevail. As a result, the loss in the above example will be recognized as soon as it is anticipated.

Thus, the four fundamental accounting concepts are:

- the matching principle;
- the going concern;
- consistency;
- prudence.

In addition, there are a number of measurement rules that are adopted universally. These are described in the next section.

14.3.2   Measurement rules

Units of measurement

The information included in a set of accounts is expressed in monetary units – pounds, dollars, euros and so on. It is therefore impossible to include the value of certain components of a business in its accounts. For example, the value of a skilled workforce would be difficult to quantify in monetary units. The wages paid to them, on the other hand, can be measured easily in these terms.

Historical cost convention

Transactions should be included in the accounts at their original cost. If a machine were purchased for £9000 three years ago, then it will be recorded in the accounts at a cost of £9000 irrespective of what that machine would cost to buy now (see also section 13.2.1).

Accounting period

The convention is that businesses will draw up accounts for a period of one year for submission to the taxation authorities and for circulation to the shareholders. This period is usually too long for internal management purposes, therefore, depending on the size and nature of the business, internal accounts may be produced on a monthly or quarterly basis.

Business entity

For the production of accounts the business entity must be isolated and defined. For a company this is relatively straightforward, since the operations of the business tend to be remote from the owners of that business. However, with sole traders and partnerships the process of definition and isolation requires scrupulous attention in order to separate the business operations from personal affairs.

14.3.3   Accounting standards

These concepts and measurement rules ensure that there is consistency and comparability in the financial information provided in the accounts by business entities. This consistency and comparability are further enhanced by the rules and guidelines of the accounting professions and by the legal framework, especially with respect to limited companies.

Most countries have developed their own accounting standards. These standards lay down a basis for the methods of accounting to be adopted in dealing with specific items in the accounts. For example, there is a standard that sets out the basis for valuing and recording stocks.

The International Accounting Standards Committee (IASC) has also developed its own set of accounting standards in order to provide for a reasonable degree of comparability between practices in different countries.

While individual countries tend to draft their own standards to be generally in line with those of the IASC, there are still differences in accounting methods that must be considered when comparing company accounts in different countries.

All standards, irrespective of the country concerned, are drafted to give general guidelines for accounting treatment. Indeed, some guidelines present a number of alternative methods from which to choose. They are, therefore, open to interpretation and allow scope for professional judgement in setting the accounting policies of a company.

14.3.4   Summarizing fundamental practices

In summary, different sets of accounts will always be comparable and will have a lot in common. Accounting standards and the legal framework will ensure that this is so. However, it is important to be aware that accounts will be prepared and interpreted in different ways. This can have a marked effect on the apparent financial position of a company and this should be taken into account when comparing one company with another, or in appraising a company in which one may wish to invest.

14.4   PRESENTATION OF FINANCIAL INFORMATION

The two most commonly used statements of account are *financial accounts* and *management accounts*. The differences between these are examined below.

14.4.1   Financial accounts

The vast majority of companies in the UK are required by law to produce a set of financial accounts on an annual basis. The contents of these accounts are to a large extent determined by the Companies Act 1985 and by Accounting Standards. The accounts are produced primarily for the shareholders, but will also be put on public record and subjected to independent scrutiny by the company's auditors. The readership is thus very wide and a great deal of care is taken to ensure that the financial position of the company is presented in the most accurate and most favourable manner.

As the contents are governed by the regulatory framework, the presentation of published financial statements from different companies will have a lot in common. Each will contain a balance sheet, a profit and loss account, a statement of the accounting policies adopted and notes providing a further analysis of the data provided.

14.4.2   Management accounts

Management accounts, on the other hand, are internal documents. They may be widely read within the company or even by potential investors, but they are not available to the general public. Good management accounts are a very necessary and effective management tool for even the smallest of enterprises. They tend to be produced on a monthly basis and are, therefore, much more up to date than financial accounts.

The level of detail varies greatly from company to company. Typically, management accounts will show the results for the month and the cumulative results for the year to date. These results are then compared with the budget for the year and with the actual results from the previous year, as illustrated in Table 14.1. It is also likely that separate figures will be prepared for each division of the company or even for each product line if the company spends large amounts of money manufacturing a small number of products.

The company's performance can now be assessed on a monthly basis and compared to the budgets set for the different divisions at the beginning of the year. Comparisons can also be made to the previous year's performance. Large variances will lead to investigation and corrective action taken sooner rather than later. In Table 14.1, the YTD variance for sales on the previous year is substantial but positive and therefore beneficial; this seems to have resulted in a considerable increase in labour costs in the past year, but this was expected, as the budget indicates. There has also been a large increase in administrative salaries (possibly something to be watched for the future) but, again, this was anticipated in the budget. These accounts would not give the management serious cause for concern.

The illustration for Division D of the Campbell Engineering Company shows the current trading position. Management accounts may also include a balance

Table 14.1 Campbell Engineering Company, Division D. April Management Accounts. Trading and Profit and Loss Account

	Actual		Budget		Prior year		Variance YTD**	
	Month	YTD	Month	YTD	Month	YTD	Budget	Prior year
	£'000	£'000	£'000	£'000	£'000	£'000	£'000	£'000
Sales	2560	11 500	2600	11 000	2050	9500	500	2000
Direct costs								
Materials	1105	5050	1200	4800	980	4320	250	730
Labour	770	3660	775	3500	515	2740	160	920
Power	130	390	100	400	140	550	(10)*	(160)
Transport	95	350	85	340	95	400	10	(50)
Gross profit	460	2050	440	1960	320	1490	90	560
Indirect costs								
Selling and marketing	27	208	28	112	30	150	96	58
Admin. salaries	115	460	110	440	90	220	20	240
Rent and rates	58	232	58	232	50	210	–	22
Net profit	260	1150	244	1176	150	910	(26)	240

*negative values are shown (. . .).
**YTD = year to date.

sheet and a cash-flow projection. The latter may be particularly important for companies that have loans to repay or an overdraft facility. The management of working capital will be discussed in more detail later; at this stage it is necessary to understand that it would be almost impossible to manage working capital without management accounts.

14.5   THE PRIMARY STATEMENTS

It should be clear at this stage that businesses prepare accounts for a number of reasons and that there are considerable differences in the ways in which financial information can be presented. However, any set of accounts should always contain a *profit and loss account* and a *balance sheet*. These two primary statements will be examined in detail below.

14.5.1   Profit and loss account

The profit and loss account shows the financial results for the company for a particular period of time, for example one year. It records the total sales made for the period and the costs incurred over the same period. Sales less costs results in either a profit or a loss. This profit or loss is the company's financial result. An example of a profit and loss account is shown in Table 14.2.

Table 14.2 Roudsari Engineering Limited
Profit and Loss Account for the year ended 30 June 2004

	£'000	£'000
Sales		900
Cost of sales		(450)
Gross profit		450
Overhead expenses	160	
Finance charges	40	
Depreciation	100	
		(300)
Net operating profit		150
Taxation	40	
Dividends	50	
		(90)
Retained profit for the year		60
Retained profit brought forward		35
Retained profit carried forward[*]		95

[*]carried forward to the balance sheet, see Table 14.4

Working through the profit and loss account:

(1) Sales for the year ended 30 June 2004 will comprise income arising to the company from all goods and/or services sold for the period from 1 July 2003 to 30 June 2004.

(2) Costs of sales, as the description implies, includes all costs directly attributable to the sales made in the period. Examples of these costs would include production materials, labour, power, storage and delivery costs.

(3) Sales less these costs results in the gross profit. To obtain a true measure of the gross profit for the period it is most important to include all direct costs incurred in making the sales. This is the matching principle discussed in section 14.3.1. The cost of sales figure is therefore calculated as shown in Table 14.3.

Opening stocks plus purchases made in the year less closing stocks equals the cost of raw materials for the sales made in that year. Opening stocks that were purchased *last* year form part of *this* year's costs, but closing stocks purchased this year do not. This makes sense because the opening stocks at the beginning of the year were sold this year and are included in sales, while this year's closing stocks will be part of next year's sales.

(4) From gross profit other expenses are deducted in arriving at the net profit retained for the year. These include overheads, finance charges, taxation and dividends. Taxation is payable on the net operating profit earned, to the Inland Revenue in the UK or its equivalent elsewhere. Finance charges would normally mean interest payments to banks and other financial institutions. Dividends are payments made to the shareholders of the company. They represent a return on the capital that they have invested in the business.

(5) The final profit retained for the year is added to the profits retained in previous years and carried forward to the balance sheet (see section 14.5.3).

Table 14.3   Roudsari Engineering Limited
Trading Account for the year ended 30 June 2004

	£'000	£'000
Sales		900
less: *Cost of sales*		
Opening stocks	42	
Purchase of raw materials	260	
	302	
less closing stocks	(54)	
	248	
Direct labour costs	180	
Storage costs	12	
Delivery costs	10	
		450
Gross profit		450

The majority of companies do not distribute all of their profits to the shareholders. Normally they will retain some of the profit within the business. Retained profits will enable the business to grow, by reinvesting in capital equipment, in new product lines or in employing extra people. Without such reinvestment a business would be unlikely to grow significantly or keep up with technological change.

14.5.2   Capital and revenue expenditure

Before examining the balance sheet (section 14.5.3), the differences between capital and revenue expenditure will be explained.

Capital expenditure is accounted for in the balance sheet and includes purchases of plant and machinery, computers and premises. Capital expenditure is defined as providing a benefit to the business over more than one accounting period. It is, of course, reasonable to apportion these costs over the periods of their use and hence match these costs with the sales generated in each period. This is what is done, and the process is called *depreciation*. Depreciation is examined later in the chapter.

The benefits of revenue expenditure, on the other hand, are normally realized in one accounting period; the payment of rent on the company's premises will be an annual charge for an annual benefit and the sum will be accounted for in the year in which it is levied. Revenue expenditure is accounted for in the profit and loss account as a charge against sales.

14.5.3   The balance sheet

The balance sheet is a statement of the company's assets and its liabilities at a particular point in time. It provides a snapshot of the financial position and is usually drawn up at the end of the year covered by the profit and loss account. The term *assets* is used to describe what the business owns and *liabilities* means what the business owes to others.

Table 14.4   Roudsari Engineering Limited
Balance Sheet at 30 June 2004

	£'000	£'000
Fixed assets		
Tangible fixed assets		700
Intangible fixed assets		20
		720
Current assets		
Stocks	54	
Trade debtors	306	
Other debtors	40	
Prepayments	20	
Cash at bank and in hand	170	
	590	
Current liabilities		
Trade creditors	(175)	
Other creditors (dividends + taxation)	(90)	
Accruals	(25)	
	(290)	
Net current assets		300
		1020
Long term liabilities		
Bank loan		(175)
Net assets		845
Represented by:		
Share capital		750
Profit and loss account		95
		845

The fundamental equation of any balance sheet is

$$\text{total assets} = \text{total liabilities}$$

The assets are balanced by the liabilities, hence the term balance sheet. A typical balance sheet is illustrated in Table 14.4.

The equation can be expanded to

$$\text{total assets} = \text{liabilities to proprietors} + \text{external liabilities}$$

The capital invested in the business by its proprietors, whether they be sole traders, partners or shareholders, represents long-term loans to these businesses and is, therefore, a liability.

The balance sheet shows how the various categories of assets and liabilities are set out. If this balance sheet were to be represented by an equation it would be as follows:

$$\text{total assets } \textit{less} \text{ external liabilities} = \text{liabilities to proprietors}$$

Table 14.4 is typical of how balance sheets are drawn up. Note that the top half and the bottom half balance. Effectively this balance sheet shows how the proprietors' investment in the business is represented by assets and external liabilities.

Assets

Let us look first at the assets forming part of the equation above,

$$\text{total assets} = \text{fixed assets} + \text{current assets}$$

Fixed assets

Fixed assets are those assets that are held in the business for the long term and are used to generate income over a number of accounting periods. These assets would not normally be resold in the course of trade. As noted in section 14.5.2, they represent the capital expenditure of the business. Most fixed assets are tangible assets. Examples of tangibles would be land, buildings, plant, machinery, computers and office furniture. Intangible assets include goodwill, patents, trademarks and so on.

Fixed assets are recorded in the balance sheet at historic cost less depreciation. Historic cost generally equates with the purchase price of the asset, however long ago.

Depreciation

It was stated above that fixed assets are recorded at historic cost less depreciation in the balance sheet. This is referred to as *net book value*. In section 14.5.2 it was noted that an annual charge is made to the profit and loss account in respect of depreciation.

All fixed assets have a finite useful economic life and there will come a time when these assets have to be replaced. Examples of this would include photocopiers, concrete mixers and lathes, as well as the very buildings that house them. Depreciation is a measure of the gradual 'wearing out' of these assets. By making an annual charge to the profit and loss account, the level of profits available for distribution to shareholders is reduced and the money retained within the business can be used for replacements.

In calculating the annual depreciation charge three factors are taken into account:

- the cost of the asset;
- the useful economic life;
- the estimated residual value.

The only one of these factors that can be stated with any degree of certainty is the cost of the asset. The other two factors involve judgement. This exercise of judgement can lead to very different results and hence it is important to note what depreciation policies a company is adopting so that the impact on declared profits can be appreciated. The effects of exercising judgement are best illustrated by an example.

Example 14(1)

Suppose that two companies purchase identical machines, and the purchase price of each is £81 000.

Company A estimates that the machine will last for eight years and that the residual value will be £1000.

Company B estimates that the machine will only last for five years and that its residual value will be £6000.

The annual depreciation charge for Company A is as follows:

	£
Cost	81 000
less Residual value	(1000)
	80 000

Spread over the machine's economic life, this gives £10 000 p.a. for eight years.

The annual depreciation charge for Company B is as follows:

	£
Cost	81 000
less Residual value	(6000)
	75 000

Spread over the machine's economic life, this gives £15 000 p.a. for five years.

Company B has an additional £5000 charged against its profits each year and is likely to be reporting lower profits than Company A. This method is called the *straight line method* because there is an equal annual charge to the profit and loss account.

An equally common method is the *reducing balance method*. With this method the depreciation charged in the early years of an asset's life is higher than in the latter years. The depreciation charged reduces in each successive year. This method attempts to recognize that the value of most machinery and equipment falls rapidly in the first years of use and then declines at a much slower rate, and that as time passes the costs of repairs and maintenance of the asset increase. Depreciation on a straight line basis plus increasing costs of repairs would lead to an escalating charge in the profit and loss account over the life of the asset. Where this might occur, the use of the reducing balance method is considered more appropriate. The method is often used for the depreciation of motor vehicles. Similarly, special tooling that has a useful life tied to a particular product is normally written off over a short period using the reducing balance method. This is consistent with the prudence concept described in section 14.3.1.

There are two alternative methods for determining the depreciation charge on the reducing balance basis. These are shown in Examples 14(2) and 14(3).

Example 14(2)

A company purchases a motor vehicle for £13 000. The directors estimate that the vehicle will be used in the business for four years, at the end of which time it will be sold for £2500.

$$\text{annual rate of depreciation} = 1 - \left(\frac{2500}{13\,000}\right)^{1/4} = 33.8\%$$

Charge in Year 1 = 0.338 × 13 000	= 4391
Charge in Year 2 = 0.338 × (13 000 − 4391)	= 2908
Charge in Year 3 = 0.338 × (13 000 − 4391 − 2908)	= 1926
Charge in Year 4 = 0.338 × (13 000 − 4391 − 2908 − 1926)	= 1275
Total depreciation charge over four years	£10 500

Example 14(3)

For the same conditions as in Example 14(2), the sum-of-the-digits method (widely used in the USA) gives the following charges.

$$\text{Charge for Year 1} = (13\,000 - 2500) \times \frac{4}{(4 + 3 + 2 + 1)} = 4200$$

$$\text{Charge for Year 2} = (13\,000 - 2500) \times \frac{3}{(4 + 3 + 2 + 1)} = 3150$$

$$\text{Charge for Year 3} = (13\,000 - 2500) \times \frac{2}{(4 + 3 + 2 + 1)} = 2100$$

$$\text{Charge for Year 4} = (13\,000 - 2500) \times \frac{1}{(4 + 3 + 2 + 1)} = 1050$$

Total depreciation charge over four years

$$= 4200 + 3150 + 2100 + 1050 = 10\,500$$

Figure 14.1 shows the net book value of the vehicle against time for the two methods above and for the straight line depreciation method.

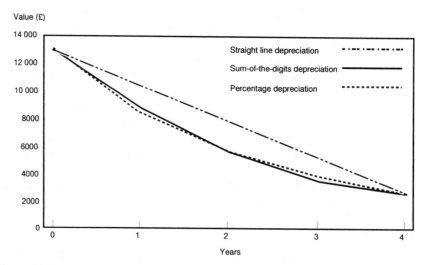

Figure 14.1

A third method of depreciation is the *usage method*. In this method the annual depreciation charge is calculated taking into account the estimated number of units that the asset (machine) will produce in its working life, or the number of hours or miles for which the asset will be productive. This method is commonly used in the depreciation of commercial aircraft.

Example 14(4)

A company purchases an aeroplane for £6.6m. The manufacturer has advised the directors that the plane will fly for up to 40 000 hours during its useful economic life.

The directors estimate that the flying hours over the next three years will be as follows:

	Hours
Year 1	5800
Year 2	4700
Year 3	5200

Residual value is estimated at £1.2m.

$$\text{Charge for Year 1} = (6.6 - 1.2) \times \frac{5800}{40\,000} = £0.783\text{m}$$

$$\text{Charge for Year 2} = (6.6 - 1.2) \times \frac{4700}{40\,000} = £0.635\text{m}$$

$$\text{Charge for Year 3} = (6.6 - 1.2) \times \frac{5200}{40\,000} = £0.702\text{m}$$

By this means, the depreciation charges will reflect the sales being generated by the machine, or aircraft, as in this case. This should provide a better measure of the profitability of operating the asset.

The depreciation policy adopted by a company should be the one that is most appropriate for the type of asset and its use, so that depreciation charges are allocated fairly across accounting periods. The original estimates of the useful economic life of each asset should be periodically reviewed to make sure that, as time passes, they are still appropriate. This is especially important for companies that use equipment at the forefront of technological development, which can, so quickly, become obsolete and have its economic life sharply curtailed.

Net book value

Returning to the net book value of an asset, this is calculated as the asset's cost less the accumulated annual depreciation charge. Consider the aircraft in Example 14(4). These net book values would be recorded in the balance sheet at the end of Year 1, 2 and 3 respectively, as shown in Table 14.5.

Current assets

Current assets are assets that are retained within the business for a relatively short period, usually for less than one year. They are readily convertible into

Table 14.5

	Cost £'000	Cumulative depreciation £'000	Net book value £'000
Year 1	6600	783	5817
Year 2	6600	1418	5182
Year 3	6600	2120	4480

cash at short notice. Current assets on the balance sheet would usually include the following:

(1) *Cash* in hand and in bank accounts.
(2) *Debtors*; that is, trade debtors and other debtors. Practically all businesses, with the exception of retailers, make their sales on credit. There is therefore a time lag between making a sale to the customer and receiving the cash. During this period the customer becomes a trade debtor. The asset in the books of the company is in recognition of the right to acquire cash from the customer at a later date. Other debtors arise from nontrading transactions, for example the sale of a fixed asset at the end of its useful economic life.
(3) *Stocks*: Stocks comprise raw materials, work in progress and finished goods held for resale. Stocks will be converted into sales and then into cash in the short term. Repair, maintenance and operating supplies, which may not be converted immediately into sales, are also stocks and are recorded as current assets. Alternative methods of valuing stocks are considered in Chapter 13, section 13.2.1.
(4) *Prepayments*: Prepayments are expenses that have been paid by the business in advance and do not relate to the current accounting period. Rent is generally paid in advance and recorded as a prepayment.

Liabilities

Liabilities represent the amounts that a business owes to other parties, both external parties and the proprietors. As with assets, liabilities are subdivided according to a time frame. The parties to whom these amounts are due are known as the creditors of the company.

$$\text{total liabilities} = \text{long-term liabilities} + \text{current liabilities}$$

Long-term liabilities

These liabilities are due for payment one year or more from the date of the balance sheet. Examples would be bank loans or hire purchase contracts. Long-term loans generally require security in the form of a mortgage over the company's assets.

Liabilities to proprietors are also long-term liabilities, but they are not secured on the assets of the company. These liabilities are normally recorded in the lower section of the balance sheet. Their composition depends on the type of business.

For partnerships and sole traders, the proprietors' liability is represented by a capital account (one for each partner in the case of a partnership). This would include the capital introduced into the business at its establishment plus all retained profits to date.

In the case of a company, the liability is represented by share capital plus retained profits. Share capital is also capital introduced by the owners to establish the business.

Any capital subsequently introduced would be added to the capital account for sole traders or partnerships or would increase the share capital of a company.

Current liabilities

These liabilities are due for payment within one year of the date of the balance sheet. Current liabilities include the following:

(1) *Trade creditors*: These are amounts owed to other parties for goods or services supplied to the company. For example, a manufacturing company will purchase its raw materials from a number of suppliers on credit terms.

(2) *Other creditors*: These include amounts due for taxation and dividends payable to shareholders, all payable within the year.

(3) *Bank overdraft*: Overdrafts provide a form of short-term finance for companies. They are arranged in advance with a company's bank to allow for short-term cash shortages. (Figure 17.8 in Chapter 17 could be used by a contractor and his or her bank to plan the overdraft provision for the duration of a project.) Technically, bank overdrafts are repayable on demand and must therefore be classified as a current liability.

(4) *Accruals*: Accruals represent a provision in the accounts for expenses relating to the current accounting period for which an invoice has not yet been received at the time of drawing up the accounts. An example of this is an electricity bill. If the meter is read on the last day of the accounting period there will be a time lag before an invoice is raised by the electricity company and despatched to the user. Where the bill is expected to be substantial, it would be sensible for the user to make provision and this is how it is done.

All the major components of a profit and loss account and a balance sheet have now been covered. There will sometimes be other items that have not been discussed in this chapter. The guiding principle is to determine whether an item represents income/costs or an asset/liability. If it is one of the former it should appear in the profit and loss account, if either of the latter it should appear in the balance sheet.

14.6 BOOK-KEEPING

The profit and loss account and the balance sheet present only a brief summary of the trading performance of a company and an outline of its assets and liabilities. The two statements are in fact a summary of every individual transaction the company has entered into. The process whereby these individual transactions are recorded and then summarized into profit and loss account and balance sheet is called *book-keeping*. Without this systematic recording of every transaction, it would be impossible for a company to function and to produce satisfactory results.

Book-keeping involves a process known as double entry, whereby every transaction is represented by a debit and a corresponding credit. Detailed instruction in double entry bookkeeping can be found in Whitehead [1].

Businesses spend large sums of money employing accountants and teams of accounting staff to maintain the accounting records. Indeed, all businesses are subject to some degree of statutory regulation, which dictates that accounting records be maintained. The taxation authorities, for example, are concerned that businesses are able to produce accounts that will form the basis for assessing how much they should be taxed.

It is also the case that the owners of a company will want to ensure that a proper account of their enterprise and investment is made available to them at regular intervals. As a result, complex financial reporting systems have to be put in place to safeguard the shareholders' investment. Good financial systems will help to ensure that:

- all stocks sold to customers are recorded as a sale;
- all cash received by the company is banked;
- all fixed assets purchased are secure and can be located;
- staff do not receive pay for more hours than they have worked;
- the possibilities for fraud and other financial loss are minimized;

and so on.

14.7 THE DIFFERENCES BETWEEN CASH AND PROFIT

Before moving on from the detail of the primary statements of accounts, consideration needs to be given to the difference between the cash and profits generated by a business.

Over the entire life of a business, total profits will equate to the total increase in cash balances. However, in individual accounting periods cash generated will not equate with profits earned. Consider the sale of a unit of production. As soon as the item is sold to a customer the profit earned on that sale can be recognized in the profit and loss account, but it is unlikely that the cash will be received on that date. As already observed, customers may take some time to settle up, so the corresponding movement in the cash balance will not occur in the same period. In addition, the company may not yet have paid the supplier for the materials used to produce the unit sold.

Example 14(5)

A company sells 10 units for £250 per unit during September 2004. The cost of producing each unit during August is £180, comprising £80 of labour costs and £100 of raw materials.

The company's cash balance on 1 August 2004 is £9800. The wages are paid on 30 August 2004 and the supplier is paid on 30 September 2004. The customer pays for the goods in October 2004. The company's cash balance moves as shown in Table 14.6.

Table 14.6

Month	Transaction receipt/payment	£	Cash balance at end of month £	Cumulative movement in cash balance £
August	Opening balance	–	9800	–
August	Wages paid	(800)	9000	(800)
September	Supplier paid	(1000)	8000	(1800)
October	Customer pays	2500	10 500	700

The profit on the sale of £700 is recorded in the profit and loss account for September 2004, which has a beneficial effect on the company's result in that month. On the other hand, the cash balance has reduced by £1800 by September 2004, so far as this transaction is concerned. It is not until October 2004 that the profit and cash movements equalize.

Profit is the most common measure of the success of a business, particularly by external parties. However, the generation of cash surpluses, even in the short term, must not be overlooked. Some very profitable businesses fail because they are so concerned with increasing sales and profits that they run out of cash! This situation can arise as a result of overtrading and frequently results from poor management of the working capital cycle.

14.8   THE WORKING CAPITAL CYCLE

The capital introduced into a business can be used to provide fixed assets or working capital. Working capital is essentially the current assets and current liabilities of the business. It comprises cash, stocks, debtors and creditors. These assets and liabilities are short term and are being continually converted from one into the other. The process is called the working capital cycle and can be represented as in Figure 14.2. The diagram shows the cycle for a manufacturing organization, but the cycle is equally important for consultancies and service providers. Such organizations incur costs, such as salaries, travel, telecommunications, rates (local authority taxation) and rent, which are not recovered until the invoices for the delivered advice or service are paid.

Starting at the top of the cycle and moving clockwise, the following occurs. A liability is incurred with the trade creditors of the business in purchasing raw materials. The raw materials are then processed to produce the finished goods that the company will place in stock until an order is received from a customer. When a sale is made on credit to a customer a trade debtor is created. When the

Figure 14.2   Working capital cycle

customer pays, the trade debtor is replaced with cash that can be used to pay off the trade creditor. The cycle then repeats itself.

A business must have enough cash to pay suppliers, wages and overheads when they are due (Figure 1.6 in Chapter 1 shows the cash flows that must be funded before payments are received from customers). At the same time there must be sufficient stocks on hand to be able to meet the demand from customers as and when orders arise. (In the case of service companies, such as consultants, the only 'stocks' might be stationery and computer consumables. These stocks have to be funded, and suppliers paid, in the same way as in any manufacturing company.) If the supply of working capital were unlimited, these requirements could be met without difficulty. However, a business never has an unlimited supply of working capital. The key to meeting both the need for cash and the need for sufficient stocks is to strike a balance between making profits and retaining enough free cash.

Accountants will give the following advice on the process of managing working capital:

(1)   Do not allow stock levels to become too high (see Chapter 20):
- valuable cash resources are tied up in stock;
- the stock may become obsolete or perish;
- warehousing costs can be high;
- the interest-earning potential of cash is lost.
(2)   Do not allow surplus cash to sit in a low-interest bank account:
- expand the business with it;
- invest in new plant and machinery;
- invest in a higher interest-earning account or in stocks and shares.
(3)   Try to take the same number of days to pay your suppliers as your customers take to pay you:
- do not act as a source of short-term finance for your customers;
- encourage early settlement by offering discounts for prompt payment;

- manage your debtors – invoice promptly and accurately; follow up on overdue payments; use Pareto analysis[1] to find out the major reasons for late payments.

14.9   INTERPRETING ACCOUNTS

The numbers shown in the profit and loss account and balance sheet, to be of some use to managers and/or owners, need to be analysed and interpreted. In section 14.4.2 the use of management accounts in running a business was discussed. The actual results for the period were compared with a budget set at the beginning of the year and with the results of the previous year. This process of reviewing and analysing results is one of the best ways of highlighting the strengths and weaknesses of a business.

Analyses are also carried out on financial accounts and industry/sector comparisons made. These are of particular importance to the shareholders.

14.10   FINANCIAL RATIOS

Financial ratios are commonly used as a means of analysing and interpreting the financial information provided in accounts. They can be applied to management accounts and to financial accounts, and are used to highlight the strengths and weaknesses in a company's financial position.

Calculation of these ratios is fairly straightforward, as shown below. However, they need to be used with care and there must be a proper basis for comparison. Comparisons of ratios are usually either made within a company to examine performance over different accounting periods or made between companies in the same business sector.

Ratios are calculated on historic data and so do not necessarily indicate what will happen in the future. Nonetheless, financial ratios can indicate potential cash-flow problems or unprofitable business operations that would affect the future of the company.

The most frequently used financial ratios will now be examined.

14.10.1   Operating ratios

Gross profit margin

This measures the percentage of profit earned on every monetary unit of sales made. It is calculated as follows:

$$\text{Gross profit margin} = \frac{\text{gross profit}}{\text{sales}} \times 100\%$$

The margin expected depends on the industry in which a company operates. In the retail industry the margin is likely to lie between 5 and 10%. The lowest

---

[1] In a Pareto analysis the outstanding debts would be listed in order, with the highest debts at the top. The top 20% of debtors will probably account for about 80% of the debt; these are the ones to pursue.

values are particularly true of the supermarket chains, which rely on a very high volume of sales to generate profits. The construction industry, with its relatively high-volume use of materials, shows similar values of gross profit margin. Manufacturing industry on the other hand, which demands a greater investment in capital equipment and in research and development, is characterized by gross profit margins in the range of 10 to 15%.

In any business, however, it is imperative that a gross profit be made in order to meet overheads, finance charges and dividends.

Return on capital employed (ROCE)

ROCE is a very important ratio and is the most commonly used measure of profitability. It is calculated as follows:

$$\text{ROCE} = \frac{\text{profit before interest and tax}}{\text{capital employed}} \times 100\%$$

and

$$\text{Capital employed} = \text{share capital} + \text{profit and loss reserves} + \text{long-term liabilities}[2]$$

The ratio provides the shareholders of the business with a measure of the return on their capital investment. The shareholders will compare this rate of return with, for example, the rate of interest they could obtain from a building society. In addition, if the investment is perceived as high risk, the shareholders will require a much higher return, and if the return is not high enough they might sell their shares to invest elsewhere.

Heavy selling of shares will result in a lower share price, which will improve the return on new shareholdings, but continuing shareholders will experience a fall in the capital value of their investment.

Asset turnover

This ratio measures the ability of a company to generate sales from its capital base. It is calculated as follows:

$$\text{Asset turnover} = \frac{\text{sales}}{\text{capital employed}}$$

If the ratio is low, intensive capital investment is indicated. If the ratio is high, the investment risk will be low.

The ratio has little value when viewed in isolation. It is the trend over successive accounting periods that is important. A rising trend would indicate that the capital invested is beginning to bring about increased sales. The ratio would also provide a useful comparator within a given industrial sector: a more successful company

---

[2]Engineers and technologists, who are often concerned with the management of physical things, should remember that 'capital employed' is also equal to fixed assets plus net current assets, as shown in the balance sheet in Table 14.4, section 14.5.3. The way in which ROCE is affected by various aspects of engineering management is shown in Chapter 21, Figure 21.3.

would expect to show a higher ratio than its competitors. As one would expect, there are considerable differences between sectors, for example:

- Amec plc = 12.7 (construction).
- Pilkington plc = 3.1 (glass manufacturer).
- GUS plc = 2.7 (retail and business services – includes Argos).

Contractors in the construction industry do not need heavy capital investment; for example, they usually hire all the construction plant needed. A manufacturer such as Pilkington has a heavy capital investment in production capacity. Figures are for 2002/2003.

Working capital turnover

This measures how quickly working capital is converted into sales. It is calculated as follows:

$$\text{Working capital turnover} = \frac{\text{sales}}{\text{current assets less current liabilities}}$$

The ratio is likely to vary considerably between different industrial sectors, as illustrated below:

- Amec plc = 15.4 (construction).
- Pilkington plc = 9.6 (glass manufacturer).
- GUS plc = 12.4 (retail and business services).

Figures are for 2002/2003.

In the retail industry the ratio should be relatively high, as indicated in the example above. This is a business where sales and hence cash should be generated quickly from stocks. Efficient contractors in the construction industry provide a service in which their material stocks are kept very low while their usage is high and their monthly income matches their outgoings, so working capital turnover should be high.

14.10.2   Liquidity ratios

Liquidity ratios measure the ability of a company to pay its debts as and when they fall due. The two most important liquidity ratios are the current ratio and the quick ratio. The latter is sometimes referred to as the 'acid test'.

Current ratio

This indicates how many times the current liabilities of a company are covered by its current assets. It is calculated as follows:

$$\text{Current ratio} = \frac{\text{current assets}}{\text{current liabilities}}$$

In a majority of businesses, current assets will exceed current liabilities, indicating that short-term liabilities are covered by assets that could be readily converted into cash. An acceptable current ratio would be about 2:1 but this is only a guide, as some very successful businesses operate on a current ratio of less than this. For a supermarket chain, its stocks and debtors are likely to be low in comparison with its current liabilities (trade creditors) but, because large sums of cash are being collected daily, it is unlikely that it would be unable to meet its liabilities when they fall due. Three examples are:

- GUS plc = 1.2 (retail and business services).
- Inchcape plc = 1.3 (automotive services, logistics, marketing).
- Kingfisher plc = 1.0 (retailing – includes B&Q and Comet).

Figures are for 2002/2003.

Quick ratio (acid test)

The quick ratio is a more immediate measure of a company's ability to settle its debts. It excludes the value of stock from the calculation, as it may not be as readily converted into cash as trade debtors would be, for example. It is calculated as follows:

$$\text{Quick ratio} = \frac{\text{current assets less stock}}{\text{current liabilities}}$$

An acceptable quick ratio would be about 1:0. If it is much less than that, the proportion of the current assets represented by stocks is high. Some examples are:

- Pilkington plc = 0.8.
- GUS plc = 0.9.
- Inchcape plc = 0.5.

Figures are for 2002/2003.

14.10.3   Working capital ratios

In section 14.8 the working capital cycle was examined and the effective management of working capital was highlighted as an important factor in the success of any business. The following ratios are commonly used as tools in this management process.

Stock turnover ratio

The stock turnover ratio indicates how many times stock is turned over in a year; that is, how many times stock is effectively replaced. It can also be adapted to indicate how many days of sales are held in stock. It is calculated as follows:

$$\text{Stock turnover} = \frac{\text{cost of sales}}{\text{average stock level for the period}}$$

and

$$\text{No. of days' stock held} = \frac{\text{average stock level for the period}}{\text{cost of sales}} \times 365^3$$

The stock turnover ratio measures how quickly goods are moving through the business. A high ratio implies an efficient operation. A low ratio indicates inefficiencies and the possibility of obsolete goods being held in stocks. The nature of the industry will be a major determinant of this ratio and the industrial sector averages should be used to assess whether a company's ratio is high or low. Some examples of stock turnover ratio are:

- GUS plc = 4.8 (retail and business services); no. of days' stock held = 76.
- Pilkington plc = 4.5 (glass manufacturer); no. of days' stock held = 81.
- Inchcape plc = 5.8 (automotive services, logistics, marketing); no. of days' stock held = 63.
- Amec = 33.6 (construction).

Figures are for 2002/2003.

The number of days of stock held should be sufficient to cover variations in usage (demand) and in deliveries (supply) plus the number of days required to convert raw materials into finished goods. World-class manufacturing activities hold only a few days' stock. A typical UK manufacturer holds between 40 and 60 days of stock. (The significance of inventory management is discussed in Chapter 5. Inventory control methods are covered in Chapter 20.) The example of Amec is provided in order to consider the case where a company carries almost no stock. Amec stocks comprise development land and work in progress, with only 20% of stocks comprising materials and goods.

Debtors turnover ratio

This ratio measures the number of days it takes to convert trade debtors into cash. It is calculated as follows:

$$\text{Debtors turnover} = \frac{\text{sales}}{\text{trade debtors}} \times 365 \text{ days}$$

In the UK, customers, other than in retailing, generally take anything between 60 and 90 days to settle their invoices. The ratio should therefore fall within this range. If it is higher than 90 days the indications are that either the company is not exerting enough pressure on its customers to pay their invoices or it is exposed to bad debts. Most companies agree credit terms with their customers. If, for example, invoices are due for settlement within 30 days of receipt of goods, the debtors turnover should be in the order of 30 days, not 90 days. By allowing customers to take longer to pay than the agreed credit period, the company is, in effect, providing its customers with cheap finance (see section 14.8).

---

[3]The latter is a financial measure and may be calculated from annual accounts, therefore the full calendar year of 365 days is used. Engineers or scientists, managing stocks of materials and components and using units of measurement such as tonnes, litres or pieces, may prefer to use *working days* to calculate usage per day, and hence the number of days' stock.

Another important point is that this ratio is calculated using trade debtors figures at a single point in time (as shown on the balance sheet). If it were calculated shortly before a large overdue debt was about to be paid, there would be a distorted measure of the usual position. It is therefore important to establish a trend in the ratio calculated over a period of time. An upward trend would indicate a deteriorating position, which would require action to bring the ratio back down to an acceptable level. Regular (e.g. monthly) review of this ratio is good management practice, especially if accompanied by an analysis of debtors over 30 days, 60 days, 90 days and so on.

Creditors turnover ratio

This ratio indicates the number of days a company is waiting before it pays its trade creditors. It is calculated as follows:

$$\text{Creditors turnover} = \frac{\text{cost of sales}}{\text{trade creditors}} \times 365 \text{ days}$$

It is likely that a company will want to take as long as possible to pay off its creditors so that valuable cash resources are kept within its working capital cycle. However, extending the credit period may mean having to forego valuable discounts for early settlement or incurring premiums for late payments and/or losing the goodwill of suppliers.

The optimum value of the ratio for a company will depend on the credit terms offered by its major suppliers and should, ideally, be in line with its debtors turnover ratio. An ideal position is reached when cash is collected from trade debtors just ahead of payments to suppliers. Where the trend in the ratio is being reviewed, an increasing ratio indicates potential short-term cash-flow problems that might lead to the company being unable to find enough cash to pay its creditors as the payments fall due.

14.10.4   Gearing ratios

Gearing ratios analyse the components of the capital invested in the business by comparing capital that is borrowed with capital provided by the shareholders.

Debt/equity ratio

This ratio is calculated as follows:

$$\text{Debt/equity ratio} = \frac{\text{interest bearing loans} + \text{preference share capital}}{\text{ordinary share capital}}$$

Preference shares carry preferential rights to dividends that must be paid before a company can pay dividends to its ordinary shareholders. If a company has issued 10% preference shares it must pay a dividend on these shares equal to 10% of their capital value, if there are sufficient profits. If that leaves no profits for distribution to ordinary shareholders, then so be it. Because preference shares have rights to a fixed rate of return they are treated like loans in calculating the debt/equity ratio.

If the ratio is greater than 1:1 the company has a high proportion of borrowing in its capital base and is described as highly geared. For most companies the ratio would be no higher than about 1.5:1. A highly geared company is a more risky company for its shareholders. This is because it must meet all interest payments and preference dividends out of profits before dividends can be paid to ordinary shareholders. Two examples are:

- Inchcape plc = 0.28 (automotive services, logistics, marketing).
- Pilkington plc = 1.3.

Figures are for 2002/2003. Inchcape plc clearly has a low gearing ratio.

Debt/assets employed ratio

This ratio indicates the proportion of a company's assets that are financed by borrowings. It also highlights the capacity that a company has for obtaining further secured borrowings. It is calculated as follows:

$$\text{Debt/assets employed} = \frac{\text{interest-bearing loans} + \text{preference share capital}}{\text{assets employed}}$$

Again, a ratio of greater than 1:1 would indicate a high level of gearing. If the ratio is less than this there is potential for more borrowing for further capital investment.

Interest cover

This ratio is used to assess whether profits are sufficient to cover the interest due on borrowings (and preference dividends if applicable). Lenders will pay particular attention to this ratio. It will be almost impossible to obtain finance if profits are insufficient to cover interest payments. The ratio is calculated as follows:

$$\text{Interest cover} = \frac{\text{profit before interest and tax}}{\text{interest payable}}$$

Lenders will require a level of cover that reflects the risk attached to the borrowings; a ratio of 2:1 should be adequate.

14.10.5 Investor ratios

The final set of ratios to be considered are the investor ratios. These ratios are very important for companies quoted on a stock exchange. They will be given careful consideration by investors choosing between companies, so they need to be acceptable to attract additional funding.

Earnings per share (EPS)

The EPS measures the amount of profit earned for a year by each share in issue. It is a good indicator of the changes in profitability of a company over a number of accounting periods. For example, the profits of a company could double from one year to the next; this might at first glance appear to be an exceptional result. However, if the company had in issue three times as many shares as it had in the prior period, the result would not be perceived to be exceptionally good and the EPS would demonstrate this. The ratio is calculated as follows:

$$\text{EPS} = \frac{\text{profit after tax and payment of preference dividends}}{\text{number of ordinary shares in issue during the year}}$$

Investors will want to see an upward trend in the earnings per share, as this indicates a growth in profitability. For example, the EPS for Inchcape plc rose from 48.4p in the year 2000 to 79.7p in 2001 to 104.5p in 2002, whereas the EPS for Amec plc for the same period fell from 19.1p to 13.7p to 3.7p.

Price/earnings ratio (P/E ratio)

This ratio expresses the current market price of a share as a multiple of its earnings. It is calculated as follows:

$$\text{P/E ratio} = \frac{\text{current market price per share}}{\text{earnings per share}}$$

P/E ratios are a common means of valuing companies. Indices of P/E ratios for different industrial sectors, compiled by analysts, provide a good indicator of a company's performance. P/E ratios can also be used to appraise a company's investment potential by making comparisons within the same sector.

Example 14(6)

A company has in issue 1.5 million ordinary shares with a nominal value of £1 each. The current market value of the shares is £6.80 each. The latest set of accounts show a profit after interest and preference dividends of £850 000. Calculate the EPS and the P/E ratio for the company.

Step 1: Calculate the EPS

$$\text{EPS} = \frac{850\,000}{1\,500\,000} = 0.57$$

Step 2: Calculate the P/E ratio

$$\text{P/E ratio} = \frac{6.80}{0.57} = 12.57$$

This would then be compared with companies in the same industrial sector to assess how well the company is performing.

14.11   LIMITATIONS ON MEASURES OF FINANCIAL PERFORMANCE

There are limitations to using accounts for assessing the performance of a business. All accounts are based on historic cost data and can fail to reflect current inflationary pressures. Certain valuable business assets are not taken into account at all. The value of a highly skilled workforce and/or a prime business location cannot be easily measured using traditional financial criteria.

When depreciation was examined, a degree of choice of method was noted that could affect reported profits. A similar degree of choice exists in measuring and apportioning overheads or evaluating stocks, as described in Chapter 13, section 13.2.

In short, accounts do not provide an absolute measure of performance, but one in which there is room for interpretation. Managers should not forget this.

14.12 SUMMARY

This chapter began by looking at the different entities used to conduct business and the basic legal framework for each. The accounts of a business were then examined in detail to establish the fundamental concepts and methods of their preparation. The use of accounts as a management tool and for financial analyses was then examined. The aim has been to provide engineers, scientists and technologists, whether currently in managerial positions, graduates aspiring to such positions or undergraduates planning to become entrepreneurs, with a basic working knowledge of the measures of financial performance. Overall, despite the reservations expressed in section 14.11, the preparation of regular, clear management and financial accounts is an essential component in the successful management of any business operation.

REFERENCE

1. Whitehead, G. M. (1997) *Book-keeping Made Simple*, 5th edn, Heinemann Educational, Oxford.

BIBLIOGRAPHY

Chadwick, L. (1998) *Management Accounting (Elements of Business)*, Thomson Learning, London.

Dyson, J. R. (2000) *Accounting for Non-Accounting Students*, FT Prentice Hall, London.

Glautier, M. W. E. and Underdown, B. (2000) *Accounting Theory and Practice*, FT Prentice Hall, London.

Glynn, J. J., Perrin, J. and Murphy, M. P. (2003) *Accounting for Managers*, Thomson Learning, London.

15

Project Investment Decisions

15.1 INTRODUCTION

From time to time, businesses need to make substantial investments in plant, equipment or buildings. It is often the case that the returns from the new investment will be small relative to the size of the investment, such that several years will elapse before the returns can repay the investment. Engineers, scientists and technologists may be called on to justify investment in new plant, new laboratories or new equipment by demonstrating that the benefits exceed the costs. In large organizations evaluations of proposed investments will be carried out by finance staff, but the engineers, scientists and technologists will need to talk to the finance people in their own language. Long-term *investment in projects* and *life-cycle costs* of capital assets are the subject matter of this chapter. The ways in which investment decisions of this nature are made and the methods of life-cycle costing will be explained. First, the notion of *payback* will be briefly considered. The remainder of the chapter is concerned with *discounted cash flow* techniques; the concepts of *present value, annual value, benefit–cost ratio* and *internal rate of return* will be examined and an explanation of how these relate to the underlying problem of *inflation* will be provided.

15.2 PAYBACK

The notion of payback is simple, and therefore popular. How long will it take to repay an initial single-sum investment? If machine X costs £1000 and earns net annual benefits of £250 per year, the payback period is four years. If machine Y costs £1200 but earns net annual benefits of £400 per year, its payback period is only three years so, despite the higher price, it would appear to be a better investment.

This, of course, is too simplistic; it ignores interest rates. If in the above example the interest rates are 10% then, for machine X, the sum of £100 interest (10% of £1000) must be paid out of the first year's benefit of £250. This leaves £150 towards paying back the initial investment of £1000. So, at the end of the first year the sum to be paid back will be £1000 − £150 = £850.

The interest payable in the second year will be 10% of the sum owed; that is, 10% of £850 = £85. This leaves £165 available for paying off the initial

Table 15.1 Payback

Year	Sum owed at start of year (A)	Total benefit = 250 Interest (B) = 0.1 × (A)	Total benefit = 250 Payback (C) = 250 − (B)	Sum owed at end of year (C) − (A)
1	1000	100	150	850
2	850	85	165	685
3	685	68.50	181.50	503.5
4	503.5	50.35	199.65	303.85
5	303.85	30.39	219.61	84.24
6	84.24	8.42	241.58	(157.34) Credit

investment. Table 15.1 shows how this process continues until Year 6, by the end of which the total initial investment has been repaid.

A similar calculation would show that for Machine Y, payback is completed within four years. This suggests again that it is the better investment; but whether this is really so in the long term must depend on other factors.

Additional factors need to be taken into consideration:

- Will the earnings for Machines X and Y remain at their given levels after the payback period? If they change, X could be the better project in the long term.
- What are the lives of Machines X and Y? If X had to be replaced every seven years and Y every five years, X might be better value.
- If X had a higher scrap value than Y, then X might be better value.
- While interest paid out on the investment has been considered, no account has been taken of interest earned on income.
- Would the investment do better if deposited in a bank?

For many managers, payback is an invaluable concept. As a matter of policy, if the payback period is short, three years (say) or less, then the investment will be worthwhile. For longer-term investments a more sophisticated approach is needed. This is where discounted cash flow should be used.

15.3 DISCOUNTED CASH FLOW

Discounted cash flow (DCF) techniques allow proper consideration of all these questions. These techniques allow rational comparisons to be made between alternative investment projects. The techniques are generally used for long-term projects, 'long-term' meaning anything from two years to 120 years or more.

If we are to choose between Project A, which has an initial cost of £100 and will earn a net annual income of £30 per year for the next ten years, and Project B, which has an initial cost of £80 and will earn a net annual income of £25 for the next five years then £27 per year for the following five years, DCF will assist us. This will be demonstrated in section 15.3.2.

The underlying premise of DCF is that money will not usually be invested in a project unless that project earns more than if the money were earning steady interest in a bank. A project is said to have negative value if it earns less than if the capital simply earned interest in a bank; a positive value if it earns more. The higher the interest rate, the more likely it is that a project will have a negative value.

DCF forms an important part of an investment decision analysis; it does not predict the future! Once a project is underway the outcome may well vary from predictions. It is unlikely, for example, that interest rates will remain constant over the duration of the project; which, among other things, makes estimating an interest rate for the DCF analysis so interesting!

15.3.1   Effect of interest rates

If £100 is invested in a bank account now and the annual interest rate is 12%, then the sum in the bank will grow as follows:

$$
\begin{array}{lll}
\text{End of Year 1} & £100 \times 1.12 & = £112.00 \\
\text{End of Year 2} & £100 \times 1.12^2 & = £125.44 \\
\text{End of Year 10} & £100 \times 1.12^{10} & = £310.58
\end{array}
$$

Having invested £100 now, the investor would be able to withdraw from the bank the sum of £310.58 after ten years. Returning to the principle of DCF in a business environment, if a company made an investment of £100 in a piece of plant or machinery now and the annual interest rate were 12%, it would need to be able to demonstrate assets and/or accumulated earnings in Year 10 to the value of at least £310.58 for the investment to be worthwhile; otherwise it might just as well have put the money in the bank.

A commercial decision to invest in plant or machinery or even buildings will, therefore, depend on whether the earnings are likely to exceed those provided by the usually more secure option of leaving the money in the bank. A commercial enterprise does not need to concern itself with the absolute value of its earnings but with the value of its earnings relative to the 'do-nothing' policy of leaving the money in the bank. This leads to the idea of equivalence. If interest rates are 12% and the project investment analysis shows earnings of £310.58 in ten years' time, then that sum can be considered equivalent to £100 earned now. £100 is said to be the *present value* of the sum of £310.58 earned at Year 10 when the interest rate is 12%. If an investment is made of £100 now and a benefit of £310.58 is earned in Year 10, then the *net present value* of the investment is given by:

$$
\text{net present value} = \text{present value of benefits} - \text{initial cost} = £100 - £100
$$
$$
\therefore \text{net present value} = \text{zero}
$$

From the foregoing it should be evident that £1.00 earned in Year 10 is equivalent to

$$
£\frac{1.00}{3.1058} = £0.322
$$

earned now. This is the basis of the present value tables provided in Appendix 4. (The use of these tables will be demonstrated later.) In deciding whether to invest in a project or not, the investment analyst determines whether the net present value is greater than zero. If it is, the project should be worthwhile, for it provides a better return than the bank.

The method of DCF, therefore, is to discount future cash flows by multiplying each future sum by the factor

$$\frac{1}{(1+i)^n}$$

where

i = rate of interest per unit time (usually per annum)

n = number of units of time (usually years) that must pass before the future sum is realized

This discounted value is the present value (or present worth) of the future sum. It is the sum that would need to be invested now at the given interest rate to attain the value of the future sum after n years.

Note that as interest rates rise the present value of future earnings falls and, therefore, the net present value of any project under consideration will be lower. This is why there is greater reluctance to make capital investments when interest rates are high.

For example, if interest rates are 15%, then £310.58 earned after ten years is equivalent to

$$£\frac{310.58}{1.15^{10}} = £76.77$$

rather than the £100 if interest rates were 12%.

15.3.2   Cash flow diagrams

It is helpful to represent cash flows for investment projects in a diagrammatic form, as shown in Figure 15.1. Upward bars represent net annual income, downward bars represent costs. Cash flow diagrams for projects A and B described earlier are shown in Figure 15.1. The diagrams show the whole *life cycle* for each project.

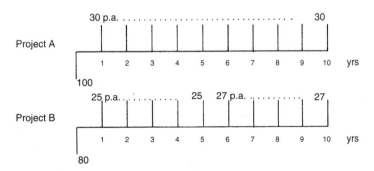

Figure 15.1   Cash flow diagrams

Reconsider projects A and B previously referred to at the beginning of section 15.3. The net present value (NPV) of each project will now be calculated. Project A

$$\text{PV of earnings} = \frac{30}{(1.12)} + \frac{30}{(1.12)^2} + \cdots + \frac{30}{(1.12)^{10}} = 169.51$$

Project B

$$PV \text{ of earnings} = \frac{25}{(1.12)} + \cdots + \frac{25}{(1.12)^5} + \frac{27}{(1.12)^6} + \cdots + \frac{27}{(1.12)^{10}} = 145.35$$

The two results are summarized below.

Project A	present value of earnings	= £169.51
Project B	present value of earnings	= £145.35

If the initial cost of each project is subtracted from the present value of the earnings, the NPV of each project is found.

Project A	NPV = 169.51 − 100	= £69.51
Project B	NPV = 145.31 − 80	= £65.35

A similar analysis at an interest rate of 4% gives:

Project A	NPV = 243.33 − 100	= £143.33
Project B	NPV = 210.09 − 80	= £130.09

A similar analysis at an interest rate of 20% gives:

Project A	NPV = 125.78 − 100	= £25.78
Project B	NPV = 107.22 − 80	= £27.22

As interest rates increase, Project B eventually becomes preferable to Project A and both projects decline in value, eventually reaching negative values. Project A has zero NPV at an interest rate of approximately 27%; this is known as the *internal rate of return* (IRR) of the project, or the *yield*. The earnings of Project A are equivalent to earning interest at a rate of 27% on the initial investment of £100. There will be more on the concept of IRR later in section 15.4.

Example 15(1)

The cash flow diagrams in Figure 15.2 show the whole life cycle of each of Projects C and D. (Note: in some of the following examples the term net present worth has replaced net present value. NPW = NPV. The purpose behind this is to familiarize the reader with both expressions.)
Project C
Present value of benefits

$$= \frac{50}{1.10} + \frac{40}{1.10^2} + \frac{30}{1.10^3} = 101.05$$

Net present value = 101.05 − 100 = 1.05.

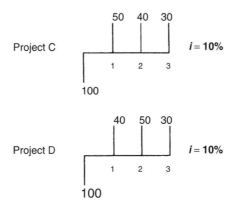

Figure 15.2 Cash flow diagrams

Project D
Present worth of benefits

$$= \frac{40}{1.10} + \frac{50}{1.10^2} + \frac{30}{1.10^3} = 100.22$$

Net present worth (NPW) = 100.22 − 100 = 0.22.

The same calculations can be made more easily by using the *present value tables* in Appendix 4, Table A4.1, as demonstrated below.

Project C

$$NPW = 50 \times 0.9091 + 40 \times 0.8264 + 30 \times 0.7513 - 100 = 1.05$$

The factor 0.9091 is found in the present value tables in the 10% column at row Year 1.
The factor 0.8264 is found in the present value tables in the 10% column at row Year 2.
The factor 0.7513 is found in the present value tables in the 10% column at row Year 3.

Project D

$$NPV = 40 \times 0.9091 + 50 \times 0.8264 + 30 \times 0.7513 - 100 = 0.22$$

The factors remain the same as for Project C above, only the benefits change.

The only difference between the two projects is the chronological order in which the benefits of 50 and 40 are obtained. It can be observed that bringing forward in time high-value benefits increases the net present value.

15.3.3 Annuities

Many investments will result in equal annual benefits for the life of the project. An annuity is a series of equal annual payments. The benefits for Project A above form an annuity. The present value of an annuity can be determined swiftly using the *present value of annuity tables* (Table A4.2) provided in Appendix 4.

The present value of £1 per annum for n years at i% interest is given by the formula

$$\frac{1}{(1+i)} + \frac{1}{(1+i)^2} + \cdots + \frac{1}{(1+i)^n} = \frac{1-(1+i)^{-n}}{i}$$

This is the basis of the present value of annuity tables (Table A4.2).

Using these tables the net present value of Project A in Figure 15.1 is given by

$$NPV = 30 \times 5.6502 - 100 = 69.51$$

The factor 5.6502 is found in the present value of annuity tables in the 12% column at row 10 years.

Example 15(2)

A project earns £25 000 p.a. for 20 years. Determine the present value of these earnings at

(a) $i = 5\%$;
(b) $i = 15\%$;

where $i =$ interest rate selected for discounting.

(a) $PV = 25\,000 \times 12.462 = £311\,550$
(b) $PV = 25\,000 \times 6.2593 = £156\,482$

Example 15(3)

A project earns £25 000 p.a. for 30 years. Determine the present value of these earnings at

(a) $i = 5\%$;
(b) $i = 15\%$.

(a) $PV = 25\,000 \times 15.372 = £384\,300$
(b) $PV = 25\,000 \times 6.5660 = £164\,150$

Example 15(4)

A project earns £25 000 p.a. for 50 years. Determine the present value of these earnings at $i = 15\%$.

$$PV = 25\,000 \times 6.6605 = £166\,513$$

Observe how, at the higher interest rate, adding 10 or even 30 years to a project's life has very little effect on the present value. Indeed, earning £25 000 p.a. for ever at an interest rate of 15% results in a present value of £166 667 (see section 15.9.2). Analyses that use DCF techniques, particularly at high interest

Table 15.2   PV of 1 million p.a. for *n* years

	$n = 40$	$n = 80$	$n = 120$
$i = 3\%$	23.11 m	30.20 m	32.37 m
$i = 5\%$	17.16 m	19.60 m	19.94 m
$i = 12\%$	8.24 m	8.33 m	8.33 m

rates, have the unfortunate effect of rendering worthless massive benefits to future generations (see Table 15.2).

15.3.4   Equivalent annual value

Large projects operating over many years, which take several years to construct and require expensive replacement of components (e.g. turbines for a power station) at long intervals, are often evaluated in terms of their *equivalent annual value* (EAV) rather than in terms of their NPV. The EAV of a project is found by converting its NPV into a series of equal payments, payable at the end of each year; in other words, an annuity.

If £25 000 p.a. for 50 years has a present value of £166 513 at a 15% interest rate, then a payment now of £166 513 has an equivalent annual value of £25 000 over a 50-year period. For large projects, perhaps constructed over two or three years and for which expensive components may need replacing, say, every 30 years, estimates of the initial capital cost and of the replacement costs can be made, converted into present values then further converted into an *equivalent annual cost* (EAC). This value will represent the minimum net annual revenue necessary to cover these capital costs. The amount by which the net annual revenue exceeds the EAC will be the EAV of the project. In choosing between alternative hydroelectric power schemes, for example, EAV is more likely to be used than NPV.

Individuals are more likely to encounter the concept of EAV when arranging a mortgage. An annual mortgage repayment is simply the EAV of the sum borrowed. Mortgage lenders set the interest rates at an appropriate level to cover costs and profit. The *annual value tables* (Table A4.3) in Appendix 4 show the equivalent annual value of a single payment now of £1.00 for a range of interest rates and project durations. The values in the annual value tables are the reciprocal of the values in the present value of annuity tables.

Example 15(5)

(1) I intend to take out a mortgage for £80 000. Determine the annual repayments if the period of the loan is
   (a)   25 years;
   (b)   30 years;
   and the interest rate is 7%.
   (a)   Annual repayment = EAV = £80 000 × 0.0858 = £6864. The factor 0.0858 is found in the annual value tables (Table A4.3) in the 7% column and the 25 year row.

(b)   Annual repayment = £80 000 × 0.0806 = £6448. The factor 0.0806 is found in the annual value tables (Table A4.3) in the 7% column and the 30 year row.

(2)   I intend to take out a mortgage for £80 000. Determine the annual repayments if the period of the loan is 25 years and the interest rate is 14%.

$$\text{annual repayment} = £80\,000 \times 0.1455 = £11\,640$$

Observe how my mortgage repayments have almost doubled with the doubling of the interest rate.

(3)   I can only afford annual repayments of £11 240 p.a. How long will the mortgage period be if I borrow £80 000 at 14% interest?

$$\text{annual repayment} = £80\,000 \times \text{factor} = £11\,240$$
$$\text{factor} = 0.1405$$

A search of the annual value tables (Table A4.3) for the factor 0.1405 in the 14% column shows that the loan period will have to be extended to about 43 years! This looks like an unaffordable mortgage.

During the life of a mortgage, as interest rates change, the mortgage lender determines how much of the capital has been repaid. From this the lender calculates the outstanding capital debt and finds the EAV of this sum over the remainder of the period, thus determining the revised annual payments.

Financial planners for companies intending to invest in premises will carry out similar analyses using the same methods as used by individuals.

Example 15(6)

An engineering company took out a 25 year mortgage for £800 000 one year ago, at an interest rate of 7%. A single annual payment of £68 640[1] has been made. The interest rate now changes to 8.5%. What will be the revised annual repayments?

Interest payable in year 1 = 0.07 × 800 000 = £56 000

Capital repaid = 68 640 − 56 000 = £12 640

Outstanding debt = 800 000 − 12 640 = £787 360.

EAV of £787 360 over 24 years at 8.5% interest = 787 360 × 0.09897

$$= £77\,925$$

The factor 0.09897 is the inverse of 10.1041 which is found in the 8.5% column, row 24 years of Table A4.2.

The new annual repayments will be £77 925. Alternatively, if the business could not afford this and if the lender agreed, the mortgage could be extended for a further 45 years, as shown below.

EAV of £787 360 over 40 years at 8.5% interest = 787 360 × 0.08838

$$= £69\,589$$

---

[1] See annual value tables (Table A4.3), 7% column, row 25 years for factor = 0.0858, therefore payment = 0.0858 × 800 000 = 68 640.

A modest increase in interest rates combined with an inability to increase repayments can result in a very unpleasant increase in the period of the loan!

15.3.5 Machine replacement

EAVs are particularly useful when deciding on a replacement policy for machinery or equipment. Where two alternative machines that do a similar job produce modestly different benefits, have different lives, different capital costs and different scrap values, EAVs can be used to determine which machine is the better investment. Assume for the moment that there is no inflation. Later on, the issue of inflation will be examined.

Example 15(7)

Machine M1

$$\text{Initial cost } C_0 = 200 \qquad \text{Scrap value} = 80$$

$$\text{Annual benefits} = 84 \qquad \text{Life} = 3 \text{ years}$$

Machine M2

$$\text{Initial cost } C_0 = 270 \qquad \text{Scrap value} = 70$$

$$\text{Annual benefits} = 90 \qquad \text{Life} = 5 \text{ years}$$

It would be wrong to find the NPV of each machine and compare the two values. The stated intention is to replace the machines repeatedly with similar models that will continue to do the job required of them.[2] The obvious approach is to consider the machines over a common period. In this case 15 years is the lowest common multiple of the two lives and could be chosen. The cash flow diagrams are shown in Figures 15.3 and 15.4.

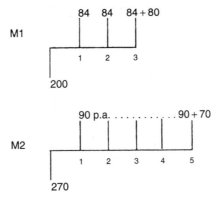

Figure 15.3   Cash flow diagrams for machines M1 and M2

---

[2] These financial evaluations make no allowance for changing technology. Where such changes are probable, engineers and scientists will need to exercise judgement before putting any reliance on calculations. In Chapter 14, section 14.5.3, similar problems arise with depreciation.

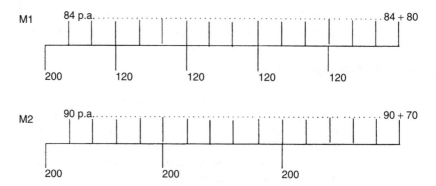

Figure 15.4 15-year cash flow diagrams for M1

The values of 120 for M1 and 200 for M2 in Figure 15.4 represent the replacement cost minus the scrap value for each 'project'.

The NPV of each scheme at a discount rate of 10% can now be found.

M1 $\quad$ NPV $= 84 \times 7.6061 - 120(0.7513 + 0.5645 + 0.4241 + 0.3186)$

$\qquad + 80 \times 0.2394 - 200 = 211.04$

M2 $\quad$ NPV $= 90 \times 7.6061 - 200(0.6209 + 0.3855) + 70 \times 0.2394 - 270$

$\qquad = 230.03$

Thus M2 is marginally better than M1 at the selected discount rate.

This method could become a little tedious. If three machines with, say, lives of three years, five years and seven years were being compared, cash flow diagrams of 105 years' duration would have to be drawn in order to find the NPVs of each machine! In fact, NPVs do not need to be compared. All that is needed is to compare the EAVs of the three-, five- and seven-year projects.

The following analysis should show that a comparison of EAVs is sufficient. Reconsidering machines M1 and M2 on Figure 15.3 and using a discount rate of 10%:

M1

$$NPV = 84 \times 2.4869 + 80 \times 0.7513 - 200 = 69.00$$

The factor 2.4869 is taken from Table A4.2 for 3 years at $i = 10\%$
The factor 0.7513 is taken from Table A4.1 for 3 years at $i = 10\%$

$$EAV = 69.00 \times 0.4021 = 27.75$$

The factor 0.4021 is taken from Table A4.3 for 3 years at $i = 10\%$

Alternatively,

$$EAV = 84 - 200 \times 0.4021 + 80 \times 0.7513 \times 0.4021 = 27.75$$

M2

$$EAV = 90 - 270 \times 0.2638 + 70 \times 0.6209 \times 0.2638 = 30.24$$

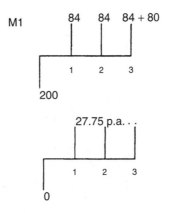

Figure 15.5   Equivalent cash flow diagrams for M1

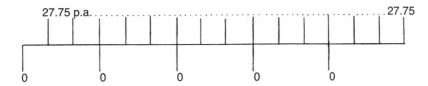

Figure 15.6   15-year equivalent cash flow diagram for M1

M2, as before, is marginally better than M1 at the selected interest rate. The reason why this comparison of EAVs is sufficient is as follows.

Finding an EAV provides a different representation of the cash flow diagram, as shown in Figure 15.5. If this cash flow diagram is repeated five times the cash flow diagram for the 15-year period (the lowest common multiple) is obtained as shown in Figure 15.6.

Therefore, finding the EAV of each 'project' gives a direct comparison over the period of the lowest common multiple or, for that matter, indefinitely. This method should be used for all comparisons of projects having lives of different durations.

15.4   INTERNAL RATE OF RETURN

Another approach to comparing alternative investment projects is by means of their internal rates of return. The *internal rate of return* (IRR) of a project is the rate of interest at which the NPV is zero. This is also known as the *yield*. An investor might decide that the highest yield should be the determining criterion in project selection, rather than NPV or EAV. In most cases the IRR of a project cannot be calculated save by an iterative method. However, it can be found by plotting values of NPV against interest rate, as shown in Figure 15.7. The curve crosses the axis at an interest rate of 16.5%, therefore the IRR = 16.5%.

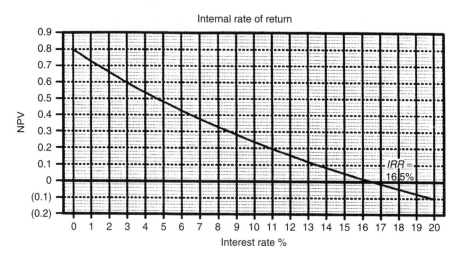

Figure 15.7   Internal rate of return

There are two ways of looking at the IRR. For the project described in Figure 15.7:

- If £1m is borrowed at 16.5% the benefits over the six years will completely repay both capital and interest.
- The accumulative benefits over the six years at an interest rate of 16.5% would equal the sum obtained by investing £1m in a bank to earn 16.5% over a six-year period.

In the case in which an investment project has an initial cost followed by a uniform series of annual payments, the IRR can be determined quite easily.

Example 15(8)

A project has an initial cost of £5m and equal annual net benefits of £962 000 for 24 years.

At the IRR, the present value of the benefits equals the initial cost

$$962\,000 \times \text{factor} = 5\,000\,000$$

$$\text{factor} = 5.1975$$

Examination of the present value of annuity tables (Table A4.2) for a 24 year annuity shows an IRR of approximately 19%.

In Figure 15.8, the NPVs for three projects, A, B and C, have been plotted against interest rates. Observe that the project with the highest yield (IRR) does not necessarily have the highest NPV across the range of interest rates. Observe

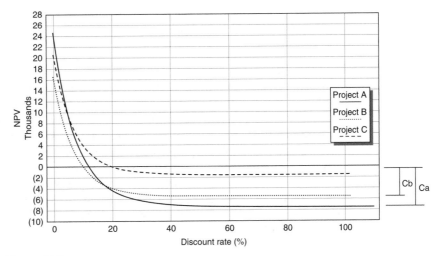

Figure 15.8

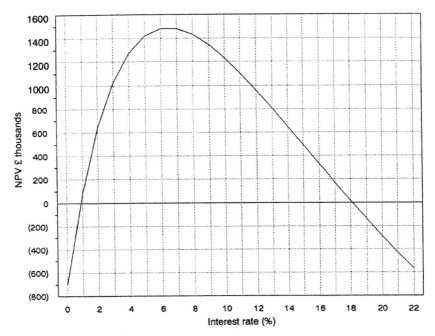

Figure 15.9   Project with two internal rates of return

also, as the interest rates become very high, that the value of NPV tends towards the initial cost of the project.

$$C_a = \text{initial cost of Project A}$$

$$C_b = \text{initial cost of Project B}$$

In Figure 15.8 the only difference between Project B and Project C is the initial cost. The initial cost of Project C is £4000 less than the initial cost of

Project B. The distance on the vertical axis between the curves of Projects B and C is £4000 at all points along the curve.

It is possible for a project to have two internal rates of return. A project that has a very large late cost, so large that at zero interest rates the total costs exceed the total benefits, can give rise to the curious effect shown in Figure 15.9.

Examples of projects of this nature might include nuclear facilities in which the decommissioning costs can exceed the initial costs and, at a more modest level, refrigeration plant, the disposal costs of which might have to be very high to ensure that no harmful CFC gases leak to the environment.

The assumption so far has been that the costs and benefits are known. In fact, they will always have to be derived. A project evaluation necessitates a determination of the costs and of the benefits. Exercise 9 (for Chapter 15) on the website (www.wileyeurope.com/go/chelsom) provides a comprehensive project evaluation.

15.5   COST–BENEFIT ANALYSIS AND LIFE-CYCLE COSTING

15.5.1   Benefit-to-cost ratios

Projects may be assessed by comparing their *benefit-to-cost ratios* (B/C).

Let B = present value of benefits
C = present value of costs
then if B/C > 1 the investment is worthwhile.

While numerous investments might be worthwhile, there will almost certainly be limits on both the capital available for investment and on revenue available for future annual expenditure. Choices will need to be made between alternative projects. If this is so, then those projects that provide the best value for money will be chosen. The B/C ratio provides one measure of value for money; the higher the ratio, the greater the value for money.

Example 15(9)

Consider two projects X and Y, each having an initial cost, C_x and C_y respectively, and each generating a series of benefits over a period of ten years.

$$\text{Let } C_x = 10\,000 \qquad C_y = 12\,000$$

Assume that at a discount rate of 8% the present value of the benefits for each project is as follows:

$$B_x = 15\,000 \quad B_y = 17\,500$$

$$\text{then } B_x/C_x = 1.5 \text{ and } B_y/C_y = 1.46$$

The B/C ratio for Project X is higher than the B/C ratio for project Y, but

$$NPV_x = 5000 \quad NPV_y = 5500$$

A project with a higher B/C ratio does not necessarily have the higher NPV.

This can be thought of in a different way by examining the incremental project Y − X.

$$\frac{B_{y-x}}{C_{y-x}} = \frac{2500}{2000} = 1.25$$

This shows that for every additional £1 invested in Y over X a benefit of £1.25 is earned. Thus, if a choice were being made between the two projects Y should be chosen.

Benefit/cost ratios are not only indicators of value for money, they also provide a good discipline for long-term project assessment in that they should ensure that much closer attention is given to future costs as well as to future benefits.

15.5.2   Cost–benefit analysis (CBA) and life-cycle costing (LCC)

Cost–benefit analysis or CBA examines future annual benefits and future annual costs, not merely the net annual benefits (or costs). For example, a project that earns annual benefits of £1m p.a. and incurs annual costs of £950 000 could be described by a cash-flow diagram as either an annuity of net benefits of £50 000 p.a., which gives no indication of the costs of operation, or as annuities of £1m benefits and £950 000 costs, which does.

Cost–benefit analysis is often used in *life-cycle costing* (LCC). LCC is another term for methods of analysis that take into account the whole life cost of a project. For example, if a bridge is to be built, the usual procedure is to receive tenders for the construction (or the design and construction) of the bridge and let the contract to the lowest bidder. So, the decision is made solely on the basis of lowest initial cost. This is how most large public works contracts and contracts for large buildings are let. No account is taken of the maintenance, repair and replacement costs of the project. No account is taken of the decommissioning costs of the project at the end of its life. For many projects this may be perfectly reasonable, where running costs are very small in relation to the initial capital cost and/or where the running costs for the alternative schemes are of the same magnitude.

The importance of LCC has in recent years come more to the fore because the physical rates of decay of concrete and steel structures have been in many cases much faster than originally anticipated. Bridge structures with design lives of 120 years have experienced such high levels of deterioration that major components are being replaced after 20 or 30 years. While the direct costs of installing new components may not be particularly high, the indirect costs that arise from delays and rerouting of traffic can be very high indeed. Where this is so, a structure having a much higher capital cost and a far lower frequency of major replacement and repair could have the lowest LCC and, therefore, should be selected. Similar arguments can be put forward when choosing between alternative building designs in which major elements such as external cladding systems or air-conditioning systems may need replacement after only 25 or 30 years.

The term cost–benefit analysis is more strictly used for public-sector projects where government funding is invested in some asset for society, which will not only provide society with economic benefits but may also provide benefits that are not directly quantifiable in monetary terms. When used in this way, CBA attempts to assess nonquantifiable costs as well. The purpose of such analyses is to assist government in choosing between alternative public-sector projects. Difficulties arise when CBA is used as an absolute, rather than a comparative measure to determine whether the project selected should be allowed to go ahead or not. For example, it seems quite reasonable to use CBA when comparing alternative highway schemes, but to claim that the analyses provide absolute values of the schemes to the community is highly questionable. Public roads in the UK do not generate income but they do result in savings to the community. Therefore, if the government has limited capital for spending on new highways, it makes sense to spend on those highway projects with the highest B/C ratios. Table 15.3 shows the criteria usually used in cost–benefit analysis for highways.

In the UK the Department of Environment, Transport and the Regions (DETR) uses the COBA program to analyse planned trunk road schemes. This is a very sophisticated program in terms of traffic and accident analyses (although it is not without its critics). However, it is less successful in its provisions for the evaluation of environmental impact.

If traffic flows are heavy, savings in journey time and savings in accidents can give rise to very large benefits. In the UK these are usually discounted over a period of 30 years at an 8% interest rate. Two further assumptions are often made: these are that traffic volumes will increase and that the national economy will also improve over the period, so that the value of journey time savings will increase at some notional annual rate. These two increases are used to offset the relatively high discount rate, thus providing the projects with higher benefit/cost ratios than otherwise.

A criticism of this approach is that certain, very real costs are seldom quantified, since they are only quantifiable in terms of compensation to third parties (e.g. landowners) and such evaluations are notoriously difficult. These costs include effects of the new highway on nonusers and pedestrians and effects on the environment, including:

- noise;
- vibration;
- pollution;
- changes in ground water levels;
- visual impact;
- community severance.

A further criticism in the UK is that, while CBA is used in this narrow sense to choose between alternative highway schemes, it is not used to choose between

Table 15.3   Cost–benefit analysis for roads

Costs	Benefits
Capital costs	Journey time savings (man-hours, vehicle hours)
Maintenance costs	Maintenance costs saved on existing routes
Operating costs	Operating costs saved
	Savings in accidents

alternative transport systems, for example whether to invest in more railways or in more roads.

Cost–benefit analysis, with the exception of highways, has not been widely used in the UK in recent years. This has been a consequence of both its general unacceptability to the British Government and the very real difficulties facing those who attempt to use it.

This section has provided the reader with a very brief introduction to CBA. A useful practical study by means of case studies is *Cost–Benefit Analysis*, edited by Richard Layard and Stephen Glaister [1].

15.6   COMPARING DCF AND ACCOUNTING METHODS

At this point it may be useful to reflect on the DCF approach by comparing the method with normal accounting practice. The following example demonstrates that the two alternative methods give the same result.

When the net present value of a project is calculated, this is done by discounting future benefits to a present value and subtracting the initial cost. This process means that there is no need to calculate the interest earned or the interest paid in each year. DCF gives the value *now* of a project for a given rate of interest, not the sum of money accrued at the end of the project or, indeed, the accumulative debts/credits over the life of the project.

Example 15(10)

For the cash-flow diagram in Figure 15.10, using a discount rate of 15%, the NPV can be found:

$$NPV = 500 \times 0.8696 + 600 \times 0.7561 + 400 \times 0.6575 - 1000 = 151.46$$

Using normal accounting methods as developed in Table 15.4, the balance in the account after three years is £230.37, as shown. The interest rate of 15% is assumed to apply equally to the debt and to the benefits (see section 15.7 for a justification of this assumption).

With an interest rate of 15% the sum of

$$\frac{230.37}{1.15^3} = 151.46$$

needs to be invested at Year 0 to obtain 230.37 by the end of Year 3.

Both methods give the same result. The difference is that DCF compares investment projects in *present value* or *annual value* terms, while the traditional accounting method looks at the *future* worth.

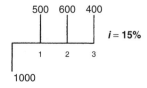

Figure 15.10   Cash flow diagram

Table 15.4   Future worth

At end of	Accumulative debt	Accumulative benefit
Year 0	1000	0
Year 1	1000×1.15	500
Year 2	1000×1.15^2	$500 \times 1.15 + 600$
Year 3	1000×1.15^3	$500 \times 1.15^2 + 600 \times 1.15 + 400$
	$= 1520.88$	$= 1751.25$
	balance $= 1751.25 - 1520.88 = 230.37$	

15.7   PERFECT CAPITAL MARKET

The analyses so far have made, implicitly, an underlying assumption, which needs some explanation and justification. The assumption is that the rate of interest payable on a debt is the same as the rate of interest earned on cash in the bank. This situation is known as the perfect capital market.

How can the perfect capital market be justified? Individuals, certainly, do not experience a perfect capital market. When a person borrows from a bank, he or she does so at a higher rate of interest than when he or she deposits savings in the bank. Businesses are treated in a similar fashion. When a business borrows money, the interest rate applied to the loan is greater than the interest rate that applies to any cash reserves that the business might have. Even so, the assumption of a perfect capital market is a reasonable one, as the following analysis attempts to show.

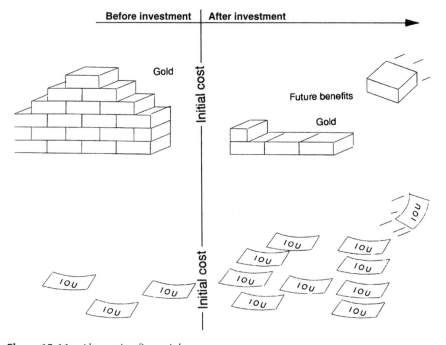

Figure 15.11   Alternative financial arrangements

Figure 15.11 represents two alternative businesses. Assume that the bank pays interest to its account holders of 8% p.a. and charges interest at a rate of 12% on borrowings.

In the first case the business has large reserves of cash (represented here as blocks of gold) deposited in the bank. A substantial investment is made, e.g. buying a factory (this is shown here by the term *initial cost*). This reduces the cash reserves; the effect of this is to lose the 8% interest that would have been earned by the depositor on the money used to buy the factory. As time passes and benefits are paid into the account the benefits will earn interest at 8%, so the one rate of interest applies to both the initial cost and to the benefits.

In the second case the business is already in debt (represented here as a pile of IOU notes) to the bank. A substantial investment is made by increasing this debt, and interest is paid on this debt at 12%. As benefits are credited to the account, the level of debt falls, so the benefits reduce the amount of interest being paid at 12%; so, once again, the one interest rate (12%) applies to both the initial cost and to the benefits.

Most businesses are net borrowers. That is, their borrowings far exceed their reserves. As time passes and particular loans are repaid, businesses will seek to expand their operations or sustain them by further loans, thus maintaining a net level of borrowing.

15.8   CAPITAL RATIONING

It has been noted that B/C ratios are indicative of value for money. However, the value of C represents the present value of all the costs of the project, present and future. Most commercial investors will be especially concerned with the initial capital cost. If several projects are available as investment opportunities and the capital available is insufficient for them all, then a simple way to proceed is to select those projects with the highest NPW/C_0 ratio, where $C_0 =$ initial cost.

Example 15(11)

Projects P1 to P10 have been analysed at a discount rate of 8%. The results of the analysis are shown in Table 15.5. In the right-hand half of the table the projects have been placed in an order of declining NPW/C_0 ratio.

Table 15.5   Capital rationing

Project	NPW	C_0	Project	NPW/C_0	ΣC_0
P1	560	1400	P1	0.40	1400
P2	540	1500	P6	0.40	2200
P3	400	2300	P2	0.36	3700
P4	380	1700	P8	0.30	4700
P5	360	1800	P4	0.22	6400
P6	320	800	P9	0.21	7600
P7	320	2000	P5	0.20	9400
P8	300	1000	P3	0.174	11 700
P9	250	1200	P10	0.167	12 300
P10	100	600	P7	0.16	14 300

If the capital is limited to £9400, Projects P1, P6, P2, P8, P4, P9 and P5 should be selected.

This simple approach does have its limitations. For example, if the capital limit were £10 000, Project P10 could also be included without exceeding the capital limit, but this is not immediately clear from the table. The method works well when the initial costs and NPWs of the projects under review are of the same order. A quick check on the difference between the capital limit and the accumulative cost should always be made to see whether a project further down the list could be included.

15.9   INFLATION

The introduction of the concept of *inflation*, and the matter of handling it, when applying DCF often causes considerable confusion. Confusion arises over the meaning of inflation, how it is taken into account and why, sometimes, it is apparently not taken into account. First, the meaning of inflation itself must be examined.

Inflation in prices is a consequence of a fall in the value of money in relation to the value of the goods being sold. If a new camera costs £100 this year and exactly one year later the identical model costs £105, then cameras (or this model) are experiencing 5% inflation. The example below shows the relationship between annual interest rate and annual rate of inflation.

Example 15(12)

This year I buy 100 kg of widgets for £103. How many can I buy for £206? Answer: 200 kg.

How many can I buy for £112? Answer: $100 \times 112/103 = 108.7$ kg.

One year ago I bought 100 kg of widgets for £100. What has been the annual rate of inflation over the last year? Answer: $(103 - 100)/100 = 0.03 = 3\%$.

Last year I invested £100 in the bank at an interest rate of 12%. How much money do I have in the bank one year later? Answer: £112.

How many widgets will my £112 buy? Answer: 108.7 kg (see above).

By how much has my investment increased in *real terms* (purchasing power)? Answer: $(108.7 - 100)/100 = 0.087 = 8.7\%$.

So, what has been the *real* interest rate? Answer: 8.7%.

$$(1 + r) = \frac{(1 + m)}{(1 + p)}$$

where r = real interest rate
  m = money interest rate (what the bank pays)
  p = rate of inflation

A common measure of inflation in the UK is the Retail Price Index (RPI). (See also Chapter 13, section 13.5.2.) This measures the changes in prices over a range of consumer goods and services as well as housing, applying a weighted formula to arrive at a monthly figure.[3] The formulae by which these indices are calculated are adjusted from time to time to take into account historical changes in how society spends its money. When domestic coal was an important part of every family budget, its price had to be included in the formula. Today these indices include such items as the price of skiing holidays, with an appropriate weighting, of course. Each time the formula is adjusted a new base date is established and the RPI for the new base date is set at 100.00.

To determine the 'current' rate of inflation, the RPI now is compared with that of 12 months previously, as illustrated below. Thus the 'current' rate of inflation is always a measure of past inflation!

Example 15(13)

(1)   The RPI now is 142.6, 12 months ago it was 134.5. What has been the rate of inflation?

$$142.6/134.05 = 1.06 \text{ Answer} : 6\%$$

(2)   The RPI now is 142.6, four years ago it was 119.6. What has been the average rate of inflation over the last four years?

$$\frac{142.6}{119.6} = 1.1923 = (1 + p)^4$$

therefore $p = 0.045 = 4.5\%$
therefore, the average rate of inflation over the last four years has been 4.5%.

Investment projects can be analysed in *real* terms or in *money* terms. Analysis in real terms means in terms of current purchasing power (current prices). Thus, the statement that the real net benefits at the end of Year 4 are £1000 means that the net benefits will have the same purchasing power in Year 4 as £1000 has today. This carries with it the implicit assumption that future money benefits are expected to keep abreast of the rate of inflation. No long-term plan could afford any other provision!

However, if the analysis of a project is conducted in *money* terms, and the net benefits at the end of Year 4, say, are stated to be £1000, then this means that £1000 is the actual sum of money that is expected to be earned; that future earnings are 'fixed' and that they will not rise with inflation. This could be a fixed-price contract. An analysis for such a project must use money interest rates.

Whether alternative investment projects are analysed in real terms or money terms, a consistent approach is essential. Thus, if the analysis is conducted in

---

[3]In December 2003 the British Government introduced an alternative index, the Consumer Prices Index, which excludes all housing costs (house prices and council taxes) from the equation. This gave a measure of UK inflation almost 1% lower than that obtained using the RPI. Since housing costs represent one of the greatest expenses of every household, the RPI must provide a more reliable measure of inflation as experienced by the individual; the new Consumer Prices Index would appear to be a measure more useful to those who try to control the economy than to those who simply are subject to it. Similar indices operate throughout Europe.

money terms, then money costs, money benefits and a money interest rate must be used. If the analysis is in real terms, then real costs, real benefits and real interest rates must be used. What is a real interest rate? In the earlier example the real interest rate was 8.7%. As has been previously observed, the real interest rate is something that is derived, not what the banks advertise; what they advertise is the money interest rate.

When should the analysis be conducted in money terms and when in real terms? If the projects under consideration are relatively short term – that is, two or three years – they can be analysed in money terms. If the projects are long term, then it is far better to analyse them in real terms. The reason for this is that it is impossible to predict inflation rates over the next, say, 20 years, therefore there is no way in which money income for any of the later years can be predicted safely. But engineers, scientists and technologists can predict operating costs and revenues at current prices for a project, therefore they can predict income in these terms; that is, real terms. A good example of this would be a power station. If demand and hence output can be predicted for Year 20, or any other year, then it is certainly possible to estimate the net revenue for that year at current prices. So, the only way to analyse this investment is to use real values at real discount rates.

There remains the difficulty of selecting a real interest rate for discounting future cash flows. Historically, the UK real (Bank of England) base rates have varied between about 1.5% and 8%, although in the 1970s negative values occurred when the inflation rate exceeded interest rates. The cost of borrowing does depend on the need and the nature of the business, but an assumption that interest rates selected for the purposes of discounting should be about 3% above the base rate does not seem inappropriate for UK projects. Government departments have in the past used 7% or 8% for discounting long-term projects, which seems reasonable. If one were investing outside the UK it would be useful to have knowledge of recent historical inflation rates as well as interest rates for the country where the investment might be made. Figure 15.12 shows interest and inflation rates for the UK for the period 1986 to 1992. The real interest rate has been determined using the equation in Example 15(13). This period has been selected because base rates were high and inflation rates unsteady. The graphs provide a clear demonstration of how the real interest rate has fluctuated over the period.

For short-term projects, current (i.e. money) interest rates and current inflation rates should be used.

Example 15(14)

Shown in Table 15.6 are the estimated cash flows for a project at current prices (that is, in real terms). Assume that the current rate of inflation is 5%. The lender advises that the borrowing rate is currently 12% and that it is unlikely to change by much over the three-year period. From this data, the future benefits in money terms have been calculated as shown in Table 15.7.

From this the NPW can be found:

$$\text{NPW} = 525 \times 0.8929 + 441 \times 0.7972 + 695 \times 0.7118 - 900 = 415$$

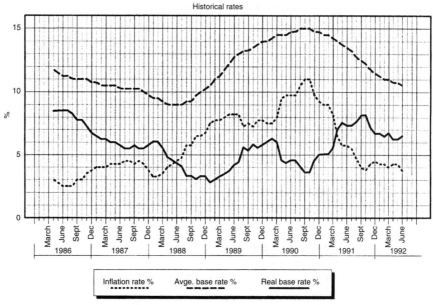

Figure 15.12   UK interest and inflation rates (1986 to 1992)

Table 15.6   Cash flows at current prices

Year	Outflow (cost)	Inflow (benefits)
0	900	
1		500
2		400
3		600

Table 15.7   Cash flows in money

Year	Outflow (cost)	Money inflow (benefits)
0	900	
1		$500 \times 1.05 = 525$
2		$400 \times 1.05^2 = 441$
3		$600 \times 1.05^3 = 695$

The factors 0.8929, 0.7972 and 0.7118 have been taken from Table A4.1 for an interest rate of 12%.

Now, by analysing the project in real terms, using the real rate of interest, which is

$$\frac{(1 + 0.12)}{(1 + 0.05)} - 1.00 = 0.067$$

that is, 6.7%, the NPW is found:

$$NPW = \frac{500}{1.067} + \frac{400}{1.067^2} + \frac{600}{1.067^3} - 900 = 415$$

$$\mathbf{NPW(real) = NPW(money)}$$

This is a valuable finding; whether a project is analysed in real terms or in money terms, the result will be the same.

15.10 PAST COSTS AND PROJECTS WITH INDEFINITE LIVES

This section explains how past costs should be dealt with and considers projects with indefinite lives.

15.10.1 Past costs are sunk costs

Two years ago we invested £28k in Project A. Since then nothing has happened; there have been no earnings, no costs. If we invest £70k *now* in this project it will earn £24k p.a. for the next 10 years. Alternatively, if we invest £85k now in Project B it will earn £26k p.a. for the same period. If the interest rate is 10%, which should we choose? Cash-flow diagrams are shown in Figure 15.13.

$$NPW_a = 24 \times 6.1446 - 70 = £77.46k$$
$$NPW_b = 26 \times 6.1446 - 85 = £74.47k$$

The earlier investment of £28k is irrelevant. Past costs are sunk costs and are common to both projects; therefore they can be ignored (if we proceed with B we have still incurred the £28k). Choose Project A.

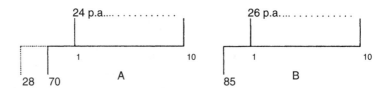

Figure 15.13 Cash flow diagrams

15.10.2 Projects with infinite/indefinite lives

(1) Determine the net equivalent annual value (NEAV) of a project that has an initial cost of 100 monetary units and earns benefits of 20 p.a. indefinitely. Use $i = 12\%$.

Assume indefinitely means that the number of years is very large. Consider the formula

$$\text{EAV of } C_0 = C_0 \frac{i}{\left(1 - \dfrac{1}{(1+i)^n}\right)}$$

where C_0 = initial cost

$$\text{when } n \text{ is very large} \frac{1}{(1+i)^n} \text{ tends to zero,}$$

and, therefore,

$$\text{EAV of } C_0 = C_0 \times i = 100 \times 0.12 = 12$$

therefore

$$\text{NEAV} = 20 - 12 = 8$$

(2) Determine the NPW of the project in (1) above. Use $i = 12\%$. In (1) an initial cost of 100 has been converted into equal annual repayments of 12 for an infinite period when $i = 12\%$, that is:

$$\text{annuity payments} = i \times \text{present value (of cost)}$$

therefore,

$$\text{present value} = \frac{(\text{annuity})}{i}$$

$$\text{NPW} = 20/0.12 - 100 = 66.67 \text{ or NPW} = 8/0.12 = 66.67$$

In short, the present value of an indefinite stream of equal annual benefits or costs is found simply by dividing the annual sum by the interest rate. Conversely, a single sum now can be converted into an annual value for ever by multiplying the sum by the interest rate. Such simplicity is welcome.

15.11   REVIEW

The foregoing sections have covered a considerable amount of material that, although simple in terms of numerical procedures, often gives rise to conceptual difficulties. Because of this, a brief review of the study material is provided in advance of the summary.

An investment involves early costs and later benefits that usually arise at periodic intervals. During its life the investment has a changing value in recognition of the predicted future net benefits.

Payback is a useful but simplistic method of assessing short-term investment projects.

DCF makes the assumption of a perfect capital market. This allows the investor to use NPW as an unambiguous parameter of gain or loss, permitting a set of projects to be considered independently of whether the investing enterprise finances its operations by borrowing or by using its own capital.

Projects may be assessed in several ways: NPW, EAV, B/C ratio, IRR, NPW/C_0 ratio. The method adopted will depend on the priorities of the investor. If the intention is to maximize present worth, then NPW should be used. If the intention is to minimize annual costs or to maximize net annual benefits, then the use of EAVs would be appropriate. For many long-term projects the investor will be far more concerned about the regular (annual) distribution of revenue than about any present value of the investment, therefore the EAVs of such projects will provide far more useful information. If the intention is to take careful account of future costs and benefits, rather than merely net benefits, in order to ascertain value for money or in order to assess the true nature of the costs and benefits, then cost–benefit analysis should be used. Where the analyst needs to compare yields, the IRRs of alternative investments can be found, and where the principal criterion for an investment decision is to obtain maximum value for capital employed at a given interest rate, then the NPW/C_0 ratios should be examined.

In drawing up a cash-flow diagram the investment analyst is making certain implicit assumptions of which he or she should be acutely aware. These are:

- certainty about the initial cost;
- certainty about future costs and future benefits;
- certainty about project lifetime;
- perfect capital market;
- nonvariable interest rates;
- how inflation will be taken into account.

The following questions should illustrate these points:

(1)   When a builder is asked to tender for the construction of a new laboratory building, does the owner really believe that the tender sum is what he or she will eventually pay for the factory?

(2)   When future operational and maintenance costs for this building are estimated, how confident can the owner be that these will be the actual costs?

(3)   When future profits/income from a project are estimated, can the owner be confident that these will be achieved?

(4)   Why has a 30-year lifetime for the laboratory or a 10-year life for a given piece of equipment been selected?

(5)   How can the interest rate(s) selected for discounting be justified?

In view of these uncertainties, engineers, scientists and technologists as well as investment analysts should avoid evaluating projects to several decimal places (!) and, when the quantitative differences between projects are small, they should pay careful heed to the nonquantifiable attributes of the projects.

A method of dealing with these uncertainties is *sensitivity analysis*. Sensitivity analysis examines the effects of changing costs, benefits, project lifetimes and discount rates to determine how sensitive the decision is to such changes. It may be that a change in lifetime from 20 to 30 years has far less effect than a 5% change in initial cost for a given project; if that were the case, a careful review of the estimating procedures by which the initial cost has been obtained and/or a review of the contractual arrangements for commissioning the project would be of far greater value than worrying about how long the investment would last. The process of sensitivity analysis will usually involve setting up a spreadsheet for the project by which the effects of changing a variety of parameters can be examined rapidly and thoroughly.

15.12   EXERCISES

Mastery of this material will only be achieved through practice. Some simple exercises are provided on the website (www.wileyeurope.com/go/chelsom) in which students are asked to determine NPVs, B/C ratios and IRRs for a number of different cash-flow diagrams. These are followed by a collection of more interesting and more exacting exercises that have the aim of improving the student's understanding of the concepts.

15.13   SUMMARY

This chapter has discussed the utility of payback as a means of investment decision analysis, and studied in some detail discounted cash flow techniques. The concepts of net present value, equivalent annual value, benefit–cost ratio and internal rate of return have been explained. The presumption of a perfect capital market has been justified. Alternative ways of dealing with inflation have been set out: either by analysing projects in money terms or in real terms. The purposes of life-cycle costing and cost–benefit analysis were discussed. Finally, methods of dealing with past costs and indefinite lives have been explained.

REFERENCE

1.   Layard, R. and Glaister, S. (eds) (1994) *Cost–Benefit Analysis*, 2nd edn, Cambridge University Press, Cambridge.

BIBLIOGRAPHY

Brent, R. J. (1997) *Applied Cost–Benefit Analysis*, Edward Elgar, London.

Lanigan, M. (1992) *Engineers in Business*, Chapter 11, Addison-Wesley, Reading.

Mott, G. (1999) *Accounting for Non-Accountants: A Manual for Managers and Students*, 5th edn, Kogan Page, London.

Thuesen, G. J. and Fabrycky, W. J. (1989) *Engineering Economy*, 7th edn, Prentice-Hall, London.

Wiggins, K., (2000) Discounted Cash Flow (CPD study pack), College of Estate Management, Reading.

16

Maintenance Management

16.1 INTRODUCTION

In many sectors of industry, maintenance has been regarded as a necessary evil and often has been carried out in an unplanned and reactive way. It has frequently lagged behind other areas of industrial management in the application of formal techniques and/or computing technology. Yet expenditure on maintenance can be a significant factor in a company's profitability. In manufacturing, maintenance typically accounts for between 2 and 10% of income, and in transport of goods by road or in civil aviation for up to 24%. Maintenance costs can represent a significant element in the life-cycle costs of some projects, as indicated in Chapter 15. In the UK, national expenditure on maintenance is about 5% of the value of the sales of goods and services and exceeds the amount invested in new plant and equipment.

Modern management practice regards maintenance as an integral function in achieving efficient and productive operations and high-quality products, and high levels of equipment reliability are increasingly demanded by 'lean manufacturing' and 'just-in-time' operations (see Chapter 20). This chapter outlines the objectives of maintenance, the concept of failure rate, the design of maintenance systems and the provisions for maintenance strategy and maintenance planning.

16.2 OBJECTIVES OF MAINTENANCE

The need for maintenance in most engineering operations is self-evident. Without maintenance, the plant and equipment used will not survive over the required life of the system without degradation or failure. The design and operation of a maintenance system must usually meet one of two objectives:

- Minimize the chance of failure where such failure would have undesirable consequences (e.g. environmental damage or reduced safety).
- Minimize overall cost or maximize overall profit of an operation. This requires striking a balance between the cost of setting up and running the maintenance operation and savings generated by increased efficiency, prevention of downtime and so on.

Maintenance may also take place for other reasons such as corporate image, for example, cleaning the windows of an office building or repainting an aircraft.

The objectives of a maintenance programme may be expressed formally in terms of maximizing plant effectiveness. A total productive maintenance (TPM) strategy attempts to maximize overall plant effectiveness (OPE). This is defined as follows:

$$OPE = \text{Availability} \times \text{Efficiency} \times \text{Quality}$$

where

Availability = the proportion of time the plant is available to do its job, expressed as a percentage

Efficiency = the 'output' of the plant as a percentage of its maximum output

Quality = the percentage of products that meet the required specification

In maximizing this measure, it is implied that full output of the plant (Availability × Efficiency) is required. If it is not, maximizing OPE may not be cost-effective and the objective of maintenance should then be to achieve the required output and quality in the most cost-effective way.

Maintenance will affect each of the constituents of OPE. Ultimately, a balance must be struck between the cost of achieving a high OPE and the benefits it brings. Decisions about directing maintenance efforts to availability, efficiency or quality must be made, again on a cost-effective basis.

16.3   PLANT DETERIORATION AND FAILURE

A knowledge of expected failure rates is fundamental to the design of a maintenance operation. A review of some key concepts is given here. There is an extensive literature on reliability analysis; some titles [1-3] are suggested at the end of this chapter.

16.3.1   Failure rate

The failure rate (or hazard rate, $H(t)$) for components, equipment or plant is normally expressed in failures per hour (or failures per thousand or even per million hours). The figures represent an average for the type of equipment, and the expected number of failures in any time period can only be expressed in terms of probabilities. Failure rates can be used to estimate the number of failures expected in the plant and give an initial indication of the maintenance effort required.

Failure rates are a function of time, typically exhibiting the well-known 'bathtub' curve shown in Figure 16.1. This is characteristic of many types of electrical and mechanical engineering equipment, as well as of people!

Three phases may be observed during the life of a component, although in practice the boundaries are often ill-defined.

In *Phase 1*, the wear-in phase, the failure rate is initially high but it reduces over a short time. Failures in this phase are due to flaws and weaknesses in materials or faults in manufacture. Such failures can be minimized by use of defect-prevention techniques such as failure mode and effect analysis during equipment and process design – see Chapter 4 and Appendix 2. The wear-in characteristic has implications on the provision of warranties for new equipment,

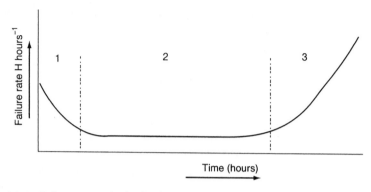

Figure 16.1   Failure rate – the bathtub curve

but of greater significance is that it will result in a high initial failure rate. On installations of a critical nature, an initial testing programme, often under severe conditions, is often undertaken to reveal Phase 1 failures before the system is put on line. The implication for maintenance is that engineers should be prepared for a heavy maintenance effort on newly commissioned plant. The high failure rate of new equipment is not generally appreciated, which can lead to image problems. The press devoted much coverage in 2003 and 2004 to the new rolling stock on Connex South East trains, which were initially breaking down more frequently than the old stock they were replacing. It is also worth noting that wear-in characteristics can be evident after maintenance, perhaps to a less pronounced degree, as a result of new parts being fitted or of faulty workmanship in repair.

In *Phase 2*, the useful-life phase, the failure rate is low and approximately constant, representing random failures. The length of this phase, sometimes referred to as durability, will vary with the type of equipment. In this phase the mean time to failure is the reciprocal of the (constant) failure rate.

In *Phase 3*, the wear-out phase, the failure rate increases due to ageing, wear, erosion, corrosion and related processes. The rate at which the failure rate increases varies with the type of equipment. For much equipment, especially electronics, no wear-out is observed within the working life and there is some debate about whether it occurs. Wear-out has an influence on maintenance policy; if an item has a distinct wear-out characteristic it can be replaced before failure is likely, for example the brake pads on a car.

16.3.2   Reliability

Reliability $R(t)$ is defined as:

'the probability that an item will perform a required function under stated conditions for a stated period of time.'

It is related to failure rate by the relationship

$$R = e^{-\int Hdt} \text{ (see Figure 16.2)}$$

Initially the reliability should be 1.0 and eventually it will tend to zero, as all engineering systems will eventually fail or deteriorate to such an extent that

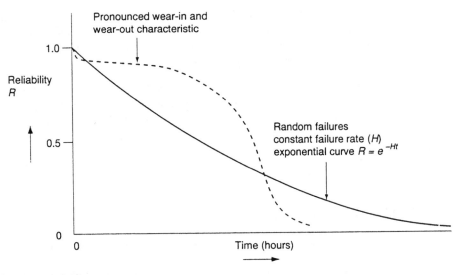

Figure 16.2   Reliability curves

they can no longer perform their intended function. Figure 16.2 shows typical reliability functions. The area beneath the reliability curve is the mean time to failure of the item.

16.3.3   Data sources

Obtaining failure rate data requires testing of many components over long periods of time. This is easy enough for inexpensive items made in large numbers, such as electronic components and ball bearings. For expensive equipment, data are obtained from in-service failure records, since a testing programme could be expensive and would take too long. For components with a very low failure rate, accelerated testing is usually employed; for example, testing of electronic components is often undertaken at a high temperature and the failure rates at normal operating temperatures calculated by means of an empirical relationship (see [1]).

Published data sources generally give failure rate figures for Phase 2 of the 'bathtub' curve; some sources are listed in the references [4–6]. In using the data one should be aware of the uncertainty associated with these quoted figures, and some sources give an indication of this. Also, the operating conditions under which the data were obtained must correspond to those of the applications being considered. Some sources [4] allow adjustments to be made for different operating conditions. Data on wear-in and wear-out characteristics are often not available. Information on failure modes is sometimes given if an item can fail in one of several different ways.

16.4   DESIGNING MAINTENANCE SYSTEMS

There is no universally accepted methodology for designing maintenance systems. Sufficient information or time for analysis is generally not available to allow a fully structured approach leading to an optimal solution. Instead, maintenance systems are designed using experience and judgement assisted by a number of formal decision aids. The two sections that follow outline a two-stage approach to the maintenance system design process:

- *Strategy:* Deciding on which level within the plant to perform maintenance, and outlining a structure that will support the maintenance.
- *Planning:* Day-to-day decisions on what maintenance tasks to perform and providing the resources to undertake these tasks.

Even here, it must be emphasized that there is not a once-through process and a degree of feedback and of reiteration will be needed.

16.5   MAINTENANCE STRATEGY

In formulating a maintenance strategy three key points have to be determined:

- At what level (see below) within the plant the maintenance is to be performed.
- What structure is needed to support the maintenance.
- What resources are needed.

To illustrate the factors that influence these decisions, consider the example of a machine tool shown in Figure 16.3.

Consider the position if a failure of the lubrication system is observed. Judgement will tell the engineers that the replacement of the whole machine (Level 1) or of the whole lubrication system (Level 2) would not be sensible. However, it is not obvious whether maintenance should take place at Level 3, 4 or 5. Factors that affect the decision are summarized below:

- *Diagnosis time:* The time taken to locate the fault. This will usually be longer at lower levels of the hierarchy. Erratic lubricant flow is a fault of the pump unit,

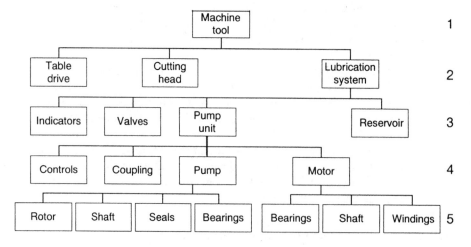

Figure 16.3   Assembly hierarchy for a machine tool

which is immediately obvious (Level 3). To determine the cause of, say, bearing failure in the motor (Level 5) will take much longer.

- *Off-line time:* Includes shutdown, isolation, removal, repair, replacement, test, start-up. In other words, the total time that the plant is unavailable as a result of the maintenance action. This usually will be lowest for maintenance at middle levels in the hierarchy (e.g. replacing a motor is quicker than replacing one of its bearings or the whole lubrication system). Unavailability of spare parts can be a factor.
- *Off-line cost:* The loss of income resulting from off-line and diagnostic time.
- *Repair cost:* Includes labour and an allowance for the capital cost of diagnostic, repair and test equipment and the space it occupies, and other overhead charges (energy, administration etc.). This cost is usually lower if maintenance is undertaken at middle levels in the hierarchy.
- *Spares cost:* The cost of spares consumed. This reduces further down the hierarchy (the cost of a bearing is clearly less than the cost of a pump unit), although the associated ordering and storage costs may at some point increase.
- *Expected failures:* The number of failures increases higher up the hierarchy. The pump unit (Level 3) fails every time a bearing or a seal fails (Level 5).

The decision must also include whether to provide a two- (or multi-) level process. It may well be prudent to replace the pump unit quickly to get the machine back in operation, and then dismantle it in a workshop to repair the cause of the failure. For equipment where it is not cost-effective to employ people with the specialist skills to carry out repairs 'in-house', units may be returned to the supplier for repair.

Analysis of the above factors will indicate at which level maintenance is most cost-effective. In practice, such decisions are made, but rarely as a result of a thorough cost analysis. More often it is a case of the application of experience and judgement in the first instance, with adjustments made during the life of the plant. The level at which maintenance is carried out may be:

- at different levels on different branches;
- different for different fault modes at the higher levels;
- different for emergency and planned maintenance.

For a plant of any complexity, the analysis clearly involves a significant effort for managers and engineers.

Once decisions on the level of maintenance have been made, a structure to support the maintenance operation must be devised. Figure 16.4 indicates a typical arrangement that may be expanded or reduced depending on the circumstances.

Simple cases will not require the centre layer (workshop), for example, computer systems in small or medium-sized companies where faulty items are either simply replaced (e.g. a keyboard) or returned to the supplier for repair. Larger operations may have an extra layer (site workshop, main base), especially where there are many operational units remote from the supplier.

Within the structure, resource requirements in terms of people (skills), tools and equipment must be identified. The allocation of tasks to people must be considered. Generally it is better that routine tasks (lubrication, cleaning, adjustment) be

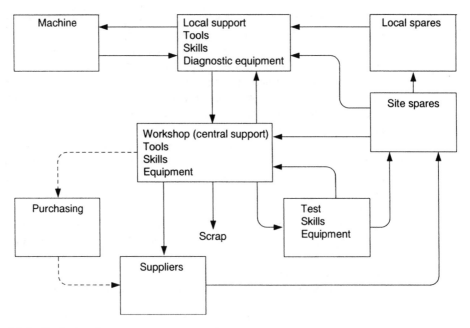

Figure 16.4   Typical maintenance support structure

made by the operator who knows the machine rather than calling on a separate person. In many industries, efforts are being made to increase the level of maintenance (and repair) carried out by work-area teams rather than by specialist maintenance personnel.

16.6   MAINTENANCE PLANNING

Within the overall maintenance structure the day-to-day planning of tasks must take place. The starting point for this planning is to decide on the basis for maintenance, which may be one of the following:

- *Operate to failure (OTF):* Where the consequence of failure is small and the time to failure is difficult to predict, this might be the most appropriate planned method of maintenance (e.g. replacing light bulbs).
- *Time-based activities:* Where maintenance to prevent failure (e.g. painting a bridge) or to replace parts that have deteriorated (spark plugs of a petrol engine) is undertaken at regular intervals of time or some related variable (miles travelled, in the case of a car). There may be statutory requirements to undertake regular inspections/tests and repairs.
- *Condition-based activities:* Where deterioration can be detected through monitoring and can be justified financially (e.g. cost of instrumentation), maintenance actions can be based on conditions.

A planned maintenance programme must be able to respond to unexpected failures that may have to be dealt with immediately (emergency maintenance) or scheduled into the daily plan.

Reliability-centred maintenance (RCM) provides the basis for a maintenance plan by considering the failure characteristics of the plant and its components and the cost and consequences of failures. A flow chart summarizing the decision process is shown in Figure 16.5. Clearly, all items in the assembly hierarchy cannot be treated independently. If a pump is being taken apart to replace the seals, it is sensible to check the bearings. Many activities have to be concentrated into major shutdown periods.

Maintenance planning needs to be a dynamic process. The time-based tasks are known well in advance, condition-based tasks shortly in advance and emergency tasks will arise with little or zero notice. Flexibility is usually achieved through backlog tasks or by moving forward or back the time-based tasks whose timing is not critical. Consideration should be given to this when creating the master schedule for the time-based work. Figure 16.6 illustrates the problem.

A further consideration in creating the plan is the availability of 'maintenance windows', possibly evenings or weekends or periods of the year when demand is usually slack. The question of opportunity maintenance also arises; when equipment is shut down and dismantled for a particular repair it may be prudent to undertake other tasks. Such actions may be planned to a limited extent.

The level, strategy and planning decisions lead in turn to requirements for spare parts, resources, diagnostic equipment and labour. Again, this will generally be a reiterative and interactive process.

Traditionally, control and information systems for the maintenance plan have been paper based. However, today computer packages are available, the more sophisticated of which are linked to purchasing and stock control functions.

Proper recording of information is the key to the operation of a planned maintenance programme. Analysis of the information can lead to improvements to the programme and greater efficiency in the use of spare parts, equipment and people. Recording is an important aspect of maintenance management, but people tend to get the work done and move on to the next task, forgetting the record keeping. Computer-based systems have helped record keeping to become more 'user-friendly', but have led to the temptation to record everything – this too must be resisted. Recorded data can be used for:

- reviewing and updating strategy;
- identifying probable areas where redesign could avoid failures and/or maintenance;
- calculating effectiveness ratios, for example, percentages of unplanned work, or maintenance expenditure as a percentage of income;
- satisfying statutory requirements for keeping maintenance records and tracing parts used.

It is usually beneficial to pre-plan as much maintenance activity as possible and to minimize unplanned work. In this way manpower and resource requirements are largely known in advance, although unexpected failures leading to unplanned work can never be entirely eliminated. The potential benefits of planned maintenance are outlined below:

- Reduces maintenance costs in the long term.
- Reduces equipment failures.

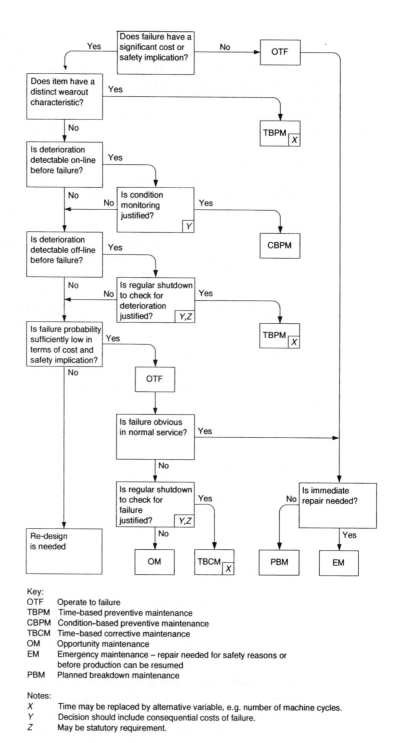

Key:
OTF   Operate to failure
TBPM   Time-based preventive maintenance
CBPM   Condition-based preventive maintenance
TBCM   Time-based corrective maintenance
OM   Opportunity maintenance
EM   Emergency maintenance – repair needed for safety reasons or
        before production can be resumed
PBM   Planned breakdown maintenance

Notes:
X   Time may be replaced by alternative variable, e.g. number of machine cycles.
Y   Decision should include consequential costs of failure.
Z   May be statutory requirement.

Figure 16.5   Maintenance policy decision chart

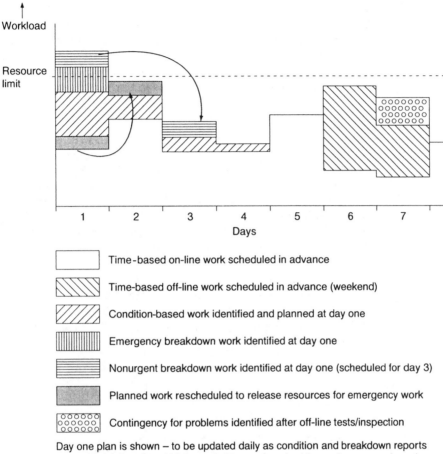

Figure 16.6   Typical maintenance resource plan

- Significantly reduces disruption caused by failure.
- Increases life of equipment.
- Improves performance of equipment.
- Improves utilization of people.
- May meet legal requirements (for example health and safety).

The inhibitors to the implementation of a planned maintenance scheme need to be recognized by management. They are largely concerned with the sometimes considerable effort and costs at start-up. These can include:

- creating an asset inventory by means of a plant survey;
- costs of designing, setting up and installing the system;
- costs of training and supervision;
- increased direct maintenance cost in the short term, as the backlog of neglect is put right.

16.7 SUMMARY

Planned maintenance will reduce the frequency of failure and the risks of unexpected shutdown to plant. This chapter has outlined the objectives of maintenance, has considered the concepts of failure rate and reliability, and the provisions for maintenance strategy and maintenance planning.

REFERENCES

1. Lewis, E. E. (1987) *Introduction to Reliability Engineering*, John Wiley & Sons, Chichester.
2. O'Connor, P. D. T. (2002) *Practical Reliability Engineering*, 4th edn, John Wiley & Sons, Chichester.
3. Smith, D. J. (2000) *Reliability, Maintainability and Risk: Practical Methods for Engineers*, Butterworth-Heinemann, Oxford.
4. U.S. Department of Defense (2002) *MIL-HDBK-217, Reliability Predictions of Electronic Equipment, Issue F* (www.itemsoft.com/mil217).
5. International Atomic Energy Agency (1988) *TECDOC-478, Reliability Data for Use in Probabilistic Safety Assessment*, IAEA, Vienna.
6. American Institute of Chemical Engineers (1989) *Process Equipment Reliability Data*, AIChE, Chicago, IL.

BIBLIOGRAPHY

Capterra, *Maintenance Software Directory*, `www.capterra.com /maintenance-management-solutions`.

Dhillon, B. S. (2002) *Engineering Maintenance: A Modern Approach*, Technonic Publishing, Lancaster, PA.

Levitt, J. (1997) *Handbook of Maintenance Management*, Industrial Press, New York.

Moubray, J. (2001) *Reliability Centred Maintenance*, Industrial Press, New York.

Nyman, D. and Levitt, J. (2002) *Maintenance Planning, Scheduling and Co-ordination*, Industrial Press, New York.

Smith, A. and Hinchcliffe, G. (2003) *Reliability Centred Maintenance: Gateway to World Class Maintenance*, Butterworth-Heinemann, Oxford.

17

Project Management

17.1   INTRODUCTION

The art of managing large projects preceded managing production processes by more than 4000 years. From the Great Pyramid (2450 BC) to the Languedoc canal (1681), from the Pont du Gard aqueduct (c. 15 BC) to the English railways of the 1840s, men and machines and materials were brought together on a massive scale for a time and then disbanded. Yet 'project management' as a recognized intellectual discipline or, more correctly, methodology has only been around for a few decades. Does this mean that the term 'project management' is merely a new name for those past practices that combined common sense and experience? This chapter will show that the formal adoption of project management methods should provide managers, working to very tight schedules, with the means to manage large and complex projects.

The aims of a project manager today are no different from those in previous centuries, namely to bring together for a finite period of time considerable resources in the form of manpower, machines and materials in order to complete the project on time, to specification and within budget. Today, however, the project manager can use a number of tools that will keep track of a multitude of different parallel tasks allowing better informed decisions to be made quickly and with confidence. Complex and concurrent engineering processes need clearly defined relationships between supplier/contractor and purchaser; the project manager must implement policies to this end. It is with these project management tools and a study of alternative relationships between purchaser and supplier that this chapter is concerned.

Although much of the material in this chapter is particularly applicable to the construction of (large) projects, the principles apply equally to manufacturing projects, installation of computer systems or even setting up, say, major one-day events for sport or entertainment. Readers will find the concepts of *concurrent (simultaneous) engineering* and *fast-track engineering*, previously discussed in Chapter 6, developed further in this chapter.

Project management is a powerful methodology, one with which all members of a project management team should be familiar. By following the methodology, young engineers, scientists and technologists should be in a strong position to take on responsibilities that might in earlier times have been assumed by their elders.

17.2   WHAT IS PROJECT MANAGEMENT?

A project is a specific, finite task in which the means for its completion must be created for its duration. Each project will be unique. It will be dependent on parent organizations for its resources, especially people. Each project, almost inevitably, will be subject to change, which may mean a departure from plan, programme and budget. Project management is the overall planning, implementation, control and coordination of a project from inception to completion to meet defined needs to the required standards, within time and to budget.

17.2.1   Appointment of the project manager

Historically the term 'project manager' has been used rather loosely. In the USA the person in charge of any construction site, large or small, was usually referred to as the project manager. He or she was, in fact, the manager of the construction project, and many months of work had been completed by many parties long before he or she could take up the appointment. The tasks of the project manager concern:

- project definition;
- planning;
- decision-making structures;
- monitoring and controlling.

These are considered in sections 17.2.2 to 17.2.5.

Project management means the management of the whole process. Therefore, if continuity is desired, the appointment of the project manager should be for this same period of time. Where the project involves a joint venture, the project manager may be selected from one of the participating companies. Where the future owner of the project is a developer of industrial, commercial or infrastructure projects, the owner may appoint one of his own staff to be project manager. Where the project is a one-off development, the future owner will probably hire a project manager specifically to manage that project. In this chapter the term project manager means *the owner's project manager*, selected to manage the project on the owner's behalf. However, at the separate and lower levels of a project, the managers who are responsible for planning, implementing and controlling operations at their particular level will be carrying out a project management role too. For large and medium-sized projects, the project manager will be supported by a project management team. In this chapter the term project manager means both the individual and the team.

In view of the considerable responsibility entrusted to the project manager, selection, training and authority are of great importance. The Latham Report [1] recommends that the terms of appointment and duties should be clearly defined. The project manager should be given the necessary authority to ensure that the work is carried out satisfactorily through to completion without frequent reference to the owner.

17.2.2   Project definition

Morris and Hough [2] assert that a key element in the success of a project is project definition. A project must be properly defined and resources should not be fully committed to the project until the definition is shown to be workable. In this introduction to project management, the underlying assumption is that the processes of feasibility studies, financial appraisal and risk assessment have been satisfactorily completed. This is referred to as the proposal stage in some literature. Project definition is more than project specification: project definition contains the objectives of the project as well as the physical characteristics and features of the project. For a building or a power scheme the task of project definition should be relatively straightforward and, therefore, pursued with vigour by the project manager. The sooner the project is fully defined the better; an imprecise definition leaves potential for change, and changes are a major cause of delays and extra costs in projects. A project that aims to provide an all-embracing computer system for a large, multidisciplinary company will, necessarily, involve a gradual approach, from a broad definition through a number of more closely defined objectives to a final set of carefully developed specifications. In this case the project manager must maintain continuous pressure on those whose responsibility it is to agree the specification.

17.2.3   Planning

Project planning is often confused with project management. Successful project management requires a well-thought-out plan, but that is merely the beginning. Project planning must be followed by project control. Project management software is largely planning and monitoring software. Project planning means:

- planning methods;
- choosing between 'in-house' services and external suppliers/contractors;
- deciding on cash flow (the cost plan or budget);
- deciding on the schedule of operations (the timing plan).

Project management software[1] allows an iterative approach to developing the plan. From an initial draft plan, data on the sequence and duration of operations (activities), on resources and on costs are input into the computer. Analysis of the output leads to modification of the plan, for example, to improve resource management, until a satisfactory plan is obtained. This iterative approach is further illustrated in section 17.3.5.

17.2.4   Decision making

The project manager must ensure that decisions are made on time and that they have no adverse effect on the budget. To achieve this in scientific and engineering projects, the project manager must establish structures for technical and managerial decision making. This means that he or she must determine by whom and at what level different kinds of technical decisions have to be made and by whom and at what level decisions on expenditure, placing orders, entering into contracts, programming and time limits have to be made.

---

[1] A discussion of project management software would warrant a chapter of its own, and would become dated fairly swiftly. The Bibliography includes one book on project management software.

17.2.5   Monitoring and controlling

As the work proceeds it must be measured against the plan. A computerized plan will permit computerized monitoring, providing extensive data on physical progress, resource use and costs. This is, of course, conditional on the input of these data. A significant feature of modern project management is the amount of paperwork and/or direct data input, often on a weekly basis, that must be provided. This is the price for faster projects and more effective control; if these goals are not realized a lot of time and money is being wasted. Monitoring, therefore, needs the wise application of input and careful control over the nature and extent of output and its appropriate distribution to different levels of responsibility. Computerized monitoring should provide the project manager with a measure of any departure from the plan and he or she will be able to examine the potential effects of any remedial measures such as the reallocation of resources or alternative sequencing before they are implemented.

17.3   PROJECT PLANNING

17.3.1   Method statement

First, of course, the methods of working have to be decided. This is often referred to as a *method statement*. Usually the methods to be used for most of the project will be well known and well tried. The method statement will be based on experience and the technical resources available. Only occasionally will there be alternatives of such a profound difference that the implications for the success or otherwise of the project need rigorous investigation by the project manager. It is not intended in this chapter to pursue this. However, Morris and Hough [2] observe that an important contributor to failure in major projects in the past has been the placing of undue reliance on new technology, on a large scale, when its use and development until then had been on a small scale. The development of Concorde and of the advanced gas-cooled reactors in the UK exemplify this. This does not rule out the use of new technologies on a large scale for major projects, but it does suggest that very generous contingency provisions should be allowed for in any plan relying on new technological development.

17.3.2   Work breakdown structure

The key to the project plan is a *work breakdown structure*. The planner must be able to provide a clear and accurate statement of the scope of work. A work breakdown structure (WBS) provides a rational subdivision of the work in hierarchical form down to the lowest level of discrete work packages from which estimates of resource requirements, durations, linkages and costs can be determined.

A WBS can provide the basis for a coding system by which any work package (activity) or any subgroup of activities can be identified. From the WBS a list of activities and precursor activities can be produced for the purposes of network analysis, from which programmes, resource charts and cost plans all flow. Network analysis is dealt with in Chapter 18.

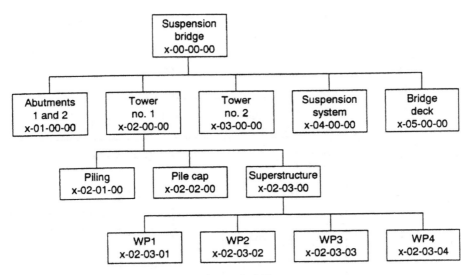

Figure 17.1   Work breakdown structure (not shown in full)

Figure 17.1 shows a WBS for the construction of a suspension bridge. The principle is that all the work at a lower level is contained within the scope of work at the higher level from which it branches. In the USA, charts of this form are sometimes referred to as 'gozinto' charts because each lower element 'goes into' the element above it. Thus the scope of work of x-02-00-00 includes each of x-02-01-00, x-02-02-00 and x-02-03-00. An objective of a WBS is clarity, which suggests a maximum subdivision into four or five elements and as many levels. A WBS in which each element at one level was subdivided into four elements at the next level down would result in 256 work packages if there were five levels to the structure!

17.3.3   Cost plan

A WBS provides a useful means of estimating the cost of a project. Estimates of costs are entered at the lowest level and 'rolled up' to successive levels. Once work is in progress the actual costs are entered at the lowest levels and rolled up to the higher levels. In presentational terms this works very well. Figure 17.2 shows a typical hierarchy of cost estimates by which a cost plan is obtained or by which actual costs can be measured.

17.3.4   Networks and bar charts

From the WBS a 'milestone' schedule will be produced identifying the several major milestones that must be achieved on the way to completion. With the aid of this schedule and from the WBS, individual activities and subactivities (work packages) are obtained. From an examination of the relationships between individual activities, the whole complex network of relationships between activities can be obtained. (This is developed in Chapter 18, section 18.2 and in Appendix 3, Case

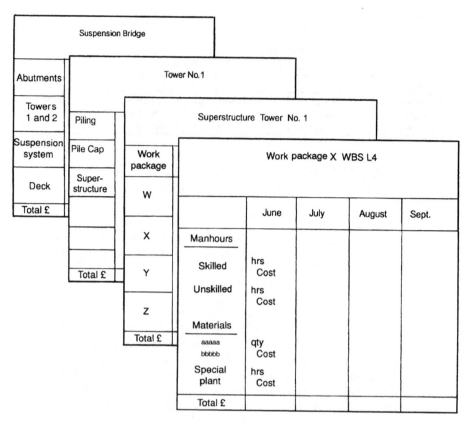

Figure 17.2   Cost plan

study: Developing a network.) The planners must then estimate the durations of the individual activities. This can be done either by examining the external time constraints, deciding how long an activity should take and applying the resources to meet that duration, or by examining the resources that are available or would be practicable for that activity and estimating how long it would take to complete. In each case the planners have the problem of estimating the productivity of the resources in the given conditions.

Since each project is unique and the problems likely to be encountered are somewhat different from those of previous projects, this is where the art (rather than the science) of project planning comes to the fore. Once the durations and linkages of all activities have been estimated, a network can be drawn from which a bar chart (or Gantt chart) can be obtained. A bar chart is by far the clearest means of representing a programme of work. Bar charts, too, can be developed for different levels of management showing successive levels of detail, based on the WBS. Figure 17.3 illustrates a long-term programme.

17.3.5   Resource charts and curves

From the bar chart, resource charts can be obtained. For each activity the planners identify the resources needed; for example manpower per day, cranes per day,

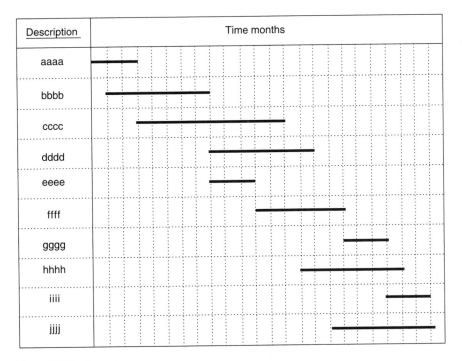

Figure 17.3   Long-term programme

quantities of materials. Manpower and cranes are nonconsumable resources; that is, using a crane for one day does not prevent its use on the next. Materials are consumable resources. By aggregating resource requirements across all activities shown on the bar chart on each day of the project, a daily resource chart for each resource can be obtained. Figure 17.4 illustrates how a resource chart is derived from a bar chart.

Resource levelling or resource smoothing is often used to eliminate undesirable peaks in demand for certain resources. If, for example, the daily requirement for welders throughout the project lay between 4 and 6, but for one week the demand rose to 15 because of the concurrence of a number of activities, then the sensible approach would be to reschedule the activities to remove that peak. If any of these activities were not on the *critical path*, then rescheduling might have no effect on the overall duration of the project. The critical path is the sequence of interdependent tasks whose total duration determines the overall length of the project. (See Chapter 18 for a full account of networks and critical path analysis.) If the number of welders could not be reduced sufficiently without rescheduling critical path activities, thus making the project longer, the planner would need to weigh up the comparative disadvantages of bringing in nine extra skilled personnel for one week against the project finishing marginally later. Project planning software usually provides a facility for resource smoothing or resource optimization, which allows the project manager to observe the effect and come to a decision.

The ideal resource chart for labour is one in which the labour demand rises to a plateau at which it remains for most of the project, dropping rapidly at the end. This affords a rational recruitment programme followed much later by a properly

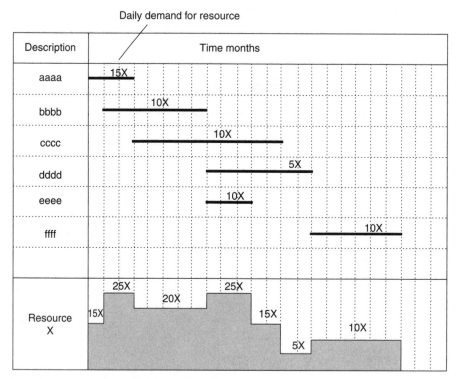

Figure 17.4   Resource chart derived from a bar chart

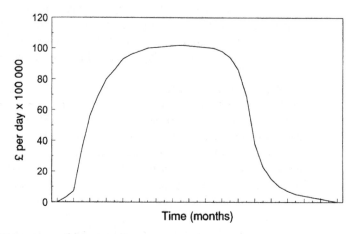

Figure 17.5   A monthly expenditure curve

planned reduction. A labour demand chart with peaks then troughs then further peaks and so on would be highly undesirable, leading to industrial relations problems (as layoffs were followed by recruitment followed by more layoffs) or the retention of unnecessary labour, or relocation of some of the workforce for short periods of time, causing much inconvenience. Projects have succeeded in which such fluctuations in labour demand have been experienced, but usually only

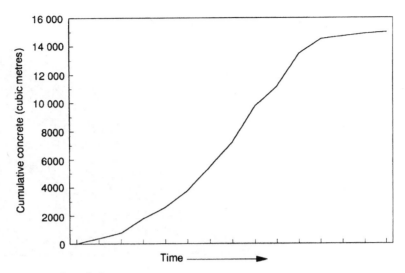

Figure 17.6   Cumulative resources curve

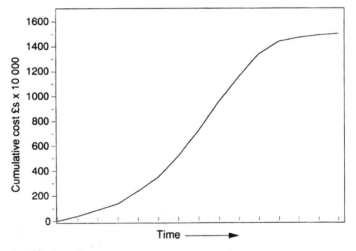

Figure 17.7   The S-curve

where most of the labour is subcontracted. Labour-only subcontractors, serving a range of projects, are better able to move their labour force between projects to suit these fluctuations. For other resources such as plant and equipment it may be possible to hire in the extra resource for short periods (although short-term hire may be more expensive).

In short, resource charts are of particular value to the project manager in respect of labour, facilitating resource smoothing. A fluctuating demand for materials and/or equipment is of less concern but can be indicative of poor planning, therefore where such fluctuations arise, a careful examination of the programme should be made.

Since each resource represents a cost, the summation of the costs of all the resources being utilized on a daily basis will provide the project manager with

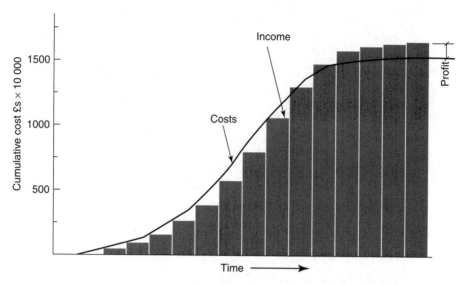

Figure 17.8   Level of borrowing during a project and anticipated profit

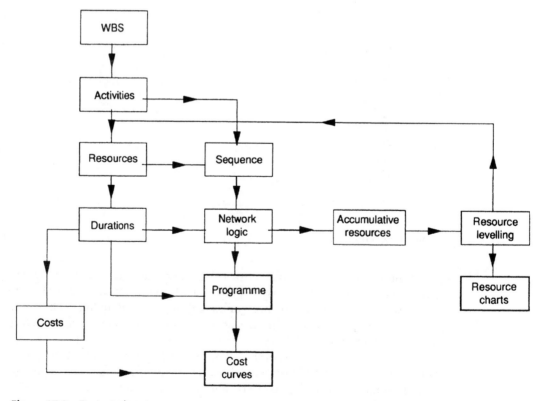

Figure 17.9   Project planning processes

a daily or weekly expenditure curve, as shown in Figure 17.5. The shape of the curve is typical of large projects. The low level of expenditure at the end would represent the period of making good, testing and commissioning. Cumulative resource and cost curves may also be obtained; see Figures 17.6 and 17.7.

Generally, cumulative curves are only of value for consumable resources, such as concrete for a construction site, and for controlling costs. The so-called S-curve in Figure 17.7 is typical for large projects. This S-curve is used to analyse progress (in terms of costs and time), as will be seen later in this chapter. An S-curve will also be used by a contractor/supplier to determine the level of borrowing during a project and the anticipated profit at the end. This is illustrated in Figure 17.8.

17.3.6   Summary of project planning

A work breakdown structure (WBS) provides a logical subdivision of the work from which programmes, resource charts and cost plans can be developed and presented at appropriate levels of operation. A primary function of a WBS is to identify the work packages; from the durations, resources and costs of each work package the programme, resource charts and cost curves for the whole project will be developed. This process is illustrated in Figure 17.9.

17.4   PROJECT MONITORING AND CONTROL

The project manager must:

- measure performance;
- control change;
- minimize delays;
- minimize extra costs;
- control and coordinate suppliers/contractors.

The first four of these will be examined in turn; control and coordination of suppliers and contractors will form the final part of this chapter.

17.4.1   Measuring performance

Performance on large projects will be measured at regular intervals; the period chosen is frequently that of a calendar month. *Monthly meetings* attended by the project manager and representatives of the current suppliers/contractors allow coordination problems to be resolved and can be a very effective means of putting pressure on anyone whose performance in the recent past has been unsatisfactory. Monthly meetings can be the single most effective driving force in a project and their value should not be underrated. A young scientist or engineer, responsible for part of a project, attending such a meeting would be expected to have up-to-date information on progress and costs, be able to explain why there had been any departure from the plan, and be able to predict performance during the coming month(s), making recommendations for remedial measures where necessary.

Performance needs to be measured in terms of time and cost. This is commonly done by monitoring performance against a bar chart and a cumulative cost curve.

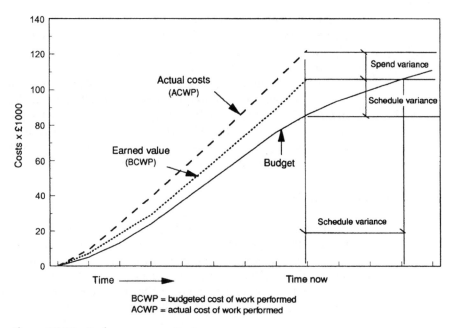

Figure 17.10   Performance monitoring curves

Earned value analysis provides a measure of performance in which time and cost form an integral whole. The method also requires that, at each measurement date, forecasts for the expected date of completion and the expected cost at completion be made. The project manager is not merely provided with a rear view mirror to drive by, but a view along the road ahead as well.

In Figure 17.10 the budgeted cost curve (baseline), the actual cumulative costs and the *earned value* of the completed work are recorded. This earned value is the budgeted cost of the work performed (BCWP) to date. From Figure 17.10, measures of the spend variance and the schedule variance can be read. The spend variance shows the difference between what should have been spent on that amount of work and what was actually spent. The schedule variance gives a measure of how far ahead of schedule the project is both in terms of time and in terms of programmed spending. The implications are that for this project getting ahead of schedule has been achieved at a price, namely the spend variance.

The following performance ratios are used in earned value analysis:

$$\text{cost performance index} = \frac{\text{ACWP}}{\text{BCWP}} \times 100$$

$$\text{cost variance} = \frac{(\text{ACWP} - \text{BCWP})}{\text{BCWP}} \times 100$$

$$\text{schedule variance} = \frac{(\text{BCWP} - \text{BCWS})}{\text{BCWS}} \times 100$$

$$\text{schedule performance index} = \frac{\text{BCWP}}{\text{BCWS}} \times 100$$

$$\text{cash-flow variance} = \frac{(\text{ACWP} - \text{BCWS})}{\text{BCWS}} \times 100$$

where

  ACWP is actual cost of work performed
  BCWP is budgeted cost of work performed
  BCWS is budgeted cost of work scheduled

The *cost performance index* tells the project manager by how much project costs are above or below budget.

The *cost variance* provides useful information to the estimators. If the cost variance is within 2 or 3% they have done a good job and they can use the same data for estimating costs of new projects. If the cost variance is of the order of 10% or 15%, then the estimators will need to reexamine their cost data and decide whether the problem lies with productivity on the current project alone or whether the data need to be revised for future projects.

The *schedule variance* provides similar feedback to the estimators on rates of progress, allowing them to confirm or modify their current data on rates of progress for similar kinds of operation.

The *schedule performance* index tells the project manager by how much the project, in general, is ahead of or behind schedule.

The *cash-flow variance* tells the project manager how far cash flows have strayed from programme. This is most important, as it may mean that the project owner (purchaser) will have to go to his or her lenders and reschedule the financing of the whole project.

These performance indices need to be used in conjunction with one another; each one taken in isolation is of very limited value. The indices may be used at several levels of a WBS by the managers responsible at those several levels. Work package managers should be required at each measurement date to give two additional pieces of information for that package:

• the estimated cost to completion;
• the estimated time to completion.

From this the estimated cost to completion of the whole project can be 'rolled up' through the work breakdown structure, and revised completion date(s) of the project and/or of its several elements can be estimated.

$$\text{Estimated cost at completion (EAC)} = \text{ACWP} + \Sigma(\text{ECW})$$

where ECW = estimated cost to completion for each work package

The project manager can provide a further measure of progress towards completion of the project from the equation

$$\frac{\text{ACWP}}{\text{EAC}} \times 100 = \% \text{ complete}$$

Example 17(1)

Sixteen months into a two-year £24m project

$$\text{actual cost of work performed (ACWP)} = £17.8\text{m}$$

$$\text{budgeted cost of work performed (BCWP)} = £18.3\text{m}$$

$$\text{budgeted cost of work scheduled (BCWS)} = £19.2\text{m}$$

$$\text{estimated cost to completion } \Sigma(\text{ECW}) = £7.2\text{m}$$

What is the predicted position for completion?

Solution

First the EAC
$$\text{EAC} = 17.8 + 7.2 = £25.0\text{m}$$

(the project is forecast to overrun by £1.0m).
   Then progress:

$$\text{schedule performance index} = \frac{18.3}{19.2} \times 100 = 95.3\%$$

This shows that progress is unsatisfactory: the project is behind schedule by 4.7% of 16 months; that is, approximately two-thirds of a month behind schedule.
   The percentage complete can also be found:

$$\% \text{ complete} = \frac{17.8}{25} \times 100 = 71.2$$

At this stage the project should be

$$\frac{19.2}{24} \times 100 = 80\% \text{ complete}$$

which indicates that progress is far less satisfactory than the schedule performance index would suggest.
   The cost performance index appears satisfactory

$$\text{cost performance index} = \frac{17.8}{18.3} \times 100 = 97.3\%$$

but in view of the estimated cost to completion, it appears that an underspend to date may result in extra costs over the remainder of the project. The project manager will need to seek further information on the reasons for the revised estimates and, at the same time, determine whether progress can be accelerated to bring the project back on schedule.

17.4.2   Controlling change

Probably the greatest single cause of delay and extra expense in most projects is change. Construction contracts very often contain provisions for change. Conditions of contract provide for variations in the scope of work and set out, usually by reference to bills of quantities (see section 17.7.2), the methods by which the extra costs can be valued and charged. Long experience in construction has taught its practitioners that on most projects problems arise that can only be overcome by changes in design and/or construction methods. The unpredictability of ground conditions is one very good reason for this. Since many construction contracts may last for several years, it is possible that the owner's/purchaser's needs might change. Unfortunately, a system that facilitates change can have the effect of encouraging change. Many small changes may be authorized by the owner or project manager, the accumulative effect of which is to disturb the regular progress of the work and cause considerable delay and expense. Therefore, the project manager for a large project must establish firm change control procedures. This can be achieved by:

- using *change request forms*;
- establishing a *change control board*.

Numbered change request forms, suitably coded to match the WBS, should identify who initiated the request, the nature of the change sought, the reason for the change, an estimate of the cost of the change and any revisions in programme. If, for example, a work package manager made the request, then one would expect him or her to provide the actual cost of work performed, the budgeted costs of work performed and scheduled, and the estimated time and cost to complete the work package. The project manager would define the levels of authority in terms of scope of work and costs. The change control board, which would meet regularly, would monitor all minor changes (having defined the limits of minor change) and would be the sole authority for major changes.

17.4.3   Minimizing delay

The project manager should anticipate the several possible causes of delay and have policies in place to mitigate the effects that such delays would have on the overall programme. A major cause of delay is change. Other causes are the late supply of technical information, late deliveries of material and equipment and underestimation of the time needed for some activities. Rigorous programme controls on the design process should resolve the first, close liaison with suppliers (see Chapter 21) should address the second, and better feedback from projects to estimators and planners might prevent the third. Re-examination of the project network and a reallocation of resources might be required where a serious underestimation of the time needed has occurred.

The studies of Morris and Hough [2] have shown that 'external forces' have seriously delayed major projects and that, too often, these external forces could have been anticipated and a contingency strategy put in place. These external forces have, typically, been political, environmental or social influences, the seeds of which were sown during the inception stages of the project. One of the most laudable features of the Thames Barrier Project in London, completed in the 1980s, was the extensive consultation with formal and informal bodies to obtain a consensus agreement that would meet the commercial, political, social and environmental needs of a widespread community. During the construction of Hartlepool nuclear power station in the early 1970s, the Inspectorate of Nuclear Installations introduced a new requirement that the reactors be provided with tertiary emergency shutdown facilities at the time when work on the reactors was already underway. The direct delays to the programme were of the order of 18 months; the indirect effects of such delays are often incalculable. The decision to introduce a tertiary shutdown facility could not have been totally unexpected since the Inspectorate worked closely with the nuclear power generators; project management strategies could have been prepared that would have mitigated the effects of these changes.

An alternative or complementary approach is to provide incentives for early completion. Bonus payments can be helpful as long as there are strict controls to ensure that the contractor/supplier does not sacrifice quality (or operate unsafely) in order to save time. In the public sector a system of 'lane rental' has been used by the Department of Transport in the UK. The principle is that a contractor appointed to carry out repair work or modifications on an existing highway is charged a weekly fee for the 'rental' of the traffic lanes that are out of use. The shorter the duration of the work, the more money is saved on the contract cost; in effect, a bonus for early completion.

17.4.4   Controlling extra costs

A project may cost more because of additional work, or lower productivity than planned, or inflation. The issue of additional work has already been discussed in the preceding section. Low productivity in one area may sometimes be counterbalanced by higher productivity elsewhere. If low productivity is experienced on a major expensive activity, costs inevitably will rise. A change in the budget by reducing later expenditure might help. The estimators, certainly, ought to be advised (see section 17.4.1). A gradual improvement in productivity through learning by experience might mitigate the effects. This is where employing a team with wide and appropriate experience is so important.

Experience is important too in those projects where the physical constructs of several engineering and scientific disciplines meet to form an integral whole. A design may lack 'buildability'. A lack of buildability is usually a consequence of the designers and detailers having insufficient experience in installation and/or construction. There is much virtue in employing designers who are experienced in the installation or construction of the work in hand. There is even greater virtue in employing designers seconded from the companies that are going to install or construct this part of the project. These designers should be employed both to advise on planning the work (the method statement) and to give advice on 'buildability'. At the design stage the cost of redesign is minimal compared to the cost of delays consequent on a redesign once the work is underway. At the interface between engineering disciplines there is an even greater risk of such problems. These problems can be avoided if either all drawings are submitted to a working group whose task is to examine, comment on and approve details at the interface, or computer-aided design (CAD) software with a 'walk-through' facility is used so that these interfaces can be looked at from all angles, or where 'mock-ups' or prototypes are assembled.

Manufacturers (for example the motor industry and the aircraft industry) have also experienced similar problems of 'buildability' and incompatibility at the engineering interface. As explained in Chapter 6, the solution has been *concurrent (simultaneous) engineering*, in which the designers and suppliers are brought closer (physically and in terms of time and task) to the point of assembly.

The most commonly proffered reason for cost overruns is inflation. Projects of the 1970s and 1980s show costs spiralling uncontrollably upwards. In inflationary times, can the project manager exercise any control over rising costs? Does it even matter, since, presumably, the completed facility will reap benefits at inflated prices anyway? The major difficulty is one of confidence. The purchaser will be borrowing from the banks or using up cash reserves at an alarming rate. The banks may lose confidence in the project and threaten to withhold funding. The original estimates (January 1986) for the cost of the Channel Tunnel between Dover and Calais came to a grand total of £5.4bn, including £0.9bn for inflation and £1bn for contingencies. In February 1990, by which time the estimated cost to completion had risen to almost £7.5bn, the group of some 200 bankers funding this project almost withdrew their support from Eurotunnel (the owners and operators) halfway through the project. Thereafter, the performance of the project managers for Eurotunnel was monitored as closely by the bankers as they, the project managers, monitored the project. This did not make life easy for the project managers! (The final cost was close to £10bn.)

Usually the effects of inflation are not uniform. A government's measure of inflation (the Retail Price Index[2] in the UK), in which certain consumer prices are measured and weighted into a formula, is unlikely to reflect the effects of inflation on a project. Wage inflation is usually greater than consumer price inflation, and materials may experience completely different rates of inflation, as in the 1970s when the prices of oil and its derivatives increased fivefold in a few months. What strategies should the project manager have in place to counter the effects of inflation? One solution is to enter into fixed-price contracts with suppliers and contractors, such that no increased payments will be made where the suppliers' cost increases are a consequence of inflation. This hard-nosed approach places all the risk on the suppliers, who should, according to this philosophy, be able to control their own costs. Unfortunately, this is seldom the case, particularly in inflationary times when employers will be equally unable to keep wage levels and the prices paid to their suppliers down. Fixed-price contracts should only be used for short periods of time, not exceeding two years, even when inflation is low. Effective cost control on, say, a four-year contract can be exercised by releasing a succession of short fixed-price contracts, none of which is likely to be put at serious risk by inflation. As tenders are received for each new contract the growing costs can be reviewed, the budget revised and estimates of future income examined in a rational manner; the process will be under control.

The alternative, where short contracts are inappropriate, is to use price fluctuation clauses within the contract by which the sums payable to the supplier/contractor are adjusted according to a previously agreed formula when the supplier's costs rise or fall. Since most industries monitor the rates of inflation within that industry, the development of a suitable formula or the direct use of an industry formula is a relatively simple matter. This method is reasonably fair to the supplier and means that at the time of tendering he or she does not have to try to guess the future rates of inflation and build them into the price.

In summary, cost control policies should make provision for:

- minimizing change;
- realistic and carefully reviewed estimates of productivity;
- reviews of buildability;

---

[2] See Chapter 15 for further information on measuring inflation.

- interface management;
- short-term fixed-price contracts;
- price-fluctuation clause agreements for long-term contracts.

17.5 CONTROLLING AND COORDINATING SUPPLIERS, CONTRACTORS AND SUBCONTRACTORS

Sections 17.5 to 17.9 will consider the management of the relationships between the project owner and his or her suppliers. (See also Chapter 19 for special consideration of contractual relationships in the construction industry.)

Depending on the size and nature of the project, the following questions may have to be considered.

(1) Should the work be done 'in-house' by the project owner's workforce or by external consultants, suppliers and so on?

(2) Should design and manufacture/construction be separate activities?

(3) How should suppliers be paid? By fixed-price contracts or contracts with price-fluctuation clauses (variable price contracts)? By lump-sum contracts, measurement contracts or cost-reimbursement contracts?

(4) How should suppliers be selected? By selective or open tendering? By lowest tender? By negotiation? By 'partnering'?

The answers to these questions will influence the procurement strategy adopted by the project manager. Some of the issues are discussed in sections 17.5.1 to 17.5.3.

17.5.1 How should the work be distributed?

Usually the answer to question (1) above should be 'no'. The project owner's business will not normally be that of project management and execution, therefore it is unlikely that he or she will have the skills to carry out the work. Specialists who have developed their expertise within one field should be in a better position than the project owner to produce the work to specification, on time and at a competitive rate. Even if the project owner has the skills, carrying out the project would mean a substantial reduction in normal business activity, which, from the long-term point of view of maintaining and developing one's market, might be unwise.

Whether the work should be shared between in-house and external providers will depend on the skills and availability in-house. Even where skills are available, it would be sensible to obtain competitive quotations for the work, as outside businesses may have the competitive edge in the tasks to be done. In the UK, local authorities are required by law to seek competitive tenders from contractors for work for which the local authority has both the resources and the skills. In large companies the decision on how much of the project should be carried out by the project owner must depend on the intended project management structure, how dependent this structure will be on the parent organization, and how much dependence can be placed on the functional departments, whose main objectives will be to meet the business needs of the company and not the needs of the project.

17.5.2   Subcontracting

The most common reason for appointing a subcontractor is that the subcontractor has specialist skills not available to the main contractors and suppliers. Generally, contractors should be free to subcontract parts of their work to others if they wish. The project manager can always, by a term in the contract, reserve to himself or herself the right to approve/refuse the contractor's choice of subcontractor. It is not uncommon in the UK to adopt a system of nomination, by which the project manager requires a main contractor to enter into a subcontract with a given supplier or specialist; this can be particularly valuable where the specialist provides a full service from design through manufacture to installation and completion. In law, the principle of privity of contract applies by which the project owner cannot sue a subcontractor for unsatisfactory performance; it is the contractor who is responsible to the project owner for the work carried out by his or her subcontractors. Therefore, it is most important that the contract between the project owner and the contractor clearly defines the contractor's obligations and does not release him or her from any responsibility should the subcontractor default. Because of the principle of privity of contract, nomination has led to some legal/contractual difficulties, especially in the UK construction industry where it is frequently used.

17.5.3   Separation of design and manufacture/construction

Construction projects can provide useful illustrations of project management. Figure 17.11 illustrates how construction projects are typically managed. The project has two phases: the design phase and the construction phase.

The engineer is responsible for design, the contractor for construction. During construction the engineer, on behalf of the promoter (project owner), supervises

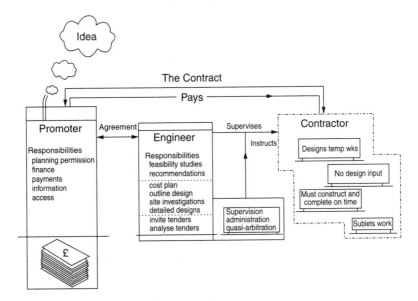

Figure 17.11   Construction project procedures

Figure 17.12   Over-the-walls engineering

the contractor's work. The role of the engineer as project manager is not as clear as it might be.

In most construction work the contractor is selected by competitive tender, the successful tenderer being the lowest bidder able to meet all the requirements of time and specification.

If all the design work precedes all the construction work, the project must take longer than if some of the work can be done concurrently. If the contractor has no involvement until all the design is complete, there is no check on buildability or project 'interfaces'. If the contractor has no input at the design stage, then the overall plan of the work is dictated by decisions made by the engineer, for the nature of the design may determine the project plan. The contractor is usually only given six weeks in which to prepare a tender; in this time he or she must provide an outline plan of the work and price it. Surely, the planning of a large project of several years' duration warrants more attention to planning than this? Shouldn't the planning of the design and of the construction be an integral activity?

What has been demonstrated here is that, in construction and, indeed, for many large projects, 'over-the-walls' engineering is the norm, when greater consideration might be given to concurrent engineering. Figure 17.12 provides a graphic illustration of 'over-the-walls' engineering; section 17.6 will consider some of the alternatives. (Over-the-walls engineering was also discussed in Chapter 6.)

17.6   APPROACHES TO CONCURRENT ENGINEERING IN CONSTRUCTION PROJECTS

Alternative methods of project management in the construction industry have become quite widespread in recent years. They have been used largely to overcome the very frequent problem of projects taking far longer than originally estimated with no apparent redress for the owner. The most important of these alternative methods are considered below.

17.6.1   Design-and-construct contracts

In the construction industry, *design-and-construct contracts* have been developed and quite widely used in the UK. Under a design-and-construct contract the contractor carries out the design as well as the construction. This has several advantages:

- The design, almost by definition, will be 'buildable'.
- Design and construction planning form an integral process.
- Construction work can commence long before the design is complete (early start means an early finish).

There are, of course, disadvantages:

- How does the project owner obtain strictly comparable competitive bids?
- Who pays for the expensive tendering process in which contractors have to develop outline designs and price them, knowing that their design may be rejected on grounds of client need, if not on cost?
- Contractors do not employ a wide range of design expertise, therefore, critics might argue, this approach only works well where designs are standard, unimaginative and straightforward.
- There is no independent supervision of the contractor's construction work by the engineer/architect.[3]

Design-and-construct contracts have been widely and successfully used in the chemical and oil industries, in the provision of 'turnkey' contracts for power stations, waste-water treatment plants and so on. (A turnkey contract is one in which the contractor undertakes full responsibility for the design, procurement, construction, installation, testing and commission of the project.) These are especially successful where the contractor appointed has considerable experience in the field.

Two other methods of project management have been widely used in the building industry in the UK in the last two decades: *management contracting* (MC) and *contract (or construction) management* (CM). Each has the same objectives: to bring the design process into parallel with the construction process (*concurrent engineering*) and to shorten the project duration (*fast tracking*).

17.6.2   Management contracting (MC)

In management contracting the promoter (project owner) appoints a firm of contractors to carry out the project management for the construction of the project. The management contractor (MCr) manages the project, does no physical construction work, and usually carries out the supervision normally done by the engineer/architect's representative. A fee plus direct costs is paid for this service. Once authorized by the promoter, the MCr appoints subcontractors progressively for discrete work packages: earthworks, foundations, superstructure, cladding,

---

[3] See further discussion of this issue in Chapter 19.

air-conditioning, lifts, ceilings and so on. The design work, carried out by consulting engineers, architects, subcontractors and suppliers, is also produced in work packages allowing a scheduled release of information to meet the construction programme; there is a phased tendering programme and a progressive appointment of subcontractors. A strong discipline must be exercised over the design process: designers will be put under pressure to meet contractors' deadlines.

Where fast tracking is intended, the design process is followed very closely by the supply/manufacture, installation and construction processes. By this means the owner will earn revenue from the facility much sooner. There are risks, however. Any necessary design changes may result in expensive modification of components or even their rejection. Last-minute problems may result in expensive delays, while incompatibilities at the engineering interfaces are resolved, and genuine mistakes may be made by personnel under pressure.

The total fee payable to the MCr will be between 5% and 10% of the cost of the project. This is a large extra cost and is unlikely to be worthwhile unless completion to time and to budget are critical and the MCr improves considerably the likelihood of completing the project to time and budget. The promoter pays the MCr for all the subcontractors' work; the MCr then pays the subcontractors. This cash flow provides the MCr with useful opportunities for short-term money investments. The proclaimed advantages of MC include:

- Completion on time and to budget.
- Permits an early start, hence an early finish.
- Promoter and contractor do not have an adversarial relationship.
- Buildability (the MCr reviews all designs for buildability).
- Planning and design decisions can be integrated.
- Individual subcontracts are relatively short, so can be fixed price.

Observed disadvantages have been:

- The MCr minimizes his or her contractual risk, so that when a subcontractor's performance is unsatisfactory the promoter finds that there is no redress from the MCr, and he or she can bring no action against the subcontractor/supplier.
- The price for the whole project is not known until all of the many subcontracts have been let.

17.6.3   Contract/construction management

In contract/construction management the procedures are similar to those of MC. The most important difference is that the contract/construction manager (CMr) does not enter into contracts with the various contractors. The promoter enters into contracts with each contractor whose work is controlled and supervised by the CMr. The CMr only receives a fee plus direct costs. The CMr can be appointed well before any site work begins, to manage the design and planning process. A proclaimed advantage of CM is that all contractors feel part of a team rather than being at the lower level of a project hierarchy. The great advantage that this approach has over MC is that if a contractor's performance is unsatisfactory, the promoter can use the contract terms to protect his or her interests. The risks in fast tracking are the same as those for MC.

CM and MC are, therefore, alternative forms of project management. It is quite possible for a promoter to adopt CM on one contract and MC on another. At No. 1 Canada Square (the tallest building in the UK) in the Canary Wharf development in London, the developers Olympia and York started the tower block under a management contract, but approximately halfway through the contract changed to contract management using the developer's own team of project managers for the contract management. This was not so much a reflection of the developers' view of management contracting as a commentary on their relationship with their management contractor. The successful transition from MC to CM halfway through the project was a direct consequence of the similarity between the two approaches to project management.

17.7   HOW SHOULD CONTRACTORS/SUPPLIERS BE PAID?

Contractors will usually base their tenders for work either on labour costs per hour (or day or week) plus materials, or all-in costs per unit quantity of the work performed.

17.7.1   Lump-sum contracts

A tender for design services usually includes a sum to cover the estimated number of hours to carry out the work, applying different rates for the time spent by engineers, by technicians and so on. Sums to cover overhead and profit are added. Contracts of this nature are usually *lump-sum* contracts, in which 'lump sums' are paid according to an 'activity schedule' as specified work packages are completed. The risk lies with the tenderer, who has to estimate how much resources (and time in this case) are needed to meet the objectives.

In large projects different parts of the work may be paid for in different ways. The design, manufacture, installation, commissioning and testing of large mechanical or electrical plant are usually carried out on a lump-sum basis, while the building and/or structure to house or support the plant could be carried out using a measurement contract (see also Chapter 19, sections 19.4.1 to 19.4.3).

17.7.2   Measurement contracts

In *measurement contracts*, bills of quantities are drawn up that identify a large range of work items and state the amount of work to be done in terms of a quantity, for example cubic metres of excavation, or metre lengths of pipe. The contractor enters a rate per unit quantity for each work item (see Figure 17.13 for a typical extract from a bill of quantities). For each item the contractor must include for labour, plant and material costs, as well as for overhead and profit. This might seem difficult at first glance, but the construction industry has traditionally operated in this way and, therefore, has a sophisticated and up-to-date record of costs and prices in this form. As the work proceeds, the contractor is paid, usually on a monthly basis, for the quantities of work carried out. In construction work the final quantities often differ from the original estimated

Item no.	Description	Unit	Quantity	Rate £/unit	Amount £
	EXCAVATION				
E323	Excavation for foundations 0.5 - 1m max depth	cu. m	450	6.50	2925.00
E532	Disposal of excavated material	cu. m	450	1.50	675.00
	IN SITU CONCRETE				
	Reinforced concrete				
F623	To bases 300 – 500 mm thick	cu. m	100	65.00	6500.00
F642	To walls 150 – 300 mm thick	cu. m	60	75.00	4500.00
	PRECAST CONCRETE				
	Beams				
H136.1	Prestressed post-tensioned 7 – 10 m long Type A1 for central span	nr.	4	5000.00	20 000.00
H136.2	Prestressed post-tensioned 7 – 10 m long Type A2 for north span	nr.	4	7000.00	28 000.00
H136.3	Prestressed post-tensioned 7 – 10 m long Type A3 for south span	nr.	4	8000.00	32 000.00
				page total	

Figure 17.13   Extract from a bill of quantities

quantities, so the contractor is paid for the work done rather than the work originally estimated.

Measurement contracts are best suited to work in which there is a tradition of measurement, where changes in quantities are anticipated as the work proceeds and the physical characteristics and dimensions of the work are designed in some detail prior to the work being carried out. Lump-sum contracts are best suited for work in which the performance of the finished project is all important, but where every single detailed dimension is of less consequence. In the USA the use of lump-sum contracts for construction work, especially buildings, is fairly common. When so used, there is much less opportunity for change as the work proceeds; the contractor has tendered to the specification and drawings provided, so even small changes will be seen as changes in scope and new prices will have to be renegotiated or the work paid for at rates that will favour the contractor. This certainly exercises a formidable prejudice against change! The Sears Tower building in Chicago, at that time (1970s) the tallest as well as one of the largest buildings in the world, was constructed using, in the main, lump-sum contracts. At No. 1 Canada Square, referred to above, most of the separate contracts were let as lump-sum contracts.

Both lump-sum and measurement contracts require much of the design work to be complete before a contractor can tender for the work. However, a *cost-reimbursement contract* is ideally suited for projects in which the owners want construction work to begin while design work is still underway (concurrent engineering).

17.7.3 Cost-reimbursement contracts

A cost-reimbursement contract is one in which the contractor is paid his or her direct costs plus an additional sum for overheads and for profit. As the work proceeds, the contractor must provide evidence of his or her direct costs by invoices for materials and components, records of wages paid, records of plant and equipment hired. Thus, a contractor can be instructed to attend at a given location with certain items of plant and a small labour force and be put to work immediately. Even if the drawings are insufficient, even if operations are delayed as a result, the contractor will be paid by the hour for plant, labour and materials. The risk lies entirely with the project owner. Why, therefore, should a project owner wish to use this form of contract?

It is ideal for a project in which the nature of the work is far from clear at the start and only gradually becomes clear as the work proceeds. It is especially suitable for projects in which there is a considerable element of risk, in which, until the work is in progress, the actual techniques have not been established. An example of this would be the welding of 75 mm thick high-strength steel for the lining of a reactor, where very high standards of welding are specified. Until nondestructive tests (and destructive tests) of sample welds and actual welds on the lining have been conducted for a variety of welding techniques, the approved procedures cannot be established, therefore neither lump-sum nor measurement pricing would be appropriate.

Project owners will only use cost-reimbursement contracts where they have to. Where they are used it is most important that the contractor's operations are firmly supervised. The project manager will want assurances that the resources (the manpower and machinery) are being used effectively and efficiently all the time and that all the materials purchased are actually being used on the project! The project managers will take a much closer interest in the contractor's paperwork than for other kinds of contract. A difficulty in this kind of contract is to provide the contractor with an incentive to keep costs down. This is certainly not achieved by paying the contractor on the basis of a cost plus a percentage fee, for, in this case, the greater the cost, the greater the fee to the contractor.

Alternative approaches include:

- cost plus a fixed fee (as costs rise, the fee as a proportion of total payment falls);
- cost plus a sliding fee (as the costs rise, the fee as a percentage falls, e.g. 7% for first £5m, 6% for next £1m, 5% for next £1m);
- target cost plus fee.

In a target contract, a target cost plus a fee are set, for example £100m cost plus a £10m fee. If the contractor carries out the work for £98m, then a saving of £2m has been made on the target; this saving is shared between contractor and project owner. If the share agreement were 50:50, then the contractor would be paid £99m plus the £10m fee. If, however, the contractor carried out the work for £102m, he would have to share the excess with the project owner and would only be paid £101m plus the £10m fee. Once the cost of the work carried out reached £120m, the contractor would, in effect, be carrying out the work at cost. Above this level the contractor would not be penalized further and would continue to be paid at cost; his or her incentive for efficient operations would be gone. There is one difficulty with this method – how is the target cost set? The main reason for using this kind of contract is its ability to cope with projects where the exact scope of work is unknown and hence the cost is unknown. A target contract is far more effective where the scope of work is fully understood and where the project owner is anxious to save money and prepared to provide the contractor with an appropriate incentive.

17.8   HOW SHOULD SUPPLIERS/CONTRACTORS BE SELECTED?

The alternatives in any project are:

- by open tendering;
- by selected tendering;
- by negotiation.

In each of the first two there will often be an element of post-tender negotiation. Negotiation, itself, can lead to a number of alternative contractual arrangements.

In competitive tendering, whether 'open' or 'selected', the project manager sends the contract documents (drawings, specifications, conditions of contract etc.) to interested suppliers/contractors. When a contractor returns the documents, duly completed, this is the tender or offer. Legally, once that offer has been accepted a contract has been made.

It is far easier to assess tenders when they are all on exactly the same basis. Parity of tendering is important for publicly accountable bodies, which will have strict procedures for the receipt and analysis of tenders. Some public bodies, as a matter of principle, will not consider tenders that propose alternative solutions for carrying out the project, because any alternative is not strictly comparable with tenders conforming to the specified project solution. Whether this approach provides the public with the best service and value is questionable; certainly there will be occasions when potentially good ideas by suppliers/contractors will not be heard.

Open tendering is when the work is advertised widely and any interested party may apply for the contract documents and then tender. *Selected tendering* is where the project manager invites three or four potential suppliers/contractors to tender. These tenderers will usually have been required to provide evidence of their experience, technical expertise and financial position. In the UK selected tendering has long been favoured for most projects; however, for large public works and for all large infrastructure developments in transport, water, energy and telecommunications, overriding EU legislation demands wider advertisement across the EU for pre-qualification than many purchasers would prefer. The advantages and disadvantages of selected tendering are given below.

Advantages of selective tendering:

- Contract documents are only sent to a few contractors.
- Recipient contractors know that the chances of being awarded the contract are reasonably high (1 in 4, say) and, therefore, their tendering costs, which must be recouped from their customers, are lower than if, say, every tender had a 1 in 20 chance of success.
- Analysis of tenders by the project manager does not take long.
- Pre-qualification of tenderers ensures they have the technical ability and financial capacity to carry out the work.

The disadvantages of selective tendering are that the process does not test the open market for value or performance or innovation. There is also the risk that a small group of contractors or suppliers might, over a period of time, be encouraged by the system to form a cartel.

The alternative to competitive tendering is *negotiation*. Normally, public bodies will not use negotiation unless the work is so specialized that there is only one known contractor who can do the work. Private companies are usually at liberty to deal with other companies as they wish. There can be very good reasons for approaching a company, asking it to carry out some work and, in advance of that work, negotiating the methods and rates of payment. Lump sum, measurement or cost reimbursement could be used. Negotiation is likely to be adopted where the two parties have worked well together in the past, have a good working relationship, trust one another and do not adopt an adversarial stance when problems arise.

In the past, negotiation has been quite widely used on an *ad hoc* basis by a range of project owners, from building developers to purchasers of high-technology advanced weapons systems. For these relationships to work well it is usually important that the project owner has considerable technical expertise in the field, to allow negotiations to proceed with confidence. In the offshore industry some customers and suppliers have entered into *alliances* in which the two parties have negotiated a form of target-price contract providing financial incentives for all parties to work towards a common goal, and in which the project management team comprises engineers and managers from customer and suppliers whose aim is to solve problems on the basis of trust rather than by resorting to contractual positions.

Where the project owner has a long-term commitment to project development, for example, an oil company with intentions to develop offshore oil fields around the world, entering into long-term *partnering agreements* with suppliers and contractors may have a number of advantages. The cost of constructing an offshore platform and its complementary installations is very high and the potential losses that can arise from late provision or unsatisfactory performance within the first few weeks or months of production are huge. Realistic levels for liquidated damages for late completion, which would provide the purchaser with adequate compensation, would be so great that no contractor could afford to pay them, so the risk of late completion has been born largely by the purchaser. Therefore, historically, the oil companies took a close interest in everything that their suppliers/contractors did, but at the same time they insisted that only the lowest bid would be accepted. This had the effect that even experienced contractors, at best, enjoyed only a tenuous relationship with the oil companies, which inhibited long-term investment in skilled personnel, engineering, information systems or plant and equipment that would improve the provision of these services.

In recent years, the oil companies have adopted very different policies with a view to reducing their operating costs by as much as 30%. *Partnering* is one such approach. Partnering is a long-term commitment between purchaser and supplier/contractor in which two or more parties work together to reduce their costs in the long term. Partnering has been established in the USA for some years, particularly in the processing and chemical industries, and more recently in Europe in the motor industry, where closer customer/supplier relationships have permitted short development times, high quality standards and competitive pricing. This practice has now become more common in the UK construction industry among clients of the industry who have a long-term commitment to development, for example the water industry. The use of partnering agreements was commended by Sir John Egan in his report on the construction industry published in 1998 [3] and has subsequently been taken up by a number of large commercial developers (see Chapter 19, section 19.5).

In the offshore industry, an alternative form of partnering has sometimes been used in which the responsibility for design, detailed engineering, procurement and construction management has been given to one company, which forms a team with the customer. Formal agreements for four or five years have been entered into, with the understanding that it is the intention of both parties to extend the agreement as the agreement period reaches its end. The project management team comprises engineers and managers from customer and contractor. The relationship is based on trust, a dedication to common goals and a mutual understanding of each other's expectations and values. A National Economic Development Council report [4] concluded that the conditions for successful partnering are that there should be:

- a significant long-term core programme;
- careful selection of the right partner;
- mutual trust and confidence between partners;
- commitment to a long-term relationship;
- preparedness to adopt each other's requirements;
- willingness to accept and learn from each other's mistakes;
- a compatibility of culture between customer and contractor.

Where there is mutual trust and compatibility of culture, the parties are unlikely to adopt adversarial stances and resort to the contract when things go wrong. Sorting the problem out together must be a better way. Lawyers, it should be noted, are not generally so sanguine about these practices.

17.9 SUMMARY

This chapter examined the role of the project manager in project definition, planning, decision making, monitoring and controlling the project. The importance of work breakdown structures was identified and earned value analysis was introduced. Change as a major cause of delay was discussed and the means of controlling delays and extra costs. Finally, the contractual methods by which contractors and suppliers can be appointed and controlled were discussed.

REFERENCES

1. Latham, M. (1994) *Constructing the Team*, HMSO, London.
2. Morris, P. and Hough, G. (1987) *The Anatomy of Major Projects*, John Wiley & Sons, Chichester.
3. Egan, J. (1998) *Rethinking Construction*, HMSO, London.
4. NEDC Construction Industry Sector Group (1991) *Partnering: Contracting without Conflict*, NEDC, London.

BIBLIOGRAPHY

Burke, R. (2003) *Project Management: Planning and Control Techniques*, John Wiley & Sons, Chichester.

Cotterell, M. and Hughes, R. (2002) *Software Project Management*, McGraw-Hill Education – Europe, Maidenhead.

Meredith, J. R. (2003) *Project Management: A Managerial Approach*, John Wiley & Sons, Chichester.

18

Networks for Projects

18.1 INTRODUCTION

It has been shown in Chapter 17 that the bar chart provides an excellent visual representation of a project programme. As a means of communication, the bar chart is equally effective whether representing high-level programmes for senior management or closely detailed programmes for those who carry out the work.

For simple projects with 20 or 30 activities, the production of a bar chart is a relatively straightforward task. A bar chart so constructed will provide a useful means of monitoring progress; late-running activities can be identified and the effects on other activities understood, so that measures can be implemented by the project manager to control progress. For complex projects in which many activities are interdependent, possibly across a number of disciplines, it will be difficult to formulate a bar chart that meets all the conditions and dependencies, and interpretation of a bar chart in which several activities are not running on schedule may become very difficult. In these circumstances analytical tools are necessary that will (a) provide a bar chart that meets all the dependencies and (b) facilitate dynamic analyses as the project proceeds. *Network analysis* is just such a tool.

In this chapter the principles of network analysis are set out and *network diagrams* developed. Two alternative forms of network diagram exist: the activity-on-arrow diagram and the activity-on-node diagram. The *activity-on-node* diagram has been selected for study in preference to the *activity-on-arrow* diagram because of its inherent superiority.

The application of *PERT* (Programme Evaluation Review Technique) to networks is considered in the latter part of the chapter.

18.2 NETWORK DIAGRAMS

The starting point for the construction of a network diagram is a list of activities and their precursor (immediately preceding) activities. The starting point for developing a list of activities is a work breakdown structure as described in Chapter 17. Perhaps the most difficult part is the determination of the precursor activity(ies) for each activity. Too often courses on network analysis give no consideration to this at all. Usually the student starts with a list of activities and precursor activities and is asked to develop the network diagram, and from it determine the critical path. In this chapter the approach will be to take as a starting point lists of activities and precursor activities and to develop and analyse the resulting network diagrams. However, in the Appendix 3 Case study: *Developing a Network* there is a model approach to network analysis, starting with a description of the project activities, from which is developed first a list of activities and precursor activities and finally a network.

18.2.1   Activity-on-node diagrams

In this system an activity is represented within a node, usually drawn as a rectangular box. Activities are linked together by arrows to show the order of precedence. Consider the construction of a column comprising three activities:

X – excavate for foundations
F – construct foundations
C – construct column on foundations

The network would be displayed as in Figure 18.1. An activity preceded by two activities and succeeded by two activities would be represented as in Figure 18.2.

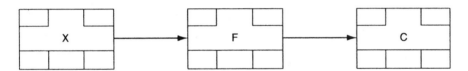

Figure 18.1   Three activities linked by arrows to show order or precedence

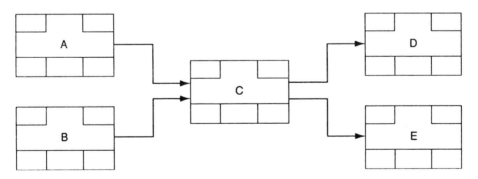

Figure 18.2   An activity preceded by two activities and succeeded by two activities

Example 18(1)

A network will be developed for the project represented by the data in Table 18.1.

Some practitioners prefer to show a start node for a project and an end node for the project, others see these as unnecessary. In this chapter start and end nodes will be included, although the authors have no preference in the matter.

Table 18.1

Activity	Duration (weeks)	Precursor activity(ies)
A	4	–
B	3	–
C	6	A, B
D	1	B
E	7	D
F	2	C
G	5	C, E
H	8	E
J	4	G
K	5	F, G
L	6	J, H
M	3	L, K

The start node will be followed by the nodes for A and B, D will follow B, and C will follow A and B, as shown in Figure 18.3.

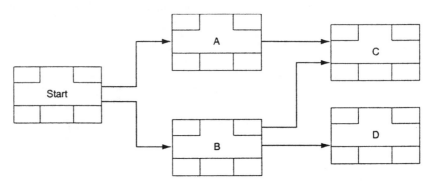

Figure 18.3   Developing the network logic

An activity-on-node network is much easier to construct than an activity-on-arrow network. There are no dummy arrows and the network can be built up methodically without difficulty. The final network is shown in Figure 18.4; durations of activities are also shown.

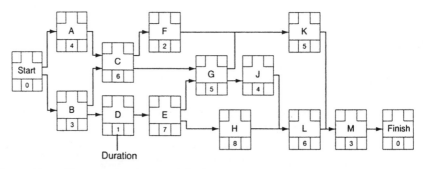

Figure 18.4   The completed network logic

The network can now be analysed to determine its duration. On each node there is a space available for the following:

Early start   the earliest time at which the activity can start
Early finish   the earliest time at which the activity can finish
Late start   the latest time at which the activity can start
Late finish   the latest time at which the activity can finish

In Figure 18.5 the earliest start and finish times have been entered. Where an activity is preceded by two activities, the larger figure for the early start must be selected. Thus the early start time for activity G is 11, not 10.

The early finish time for Activity M is 29 weeks; this therefore is the duration of the project. Therefore the latest date for completion of the project (29 weeks) can now be entered. Once this has been done one can work backwards from the finish to obtain the late finish and late start times for each activity; these have been entered in Figure 18.6.

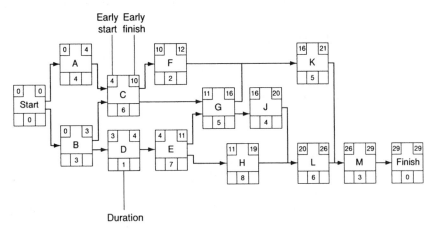

Figure 18.5

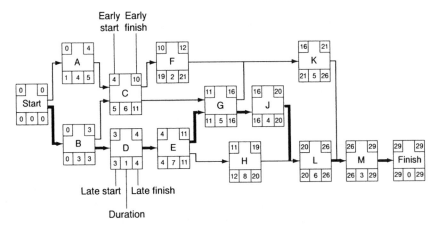

Figure 18.6

Where two or more activities follow one, as in the case of J and K both being preceded by G, the lower figure must be selected for the late finish time of G. Thus G cannot finish any later than week 16, otherwise the duration of the project would be extended. The *critical path* is shown by means of a thicker arrow. If the duration of any activity on the critical path is increased, then the project duration will be increased by the same amount. Activities not on the critical path have *float*. The float for noncritical activities is given by

late finish – duration – early start

The values of float for each activity are shown in Table 18.2. Activities with zero float are on the critical path. From this the project manager could observe, say, that a serious delay of eight weeks on Activity F would have no effect on the project duration, nor would speeding up Activity H have any useful effect. Likewise, if Activities G and J were accelerated to be completed in seven weeks total instead of nine weeks, the project duration would only be reduced by one week unless Activity H were accelerated by one week also. A delay to any activity on the critical path would delay the project by the same amount.

Table 18.2   Float

Activity	A	B	C	D	E	F	G	H	J	K	L	M
Total float	1	0	1	0	0	9	0	1	0	5	0	0

The arrows which link nodes can have durations. Activity-on-node diagrams in which finish-to-start (F–S) links have durations, and in which there are start-to-start (S–S) and finish-to-finish (F–F) links, will now be considered.

The nature and duration of each link will be shown against each arrow. Figure 18.7 illustrates the case in which Activity F cannot start until one week after the finish of Activity X, and Activity C cannot commence until two weeks after the completion of Activity F.

Start-to-start links and finish-to-finish links can be represented in a similar fashion. Where both links exist between a pair of activities, both links must be shown as separate arrows. This is because it is possible for the critical path to pass through one link and not the other.

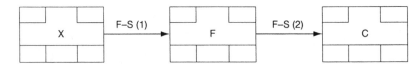

Figure 18.7

Example 18(2)

A project consists of Activities A, B and C with links as shown in Table 18.3.

Figure 18.8 is the activity-on-node diagram for the project. The activity times have been entered. The early start time of Activity C is 10 because it can start

Table 18.3

Activity	Duration (weeks)	Precursor activity	Link type	Link duration
A	4	–		
B	24	A	F–S	0
C	12	B	S–S	6
			F–F	3

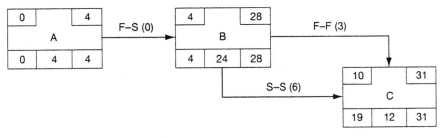

Figure 18.8

six weeks after the start of Activity B; however, the late start time of Activity C is week 19, thus the critical path passes through the finish-to-finish link but not through the start-to-start link.

Example 18(3)

Table 18.4 provides data on a simple project. Draw the network and find the critical path.

Figure 18.9 shows the completed network. Construction of the diagram logic is straightforward; the only difficulties that might arise are those concerning

Table 18.4

Activity	Duration (weeks)	Precursor activity	Link type	Link duration
A	4	–	F–S	0
B	3	–	F–S	0
C	6	A, B	F–S	0
D	1	B	F–S	6
E	12	D	S–S	4
F	2	C	F–S	0
G	5	E, F	F–S	0
H	8	E	S–S	3
			F–F	2
J	4	G	F–S	1
K	5	F, G	F–S	0
L	6	J, H	F–S	0
M	3	L, K	F–S	0

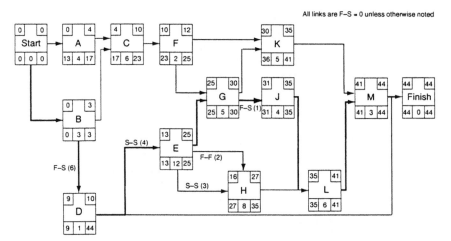

Figure 18.9

some of the activity times. The early finish time for Activity H must be Week 27 not Week 24 (16 + 8), because the finish-to-finish link of two weeks between Activity E and Activity H imposes this condition. The late finish time for Activity D is Week 44, for, although the start of this activity is on the critical path, there is no condition placed on its completion at all, other than to finish within the duration of the project. It should be noted that if the start and finish nodes had been omitted from the diagram, Activity D would have represented a finish activity, as would Activity M. This form of network construction – that is, omitting start and finish nodes – allows sectional completion; the project manager could require Activity D to be completed by Week 10 and handed over to the owner. In this case the late finish time for Activity D would be Week 10, not Week 44.

The bar chart can now be drawn. It is shown in Figure 18.10. All activities with float have been shown as starting at their earliest start time except Activity H. It would be unwise to start Activity H earlier than Week 19. Activity D is shown to be on the critical path; this is because its commencement is on the critical path although its finish is not. Since its duration is only one week it would be difficult to show it any other way (on the timescale provided).

18.2.2 Draft networks for large projects

When a large project is being planned, planners and estimators for different parts of the project may provide information on rates of production and on the links between activities that are at odds with one another, although this may not be immediately apparent. Construction and analysis of a first draft network should show the planners these contradictions and allow them to be resolved in the final draft plan.

Example 18(4)

As part of a large project a length of welded steel pipeline is to be installed in a trench. Excavation of the trench will be carried out by a civil engineering

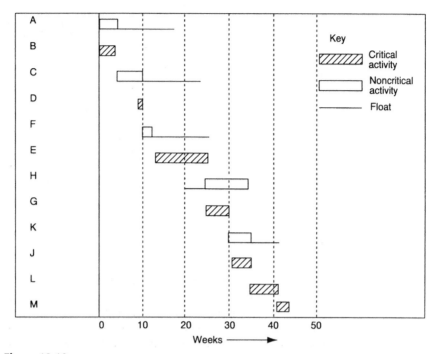

Figure 18.10

group. Placing sections of the pipe into the trench will be carried out by the suppliers of the short lengths of the fabricated pipe. Welding will be carried out by a mechanical engineering group. Nondestructive testing (NDT) of the welds will be carried out by independent assessors for statutory purposes and, once this has been finished, the civil engineering group will place backfill around the pipe. The different groups have been asked to provide estimates of their rates of progress. The pipeline is 120 km long. Each group assumes that it has one team working, starting at one end and proceeding to the other. The initial plan provides for all operations to be at least 4 km apart. These data have been analysed and are presented in Table 18.5.

Table 18.5

Group	Activity	Estimated rate of working (km/day)	Duration (days)
Civil engineering	X – excavation	2	60
Supplier	P – placing	4	30
Mechanical engineering	W – welding	1	120
NDT assessors	N – NDT	2	60
Civil engineering	B – backfilling	4	30

The activity-on-node diagram for this subproject is shown in Figure 18.11. The first thing that should be observed is that the critical path can pass through

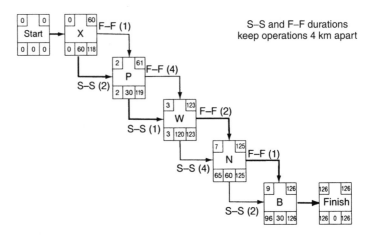

Figure 18.11

a S–S link between two activities without passing through the corresponding F–F link and vice versa. Therefore, in constructing an activity-on-node diagram one should always draw the two links, where they arise, separately.

Analysis of the network shows an undesirable feature. Activity P (placing) is a *stretch* activity. It need take only 30 days, yet it cannot be completed in less than 59 days. This is because the start of the activity is on the critical path, therefore it cannot be delayed, yet it cannot be completed until after the completion of Activity X (excavation). This would be highly undesirable from the supplier's point of view. Either a second draft plan should be prepared in which, for example, the excavation process were accelerated to be completed in 30 days, or the suppliers should be consulted to determine whether they would be satisfied with the draft plan. This would be most important in order to avoid any claims by the supplier that his or her work was being delayed once the work was underway.

It can be seen that production of a draft network is only a first stage in the planning process that can yield useful information. Further development of the plan may be necessary in which the nature of links between activities, their resourcing and, possibly, the method statement may need to be reassessed.

18.3   PROGRAMME EVALUATION AND REVIEW TECHNIQUE (PERT)

A chapter on networks would be incomplete without an introduction to PERT. This technique was developed by the US Navy between about 1958 and 1960 for the Polaris submarine construction programme. PERT takes into account the difficulty of estimating the durations of individual activities. In research and development projects in particular, the estimators may be unwilling to provide a single value for a duration but may be persuaded to provide a *most optimistic duration*, a *most pessimistic duration* and a *most likely duration* for that activity. If these figures are interpreted to be part of a frequency distribution then an expected duration of the activity, and hence of the whole project, can be determined. PERT uses a beta frequency distribution (this is a skewed distribution, which, it is claimed, provides a satisfactory model of reality and is easy to use) and from this the *expected mean duration* for an activity and the *standard deviation* of the distribution can be found from the formulae

$$t_e = \frac{a + 4m + b}{6}$$

$$s = \frac{b - a}{6}$$

where
t_e = expected mean duration
a = most optimistic duration
b = most pessimistic duration
m = most likely duration
s = standard deviation of the distribution

A comprehensive explanation of PERT is to be found in Moder and Phillips [1], which remains a fundamental text for this subject. Moder and Phillips recommend that a and b be selected as the lower and upper ten percentiles of the hypothetical performance time distribution.

Expected mean durations are normally calculated to one decimal place of the unit of measurement of time, in order to arrive at an expected duration of the project through the critical path.

Example 18(5)

Draw the network, determine the critical path and find the expected mean duration for the subproject defined by Table 18.6.

The value of the expected mean duration for each activity can now be found.

For Activity A

$$t_e = \frac{3 + 4 \times 5 + 9}{6} = 5.3$$

Table 18.6

Activity	a	m	b	Precursor activity	Link type	Link duration
A	3	5	9	–	F–S	0
B	4	6	8	–	F–S	0
C	5	8	10	A, B	F–S	0
D	3	6	9	B	F–S	6
E	6	9	15	D	S–S	4
F	3	4	5	C	F–S	0
G	8	12	15	E, F	F–S	0
H	2	6	8	E	S–S	3
					F–F	2
J	4	7	9	G	F–S	1
K	3	5	10	F, G	F–S	0
L	7	9	11	J, H	F–S	0
M	10	12	15	L, K	F–S	0

Table 18.7

Activity	A	B	C	D	E	F	G	H	J	K	L	M
t_e	5.3	6.0	7.8	6.0	9.5	4.0	11.8	5.7	6.8	5.5	9.0	12.2
s	1.0	0.67	0.83	1.0	1.5	0.33	1.17	1.0	0.83	1.17	0.67	0.83

The value of the expected mean duration (t_e) for each activity has been calculated in like fashion; these are shown in Table 18.7. The standard deviation (s) for each activity has also been calculated using the formula given previously. These figures have also been entered in the table.

Figure 18.12 shows the network. The values of t_e have been entered for each activity and all the earliest and latest start and finish times calculated for each activity. The critical path is shown and the expected mean duration of the project is 66.3 weeks, which for planning and programming the work would be taken as 67 weeks.

The standard deviation of the project duration can be determined as follows (in this case a normal distribution is used for summation of variances of activities).

The variance for the project duration equals the sum of the variances of the activities along the critical path. The variance for an activity equals the square of the standard deviation for that activity. The standard deviation for the project duration equals the square root of the variance.

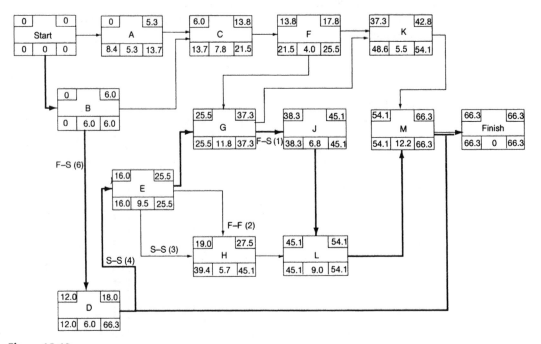

Figure 18.12

Therefore the standard deviation for the duration of the project

$$= \sqrt{0.67^2 + 1.0^2 + 1.5^2 + 1.17^2 + 0.83^2 + 0.67^2 + 0.83^2} = 2.63 \text{ weeks}$$

and probabilities can be determined for different project durations from normal distribution curves. For example, the probability that the duration will be greater than

66.3 (the mean) $+ 2 \times 2.63$ (2 × standard deviation) $= 71.6$ weeks, is 2.3%.

The probability that the duration will be greater than

66.3 (the mean) $+ 3 \times 2.63$ (3 × standard deviation) $= 74.2$ weeks, is 0.1%.

These are attributes of the normal distribution curve and are of some value in determining a range of durations for a project.

18.4   SUMMARY

Two kinds of network diagram are available: activity-on-arrow and activity-on-node. The former are mainly of historical interest and have not been considered. Activity-on-node diagrams are usually straightforward to construct and yield immediately useful data. A first draft network may, on analysis, allow the planner to perceive potential problems that can only be overcome by challenging some of the assumptions about resources or productivity or the nature of the links between activities. Two or three draft networks may be necessary before the planner can formulate the programme for the project. Network diagrams are the tools by which programmes in bar chart form can be established, and provide a most effective means of observing the effects of delayed/accelerated activities on the overall programme.

PERT is a useful technique for project planning where estimators prefer to state durations for individual activities in terms of a frequency (probability) distribution.

REFERENCE

1.   Moder, J. J. and Phillips, C. R. (1964) *Project Management with CPM and PERT*, Reinhold Publishing, New York.

BIBLIOGRAPHY

Harrison, A. (1997) *A Survival Guide to Critical Path Analysis and the Activity on Node Method*, Butterworth-Heinemann, Oxford.
O'Brien, J. (2000) *CPM in Construction Management*, McGraw-Hill, Maidenhead.

19

Project Management – Managing Construction Procurement

19.1 INTRODUCTION

Construction work, other than maintenance operations, is essentially project work. Expenditure on new construction work in the UK runs into tens of billions of pounds annually. Study of how the industry operates should yield some interesting insights into large-scale project management.

Buildings, roads, bridges, power stations, water supply systems, sewers and so on provide the infrastructure essential for all social and economic activity. Engineers are active in both the provision and use of this infrastructure. Scientists and technologists will, from time to time, commission new buildings, new manufacturing and/or processing facilities and could be responsible for the project management of the whole design and construction process. New transportation systems may be commissioned by governments or by private organizations. Experienced engineers, scientists and technologists often manage the construction procurement process, and even new graduates in these disciplines may be members of project management teams. This chapter aims to provide an appreciation of the various roles and responsibilities of the parties to the procurement process and an appreciation of relationships and forms of contract that might be encountered.

Construction procurement means buying in all the skills necessary to produce the facility, from feasibility studies, through design, manufacture, installation and construction to making the facility operational. This chapter sets out the alternative procedures available to a purchaser (or promoter) for the design and construction of a desired facility and will discuss how effective the means of procurement might be in achieving project completion swiftly and satisfactorily.

The chapter concludes by considering how some purchasers have become the main drivers in achieving the most effective project management through what is known as *best practice*, and to what extent best practice is adopted across the UK.

19.2 THE BUSINESS ENVIRONMENT[1]

The UK construction industry is characterized by large numbers of small to medium-sized contractors and only a few really large companies capable of taking

---

[1] See Chapter 2.

on most forms of construction. Of these perhaps only a handful have a world-class capability. Apart from housing developments, the construction industry is a designer-led industry. Many British firms of architects and of consulting engineers have a worldwide reputation for their excellence in investigation and design. This has meant that the industry is characterized by a separation of the design and construction processes and that endeavours by customers of the industry to bring about a closer union of the two processes will never be universally effective.

19.3   CONSTRUCTION PROCEDURES

The promoter of a construction project decides on the nature of the relationships between itself and its designers, suppliers and contractors. The promoter could, for example, enter into a single contract with one organization that commits itself to carry out preliminary studies, produce detailed designs, organize suppliers and construct and make operational the facility, and even run it, if so desired. Alternatively, the promoter may enter into separate contracts with various organizations, each of which will conduct its part of the whole process; the promoter could actively carry out some of the work itself, if it had the skill and resources. Clearly, the procedures adopted will be determined by the expertise the promoter has in both managing and participating in the construction procurement process and the resources it is willing and able to bring to bear on that process.

The alternative procedures open to a promoter are numerous and varied, but can be considered under two broad headings:

- Design is separate from construction.
- Design and construction are closely integrated.

19.3.1   Separation of design and construction

Until the 1970s this was the traditional approach to construction procurement. The procedures involve separation of:

- finance;
- preliminary studies;
- design;
- manufacture;
- installation;
- construction and making operational.

Generally, construction and making operational are one integral activity except where the facility is a major piece of process plant, for example, a chemical works or a power station, where bringing the facility into full operation may be a gradual process over some months or even years. For this reason, in order to simplify matters, the term *construction* should be taken to mean *construction and making operational* unless otherwise stated.

In traditional construction procurement the promoter obtains the finance, commissions consultants to conduct preliminary studies and make recommendations on the nature and form of the facility, employs consultants (often the same ones) to prepare outline designs and detailed designs for the facility, enters into separate contracts with manufacturers to make and, usually, install major plant or specialist service systems, and enters into a contract with a construction company to build the facility. Clearly, the processes are linear; there is no concurrent engineering. These provisions are ideally suited to a designer-led industry (see section 19.2).

The promoter remains responsible for coordination and progression between the various participants and their activities.

The promoter could be an organization that has a policy of continuous development, for example a water company, a commercial developer or a government agency. Considerable advantages can accrue if a partnering agreement is entered into between a promoter, consultants and contractors (see section 17.8). However, government agencies will usually be restrained from entry into such agreements on the grounds that the award of contracts for each and every individual project must demonstrate to the auditors a clear price advantage of the supplier over the competition.

19.3.2   Design and construction are integral activities

In this case the promoter obtains the finance, then commissions consultants to conduct preliminary studies and make recommendations. The promoter then employs a single company to carry out the design and construction of the facility, making that same company responsible for procuring the manufacture, supply, installation, assembly and so on of all plant and specialist service systems and for the construction of the facility. Partnering agreements between promoter and provider are just as appropriate for continuous development where design and construction are integral as they are for separation of design and construction.

In these contracts the purchaser appoints a project manager, who will be fully responsible for managing the design and the construction of the facility. Even the largest British contractors only have a limited design capability, therefore for design-and-build contracts the contractor frequently has to employ a firm of consulting engineers and/or architects. The procedures do allow considerable input by the constructor(s) into the design and planning of the project, which should result in savings in cost and time (see section 17.6). In these circumstances the project manager is handing over the responsibility for the design and the coordination of suppliers and contractors to others and is reduced to 'accepting' or 'approving' the designs. Technically he or she is at one remove from responsibility for the nature and quality of the facility. This will suit a purchaser who prefers a 'hands-off' approach. However, these procedures also allow for a proactive project manager, experienced in the management of construction projects, who seeks to drive down costs and to reduce construction times by influencing the performance of designers, suppliers and contractors.

It is sometimes the case that finance, design and construction are integral activities. This will arise when a promoter is unable or unwilling to raise the money for the project but will be in a position to pay for the completed project once it is in operation. Thus a facility that when operational provides a substantial income will allow the promoter to repay the project provider over a lengthy period of time, typically 15 to 20 years. Projects procured under this system are described under various appellations, such as:

- *DBOM* – Design, Build, Operate, Maintain, in which the builder designs and builds the facility, then operates it and maintains it for a specified number of years before transferring ownership to the promoter.
- *BOOT* – Build, Own, Operate, Transfer, in which the builder 'owns' the facility (because he or she paid for it), operates it for many years using the income to replenish his or her finances and, after an agreed period, transfers ownership of the facility to the promoter.
- *DBFO* – Design, Build, Finance, Operate, similar to the foregoing.

In the United Kingdom this system of procurement has been widely adopted by government since the mid-1990s for public works. The Second Severn Crossing (a high-level suspension bridge) and the Dartford Crossing (a combined high-level bridge in one direction and the under-river Dartford tunnel in the other direction) across the Thames are both good examples of this. Typically, joint ventures finance, design and build the facilities then operate and maintain them, receiving income from tolls. Risk[2] is shared between the joint venture and the government agency. Eventually the ownership of such facilities is transferred to the government, which then advertise for facilities managers (companies that specialize in operating such facilities) to bid for operating and maintaining them for another 25 years or so.

The Second Severn Crossing, completed in 1996 as a DBFO project, provides an interesting example of the risk shared between the joint venture and government [2]. As part of the agreement the joint venture took on an additional deferred debt of £62m, payable in 2013. This deferred debt was one that had arisen from the construction of the First Severn Crossing in 1966, from expensive strengthening work on the structure in the late 1980s and from an inadequate tolling regime. The joint venture comprised John Laing plc 35%, GTM-Entrepose 35%, Bank of America 15% and Barclays de Zoete Wedd (BZW) 15%, with Laing Engineering appointed as maintenance contractor and GTM-Couferoute as tolling contractor. The income from the completed project is determined by the number of vehicles that make the crossing daily and by the toll price. Who predicts the traffic volume? The government. Who sets the maximum toll price? The government. This suggests that a considerable risk lies with the joint venture. In fact, this joint venture provided equity for the project to the value of only £50 000. The European Investment Bank provided a loan of £131m. The remaining £353m was raised through the efforts of the Bank of America and BZW. It is the investors who bear the lion's share of the risk.[3] The sole risk for the government is if the new bridge fails to last beyond the period of the concession. A bridge of this kind, properly maintained, should last a century and longer without any problem.

---

[2]Risk management is outside the scope of this chapter. Large projects inevitably are liable to risks that can be very costly. Risks in construction concern health and safety, the environment and delays and extra costs. The Institution of Civil Engineers and the Faculty and Institute of Actuaries have developed a process for analysing and responding to these risks; the process is described as Risk Assessment and Management in Projects (RAMP) and is set out in *Risk Analysis and Management for Projects* [1].
[3]This is referred to as single recourse finance. In the event of the joint venture being unable to repay the debt, the investors have recourse to one asset only, the bridge.

The UK's Conservative Government of the 1990s introduced this system of procurement and referred to it as the Private Finance Initiative (PFI). The opposition Labour party firmly opposed these measures. However, when New Labour came to power it was reluctant to renege on any of the agreements entered into by its predecessors, and quickly saw the potential for providing new hospitals, roads, prisons and so on without having to resort to capital expenditure out of public funds, so the system endures and expands. The Labour Government refers to the policy as the Public Private Partnership (PPP) and the expressions PFI and PPP are both commonly used and seen to be interchangeable by most commentators. Use of PFI/PPP is controversial. Those who advocate the system argue that the private sector will design, construct and maintain a facility more efficiently because its financial success depends on completing the project within time and budget, to the agreed standards. Those who oppose PFI/PPP argue that (a) private companies borrow at higher interest rates than government, therefore the cost will be higher, and (b) the payments by government to the operators of these hospitals, prisons and so on are excessive – after all, there is no direct income from patients or prisoners – and that the companies are making unduly high profits.

19.4 CONTRACTUAL PROVISIONS

Contracts define obligations and determine relationships within a contract. This section examines a number of typical forms of contracts and agreements to illustrate the management systems that result from their adoption, and observes their place within the two broad alternative procedures for procurement.

Standard forms of contract are usually used in the construction industry. Standard forms of contract, drafted by representatives from all sides of the industries involved – that is, promoters, consultants, contractors and suppliers – have several advantages. They become familiar to all users. The way in which the courts of justice have interpreted clauses becomes widely understood, and no party to one of these contracts can claim any of the conditions to be unfair and seek to set aside those conditions by reference to the Unfair Contract Terms Act 1977. (Where large, powerful companies seek to impose onerous contract conditions on their suppliers, using their own carefully drafted contracts, it is always possible for the courts to be asked to declare such conditions unfair and release the suppliers from the onerous conditions.)

Standard forms of construction contract establish an intermediary between the purchaser of the project (known variously as the employer or purchaser) and the provider (known as the contractor). In the UK numerous standard forms of construction contract exist. Several will be referred to in the following paragraphs. Those referred to have been selected because they illustrate the kinds of contracts available and are widely used.

The intermediary is variously referred to as the engineer, the supervising officer, the project manager or, in building contracts, the architect. For simplicity's sake the intermediary will be referred to from now on as the A/E/PM. The A/E/PM is not a party to the contract but is appointed under a separate agreement by the purchaser. In each case its primary role is to ensure that the project is completed

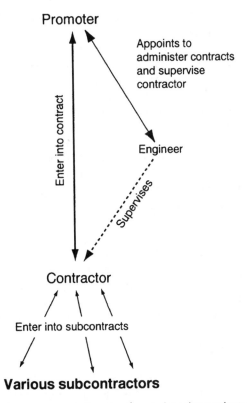

Figure 19.1   Typical management structure for engineering projects

to specification; the A/E/PM is also responsible for controlling the price paid for the project, once a contractor has been appointed, and for ensuring that the contractor finishes on time. Figure 19.1 shows a typical management structure provided by these forms of contract.

19.4.1   Building and civil engineering contracts

Standard forms of contract differ mainly in the provisions for payments to the contractor and also in their expectations of the task to be carried out by the contractor. A widely used form of contract for large buildings in the UK is the *Standard Form of Building Contract 1998 Edition, Private with Quantities* **(JCT98)** [3]; a widely respected form of civil engineering contract in the UK is the *ICE Conditions of Contract, Measurement Version*, 7ᵗʰ edn, 1999 **(ICE7)** [4]. JCT98 and ICE7 are, effectively, measurement contracts.

JCT98 obliges the contractor to 'carry out and complete the Works in compliance with the Contract Documents'. Carrying out the works can mean building and/or designing elements, subcontracting parts of the work, installing plant and equipment and so on. A large modern building is so complex with mechanically fixed cladding systems, raised floors, suspended ceilings, air-conditioning systems, security systems and so on, in which many of the installations are designed and installed by the manufacturers, that this form of wording is essential.

ICE7, on the other hand, is a typical civil engineering contract in which the assumption is that the contractor does no design and has no input into the planning process; therefore the contractor's obligations are to 'construct and complete the Works'. There is provision for design by the contractor, but only where 'express provision' is made in the contract documents.

Under JCT98 and ICE7 the architect and the engineer, respectively, are responsible for designing the project. Even though JCT98 allows for a considerable design input by the contractor, the usual purpose of this provision is to allow for the design and installation of specialist features and/or services where the expertise lies with the subcontractor and not with the architect. Because the contract is between the contractor and the employer, and the contractor is responsible for its subcontractors, under the legal principle of privity of contract the contractor must take on the contractual design responsibility even though it may carry out no design at all. Neither of these contracts is suitable for a concurrent engineering approach.

These contracts suit the procedures described in 19.2.1 where finance, design and construct are essentially separate and largely linear activities.

Alternatively, both the Joint Contracts Tribunal and the Institution of Civil Engineers have produced design-and-build forms of contract for the use of purchasers who wish to adopt the procedures set out in 19.3.2.

19.4.2   Contracts for plant

Where the contract involves the supply and installation of major plant, a contract commonly used in the UK is the *Model Form of General Conditions of Contract, for the Supply of Electrical, Electronic or Mechanical Plant – with Erection*, 1988 edition **(MF/1)** [5]. This is a lump-sum contract.

MF/1 expects the contractor to 'design, manufacture, deliver to Site, erect and test the Plant, execute the Works and carry out Tests on Completion'. The assumption is that the site is being constructed and made ready for the plant by others. This is typical of most major plant contracts, for example the installation of turbines within a hydroelectric power station. Therefore, where this contract is used, construction of the buildings to house the plant necessitates a separate construction contract. In general, in a project where several main contracts exist concurrently, the purchaser is responsible for coordinating the operations of the separate contractors, which means that the purchaser must appoint a project manager to oversee the coordination and control of the separate contractors.

MF/1 is, in effect, a design-and-build contract and provides opportunities for concurrent engineering and consequent savings in time. Being a lump-sum contract, the contractor's tender price to cover design, manufacture and installation is somewhat speculative, however well researched and informed, and, for the promoter, there can be real difficulties in choosing between tenders, although, since plant performance is usually the most important criterion, this presents less difficulty than choosing between alternative designs for, say, a bridge.

19.4.3   Contracts for complex facilities

Each one of the three contracts referred to above (JCT98, ICE7 and MF/1) is limited in its application, the first to buildings, the second to traditional civil engineering where the contractor has no input into the planning or design process, the third to the design, manufacture, installation and commissioning of plant. A new engineering contract was drafted in the 1990s, which makes it suitable for a wider range of construction projects and provides for several alternative means of payment to the contractor. It is *The Engineering and Construction Contract*, 2nd edn, 1995 **(ECC2)** [6]. ECC2 provides within its covers six alternative forms of contract, making it suitable for a whole range of construction projects. The six options, A to F, allow for a choice between a lump-sum contract, a measurement contract, a lump-sum target contract (see Chapter 17), a measurement target contract, a cost-reimbursement contract and a management contract (see Chapter 17).

ECC2 obliges the contractor 'to Provide the Works' and the expectation is that the contractor will design part of the works. ECC2 is a contract drafted by people who had used the Institution of Civil Engineers' standard conditions of contract for many years and felt that it was insufficiently flexible in terms of contractor input, as well as being a contract that too often gave rise to disputes. Further, the six options provide for a wide range of methods for paying the contractor, as already shown.

Under ECC2 the project manager will either be a senior employee of the promoter or, more usually, a member of the organization that carries out a larger part of the design. But the contract readily permits contractor involvement in design and planning of certain operations and so is useful where a concurrent engineering approach is desired. By selecting either Option E (cost-reimbursement contract) or F (management contract), ECC2 can be used where the purchaser desires greater integration between design and construction.

Another form of contract, which is used for the integral design and construction of process plant, is *The Green Book Form of Contract, Reimbursable Contracts*, 3rd edn, 2002 **(IChemE Green Book)** [7]. In this contract the contractor's obligation is to 'execute the Works'. So, whatever the employer decides constitutes 'the Works', be it design, installation, construction or testing, that is what the contractor has to do. Since this is a cost-reimbursement contract the work may develop as the contract proceeds, so what constituted 'the Works' at the beginning may be very different from what is completed by the end.

19.4.4   Role of the A/E/PM

What all these contracts have in common is the appointment of the A/E/PM. It is the A/E/PM who supervises the operations, approves payments to the contractor, on the basis of quality as well as quantity, issues instructions and grants extensions of time and additional payments. This is a powerful position, one of authority, respect and independence. The individual will be familiar with the kinds of operations necessary to complete the works and will be expected to exercise proper professional judgement, which will recognize that operating safely and meeting the required standards are more important than finishing on time and to budget. This may at times put the A/E/PM at variance with both the purchaser, who pays for their services, and the contractor, who has to follow their instructions. That makes it not always an easy role to play. These independent powers to issue instructions and grant extra payment and time are seen by some critics as a frequent cause of cost and time overruns in construction projects.

It should be clear that these provisions could give rise to a dispute between the A/E/PM (on behalf of the purchaser) and the contractor. In these standard forms it is assumed that the A/E/PM will endeavour to resolve any dispute by acting impartially between the two parties. A difficulty arises where JCT98, ICE7 and ECC2 are used. Because the A/E/PM is the designer, technical adviser and the sole authority for issuing instructions and variation orders, many of the disputes that do arise are actually differences between the A/E/PM and the contractor, so the A/E/PM has to rule on disputes for which it has some responsibility. Even so, most professional engineers, when called on to settle disputes of this nature, do so with exemplary impartiality, sometimes to the employer's consternation. However, there will be times when the contractor refuses to accept the A/E/PM's ruling and so these contracts make provision for the appointment of an adjudicator, agreed at the start of the contract, to make decisions on the rights and wrongs of a case in the event of a dispute. The purpose of such provisions is to avoid the very high costs of going to arbitration or even to the courts. This doesn't mean that cases are never taken to the courts, but it does reduce the likelihood.

Under all five standard forms of contract the contract operations are watched and supervised by the A/E/PM's site representative, although in JCT98 that representative, known as the clerk of works, is given only powers to inspect. The powers vested in the site representative should be a reflection of the difficulty and size of the operations as well as the physical distance of the site from the offices of the A/E/PM. The greater the difficulty, the larger the project and the greater the distance from head office, the greater the powers that should be vested in the site representative. In construction it is often imperative that decisions on how to proceed must be made on the spot and often at short notice. A clerk of works has no powers to issue instructions and could only refer the matter back to head office (which would be to no avail at a weekend!). Computer links can mitigate some of these problems, but only if the link is continuously manned at the head office end. Where immediate communication cannot be relied on, appointment of a resident architect provided with some or all the powers of the architect is a proper solution.

Construction sites are inherently dangerous. Responsible contractors place safe operational conditions and safe methods of working at the top of their agenda. The role of the A/E/PM's representative in observing operations ensures their strong interest in safe working, and in some contracts, e.g. ICE7, the engineer has the power to suspend operations if he/she views them as unsafe. There can be no doubt that an independent view of the safety of site operations is beneficial. Sometimes, for example, in the excavation of a tunnel through unstable ground or the construction of an earth and rockfill dam using local materials, the technical expertise of the A/E/PM is superior to that of the contractor and his/her understanding of the safety of the excavation or the fill should be paramount in any decisions on how fast or by what means to proceed. It can be argued that where the construction techniques used are a major determinant of operational safety, only those forms of contract that give independent powers to the intermediary should be used. In other words, integral design and construction may not result in the safest working environment. Self-certification of quality and safety may well save time and money but may have less desirable consequences too.

19.5   BEST PRACTICE

Partnering agreements have been widely adopted by frequent users of the construction industry. Cost-reimbursement contracts can be used as a basis for partnering agreements because their remit is for construction work where the project parameters are loosely defined. They clearly depend for their success on a considerable degree of trust between the project manager, the designer and the contractor. Today, partnering agreements, which should result in close accord between promoter, designer and contractor, are seen by many promoters as providing them with the best service from the construction industry. These agreements are fine where there is continuous development, but they are inappropriate for a one-off purchase of construction.

Critics of the UK construction industry have observed that the traditional forms of contract such as the forerunners of JCT98 and ICE7 prevented contractor involvement in the planning process and too frequently gave rise to disputes. An unfortunate tradition arose in which contractors bid low for work, then made up the shortfall by claiming extra money and extra time at every opportunity. Projects were neither completed on time or to budget. Clients of the construction industry felt ill served by architects, consultants and contractors, who seemed to thrive on this disputes culture at the expense of the client.

JCT98 and ICE7 are new editions of construction contracts that aim to reduce conflict by allowing for greater consultation between designer and contractor when problems are foreseen or encountered and a more pragmatic approach to claims for additional costs and extra time using contractor fixed-price quotations before extra work is carried out. ECC2 requires the parties to a contract to work in a spirit of mutual trust and cooperation (admittedly, a difficult legal concept) and for a similar emphasis on consultation and for fixed-price quotes before any extra work. These improvements are a direct result of the observations of Sir Michael Latham in *Constructing the Team* [8], a report commissioned by government, and because of pressure from major clients to the construction industry.

Both Sir Michael Latham and later Sir John Egan in *Rethinking Construction* [9] concluded that a key factor in improving performance within the construction industry was through client (that is, 'purchaser') involvement. Egan proposed that, under committed leadership **by the client**, the industry should focus on the customer and that the process of procurement and the team of designers, suppliers and contractors should be integrated around the product. Major clients of the industry in partnership with leading consulting engineers, architects and contractors have adopted Egan's best-practice proposals and committed themselves to a series of projects to demonstrate, through published feedback on costs, time and productivity, the efficacy of these measures.

It must be emphasized that these proposals only serve directly the interests of the major clients of the construction industry, and these major clients, apart from government, procure only a small proportion of the industry's output. It is possible that the efficiencies achieved for major clients might have some spin-off where contractors and consultants are dealing with one-off purchasers, but this will only happen if the purchaser appoints a professional and experienced A/E/PM who adopts as much of the philosophy of best practice as is practicable to manage the construction procurement.

The experience underlying Latham and Egan was largely derived from the building industry rather than civil engineering. In civil engineering where large-scale excavation and earthmoving operations are common, where fitting in with the surrounding topography requires engineering decisions as the work proceeds, where the quantities of materials such as concrete and steel can be very large and where implementation of the actual design has its own safety issues, there is a strong case for empowering the A/E/PM to issue instructions, change quantities, approve working operations and even, if necessary, to suspend the works. In these circumstances the authority provided to the engineer under an ICE7 type of contract should be preferable to self-certification. Construction is a dangerous business.

19.6   SUMMARY

This chapter considered the two alternative procedures for construction procurement – namely design separate from construction and design and construction being closely integrated – and looked at PFI/PPP contracts in the public sector. The structure of the industry was described, showing the clear separation in business terms of the designers and the contractors, which has meant that large construction projects are often managed very differently from other large projects. A number of conditions of contract were examined in order to illustrate the alternative relationships possible between promoter and provider and to consider how these forms of contract can affect project duration and project costs. Best-practice relationships were outlined and their significance for one-off purchasers and for safety were discussed.

REFERENCES

1. *Risk Analysis and Management for Projects*, 1998, Thomas Telford, London.
2. Lawrence, S. C., Payne, A. C. and Mustafa, A. (1998) *The Second Severn River Crossing*, City University Business School, London, for The Worshipful Company of Paviors, London.
3. *Standard Form of Building Contract 1998 Edition, Private with Quantities*, Joint Contracts Tribunal, London.
4. *ICE Conditions of Contract, Measurement Version*, 7th edn, 1999, Thomas Telford for the Institution of Civil Engineers, London.
5. *Model Form of General Conditions of Contract, for the Supply of Electrical, Electronic or Mechanical Plant – with Erection*, 1988 edition, Joint IMechE/IEE Committee on Model Forms of Contract, London.
6. *The Engineering and Construction Contract*, 2nd edn, 1995, Thomas Telford for the Institution of Civil Engineers, London.
7. *The Green Book Form of Contract, Reimbursable Contracts*, 3rd edn, 2002, Institution of Chemical Engineers, Rugby.
8. Latham, M. (1994) *Constructing the Team*, HMSO, London.
9. Egan, J. (1998) *Rethinking Construction*, HMSO, London.

BIBLIOGRAPHY

Bennett, J. (1998) *Seven Pillars of Partnering*, Thomas Telford, London.

Contractor's Key Guide to PFI (1998), Construction Industry Council, London.

Langford, D. (2001) *Strategic Management in Construction*, Blackwell Science, Oxford.

Scott, R. (2001) *Partnering in Europe: Incentive Based Alliances for Projects*, European Construction Institute, Loughborough.

Walker, A. (2002) *Project Management in Construction*, Blackwell Science, Oxford.

Woodward, J. (1997) *Construction Project Management: Getting It Right First Time*, Thomas Telford, London.

20

Inventory Management

20.1 INTRODUCTION

All scientists, technologists and engineers have the opportunity to practise inventory management. This may be in their domestic lives, ensuring that they do not run out of food, fuel, washing powder or squash balls. Or in their early working lives they may be given responsibility for the stationery or other vital everyday office consumables. Later they may be managers of supply and logistics systems with multimillion-dollar turnovers. At all these levels the principles of inventory control set out in this chapter can be applied.

Inventory appears in a company's financial accounts as an asset (see Chapter 14 for further details). It can provide benefits to a business, such as:

- improving availability, to assist sales;
- acting as a regulator to smooth production levels;
- protecting against variations in demand or supply that could interrupt operations;
- providing a hedge against inflation;
- reducing the effects of breakdowns.

Inventory has therefore been regarded as 'a good thing', and techniques have been developed for managing or controlling inventory to provide the benefits while minimizing inventory costs.

The first part of this chapter provides a description of inventory-related costs and the use of economic order quantity (EOQ) calculations and Pareto analysis. An example of these techniques applied to nonproduction or maintenance, repair and operating (MRO) materials is provided.

Reasons for holding inventory are then examined and the apparent benefits are questioned. This reveals that, far from being 'a good thing' that improves business performance, inventory is an obstacle to the identification and resolution of fundamental problems. The potential for just-in-time (JIT) inventory management is then discussed, using examples from the European automotive industry.

20.2   THE ELEMENTS OF INVENTORY CONTROL

20.2.1   The inventory cycle

In theory, a graph of inventory levels has a saw-tooth profile. The opening stock is reduced by steady usage or demand, and replenished by a delivery just as stock is exhausted, as shown in Figure 20.1.

20.2.2   Performance measures

An inventory or material control system scores a success if the item required is available at the time demanded. Some systems, particularly those serving random or sporadic demand, such as spare parts inventory, are measured in terms of 'first-time fill' – that is, the percentage of orders filled on demand – with 90% or higher regarded as good performance. The time taken to get the missing 10% is another measure of performance. The inventory management task in retailing is similar to spare parts provisioning, and is covered by Susan Gilchrist [1]. The article states: 'The principal element of customer service is the availability of the product... Marks and Spencer, widely seen as having the best-run supply chain in the industry, has option availability of more than 90 per cent in its stores, significantly above the industry average of about 65 per cent'.

The system scores a failure if a demand cannot be met first time – often termed a 'stock-out'. If the order is held for later supply, the unfilled demand is usually described as a 'back order'. The number and value of back orders, and their age, are also performance measures.

First-time fill can be increased and back orders reduced by holding more stock, but this can be prohibitively expensive. To meet demand between replenishments, the system will contain 'cycle stock' and, to cover variations in demand or delivery, 'buffer stock' or 'safety stock' is added. If demand is random (in statistical terms, normally distributed about the expected mean value), then safety stock would

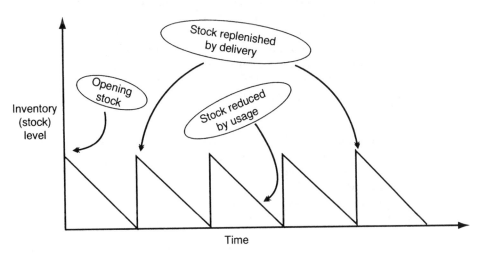

Figure 20.1   The 'saw-tooth' inventory cycle

have to be increased by about 20% to raise first-time fill from 95% to 96%. Whether this is worthwhile depends on the nature of the demand – if a stock-out were life threatening it would be worthwhile, and some managers feel the same way about stock-outs that threaten production or construction.

So, another performance measure is needed to gauge whether the first-time fill rate is achieved economically. This measure is usually:

- a *turnover rate* (demand or usage during a period, normally a year, divided by inventory held during the period); or
- the *number of days' or months' stock held* (the reciprocal of a turnover rate, see Chapter 14).

Higher turnover rates are 'good'; higher days' stock are 'bad'.

20.2.3   Acquisition costs and holding costs

The inventory control system requires two basic decisions:

- how much to order or schedule;
- when to order or schedule it.

The two factors affecting these decisions are the costs of acquiring more material and the costs of holding it.

Acquisition costs include the cost of several administrative tasks:

- Raising a requisition and getting it approved.
- Obtaining quotations from potential suppliers.
- Raising an order and getting it approved.
- Placing the order and, possibly, 'chasing' the supplier for delivery.
- Processing the receival.
- Paying the supplier.

Most of these costs are independent of the quantity ordered, and total acquisition costs can be minimized by minimizing the number of orders and deliveries. In a bureaucratic company, with many checks and controls, such as those shown in Figure 20.2, the administrative costs of acquisition can be as high as £50 per delivery, and even in a company with fewer controls these costs can be between £15 and £20 (UK figures, 1994–95. With increased efficiency offset by higher salaries, 2004 figures are probably similar.)

Holding costs comprise:

- depreciation on the buildings and equipment used to store the inventory;
- heating, lighting, cleaning and maintaining the storage area;
- security and stock checking;
- insurance;
- deterioration, damage and obsolescence;
- interest forgone or charged on the cash tied up in stock.

In total, these costs can be between 20% and 30% per annum of the average inventory value – the major element being the interest on the money tied up in

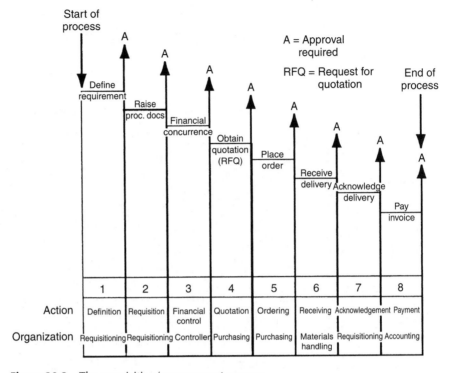

Figure 20.2   The acquisition/procurement process

stock. Holding costs can be minimized by frequent deliveries of small quantities, but this puts up the acquisition cost.

The balance of acquisition costs and holding costs to minimize total costs can be determined by calculating the 'economic order quantity' (EOQ), as described below.

20.2.4   Economic order quantity

Let A = the total acquisition cost per order
H = the annual cost of holding a unit of inventory
D = the annual demand or usage
Q = the quantity to be ordered

Then:

The number of acquisition cycles a year is D divided by Q, and the yearly acquisition cost is $\dfrac{A \times D}{Q}$

The average 'cycle' inventory is Q divided by 2 and the cost of holding inventory is $\dfrac{H \times Q}{2}$

Alfa Romeo
AUTO

PROM INDI

Luglio 1983

Il lotto economico è stato calcolato con la seguente formula:

$$\sqrt{\frac{2 \times consumo\ annuo \times costo\ ordinazione}{prezzo\ del\ materiale \times costo\ di\ giacenza}}$$

assumendo come costo di ordinazione L. 38000 e costo di giacenza 20%

NB.- Per i lotti economici tratteggiati la quantità richiesta deve essere uguale a ~ sei mesi del consumo mensile e non altro.

LOTTI ECONOMICI

foglio 1

ordine chiuso

consumo medio mensile — Prezzo del materiale (L. unità di misura)

DA	A	1÷100	101÷200	201÷300	301÷400	401÷500	501÷600	601÷700	701÷800	801÷900	901÷1000	1001÷1200	1201÷1400	1401÷1600	1601÷1800	1801÷2000	2001÷2300	2301÷2600	2601÷3000	3001÷3500
0	0.6	161	117	68	57	50	46	41	39	37	35	33	31	28	26	24	22	22	20	20
0.51	1	263	151	117	98	87	78	72	68	63	59	55	52	48	46	41	39	37	35	33
1.01	1.5	338	196	151	129	113	102	94	87	83	78	72	65	61	59	54	52	48	46	41
1.51	2	400	231	179	153	133	120	111	102	96	92	85	78	72	68	65	61	57	52	48
2.01	2.5	453	262	203	170	150	137	126	116	109	102	96	89	83	78	74	70	65	61	57
2.51	3	501	289	224	190	168	150	140	129	122	116	107	98	92	85	81	76	72	68	61
3.01	4	565	326	253	214	187	170	157	146	137	129	120	111	102	96	92	85	81	76	70
4.01	5	641	370	286	242	214	194	179	166	155	146	137	126	118	109	105	98	92	87	81
5.01	6	708	409	317	268	235	214	196	183	172	161	150	140	129	122	113	107	100	94	84
6.01	7	770	445	344	292	257	231	214	198	187	177	163	150	140	131	124	116	109	102	96
7.01	8	827	477	370	312	275	249	229	214	201	190	177	161	150	142	133	126	118	111	102
8.01	9	880	508	394	334	294	266	244	227	214	203	187	172	161	150	142	133	126	118	109
9.01	10	931	537	416	353	310	281	257	240	227	214	198	183	170	159	150	142	133	124	116
10.1	11	911	566	439	371	325	294	270	253	238	225	209	192	179	168	159	148	140	131	122
11.1	13	1041	605	469	396	349	316	290	270	253	240	228	205	192	179	170	159	148	140	131
13.1	15	1136	654	506	427	377	340	312	292	275	259	240	223	207	194	183	172	161	150	140
15.1	17	1210	699	541	458	403	364	336	312	292	277	257	238	220	207	196	183	172	161	150
17.1	20	1301	751	582	490	434	392	360	336	314	299	277	255	238	222	211	198	185	174	161
20.1	25	1424	821	641	541	477	432	397	371	347	329	305	281	262	246	233	218	205	192	179
25.1	30	1565	915	709	599	528	477	438	410	384	364	338	310	290	270	257	240	227	211	196
30.1	40	1788	1032	800	676	595	538	495	462	434	410	381	349	327	305	290	272	255	240	222
40.1	50	2027	1170	906	765	676	610	562	523	490	464	432	397	371	347	327	307	290	270	251
50.1	75	2388	1379	1068	903	796	719	663	617	580	547	508	469	436	410	386	364	340	318	296
75.1	100	2826	1631	1264	1068	942	852	783	730	685	647	602	554	517	484	458	429	403	377	351
101	125	3210	1853	1436	1218	1068	966	889	815	776	735	683	628	584	549	519	486	458	427	392
126	150	3548	2048	1587	1339	1182	1068	981	916	859	811	754	693	647	606	573	538	506	473	440
151	200	4001	2310	1789	1511	1332	1203	1107	1031	968	916	852	783	730	685	647	608	571	534	495
201	300	4780	2760	2138	1805	1591	1439	1323	1234	1158	1091	1018	935	872	820	774	726	682	639	593
301	400	5554	3206	2522	2136	1894	1703	1567	1468	1369	1295	1203	1107	1031	968	916	857	807	754	700
401	500	6410	3701	2867	2426	2126	1934	1777	1655	1552	1467	1365	1256	1171	1099	1038	974	916	857	796
501	750	7553	4361	2379	2856	2486	2267	2093	1949	1831	1731	1609	1448	1380	1286	1223	1149	1079	1009	927
751	1000	893	5169	3998	3373	2976	2692	2474	2307	2165	2047	1903	1751	1631	1533	1448	1358	1275	1195	1107
1001	1500	10682	6167	4776	4033	3553	3215	2965	2758	2585	2445	2278	2091	1919	1931	1731	1621	1526	1422	1325
1501	2000	12637	7296	5653	4771	4207	3815	3502	3259	3063	2899	2692	2474	2311	2167	2047	1931	1672	1567	1452
2001	3000	15094	8720	6756	5711	5036	4556	4186	3903	3662	3466	3215	2954	2758	2594	2442	2300	2156	2019	1873
3001	4000	17872	10318	7992	6754	5951	5395	4952	4623	4327	4091	3815	3502	3270	3052	2899	2714	2551	2381	2217
4001	5000	20264	11700	9062	7652	6736	6104	5624	5232	4913	4643	4323	3966	3766	3209	3025	2809	2708	2511	
5001	7500	23882	13758	6680	9025	7935	7194	6976	6194	5777	5472	5090	4687	4360	4017	3869	3630	3412	3183	3065

Alfa Romeo
AUTO
PROM INDI
Luglio 1983

Il lotto economico è stato calcolato con la seguente formula:

$$\sqrt{\frac{2 \times consumo\ annuo \times costo\ ordinazione}{prezzo\ del\ materiale \times costo\ di\ giacenza}}$$

assumendo come costo di ordinazione L. 30000 e costo di giacenza 20%

N.B. – Per i lotti economici tratteggiati la quantità richiesta deve essere uguale a ~ sei mesi del consumo mensile e non altro.

f.lo 2 — LOTTI ECONOMICI — *ordine chiuso*

Consumo medio mensile — Prezzo del materiale (L. unità di misura)

DA	A	3501÷4000	4001÷4500	4501÷5000	5001÷6000	6001÷7000	7001÷8000	8001÷10000	10001÷15000	15001÷20000	20001÷40000	40001÷60000	60001÷80000	80001÷100000	100001÷200000	200001÷400000	400001÷600000	600001÷1000000	1000001÷1500000	1500001÷2000000
0	0,5	17	15	15	16	13	13	11	9	9	7	4	4	4	2	2	2	2	1	1
0,51	1	31	28	26	24	22	22	20	17	15	11	9	7	7	4	2	2	2	2	1
1,01	1,5	39	37	35	33	31	28	24	22	17	13	11	9	7	7	4	2	2	2	2
1,51	2	46	44	41	37	35	33	31	26	22	15	13	11	9	7	4	4	2	3	2
2,01	2,5	52	48	46	44	39	37	33	28	24	17	13	11	11	9	7	4	2	3	2
2,51	3	57	54	50	48	44	41	37	33	26	20	15	13	11	9	7	4	4	3	3
3,01	4	65	61	57	54	50	46	41	35	31	24	17	15	13	11	7	7	4	4	3
4,01	5	74	70	65	61	57	52	48	41	35	26	20	17	15	11	9	7	4	4	3
5,01	6	83	76	72	68	63	57	52	46	37	28	22	20	17	13	9	7	7	5	4
6,01	7	89	83	78	74	68	63	57	48	41	31	24	20	17	13	11	7	7	5	4
7,01	8	96	89	85	78	72	68	61	52	44	35	26	22	20	15	11	9	7	5	4
8,01	9	102	96	89	83	76	72	65	57	48	37	28	24	20	15	11	9	7	6	5
9,01	10	107	100	96	89	83	76	69	59	50	39	31	24	22	17	11	9	7	6	5
10,1	11	113	107	100	94	85	81	74	63	52	41	31	26	24	17	13	9	7	6	5
11,1	13	122	113	107	100	92	85	78	65	57	44	33	28	24	19	13	11	9	7	6
13,1	15	131	122	116	107	98	92	85	72	61	46	35	31	26	19	15	11	9	8	6
15,1	17	140	131	124	116	107	98	89	76	65	50	39	33	28	22	15	13	9	8	7
17,1	20	150	142	133	124	113	107	96	83	70	62	41	35	31	24	17	13	11	9	7
20,1	25	166	155	148	137	126	118	107	92	76	59	46	39	33	26	17	15	11	10	8
25,1	30	183	172	161	150	140	129	118	100	85	65	50	41	37	28	20	15	13	10	8
30,1	40	207	194	183	170	157	146	133	113	96	74	57	48	41	33	22	17	13	12	10
40,1	50	233	220	207	194	177	166	150	129	109	83	63	54	48	37	26	20	15	13	11
50,1	75	277	259	244	227	209	194	179	150	129	98	76	63	57	44	31	24	20	15	13
75,1	100	327	307	290	270	249	231	209	179	150	116	89	76	65	52	37	28	22	18	15
101	125	371	347	329	305	281	262	236	203	172	131	100	85	76	69	41	33	26	20	17
126	150	410	386	364	338	316	290	264	265	190	144	111	94	83	65	46	35	28	22	19
151	200	462	434	410	381	351	327	296	253	214	163	126	107	92	72	52	39	31	25	21
201	300	582	519	490	456	419	390	353	303	256	196	150	129	113	87	61	48	37	30	26
301	400	652	613	580	538	495	462	421	358	301	231	179	150	133	102	72	57	46	36	30
401	500	729	695	654	610	568	523	477	405	342	262	201	170	150	116	81	63	50	41	34
501	750	872	820	774	719	663	617	562	477	403	310	240	201	179	140	98	76	59	48	40
751	1000	1031	970	916	852	783	730	665	565	477	364	283	240	211	163	116	89	72	57	48
1001	1500	1231	1158	1091	1018	935	872	796	676	571	436	338	286	251	196	140	107	85	68	57
1501	2000	1352	1448	1199	1111	1027	955	855	740	626	477	371	312	275	214	150	116	95	80	68
2001	3000	1744	1647	1550	1429	1325	1234	1125	959	807	617	477	403	355	277	196	150	120	95	81
3001	4000	2064	1938	1833	1707	1565	1458	1330	1131	953	730	565	477	419	323	229	179	142	113	95
4001	5000	2334	2202	2071	1929	1777	1655	1506	1280	1081	828	639	538	477	366	259	203	161	123	108
5001	7500	2758	2590	2452	2282	2093	1919	1779	1511	1275	977	754	639	562	436	307	240	190	151	124

Figure 20.3a Alfa Romeo economic order quantity (EOQ) matrix

The total annual cost, T, of acquiring and holding inventory can be calculated as

$$T = \frac{A \times D}{Q} + \frac{H \times Q}{2}$$

The value of Q to minimize T can be calculated by setting the first derivative dT/dQ to zero.

$$\frac{dT}{dQ} = \frac{-A \times D}{Q^2} + \frac{H}{2}$$

when this is zero

$$Q^2 = \frac{2A \times D}{H}$$

$$Q = \sqrt{\frac{2A \times D}{H}}$$

This is the economic order quantity (EOQ). The same formula can be used to calculate the 'economic production quantity' to balance set-up costs (a form of acquisition cost) against the cost of holding the quantity produced. In theory, EOQs could be calculated for every item to be held in stock. Alfa Romeo was doing this in the late 1980s to control its MRO inventories. Every Alfa Romeo stock control analyst was given the EOQ matrix shown in Figure 20.3.

The first two left-hand columns show ranges of average monthly usage, for example from 0 to 0.5, from 0.51 to 1 etc. The other columns show the price of the material in lira per unit of measure, for example from 1 to 100, 101 to 200 etc. For any usage up to 7500 per month, and prices ranging up to 2 million lira, the analyst could read off the EOQ. For example, the EOQ for an item with monthly usage of 5001 and price up to 100 lira was 23 882, as shown in the bottom left-hand corner. The analyst would have rounded this to a quantity that the supplier was willing to ship – say 24 000.

This degree of detail and precision is not effective or efficient. Engineers and scientists should beware of such pseudo-scientific systems where spurious accuracy is substituted for sensible analysis and consideration of the business purpose. In this case better overall results can be obtained, and fewer people employed, by using a simpler system that classifies items as fast-moving/high-value, medium-moving/medium-value and slow-moving/low-value, and applies EOQ-like rules to each class, rather than each item. The three classes are normally called A, B and C, and the system is known as ABC inventory control. An example is given in the next section.

20.2.5   An ABC inventory model

ABC inventory control is based on Pareto's law, which can be roughly interpreted as 80% of a problem comes from 20% of the possible causes. Pareto was a nineteenth-century Italian engineer who chose to study the distributions of ownership of wealth and receipt of incomes. He found that 80% of wealth was in the hands of 20% of the population. The pattern was the same for incomes – hence Pareto's law is also known as the 80/20 rule. Although established as a result of studying wealth and incomes, Pareto's law applies to many materials management situations, and is particularly relevant to inventory management. In most material flows where large numbers of different parts or items are involved, it will be found that about 80% of the value derives from about 20% of the items. This means that inventory control can be exercised efficiently by concentrating on about 20% of the items, and using simple rules to control the other 80% of items.

This is particularly true of MRO items, and since many engineers and technologists become involved in the maintenance, repair and operation of plant, equipment, machinery, buildings, ships, planes or other engineering products, an MRO example is described here.

The Pareto chart shown in Figure 20.4 is based on data from the 1980s for a transmission manufacturing plant that had been operating for over 20 years. It is probably typical of many plants of similar age in many industries, and there are general lessons to be learned from the plant's experience.

The plant stocked just over 30 000 items, which included things like oils and greases, spare parts for machines and equipment, tools for machines, hand tools, cleaning materials, safety clothing, paint, hardware, building materials, motors, valves and so on. In 20 years, a lot of material had been accumulated!

Two Pareto curves are shown. One is the accumulative share of turnover, or usage, plotted against number of items; the other is the accumulative share of inventory. Neither curve complies exactly with the 80/20 rule, but it is clear that relatively few items account for a high percentage of usage – 80% of turnover (or usage) came from less than 1000 items with individual turnover (i.e. units used per month, times unit price) of more than £30 per month, and over 90% from 3000 items with individual turnover of more than £7 per month. From the inventory curve it can be seen that 1000 items generated one third of the

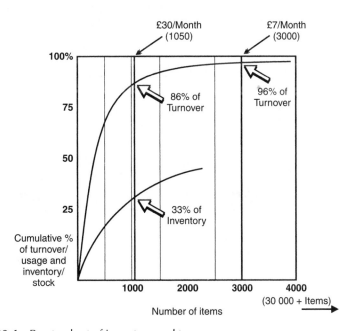

Figure 20.4   Pareto chart of inventory and turnover

inventory, and 3000 items counted for over 50%. Conversely, almost 50% of inventory resulted from holding 27 000 slow-moving (or nonmoving) items.

Annual usage was about £4m and average stock was about £1.8m – which means that the turnover rate was 2.2 times per annum or, using another measure of performance, the plant held 5.5 months' stock.

These were not considered to be 'good' performance measures and the plant improved them by applying the ABC approach. The ABC approach is to divide the inventory into three classes.

The Class A parts warrant continual attention from plant engineers, inventory analysts and purchasing to dissolve the problem. The fastest-moving 1000 items accounting for more than 80% of usage were designated as Class A. For these items it is possible to operate with less than one month's stock.

Class B parts deserve some attention, but can be managed with a simple rule or set of rules based on EOQs, and using reorder point or two-bin systems. For example, the rule could be 'when the reorder point is reached, or the first bin is empty, schedule delivery of two months' usage'. The reorder point can be as simple as a line on the bin or racks – when the physical level of stock falls below the line, it is time to reorder – and the reorder process as simple as pulling a card that carries predetermined instructions and inserting it in a reader or sending it to the supplier. Figure 20.5 illustrates the use of these two reorder rules. Although simple, these rules are efficient and effective. At its manufacturing facility in Montpellier, France, IBM has combined the two-bin system with sophisticated bar coding and direct electronic links to suppliers as part of a complete update of its operations [2].

It is not necessary to calculate EOQs precisely for each item. The system is very robust, and it may be necessary to simplify materials handling by rounding to the appropriate unit of packing – packs of 10, 12, 100 and so on. (The reorder card would show this sort of information.) In the transmission plant under consideration the next 2000 fastest-moving items, after Class A, were designated Class B and their stock levels targeted between two and four months.

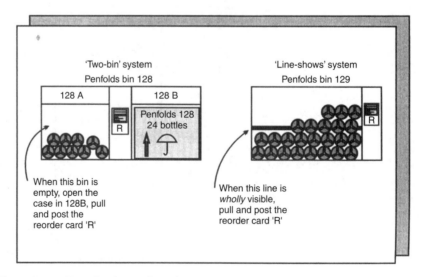

Figure 20.5   Two simple reorder rules

Class C parts do not repay detailed consideration. A very simple rule, like 'order one year's supply or £20 worth, whichever is less', will suffice. Like most inventories, the best solution for Class C items is not to have them in the first place or get rid of them fast if demand has died. In the example plant, the balance of 27 000 items was designated Class C.

A series of actions, by teams with shared values and common goals, resulted in the reductions in stock and the increased turnover rate, measured as 'turns per annum', shown in Table 20.1. Examples of the actions are given.

Some of the agreed actions are described below.

Reduce stock levels on high usage items – the mindset of the old hands (anything better than two turns a year is 'good') meant that 20 days' stock was regarded as 'tight'. Negotiation of faster response from suppliers (e.g. by using blanket orders, fax or computer links) and a change of attitude often reduced the holding to four or five days' stock. A blanket order commits the user to buy an item, or group of items, from the suppliers for a period of time – normally at least a year. During this time, the user simply issues 'releases' against the order, which can be done by mail, fax, e-mail or tied line. The order may specify delivery within 24 hours, or 24 hour-a-day availability. This cuts out the enquiry/order cycle. Chapter 21 describes blanket orders in more detail. Stocks of heating oil were virtually eliminated by changing to gas heating – 'virtually' because some protection against interruption of gas supply was needed. Stocks of hand tools were reduced to two days, by displaying the stock record via a PC link to a local stockist who replenished the plant automatically. Similar schemes were developed for standard bearings and electrical components.

Extend use of consignment stocks. It proved possible to arrange that some commodities, such as oils and greases, protective clothing and cleaning materials, were delivered to the plant, but remained the suppliers' property until the items were used. Sometimes this involved 'facility management' contracts, e.g. for lubrication systems plant-wide, or plant and equipment cleaning, or parts-inclusive maintenance contracts. In such cases, the supplier provided a service and all the materials required, which meant that the user no longer held them in inventory. (At the start of such schemes, the ideal arrangement is that the contractor takes over existing stocks.)

Identify duplication/combine usage. Items such as motors and bearings, held as replacement parts for repair of machines and equipment, were often bought initially from the equipment suppliers under the equipment suppliers' part numbers while identical parts were already in stock under the motor makers' or industry

Table 20.1   Stock levels and turn rates before and after ABC control

	Usage p.a. £m	Before		After	
		Stock £m	Turns p.a.	Stock £m	Turns p.a.
Class A (1000 items)	3.44	0.62	5.6	0.21	16.2
Class B (2000 items)	0.42	0.48	0.8	0.14	3.0
Class C (27 000 items)	0.14	0.69	0.2	0.47	0.3
All Stock (30 000 items)	4.00	1.79	2.2	0.82	4.8

numbers. By identifying duplication, sometimes with help from the bearing or motor suppliers, one or more items were deleted.

In the case of motors, the same unit with different mountings – end, top or bottom – was stocked several times, and stock was reduced by deciding to hold one or two motor units and a range of mountings. Also motors with the same ISO specification from three or four different vendors were held, and by standardizing on one manufacturer it was possible to delete the other two or three. Repaired or rewound motors were stocked, but some users always insisted on having the brand new item while hundreds of repaired items gathered dust. Once the plant manager made it clear that he was interested in lower inventories as well as achievement of schedules, the plant managed to reduce significantly stocks of new motors and old motors.

Improve scrap rates/disposals. All that was necessary was to convince the plant controller that his budget provision for obsolescence was not sacrosanct. For years he had refused to sanction disposal of 'usable' stock, which had been identified as obsolete or excess to likely all-time requirements, once the amount allowed in the budget for scrap and obsolescence was approached. This amount had been determined centrally and the instruction to the plant controller was that the amount must not be exceeded. Additional scrapping would have meant that the budget was exceeded, creating an 'adverse variance' for the plant manager to explain to his boss. Once it was clear that the boss preferred to get rid of the obsolete or surplus material (a shared goal), it was possible to reduce inactive stocks dramatically. For thousands of small-value inactive items it was necessary to introduce some rules of thumb implemented by computer programs, since the time taken for review (by engineers) could not be justified. The rules were of the form: if there has been no usage for ten years, scrap all stock; if there has been no usage for five years, scrap 75%, or down to one piece.

These were actions to correct past errors. Other improvements came by avoiding inventory, which can only happen if shared goals are established and communications are improved. It is important that the goal of inventory reduction is shared by production management, who may wish to protect against interruption of operations at almost any cost. Plant managers may fear the agony that follows a missed schedule more than the dull pain that may result from an inventory budget overrun, and will only be able to strike an appropriate balance if senior management heed the eighth of Deming's 14 obligations of top management:

> 'Reduce fear throughout the organization by encouraging open, two-way communication. The economic loss resulting from fear to ask questions or report trouble is appalling'.

For a full list of Demings 14 obligations, read *Total Quality Management* [3].

In the transmission plant, it was possible to establish a specification control activity to work towards the shared goal of inventory avoidance. The specification control activity consisted of one person with a PC and access to a database showing stock and usage information, who could check new requisitions for duplication before the duplicate item was ordered. This person was also able to check the quantity requested by the plant engineer against historical usage of like parts, and had authority to scale down excessive orders. (A better solution would have been to have the requisitioning engineer consult the database rather than make a hurried guess at the requirement, or blindly accept the supplier's recommendation.)

Another avoidance action was the determination of plant, machine and equipment maintenance requirements prior to agreement to purchase, and discussing with suppliers ways in which the equipment could be made more reliable. Equipment failure mode and effect analysis (FMEA) and study of mean time between failure (MTBF) and mean time to repair (MTTR) data can eliminate many costly 'investments' in spares [4].

The plant also achieved big savings through a policy of phasing in initial stocks of tooling and parts prone to wear. This was particularly beneficial where something new was being introduced and usage rates were uncertain – for example, tungsten carbide tools instead of high-speed steel tools. There is no point in stocking up with usage-related parts at the level forecast to be required to support full production if the climb to full production extends over six months or more. In that time actual usage can be determined and many of the items will be changed. If an item is changed, it may still be good practice to exhaust the replaced item if it is usable, rather than scrap – or, worse, keep the old and switch to the new.

This review of the ABC approach to inventory management has raised some of the general issues of inventory management, including the question: why hold inventory at all? These issues are examined in the next section.

20.3   WHY HOLD INVENTORY?

The basic reason for holding inventory is to guard against uncertainty. Some of the common areas of uncertainty and sources of variation in inventory levels are identified in this section and illustrated in Figure 20.6.

The ideal or desired situation is shown in section (a) of the graph, where the opening stock is steadily used up and is just exhausted when the replenishment arrives to restore the stock level to its starting point.

Section (b) shows what would happen if usage were greater than expected – maybe because scrap rates somewhere down the line were higher than usual, for example, or a mix of mortar had been allowed to 'go off' and more replacement material had to be called forward, or because the production supervisor thought that a 'good run' was desirable when the operation was running well. Stock would be exhausted before the replenishment delivery. This experience would probably lead to introduction of a buffer or safety stock just in case future demand again exceeded forecast.

Section (c) shows the stock levels if demand were below forecast, perhaps because operations were slow due to absenteeism, or due to the introduction of inexperienced operators, or because a tooling or die change took longer than expected. The next delivery would push stock levels above the target.

Section (d) illustrates a 'stock-out' due to a sudden surge in usage, which could happen for a variety of reasons – a record error, a production supervisor trying to get ahead of schedule, or an unexpected order could all have this effect.

Section (e) shows the effect of a shortfall in the delivery quantity – in this case it appears that prompt action was taken to secure delivery of the balance and no stock-out occurred.

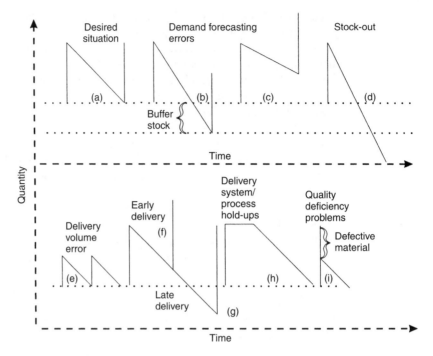

Figure 20.6   Sources of variation in inventory levels

Early delivery, as shown in section (f), results in excess stock.

Late delivery, as shown in section (g), could cause a stock-out (unless the *just-in-case* buffer stock were still there). In the 1970s, strikes at UK suppliers frequently resulted in late deliveries, which led their customers to hold high levels of safety stock or to take their business elsewhere. The effects were still evident 30 years later in the form of imports and cautious attitudes in inventory management.

Misleading information about stock levels can affect calculation of forward requirements, as shown in section (h). This could result from slow processing of receival or usage information, which would cause overscheduling or underscheduling respectively.

Section (i) illustrates one of the stock controllers' major problems – materials found to be defective on arrival, or when being processed, not only cause immediate shortages but also lead to extra stock being held so that all subsequent deliveries can be carefully checked, *just in case* there is something wrong.

The reasons for holding inventory, as just described, are really problems somewhere in the system. The problems do not occur one at a time for one item, as described, but at any time for many items, causing a complex or even chaotic situation for the inventory controller – unless the problems are resolved.

If inventory is continually used to hide problems it will rise to levels that make the business uncompetitive. Funds will be tied up in stock and allocated to building storage space, instead of being invested in new product development and more efficient equipment.

Far from being 'a good thing', inventory is increasingly seen as an 'evil' or a 'cruel sea', which hides fundamental problems, as illustrated in Figure 20.7.

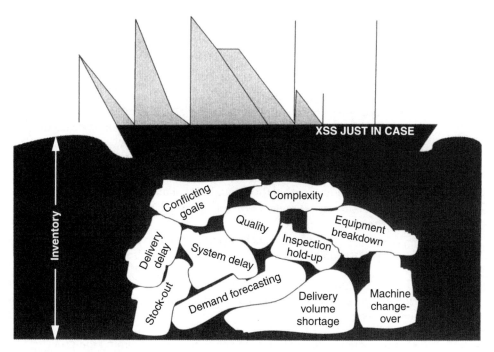

Figure 20.7   Sailing the cruel sea of inventory with its hidden rocks

As long as the rocks (the problems) are there, they are a danger to the ship (the business unit). Rather than build barrages to deepen the water (add inventory), it would be better to remove the rocks (dissolve the problems) and proceed confidently with lower levels of inventory. (Although widely used, this is not a very good analogy, since the captain (manager) does not have control over the sea level (inventory level). In reality, it is nevertheless good business practice to reduce inventories, expose the problems and dissolve them.) Having recognized inventory as an evil, 'management' may decide to eliminate it using just-in-time methods, instead of *just-in-case*. The next section describes what is involved.

20.4   JUST-IN-TIME INVENTORY MANAGEMENT

It is an appealing idea that if materials can be made to appear at the point of use just in time (JIT) to be used, then inventory can be dramatically reduced or even eliminated. The idea is so appealing that companies may be tempted to try to install JIT processes as a sort of overlay to existing systems, and without solving the fundamental problems that have led them to hold so much inventory. The first part of this section gives an indication of the difficulties of this approach to JIT. It is based on a study designed to determine the feasibility of introducing JIT to an established car assembly plant. This is Ford's plant at Saarlouis in Germany, which has a reputation for being one of Europe's best for efficiency and good product quality.

20.4.1   Attempts to 'retrofit' JIT

The study identified some of the preconditions for JIT as:

- Stable schedules.
- Reduced product complexity.
- Record integrity – plant bills of materials exactly the same as the central database, stock records adjusted promptly and with complete accuracy.
- Inbound traffic control – the plant needed to know the progress of every incoming truckload and what it contained.
- Container control – the numbers, identity and location of durable containers had to be known, so that they would be available for return to the right supplier.
- Local sourcing – suppliers of frequently required parts should be located close to the assembly plant.
- 'Parts per million' (ppm) defect rates – that is, suppliers' JIT deliveries had to be 100% usable.

This list is incomplete and other requirements will be identified later in this section. However, even the incomplete list led to identification of several preparatory tasks, which were:

- Completely replace all shipping containers for incoming material.
- Substantially modify the assembly building to create more delivery points.
- Change in-plant materials handling.
- Change plant layouts.
- Retrain plant and supplier employees.
- Introduce new in-plant information systems.
- Introduce new supplier communication systems.

The study identified, but did not resolve, another issue:

- How to deal with supplies from the company's own engine and transmission plants that were located hundreds of miles away from the assembly plant.

The cost of the changes was estimated at £12m, and the benefits were too small and uncertain to justify the investment, so the plant kept its old practices.

The study report highlighted the concern related to the engine and transmission plants, and prompted another study to determine how these plants could be run on a JIT basis.

This new study team quickly discovered that a wealth of information existed on JIT concept implementation. From this literature and their own discussions with suppliers, competitors and transportation companies, they came up with the following definition of JIT (the emphasis is their own):

'The concept by which the supply and production of *quality parts* is regulated *by next user demand with minimal inventory*

- at the supplier;
- in transit;
- in plant.'

Their list of key issues was similar to that of the earlier study, but had the following additions or differences of emphasis:

- 'Pull' material flow system (see below).
- Maximum machine 'up-time' (see below).
- Supplier selection.
- Quality the key to minimal inventory.

The concept of a *pull material flow system* is a key part of JIT, which is illustrated in Figure 20.8. By *maximum machine up-time* the study group meant that machines should be reliable, should be fitted with tools that would last at least through a shift, should be flexible and quick to change from one product derivative to another. Another JIT requirement – a flexible workforce – was not listed.

Quick tool changeover or die changeover to make a different part is an essential part of JIT methods. In section 20.2.5 above, it was shown that the EOQ formula can be used to weigh set-up (changeover) costs against holding costs. The JIT approach concentrates on reducing or eliminating the set-up cost. The results are spectacular in press shops or injection moulding or transfer line machining. Set-ups that used to take several hours, leading to 'economic' runs of several days' stock, are now performed in minutes and the economic run is a few hours' stock. The reduction in set-up times comes from redesign of tools and machines through cooperation with the equipment suppliers; from assuming that the changeover is right first time, rather than waiting while quality checks are made; from scheduling like parts in sequence; from computer control of changeovers; and from using carefully trained teams of production workers who are already on hand, instead of specialists who have to be called from a remote location.

Despite their clear understanding of JIT requirements, the Ford team was not able to install JIT production even at a new extension to the engine plant in Bridgend, South Wales that was still under construction, because:

- the project was raised without planning for JIT;
- plant layout was not consistent with JIT;

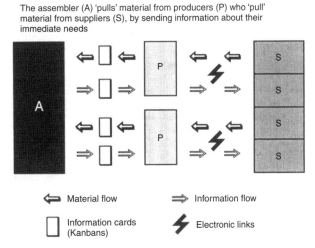

Figure 20.8 The 'pull' system of material flow

- production lines were not 'balanced' for JIT;
- sourcing actions had storage implications.

Nevertheless, some progress towards JIT was made, and Figure 20.9 shows the plan for one of the major components – the cylinder head. The figure shows Montupet Foundry, Belfast at the bottom, which was the source of castings, from which deliveries of 2500 pieces were to be made once per shift. In Phase 1, when annual output was planned to be 550 000 units, the castings were to be machined on Head Line 1, which was equipped with tooling and control systems such that it could produce all derivatives without interruption for changeover. (This is an example of 'flexible machines' as required for JIT.) In Phase 2, an additional

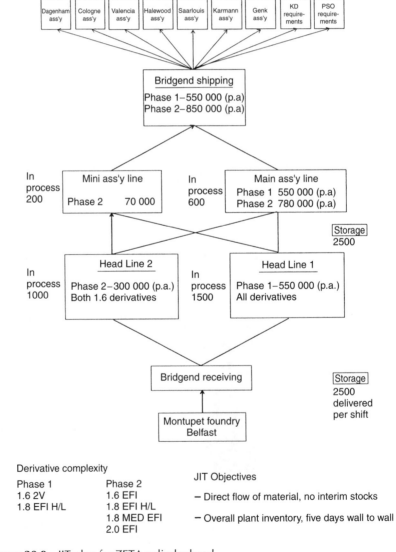

Figure 20.9 JIT plan for ZETA cylinder head

300 000 units a year were to be machined on Head Line 2, which was similarly equipped. From the machining lines, the cylinder heads were to be assembled into engines on the main assembly line in both Phase 1 and Phase 2, supplemented by a mini assembly line in Phase 2. The assembled engines were to proceed to the shipping bank, for despatch to the various destinations shown at the top of the chart. The planned numbers of cylinder heads to be stored and in process are shown against each operation. (Note the correspondence with Figure 5.5 in Chapter 5.)

The target of five days' stock (turnover rate 44 times a year, with 220 working days a year) was a major improvement on past performance, but the location of the plant in Wales, with some of its customers in Germany, Belgium and Spain and suppliers in Northern Ireland and the USA (not shown), meant that two of the JIT objectives could not be met – minimal inventory at suppliers and in transit. This was one reason why Phase 2 was installed not in Bridgend but in Cologne, with castings resourced to a nearby foundry and with customer plants less distant. (This change of plan explains all the 'were to be' phrases in the explanation of Figure 20.9.) Building the engine in two plants, each closer to its customers and suppliers, meant that the overall supply system was closer to the JIT concept.

20.4.2   A planned JIT system

This section concludes with an example of a more successful JIT application, where the requirements of JIT were part of the plan from the outset. It relates to the Nissan Manufacturing UK (NMUK) plant at Washington, near Sunderland, UK.

One factor that led Nissan to select the Washington site was its accessibility by road from the Midlands concentration of UK suppliers to the car industry. Even more important was the availability of adjoining sites where key suppliers were able to set up their own new plants. Figure 20.10 is a representation of

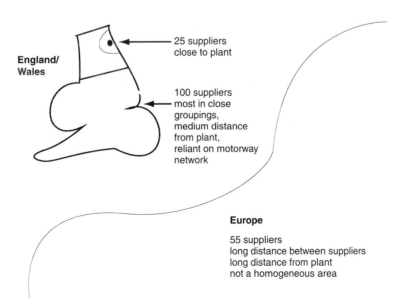

Figure 20.10   Geographical distribution of supply base of NMUK

the location of NMUK's suppliers soon after they began operations in 1986. The 25 suppliers adjoining the plant provide over 50% of the value and volume of NMUK's requirements (another example of the use of Pareto analysis in materials management). Parts and assemblies from these suppliers are delivered to the point of use in NMUK's plant at less than one hour's notice.

As production at Washington approached 100 000 units a year, NMUK revised its arrangements for deliveries from the Midlands-based suppliers. Shipment by individual suppliers led to congestion and did not provide the necessary degree of control over delivery. The NMUK plan was to go from Scenario 1 in Figure 20.11 to Scenario 2, which it achieved by schemes like that shown as pilot area solution.

In Scenario 1, which has now been superseded, each supplier shipped its own parts, to be offloaded into storage from which there were frequent deliveries of

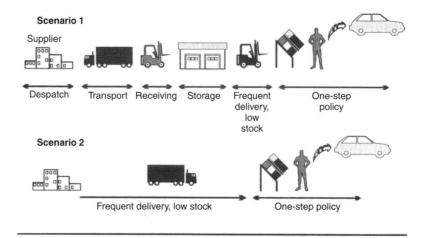

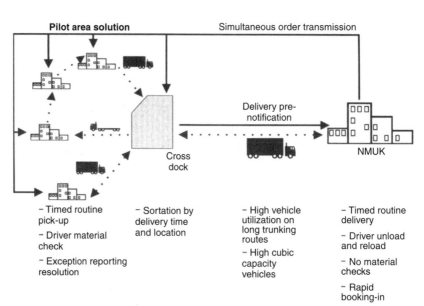

Figure 20.11 Development of NMUK's JIT system

small quantities to the production line, where material was placed 'one step' from the assembly line. In Scenario 2, which is present practice, small quantities from several suppliers are consolidated for shipment, to arrive just in time for offloading and movement direct to the production line.

In the pilot area solution diagram, information flows are shown by solid lines and material transportation flows by dotted lines. NMUK's requirements are simultaneously transmitted to suppliers through a communications network, so that each supplier can have material ready for a timed pick-up by NMUK contracted transport. The transport contractor's driver is responsible for checking the identity and quantity of the material before it is taken to a 'cross dock'. There it is sorted by required delivery times and separate delivery locations within the NMUK plant. Materials with the same delivery time are loaded to high-capacity vehicles for shipment to NMUK, and the contents of each load are pre-notified via the communications network. This allows 'one-touch' booking in of the complete shipment and unloading without further checks. Small, easy to manhandle quantities of each item are moved to within 'one step' of the production lines.[1]

The siting of NMUK's 25 key suppliers, and their inbound 'lean logistics' operations, enable them to manage with about one day's stock of materials (a turnover rate of more than 200 times a year). This is the result of careful, long-term planning – and of the opportunity to start at a 'greenfield' site, with all the worldwide experience of the parent corporation to draw on. The extent of the planning is indicated in Figure 20.12.

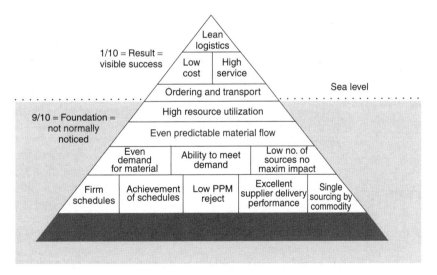

Figure 20.12 Prerequisites of lean logistics – the iceberg principle

---

[1]Increasingly, materials management systems such as that set up by Nissan are subcontracted to logistics specialists, which combine warehousing and transportation with Internet communication systems to track consignments across the globe, if necessary. In retailing, the use of electronic point-of-sale (EPOS) analysis of usage results in automatic replenishment of stocks without any action from local store management. Human intervention is still necessary to keep shelves filled and customers satisfied.

It is notable that some of the activities shown in the submerged part of the iceberg are 'enabling' actions by NMUK, such as 'firm schedules' and 'even demand for material', which assist supplier reactions such as 'excellent delivery performance'. Others indicate NMUK's sourcing policies, but the foundations of supplier performance are in the unspecified lower area, which includes the building of long-term partnerships. It was these partnerships that enabled Nissan to include suppliers in the planning of the Washington plant, and to arrange that 25 of the suppliers established new plants 'next door' to the Nissan site.

Nissan is not alone in this sort of achievement. The Saturn company, established in 1989 – GM's first new car company since 1917 – has similar performance at its integrated facility in Spring Hills near Nashville, USA. There GM has an in-house foundry, as well as engine, transmission and injection moulding facilities (yes, injection moulding – all but three body panels are plastic). Most major suppliers have been persuaded to set up nearby and, as a result, the Saturn plant holds less than one day's stock.

As with Nissan Manufacturing UK, the achievement of 'lean logistics' by the Saturn company required long-term planning of the entire manufacturing and materials management system.

The selection and development of suppliers is an important element of such long-term planning. This is dealt with in Chapter 21.

20.5 SUMMARY

This chapter covered the basics of inventory management. It showed that the questions 'How much should be ordered?' and 'When should it be ordered?' have traditionally been answered in a way that weighs acquisition costs and holding costs to minimize total inventory costs. It also showed that a better question is 'Why hold inventory at all?' since, far from solving problems, inventories often cover up problems and delay their solution. Attempts to introduce just-in-time (JIT) methods were described, showing that successful introduction requires long-term planning of the materials management system and the early involvement of suppliers.

REFERENCES

1. Gilchrist, S. (1994) 'Stores aim to reclaim buried pots of gold', *The Times*, 6 September.
2. Bradshaw, D. (1991) 'An Eastern breeze in the Med', *Financial Times*, Technology page, 25 June.
3. Peratec Ltd (1994) *Total Quality Management*, 2nd edn, Chapman and Hall, London.
4. Moubray, I. (1991) *Reliability-Centred Maintenance*, Butterworth-Heinemann, Oxford. Chapters 3 and 4 explain mean time between failures and mean time to repair, together with preventive maintenance, predictive maintenance and other maintenance management techniques.

BIBLIOGRAPHY

Just in Time, a booklet by the consultants A. T. Kearney written for the Department of Trade and Industry, is a very useful guide to JIT and its benefits. DTI publications are available through `www.dti.gov.uk/publications`.

Manufacturing Resource Planning, An Executive Guide to MRP 11, another DTI booklet, gives an overview of material and production scheduling. This is also available through `www.dti.gov.uk/publications`.

The Management of Manufacturing by E. I. Anderson, published by Addison-Wesley (1994), includes an extensive discussion of inventory management in Chapters 4 and 5, and of production planning and scheduling in Chapters 6, 7 and 8.

Optimizing plant availability is covered in *An Executive Guide to Effective Maintenance*, published by the DTI. Since reliable equipment is a precondition for JIT, this is a 'companion' to the JIT booklet cited above. It is also available through `www.dti.gov.uk/publications`.

Management
of the Supply System

21.1 INTRODUCTION

Chapters 4, 5, 6 and 20 have shown that total quality management (TQM), just-in-time (JIT) materials management, simultaneous or concurrent engineering (CE) and the management of advanced materials and new technologies all depend on early involvement of suppliers. Chapter 8 listed 'Build close, stable relationships with key suppliers of parts and services' as one of the five common strategies among world-class manufacturers. This chapter describes the interdependence of TQM, JIT and CE and how they all depend on supply system management (SSM). The concepts of a procurement team and a 'purchasing portfolio' are introduced. 'Old' methods of selecting suppliers are reviewed and the need to change these practices explained. Features of the 'new' ways of choosing suppliers and building partnerships are then outlined.

21.2 THE SUPPLY SYSTEM

21.2.1 What are suppliers required to supply?

What an organization requires from its suppliers has changed as the nature of competition has changed. In the first half of the twentieth century the primary requirement was low-cost supplies, and low cost was achieved by the application of 'Taylorist' methods to the main element of cost, which at that time was the cost of labour (see Chapter 3). By the 1950s, in addition to low cost, suppliers were increasingly required to provide goods and services of consistently high quality. They were encouraged to pursue total quality management (TQM), using the guidance and advice of 'quality gurus' such as Deming, Juran and Taguchi. Customer companies found that the task of introducing their suppliers to TQM, and in particular to the idea of defect prevention, was assisted by designing to a higher level of assembly and so reducing the number of suppliers required to participate in product and process design (see Chapter 4). This policy of working with fewer suppliers, each responsible for assemblies and subassemblies, rather than with many suppliers providing individual components and materials also simplified the materials management task and led to further cost reductions (see Chapter 5). By the late 1970s 'time' had become an important aspect of competition – the time required to bring new products to market, which was reduced by the avoidance of change using simultaneous (concurrent) engineering (see Chapter 6), and the time to respond to an order (see Chapter 20). The new challenge since the 1990s is also time related, but now it is the time to bring new materials and technologies from invention and development in the science base to be incorporated in new products and processes. Suppliers are now called on to supply revolutionary new ideas and unique new products. This entails the entry of the science base, which includes academic as well as industrial research organizations, as new links in the supply chain. Rather than forming the first link in the chain, feeding forward to materials suppliers, these new participants need to join the concurrent engineering teams made up of members from all parts of the supply chain and help to supply new low-cost, high-quality products incorporating advanced materials and technologies at a faster rate.

In summary, what suppliers are now expected to supply can be seen in the mirror of customer expectations, as stated in the electrical engineering company ABB's 1993 Annual Report:

'In all of ABB's markets, customers now expect more than top quality at a competitive price. They also want a supplier who has an intimate understanding of their business, who can anticipate changing needs and shifting markets, who can deliver fast, innovative and total system solutions that help them achieve their business goals.'

Later reports show the same view of what ABB aims to deliver, but with changes to its organization, systems and use of IT in how it delivers it.

21.2.2   Casting off the supply chain

The so-called original equipment manufacturers (OEMs), such as aircraft, oilrig, ship, car or computer producers, have to reach further down the supply chain for their concurrent engineering team partners. In similar ways, suppliers of services need to look further ahead to learn of new enabling technologies that will help them improve their 'products', processes and performance. Civil and construction engineering companies, too, are interested in new materials and new processes and fewer suppliers to assist them in developing better products that are quicker to build. This drive for innovation entails a new concept of supplier management: since all the members of the development team have to be involved concurrently, it is time to abandon the concept of a 'supply chain', which implies sequential links, and substitute the concept of the supply system. Management of the supply system is the current key to competitive advantage.

Figure 21.1 is a representation of the supply system, showing the interdependencies of the various 'players'. Recognition of the constant and concurrent effects of these interdependencies distinguishes the supply system concept from the supply chain introduced in Chapter 2.

The changes in the nature of competition referred to in section 21.2.1, and the corresponding changes in management focus or themes, are depicted in Figure 21.2.

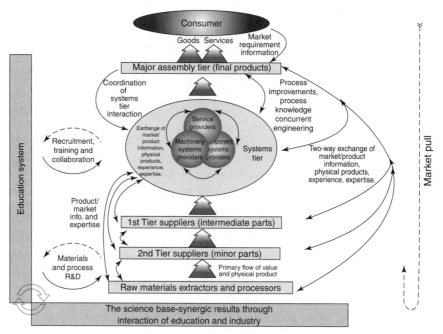

Figure 21.1   The supply system
Source: The figure was provided by Graham Clewer, who developed the idea of the supply system as a replacement for the supply chain as part of his PhD studies at the Engineering Management Centre, City University, and is reproduced with his permission

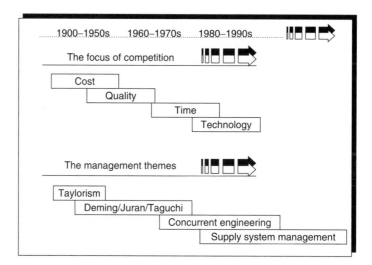

Figure 21.2   The moving edge of competition

The theme 'supply system management' (SSM) is shown overlapping 'concurrent engineering' (CE), since new relationships with suppliers and new ways of selecting them are essential for the success of CE. In this chapter, which deals with practices developed in the 1990s, the term 'concurrent engineering'

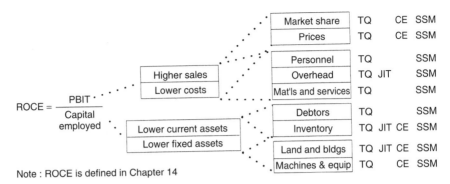

Figure 21.3   The dependence of ROCE on SSM

is used, reflecting general usage, rather than 'simultaneous engineering', which was more widely used when the concept was reintroduced in the 1980s.

This is also true of the other contributions that suppliers are expected to 'deliver' to help their customers improve their business performance. All the elements of the primary business performance indicator – return on capital employed (ROCE; see Chapter 14) – are affected by one or more of total quality (TQ), JIT and CE, and they all depend on SSM as shown in Figure 21.3.

The links between the elements are as follows:

- Higher sales result from a combination of increased market share, entry to new markets and premium prices, which all require superior quality at the time of market entry, and market entry of new products ahead of competition, which requires CE, which requires SSM.
- Lower personnel costs depend on better processes – doing things 'right first time' (TQM) – and on having suppliers do more things for their customers (outsourcing), which depends on SSM. Having fewer employees also leads to lower fixed assets – less office space and less equipment are required.
- Lower overheads can be achieved by having less inventory and equipment to finance and smaller buildings to service and finance. See below how this is linked to SSM.
- Lower costs of materials and services depend on suppliers delivering TQ and achieving TQ in their own operations, which depends on CE, which requires SSM.
- Lower debtors can be achieved by applying TQ techniques to debt management and by partnership-style relations with customers, which is part of SSM.
- Lower inventories depend on JIT, which depends on TQ and CE, which depend on SSM.
- Less inventory can lead to smaller, lower-cost buildings, as can the use of less equipment that is more fully utilized thanks to equipment and materials suppliers' TQ, which depends on CE, which depends on SSM.

Within SSM, the key elements are supplier selection and the development of long-term alliances with fewer, better suppliers. The processes of selecting suppliers are dealt with below. Rather than 'purchasing', which has become regarded as a clerical, rubber-stamping exercise, the term 'procurement' is used.

21.3   PROCUREMENT – ANOTHER TEAM GAME

Business is a team game. As in all team games, the chances of winning are improved by choosing a good partner or good team mates, so that among the team there are complementary skills, knowledge and experience. The potential of the team is realized through practice, to develop cooperation and collaboration.

Throughout this book there have been examples of the need for cooperation between companies and for collaboration between functions within companies. This is particularly true for procurement – the process of identifying the best long-term suppliers of materials, equipment, goods and services and forming alliances with them fosters the joint pursuit of shared goals. That is a task very different from the old job of buying from the cheapest source that meets immediate requirements. The new way calls for a new team within the customer company (and in the supplier company). The chart in Figure 21.4 shows the functions involved in procurement of components, materials, equipment or services; it can be used as a generic model. Note that communication between functions is electronic, as are most links with suppliers, though these are not shown.

The specification and requisitioning activities – the internal customers of purchasing – are usually the responsibility of engineers, scientists, technicians or managers in HRM, sales or finance staffs (who could also be engineers, scientists or technicians by training). The interaction between these internal customers and service providers is one of the key factors that determines the team's success and is very much dependent on the organization's *style* (see Chapter 3) and its human resource management (see Chapters 9 and 10).

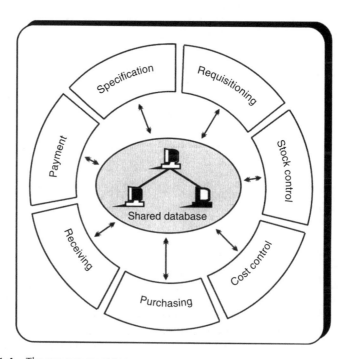

Figure 21.4   The procurement team

Not every procurement transaction warrants fielding the top team. Procurement, like all processes, benefits from Pareto analysis in order to apply effort where it will have greatest effect. A 'buy' may be important simply because of its value (i.e. its effect on life-time cost) or it may have strategic significance. Engineers and scientists have a key role in identifying those materials, products and services that have strategic significance because they may be important to the organization in ten or more years' time, as well as those that are required in the next product development cycle. They also are vital members of the team that identifies possible supplier partners. Other team members may be users (shop-floor employees), product planners, maintenance and safety personnel.

A 'portfolio' can be compiled and a plan developed for dealing with each category of 'buy'. Such a portfolio is not static and has to be updated as the business environment and corporate objectives change. An example for an entirely fictitious company is shown in Figure 21.5.

Some items do not deserve to be in the portfolio. The chart in Figure 21.6 is the typical procurement process introduced in Chapter 20, with an important added line of information. It shows across the bottom of the chart the time taken to perform each function. Figures are based on an internal Ford study of the procurement of nonproduction materials/goods and services. The number in the bottom right corner (130) is the average time in minutes to perform all the processes – not the elapsed time but the total processing time.[1] In the UK, at average salary and overhead rates, that costs about £60. This far exceeds the

Product	Risk Operations	Cost	Actions	Information	Method	Responsibility/ timescale
Strategic Benzol Cyclohexan	H	H	Precise requirements Detailed market research Long-term relationships Identify risks	Detailed market knowledge Long-term needs	Market analysis Risk analysis Simulation Price forecasts Long term	Purchasing manager Long term
Bottleneck Catalysts Metals	H	L	Volume guarantee (price premiums?) Supplier follow-up Safety stocks	Mid term needs Storage costs Good market knowledge	Negotiated agreements Mid/long term	Purchasing agent Mid/long term
Key DP equipment Motors Heating oil	L	H	Bulk buying Alternative products Spot buys?	Good market knowledge Transport costs Maintenance costs	Competitive bidding leading to extended agreement	Senior buyer
Normal Stationery Office equipment	L	L	Standardization Groupage	Market overview EOQ	Competitive bidding Short-term agreement (1 year?)	Buyer short/mid term

Figure 21.5   Purchasing portfolio analysis

[1]The almost universal use of electronic links between components of large organizations does little to reduce the process time – only the transmission time. Savings come from eliminating unnecessary repetition of the process.

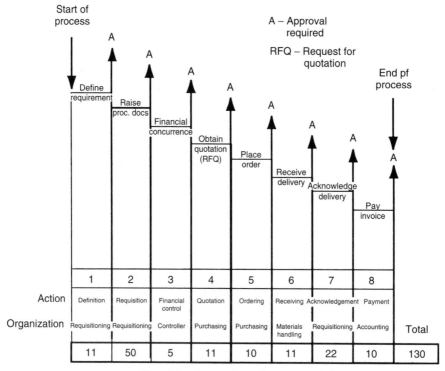

Figure 21.6 The acquisition/procurement process

saving (15%?) that the most astute buyer could achieve on a high proportion of requisitions – in fact, it exceeds the value of a high proportion of transactions.

Another Ford study of nonproduction procurement in its UK operations during 1984 showed that the average value of the more than 4000 requisitions per day was £36. Throughout the country, there were over 1000 deliveries per day, some involving more than one item. The average value per item was £153, but 40% had a value below £25 and 70% had a value below £100. The values have changed since the study was made but the pattern has not, and the implications still apply.

It does not make sense to apply the might of the full procurement process to any requirement with a cost of less than £50, or even a requirement costing £200. These needs can be met from petty cash, or through a company credit card, or through an administratively simple purchasing device, the 'blanket order', which readily lends itself to 'e-business' using the Internet. Figure 21.7 illustrates how a blanket order system operates.

Once set up, the blanket order enables any authorized user, in any of the several different locations, to obtain what they need by issuing a 'release' direct to the supplier. The supplier ships direct to each user location and consolidates all transactions covered by the blanket order into a single monthly invoice. This takes time out of actions 2, 3, 4, 5 and 8 shown in the 'procurement process' chart in Figure 21.6. In 1994, Ford Motor Company held meetings with its leading suppliers to explain its plans to change its purchasing methods and organization

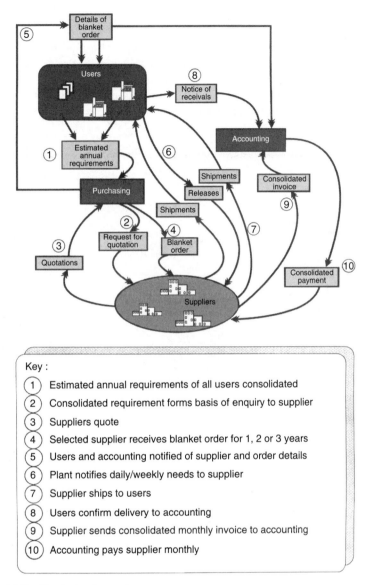

Figure 21.7 Blanket order system

on a global basis. One of the major planned method changes was to increase the use of blanket orders, since Ford's studies had shown that to process an individual purchase notice (order) cost $280, whereas processing a release against a blanket order cost $190. Again, the numbers have changed, but the relationship has not. Ten years later, the argument for using blanket orders is just as strong and becomes stronger when applied to e-commerce.

Delegated 'local' authority or centrally negotiated blanket orders can take care of many requirements that have low individual value and no strategic significance. It is the other end of the scale that demands attention. This is the 20% of items that account for 80% of the total value of purchases, or the few items that differentiate the product, or the even lower number involving new technology and/or long lead times that can make or break a project.

For machinery and equipment purchases, this process of separating the important few from the rest should include a decision about the need to use the CE approach. Although there are potential benefits in terms of improved quality and savings of time and investment cost, CE requires the application of more human resources than the conventional procurement process. Figure 21.8

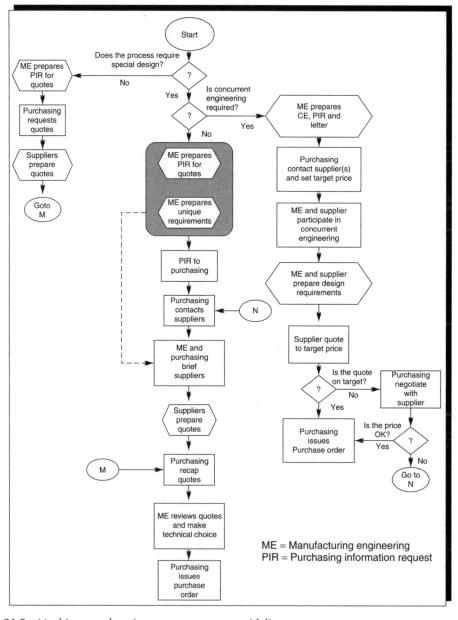

Figure 21.8   Machinery and equipment procurement guidelines

shows an algorithm that can help to determine whether a simple request for quotations would be appropriate, or whether the additional effort involved in CE should be applied.

Successful organizations make these decisions about the use of CE, and the commodities to which CE should be applied, early on. This means early enough to involve suppliers in developing concepts, designs and delivery processes (delivery in the sense of meeting all the customers' needs, not delivery by truck, though that may be part of it).

In the automotive industry early involvement means identifying the suppliers of key parts at least four years before production starts. For major changes in materials or the introduction of radical new features or processes, it may mean ten years. Some other industries, such as pharmaceuticals or aerospace, have even longer development spans, and others such as home entertainment equipment or computer peripherals have shorter spans. Since the automotive industry is somewhere in the middle and has been a leader in making changes to procurement processes, it can provide an insight into methods of supplier selection and management. The following sections describe these methods and how they have been changing.

21.4   PROCUREMENT – THE OLD WAY

The 'old' ways of procurement in the automotive industry are still widely used in other industries and by some laggard automotive companies. Engineers, scientists and technologists may therefore find themselves working with or in organizations where these methods continue to be used, and may be assisted by this section on what should be a dying art.

Three criteria have been used to make purchasing decisions in the engineering industries from the time they were founded. These are:

- cost;
- quality;
- delivery performance.

In many industries, the realization that suppliers' quality management determines their ability to reduce costs and deliver on time is fairly recent. In some, the belief that quality can be secured by inspection still persists, but generally, with a time lag of a few years, supplier quality management has changed in the same ways as their customers'. Management of quality has been covered in Chapter 4, and this section therefore concentrates on the old approach to supplier cost and delivery management, with passing reference to quality.

The table in Figure 21.9 shows the 'old' basis of sourcing decisions, for components (parts) and for machinery and equipment.

The division of responsibilities between purchase, engineering and quality departments, who may all have been pursuing different goals, encouraged 'compartmental' attitudes and 'turf protection', plus time-consuming blame allocation when things went wrong. Conflicts arose, for example, when designers wanted development agreements with suppliers that buyers thought were uncompetitive on price or had a poor delivery record.

Assessment criteria	Assessed by department...		...on the basis of	
	For components...	For machinery and equipment...	For components...	For machinery and equipment...
Ability to meet specifications	Purchasing/ quality control Product Development	Purchasing/ manufacturing engineering	Function, laboratory and dimensional tests	Demonstrations Witness Judgement
			Development programme	
Price	Purchasing	Purchasing	Competition and estimates	
Reliability	Purchasing	Purchasing	Delivery, warranty, and performance records financial appraisals and buyer visits	

Figure 21.9   The procurement process – sourcing decisions the old way

Quality was controlled by issuing detailed specifications, with extensive test requirements, against which suppliers with approved quality control systems were invited to quote. At the quotation stage, suppliers would rarely admit to any difficulty in meeting the specification, so, increasingly, price became the main criterion. This was reinforced by design cost control systems that assigned cost targets to designers, as well as by purchase cost control systems that assigned targets to buyers.

Design cost control and purchase cost control systems work in similar ways. By identifying the 20% of 'key' parts that account for 80% of product cost, the cost analysis task can be reduced to a manageable scale. For these key parts it is possible to build up a cost model. This would show what it would cost an efficient producer, using suitable equipment and processes and paying market labour rates and material prices, to make that part or assembly. Profit margins, R&D costs and other overheads can be incorporated. Once the model has been constructed, it is possible to estimate the effect of changes. These include design changes, pay awards, material cost increases, energy cost movements, exchange rate fluctuations and so on.

Such systems can be used to measure the cost performance of an individual designer or buyer, or the collective performance of a design group, a manufacturing plant or an outside supplier. In finance-driven companies, this is the principal application, with the added refinement of targeted annual cost reductions. Similarly, estimates can be aggregated to generate total product costs, and to set targets for cost reductions from one design iteration or model to the next.

Even with competitive bidding, the purchase cost estimates can be used by buyers to check each supplier's quotation and identify opportunities for negotiation. In single-bid situations, for example where a supplier has been engaged in a design contract, the estimate may be the only check on their quotation to determine whether it is reasonable or extortionate. If it is extortionate, the only remedy for the buyer is a long memory and carefully kept records.

The same information can also be used in a more constructive way, to focus shared cost-reduction effort through value analysis and the highlighting of opportunities to introduce better methods, alternative materials, alternative sources and so on.

The procurement process within which such price control mechanisms were used is illustrated in Figure 21.10. In better-run firms, the list of suppliers invited to quote was agreed between users, designers and buyers, but in many organizations commitments were made by designers or users – or worst of all by the boss – and the buyer was expected to negotiate the best deal afterwards. The sequence of actions in this version of the procurement process is as follows.

The stream of releases from product development (design engineering) is divided by pre-production control into two flows of 'buy' parts and 'make' parts, on the basis of earlier policy decisions or 'integration studies', also conducted previously.

Purchasing receives the information about buy parts and issues an enquiry or 'request for quotation' to potential suppliers. The suppliers send in their quotations, purchasing selects the 'best' bid and issues an order to the successful supplier.

Manufacturing engineering receives the information about 'make' parts and checks its capacity and capability to produce the volumes and specifications. If additional equipment or tooling is required, it issues a purchase information request (PIR) to purchasing, which goes through an enquiry/quotation process similar to the buy parts process. However, for equipment there is an added loop, to cover the review of quotations jointly by engineers and buyers – the 'recaps' in the figure. This allows for price and performance tradeoffs to be discussed, which are unlikely to be feasible for parts and assemblies. The make parts will also require

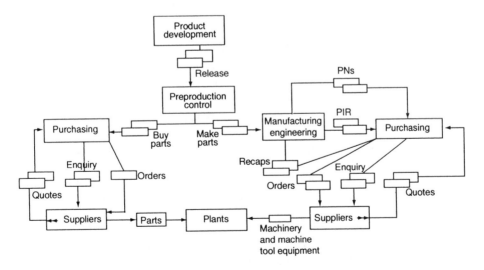

Key to abbreviations:
PN – purchase notice (an order)
PIR – purchase information request (an enquiry)
RECAP – recapitulation (a detailed comparison of the quotation)

Figure 21.10   The procurement process – purchasing/engineering/supplier links

outside purchases of materials and parts, and the buy parts process is followed for these items. Long-lead requirements of manufacturing – such as a new building, or a major building extension or modification – are bought through a similar process as equipment purchases, but the need for this is identified through the integration studies mentioned earlier, on a timescale that recognizes construction lead times.

In these 'old' procurement processes it was normally the quoted purchase price that was considered when the source was selected. For most purchases, the difference between price and lifetime cost, which recognizes running costs, maintenance costs and disposal costs (see Chapter 15), was not taken into account. This difference can be so great for long-life products that a decision based on price alone can be a bad one and, increasingly, purchasing decisions for buildings, manufacturing equipment and transportation equipment are taking account of lifetime costs. The same should apply to office equipment purchases. For example, the price of a laser printer can be as little as 10% of the cost per page of print over a three-year period of printer life [1].

The processes described above are based on manufacturing industry practice, but something similar applied, and continues to apply, in the construction industry. A consultant placed between the user and the constructor performs the same role of developing very detailed specifications, establishing a list of suitable suppliers, soliciting competitive bids, analysing responses and recommending order placement with the company quoting lowest – providing it meets the specification and lead time. The detailed specification is later used to try to control the time and cost effect of changes. (See Chapters 17 and 19.)

In the 'old' manufacturing procurement process, extensive buyers' terms and conditions were combined with the detailed specifications to protect the purchasers' interests. These were matched with similarly restrictive terms of offer and sale to protect the sellers' interests. Attempts to negotiate settlements of disputes would start with attempts to determine whose terms and conditions applied. The terms and conditions might include sections designed to assign responsibility for adherence to delivery or completion timing, as well as compliance with specification and price. Graham and Sano in *Beyond Negotiation* [2] note the contrast between western contracts, which may take 100 pages to define buyer rights and supplier responsibilities, and Japanese contracts of a page or two that set out the shared objectives of the agreement. They also observe that the Americans talk of 'concluding business deals', while the Japanese speak of 'establishing business relationships'.

It was common practice in the West to reinforce these contractual arrangements with monitoring – 'chasers', 'follow-up', site engineers, timing analysts, project control engineers or analysts (sometimes working against each other, one to control cost, the other to control timing). Extensive records were kept of back orders, past dues, slippages and delays, and alleged reasons for them. In many western industries, security of supply of parts and materials was often sought through dual or triple sourcing, as a protection against poor labour relations, quality concerns, transportation difficulties and the like. Where multisourcing failed to give this protection, it was supplemented by inventory on the user's site or at a 'neutral' warehouse.

'Transportation difficulties' and 'warehousing' are reminders of the traffic and customs function within 'supply'. (See Chapter 5.) They are tagged on here much as they were in the old ways of procurement. After a part or product had been sourced, the materials handling and traffic personnel tried to negotiate packaging and packing modes with the selected supplier, and the traffic or transportation personnel would review inbound shipping arrangements. This could only be done after the programming and inventory planning group had had its say to determine target inventories, delivery frequencies and quantities. It also needed to know prices, volumes, supplier location and reliability in terms of quality and continuity of operations to do this. Shipping was often the suppliers' responsibility, but there has been a trend to ex-works purchasing for two sets of reasons:

- Suppliers fell down on the quality of their packing and delivery performance (under pressure to reduce cost).
- Individual suppliers were constantly faced with the task of delivering uneconomic (part) loads, and using common carriers to do this added more delay and risk.

The pressure to introduce JIT concepts forced changes in transportation and packaging, as well as in inventory planning, and was one of the reasons for moving away from a separate procurement process towards integrated purchasing, logistics and planning. Other reasons are included in the next section, which describes the 'new' processes of procurement.

21.5 'NEW' PROCESSES OF PROCUREMENT – PARTNERSHIP SOURCING

21.5.1 The 'drivers' for change

As the edge of competition moved from price alone, to include quality, durability, reliability and then time to market and time to satisfy customers' orders, the old purchasing and follow-up practices were exposed as defective and destructive.

They generated internal conflicts and conflicts between buyer and seller. Removal of these conflicts required new relationships within companies and new relationships between companies. All the 'success formulae' identified in Chapter 8 stress the need for long-term partnerships with fewer suppliers. Section 21.2.1 summarized the ways in which TQ, JIT and CE drive to the same conclusion: suppliers have to be involved early in the planning of the product, and in the entire production and distribution system. These programmes – TQ, JIT and CE – require 'real-time' communications systems using computerized data exchange (CDX), which, again, is a less difficult task if there are fewer suppliers involved.

So, from many directions there are pressures to change from relationships with many suppliers who bid against each other for short-term contracts, to the practice of early identification of fewer suppliers to be treated as partners pursuing shared goals. This practice precludes the old form of competitive bidding as a means of supplier selection – at the time when suppliers join the CE teams there are no detailed specifications for them to bid against.

The chart in Figure 21.11 shows the timing of 'early sourcing' in the automotive industry in the late 1980s, when alternatives to competitive bidding as a means of supplier selection were being developed. Since then, the 48-month new vehicle development cycle has been reduced to 36 months typically, and as low as 24

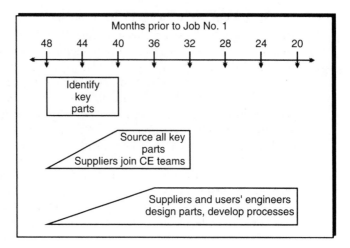

Figure 21.11   The early sourcing process

months exceptionally. However, these shorter cycles result from the policies and practices developed in the late 1980s, and these are therefore reviewed in the next sections.

21.5.2   New policies and new attitudes towards suppliers

The sourcing policies and attitudes that were widely adopted in the late 1980s had been advocated by Deming some 40 years earlier [3]. The fourth of his 14 responsibilities of management is:

> 'End the practice of awarding business on the basis of price tag alone. Purchasing must be combined with design of the product, manufacturing and sales to work with the chosen suppliers.'

Acceptance of this advice is evident in policy statements by many of the automotive companies. The following examples are from material supplied by the companies. Nissan Manufacturing UK:

> 'The principle of "Partnership" is the key to our supplier relationships. It is, however, not just a purchasing principle – it is a whole company philosophy. To work properly it needs time, patience, trust and full commitment from all parties involved in the buyer/seller relationship on both sides of the partnership . . . Partnership sourcing is a commitment . . . to a long-term relationship based on clear mutually agreed objectives to strive for world-class capability and competitiveness.'

The Rover Group:

> 'Our new standard emphasises strategic issues which are often overlooked and encourages both ownership and flexibility in achieving our common goals

... This new standard is an acknowledgement by the Rover Group that we need to form close, lasting partnerships with suppliers to maintain successful, competitive businesses into the next century.'

The Rover Group, it could be added, was one of the earliest companies in the UK to apply this strategy to the procurement of new buildings.

Chrysler Corporation:

'At Chrysler, we don't view our suppliers as generic interchangeable "vendors" whose only purpose in life is to do what we tell them to do and to never have any ideas of their own. Instead, we view our suppliers as true extensions of our company – as an integral, creative link in the value-added chain, just as we ourselves are merely a link. We call it "the extended enterprise concept." We've totally scrapped the age-old system in the auto industry of auctioning off contracts to the lowest bidder. Instead, we set what we call a "target cost" and then we work closely with trusted, pre-selected supplier partners to arrive at that target. And we don't do that by cutting profit margins, but by encouraging new and innovative ideas, and by rooting out waste and inefficiency – much of it, by the way, in our own systems.'

This last point in the statement by Chrysler is also a part of the Xerox policy (just to get away from the automotive industry for a few lines). Xerox calls it 'cooperative contracting' and it has the same results as Chrysler's approach: lower total costs, but not at the expense of supplier profit margins. The chart in Figure 21.12 illustrates the effect of the Xerox practice.

All these policy statements describe close, long-term relationships with fewer, pre-selected suppliers. The way in which these select few are chosen is described in the next section.

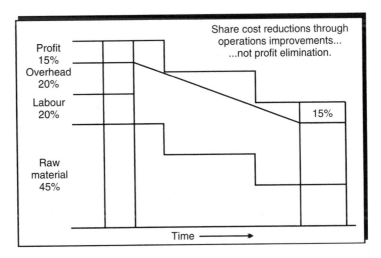

Figure 21.12   Cooperative contracting

21.5.3   Selection criteria for pre-sourcing

History may be 'bunk', as Henry Ford I is alleged to have said, but it nevertheless can be used as a guide to the future. Carefully recorded history and objective assessment form the basis of most of the supplier selection processes that have been substituted for competitive bidding. The automotive industry generally uses five headings under which supplier history is compiled, and which are used to make an overall assessment of a supplier's capability and potential. These are:

- management ability and attitude;
- quality performance;
- delivery performance;
- technical capability;
- commercial performance.

Some schemes incorporate an assessment of management under the other four headings. One of these, the Ford supply base management (SBM) system, forms the basis of the following explanation. Figure 21.13 shows the logo from the cover of Ford's SBM booklet, with the four criteria and the activities responsible for rating supplier performance for each of them.

Quality is the overriding criterion. Most producers in the automotive, aerospace, information technology, pharmaceutical and electrical and electronics industries have their own quality management systems and standards for assessment of materials and component suppliers that are more demanding than the international BS/EN/ISO 9000 standards. They have developed comprehensive measures of supplier performance over the years, and many suppliers recognize the value to their own operations of making the continuous improvements demanded

*SQE = supplier quality engineering

Figure 21.13   Supply base management rating criteria

by these quality management systems. Compliance with the customer's quality system requirements is the price of admission to the supplier selection game.

The Ford Q101/Q1 system scores supplier quality performance under three headings with weighting for each. These are:

- Supplier management attitude and commitment   20%
- Supplier quality management system   30%
- Supplier delivered quality performance   50%

Suppliers that consistently score above 80% may petition for the Ford Q1 award, and only suppliers with Q1 status are considered for new business. Based on their scores, suppliers are classified as 'preferred', 'potentially preferred' or 'short-term'. The last category indicates that a supplier will not be used beyond existing contracts.

Objective measurement of equipment suppliers' quality performance is less well developed. The issue is usually clouded by variations in material or rough part quality or tooling quality, by the effectiveness of prior or complementary operations, and by standards of equipment maintenance and operator training. Nevertheless, by considering process capability (see Chapter 4), mean time between failures, mean time to repair and so on, it is possible to reach consensus between equipment users, process designers and purchasing on a ranking of potential suppliers for each type of equipment, and to agree whether quality performance is roughly equal or markedly different. It should be possible to differentiate between 'acceptable' and 'unacceptable' suppliers, and it is quite likely that one or two 'quality preferred' suppliers can be named.

Current technical capability is a clue to future performance. Since suppliers' participation in the design process is essential for successful CE, their capability in this respect is a vital part of the selection process. The activity best able to make this assessment is the customer company's own technical staff. Purchasing has to ensure that all serious contenders are considered and that its technical colleagues' opinions are expressed in a rational, measurable way. Scoring or ranking systems can be established to evaluate a supplier's R&D investment and facilities, personnel, rate of innovation, use of CE with their own suppliers, CAD/CAM and CDX capability and so on. Absolute precision in ranking is not necessary, and is not meaningful. A useful and sufficient separation between suppliers could be achieved by assessing them as:

- Preferred – frequently provides relevant, tested innovations.
- Acceptable – sometimes provides relevant, tested innovations.
- Unacceptable – rarely provides relevant, tested innovations.

This approach can be applied equally well to materials suppliers, component suppliers and equipment suppliers. Regular, structured exchange visits between customer and supplier to develop an understanding of each other's operations, with shared, structured reports, will enable these evaluations to be built up over an extended period, so that 'off-the-shelf' selections can be made for new projects.

Measurement of delivery performance and capacity is essential. By systematically measuring supplier delivery performance and fairly analysing reasons for lateness, it is possible to build up a ranking of suppliers according to this criterion. Rating systems that include measures of supplier responsiveness to change, their planning capability, their management attitudes and commitment and so on can be used to generate scores, say on a scale of one to ten for each criterion. The scores could be weighted and aggregated to provide classifications such as:

- preferred;
- acceptable;
- unacceptable.

Early involvement of preferred or acceptable suppliers in new programmes enables them to improve their delivery performance by planning new processes and capacity in advance. Process capability and capacity and subsupplier capability can all be brought to the required levels in good time.

Component, materials and equipment suppliers must have sufficient confidence in their customers to honestly discuss their order backlogs and production and investment plans. It may be necessary for customers to support their suppliers when they make requests from their financiers for investment funds, or to modify their payment terms to assist in the funding of work in progress or new facilities.

Continuous exchange of information about current performance, forward requirements and anticipated problems is necessary to maintain deliveries on schedule, or to prepare offsetting measures. It is an important customer responsibility to constantly check demand from all user areas against supplier capacity and commitments. For example, service requirements should be included as well as current production needs, or coordinated demands from associate and subsidiary companies around the world.

Commercial performance seals the relationship. The 'right price' is essential to both supplier and customer, but may not be known when sourcing commitments are made. CE helps suppliers to improve their efficiency and reduce their costs, so that they can supply at prices that are competitive on a global basis, yet make returns on capital that are sufficient to support their own profitable development. Chrysler's 'target pricing' and Xerox's 'cooperative contracting' are ways of supporting suppliers' efforts in this respect. For 'global' sourcing, these pricing discussions will include projections of supplier productivity improvements and material cost control, taking account of inflation and exchange rate projections. Where design details of the new product are not available, the discussions may be based on 'surrogate' parts, or commodity groups rather than specific parts.

Constructive use of the information is vital for developing partnerships. Evaluation of supplier performance is a continuous process, and the results must be shared regularly with individual suppliers so that opportunities for improvement can be identified. Comparative data should be maintained by the customer's purchasing activity so that supply base capability can be assessed in the light of the purchasing portfolio of long-term needs. An example of a Ford summary of quality, delivery, technical and commercial ratings of suppliers in a commodity group is shown in Figure 21.14.

The commodity summary covers the UK only, and shows an objective for the reduction in the number of suppliers from 4 to 3. This is no longer typical. Ford, like many multinational companies, is now buying its most important requirements on a global basis, and a commodity summary would show global supplier ratings and global supplier reduction plans.

Supply Base Management

Commodity : wheel bearings

Buyer: A. N. Other   Code: E100

Objective no. suppliers 90 4 91 4 92 4 93 3

Date: April 1991   Division: BAO

Supplier name(s) Ship point Preferred long term	Code	Sales		Ratings								Remarks
		Totals ($ millions)	Commodity ($ millions)	Quality	Date	Tech	Date	Dely*	Date	Comm	Date	
W. H. Smith Romford, Essex	A100B	4.8	2.4	Q1	6/89	90	2/90	91	3/90	93	4/90	
E. H. Willis Tipton, W. Midlands	B200A	3.7	1.9	Q1	8/89	91	2/90	89	3/90	85	4/90	
Potential long term												
J. Brown Dudley, W. Midlands	C150B	2.8	2.4	85P	3/90	84	2/90	87	3/90	81		Corrective action plan Available resurvey Q3 1990
Short term												
D. Lete Runcorn, Liverpool	E250A	1.4	1.0	80A	3/90	81	2/90	76	3/90	72	4/90	Deleted with CDW27 1993

*Median Score—of Delivery Ratings

Figure 21.14 Commodity summary

21.5.4 Global procurement

Ford had developed the SBM sourcing process far enough for it to be used in 1989–90 to identify preferred suppliers to participate in the CDW27 programme, which was the code name for the one car eventually marketed as the Mondeo in Europe, and the Contour and Mystique in North America, replacing two separate and different medium car ranges.

Ford's European Automotive Operations (EAO) had been using the supply base management concept for several years prior to the start of the CDW27 programme. This had enabled EAO to combine its requirements for all its European plants, and to make a 'best buy' using information from the worldwide supply base. Similar sourcing practices had been used by North American Automotive Operations (NAAO) for its total requirements. The CDW27 project was an opportunity to optimize on a global basis, for the combined needs of the CDW27 producer plants in the USA (St Louis) and in Europe (Genk, in Belgium). To do this, Ford established temporary 'worldwide sourcing operations' that coordinated information from EAO and NAAO to compile a single list of preferred suppliers to work with the single CDW27 design and manufacturing team in Europe. Figure 21.15 illustrates the transition from 'regional' to 'global' purchasing. The 'volume savings' referred to in the figure are not economies of manufacturing scale, since most suppliers used separate facilities to serve the St Louis and Genk plants and to apply JIT methods to a limited extent. (Full JIT would have entailed location of the 20 to 30 most important suppliers close to each of the assembly plants, which was not done.) The savings came primarily from combining Ford and supplier design and process engineering capabilities to work on a single, global design and a single best manufacturing process – something the Japanese car producers have always done.

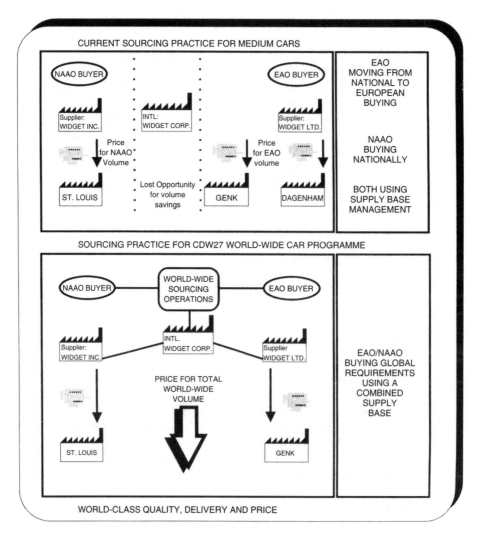

Figure 21.15   The change from regional to global sourcing at Ford

Throughout the CDW27 programme, Ford continued its policy of reducing the number of suppliers with whom it was involved. Despite being assembled in plants in two continents, the CDW27 was supported by fewer than 300 suppliers – about half the number that EAO and NAAO each used on the superseded models. For the year 2000, Ford aimed to be working with only 600 to 800 suppliers of production requirements worldwide, and with about 5000 suppliers for its 'nonproduction' requirements, instead of something like 50 000. It knew that reductions of this sort are possible – Xerox reduced its supplier base from 5000 to 400 as part of its 'leadership through quality' programme.

Global procurement by automotive, aerospace, IT and other technology- and science-based organizations is now standard practice. Establishing a global procurement and supply organization was one of the first priorities for the merged DaimlerChrysler companies, with the aim 'to enhance global coordination and accelerate the internal transfer of knowledge and technology'. As its Japanese and American competitors had done for a decade or more, DaimlerChrysler assigned a 'lead buyer' to have responsibility for coordinating all the group's requirements of those materials and products 'with global synergy potential'. A data warehouse was put in place to provide the detailed information needed for worldwide commodity management. However, since it takes years to build the relationships necessary for partnership sourcing, the geographical pattern of DaimlerChrysler purchases changed only slowly. In 2002, the $95 bn purchases by the automotive divisions were placed 48% in Europe and 48% in North America, but they were 'intensively analysing' the potential for expanded purchasing from Asia.

21.5.5   E-procurement

Seats on trains and boats and planes can be bought very efficiently via the Internet. So can seats in theatres and beds in hotels, and many other standard products and services. Buying a boat, train or plane, or building a hotel or hospital into which seats and beds may be installed, is a different matter. Few industrial purchases are made on the Internet or by e-procurement.

DaimlerChrysler's widespread use of advanced information technology for internal knowledge management in its procurement and supply organization described above is also representative of most global corporations. Its external use of IT in procurement has also been developed, but according to its latest published annual report (2002) was still relatively small. The Covisint Internet platform, established by a consortium that includes several automotive companies, was used by DaimlerChrysler to process 'approximately 700 online bidding events'. What share of total bidding events or total purchases this constituted was not disclosed.

Covisint was formed by a group of European automotive industry companies that included Ford, General Motors, DaimlerChrysler, Delphi Corporation and Lear Corporation. It provides a data-messaging hub for handling purchase orders, shipping notices and schedules. It claims to be the leader in the automotive industry for sharing business processes with suppliers and customers. The Covisint website [4] shows that since operations started in January 2001 to October 2003 the hub handled 4800 online bidding events on behalf of the founder member companies. This is a tiny share of their total bidding events and, since Covisint claims to be the industry leader in e-business, it can be inferred that online purchasing by automotive companies is still insignificant. However, from 30 November 2003 the Covisint facility was made available to users other than the founder companies, and automotive industry e-business will undoubtedly grow.

In Chapter 8 it was shown that ABB is a leader in the application of IT and has been encouraging e-business throughout its customers' supply chains by embedding its own Industrial IT product in most of the equipment it supplies. ABB reports that 'an ABB salesperson can *sit with* a customer, log into the company's website and configure and order a distribution transformer'. However, in 2003, only a small percentage of its sales was made in this way.

With the leading companies of the automotive and power industries still in the early stages of e-commerce, engineers, scientists and technologists involved in industrial procurement can still expect to *sit with* a salesperson or buyer when negotiating to buy or sell a product or service, rather than sit at a screen and keyboard.

21.6   SUMMARY

This chapter described the changes in inter-company relationships that are necessary to achieve world-class levels of business performance. It was noted that the shift of the competitive edge to include faster application of new technologies demands extension of the boundaries within which companies operate. This requires that the whole supply system is managed concurrently, with the science base as well as materials, component and equipment suppliers involved as partners at the early stages of new product and process development. A description of the 'old' methods of supplier selection by competitive bidding showed that these are inadequate for the new competitive environment, and that buyers need to change to methods that promote cooperative rather than confrontational relationships. Sourcing policies of some leading automotive companies were listed to illustrate how assemblers are seeking to establish a sense of partnership with their suppliers, and to take account of quality, delivery and technical ability as well as price when selecting their partners. Ford's supply base management system was described as an example of these sourcing processes and their use to move towards global procurement practices. It was shown that e-business is not so widespread in industrial procurement as it is in personal and domestic purchasing.

REFERENCES

1. Lloyd, C. (1993) 'Hidden costs of running laser printers', *The Sunday Times*, 14 December, based on a report by Context, a London UK research company.
2. Graham, J. L. and Sano, Y. (1989) *Smart Bargaining: Doing Business with the Japanese*, Harper and Row, New York. Chapter 7 gives extracts from typical American and Japanese contracts, and comments on the way in which they reflect different attitudes to supplier relationships.
3. Deming, W. E. (1982) *Out of the Crisis*, Cambridge University Press, Cambridge.
4. www.covisint.com.

22

Marketing

22.1 INTRODUCTION

Marketing can be defined as follows:

> 'Marketing is the management process responsible for identifying, anticipating and satisfying customer requirements profitably.' (UK Chartered Institute of Marketing)

> 'Marketing is the process of planning and executing the conception, pricing, promotion and distribution of ideas, goods and services to create exchanges that satisfy individual and organisational objectives.' (American Marketing Association)

In order to bridge the 'planning gap' between where a technical organization is and where it would like to be in terms of its products and the markets in which it operates, a strategy for marketing engineering or scientific practice and for corporate planning must be adopted. The organization has to evaluate where it is and forecast where its current policies and technologies will take it.

This projection can leave a gap between where management wishes to be and where it will be unless purposeful action is taken. The process of strategic management involves a decision about 'where we want to be', a critical evaluation of 'where we are' and the creative process of deciding the actions – policies, products, services and projects – that will bridge this planning gap. Success in this process requires a *marketing orientation*.

Successful marketing is vital to the success of any company and, therefore, is of considerable importance to the success of technical enterprises and to the engineers, scientists and technologists who work for them. This chapter will take Figure 22.1, the outline of the business planning process, as its starting point and will consider the concept of '*mission*', *methods of external and internal analysis*, the *marketing mix*, *marketing tools* and *marketing information systems*.

22.2 BUSINESS PRINCIPLES

Consider the issues that confront an extractor and distributor of natural gas. The organization may define its operation in terms of 'we supply gas'. However, such

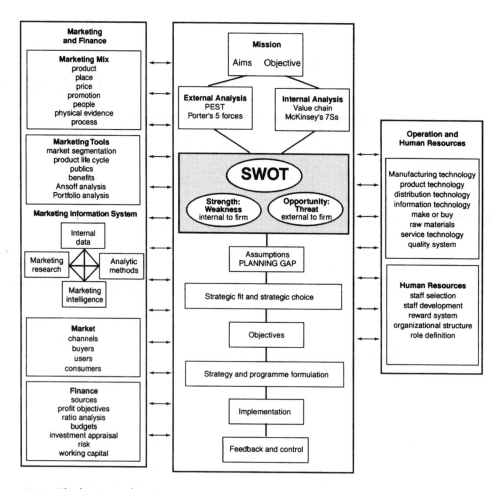

Figure 22.1   The business planning process

a company will not survive in the long term as it will not be able to evaluate its environment or serve its customers. The domestic customers may buy gas but they do not want gas, they want the benefits that the gas supplies them with. A century ago in a Victorian house the benefit from coal gas was light. In a modern household the benefits of gas are multiple, including reliability (relative freedom from supply cuts), warmth (central heating), cleanliness (ample supply of hot water) and good food (cooking).

As soon as the business is defined in *benefit* terms it becomes clear that the prime benefit in this case is the supply of energy and that competition can come from many sources. The organization is in direct competition with electricity and other suppliers of energy, and indirectly with organizations and technologies that reduce energy usage such as house insulation. The key concept is that people do not buy products: they have needs, wants and desires for benefits. A supplier cannot take a myopic view of their market or it will fail to keep up with changes in customer requirements, the external environment and competitor development. The gas company would do better to define its business as 'we supply energy for modern living'. Armed with the concept of *benefits*, the strategic planning process can now be considered.

22.3   THE BUSINESS PLANNING PROCESS: MISSION, AIMS AND OBJECTIVES

Figure 22.1 shows the outline of the business planning process. The starting point is the definition of the business in marketing-oriented terms, the *corporate mission*. Good mission statements have a number of characteristics:

- A long timescale – for a power supply company this should be looking well beyond the life of a production unit and needs to project decades, rather than years, ahead.
- A credible definition of the competitive scope of the business – the definition 'we supply gas' is very different to 'we supply energy', which would not preclude the organization entering the electricity supply market.
- A focus on a manageable number of key goals.
- A customer-oriented perspective – the mission must be defined from the customers' *benefit* viewpoint.
- It must inspire and motivate – an organization that defines its mission in terms of 'to provide a 20% return on capital' might suggest to both employees and customers that the organization's only concern is what it gets out, not the benefits it supplies, nor its values, ethical stance or legal requirements.
- It must be realistic – acknowledging the organization's resources and distinctive capabilities.

The mission defines the long-term framework that the organization will function within, but for effective management it has to be translated into operational terms. Pursuing the example, the energy-based mission for the gas company would embrace the possibility of its supplying gas to combined heat and electrical power plants for industrial, commercial and institutional organizations. In this case the mission would be amplified with an aim: 'to enter and achieve a significant market share in the supply of combined heat and power'. This provides a specific focus, but fails to give the chief project manager any organizational targets, specific objectives or aims. These might be to:

- Have a proven marketable system in one year.
- Have the first customer installed system within two years.
- Build up to a profitable sales level of ten systems a year within four years.

Objectives are quite different from aims. The relationship is the same as between a vector and direction – the former has direction and magnitude. So a good objective has to have a realistic defined target, quantitative if possible, as well as a timescale for achievement. Armed with this the chief project manager has specific goals and can proceed to environmental analysis.

A *mission statement* is most effective when used in conjunction with a vision, which is an inspirational ideal defining a preferred future for the organization. In this example it might be: 'global energy providers of the future'.

22.4   THE BUSINESS PLANNING PROCESS: THE EXTERNAL ENVIRONMENT

Two tools allow analysis of the external environment: PEST (political, economic, social and technical) or SLEPT (incorporating legislative) and Porter's five forces competition analysis [1, 2].

Figure 22.2 shows a PEST model. The detailed consideration of the marketing aspects is developed in sections 22.7 to 22.11.

22.4.1   PEST

The *political* forces acting on the organization include the specific demands of the external environment such as legislation, as well as the ideology and attitudes within the operating environment. For example, in the supply of combined heat and power, the attempt to identify realistic alternatives to energy generated via fossil fuels initially led to the development of energy from nuclear power. However, legislation, safety and efficacy concerns in the UK generated external environment pressure on energy suppliers and led to the development of wind, solar and hydroelectric power as evolving competitive alternatives.

An organization in the UK now has the ability to sell excess electrical power and this has changed the economic viability of energy suppliers. The project manager will be faced with a long list of detailed legal requirements covering the safety of the unit and environmental compliance issues (e.g. noise levels, venting of exhaust gases, etc.). The attitudes and ideology are important as they affect the degree, flexibility and speed with which factors can be resolved. In the present climate of a global commitment to reduce carbon dioxide emissions and to reduce the wasteful use of finite resources, a supportive climate exists for energy efficiency projects.

P	E	S	T
Political	Economic	Social	Technical
Legislation Ideology Attitudes	Trends Credits Income Interest rates Tax Exchange rates International trends Gross National Product Wealth distribution	Values Attitudes Work ethic Lifestyles Demographics Workforce Religion Status	Discoveries Developments Substitutes

Figure 22.2   PEST

Note that PEST and competition analysis do not provide a neat set of orthogonal vectors; these do not exist in business. All the elements, to some extent, interact; but, if the strategist does not consider the differing aspects, critical factors may be neglected and lead to project failure. In the case of energy there is a social concern for 'green' issues and the pressures on energy resources have economic consequences. These two forces result in legislation and changes in the political environment. Engineers and scientists search for innovative technical approaches to the issues. The new technological skills needed might exist in organizations that have not previously been active in the arena and new competition patterns might be established. For example, in the production of energy through wave power, hydrographers, oceanographers and other scientific experts in areas such as coastal erosion are required to work alongside engineers. For an electrical energy-based company this will be a new area of competency.

Discounted cash flow analysis (see Chapter 15) provides a specific detailed investment appraisal tool, but a more general *economic* view has to be exercised at the strategic level. Long-term trends in finance sources and interest rates are important for engineering projects with long construction periods and even longer operational lives. Where the engineer or scientist is required to design products or production facilities for the manufacture of consumer goods, trends in the shape of the population and lifestyle must be considered. As discussed in Chapter 2, engineering industries operate in international markets with global procurement and manufacturing operations, and long-term trends in economic development and exchange rates must be evaluated (also discussed in Chapter 2). In the example, the project manager will be concerned with the sources and costs of finance for customers who will purchase the units of energy and how these will balance against future energy costs and potential customer savings. In the procurement of production units, long-term judgements of relative domestic cost structures against those of overseas suppliers must be considered among the factors affecting the procurement decision.

Social issues affect all aspects of an organization. Attitudes to employment, for example, will affect how an organization structures and operates its manufacturing environment. Considerations here include work ethic, demographics and workforce skills. Consumer issues not only affect the manufacturers of consumer products but also major utilities. A general social attitude to safety and quality of life will induce a car manufacturer to introduce comprehensive safety features into its basic car design, such as side impact bars, airbags, active braking systems, power steering and safer seat design into its products to satisfy individual consumers and also to introduce energy-efficient methods of manufacture in order to sustain a positive corporate image such as 'products designed and built with care for people who care'.

Customers changing attitudes and opinions have an impact on the social requirements of a product. Customers demonstrate wants through their purchasing power, which affects the elasticity of pricing. 'People carriers' is a UK term for cars that carry more than five people. These cars target the family market and customers buy the benefit of safety as one of the key attributes of the vehicle. Poor results from New Car Assessment Programme (NCAP) tests for several people carrier manufacturers resulted in an immediate drop in sales and a customer move towards companies demonstrating safety with high NCAP ratings, such as the Renault Espace, rated 5 in its NCAP results – which is the highest rating currently attainable.

Technology provides the ability to supply benefits in entirely new ways with effects that extend far and wide. The rapid development of the Internet and its adoption into the business environment has revolutionized the way we work. Companies can operate and compete globally without geographical, financial or communications barriers. The Internet as a channel of distribution is providing an alternative to the traditional outlet-based approach. Companies such as Amazon have managed to compete effectively with high-street shops in the provision of books and music using price-cutting strategies to achieve *competitive advantage*. Amazon does not have the costs of maintaining a high-street presence or of the maintenance of extensive stocks, and passes the benefits on to the customer.

The Internet has also engendered 'instant' communication via e-mail across continents. The traditional 28-day standard reply has now been replaced with customer satisfaction surveys revealing that customers who do not receive a response to an e-mail within six hours will shop elsewhere. Responding rapidly and effectively to customer wants means maintaining customer satisfaction and provides another way for an organization to achieve *competitive advantage*. If it is an age of speed, it is also a new age of information. Customers and competitors have access to a plethora of intelligence on values, policies, products and price and can differentiate on all these factors when making purchase decisions. Where product differentiation is limited the key factor becomes price and with comparisons available at the click of a mouse, customer loyalty can be quickly lost.

For marketers the Internet is a tool to assist in the provision of service and relationships. It is a virtual marketplace and using the Internet as a channel of distribution does not mean that organizations should be reduced to faceless providers of low-cost products. The Internet provides an excellent tool for *relationship marketing*, where an organization invests in the relationship it builds with its customers to develop a long-term, mutually beneficial connection. The Internet enables the marketer to target its offering specifically to the customer.

It is only the strategically oriented engineer or scientist who can fully evaluate the commercial impact of developments, discoveries and substitutes. The whole business is affected: procurement, manufacturing operations, distribution and marketing. The pace of technological change and the rate of adoption of innovations are such that an organization's competitive position can be quickly changed (see Chapter 6).

22.4.2   Porter's five forces of competition

Figure 22.3 shows Porter's five forces of competition model [2]. To use this model the strategist has to be clear about the benefit being supplied to the market and the platform from which the analysis is conducted. The results of an analysis of air travel will be very different for an airline, an aircraft manufacturer, a travel agency or a holiday tour operator. Many engineering technologies have more than one application satisfying differing benefit needs. This is called *segmentation* and the marketing implications are discussed in section 22.7. It is necessary to complete the competition analysis for the differing benefit segments. Gas turbines were developed for air travel, but are now used for electricity generation and

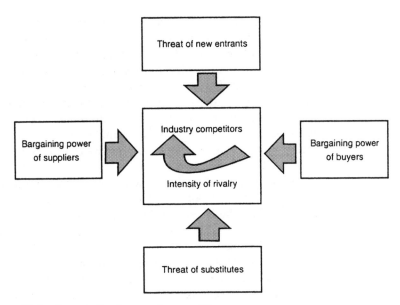

Figure 22.3 Porter's five forces of competition
Source: M. E. Porter (1990) *The Competitive Advantage of Nations,* The Free Press, London

the powering of warships. Nuclear or coal power stations compete with gas, electricity, hydroelectric and other sources of energy. Diesel engines may be used to power a ship but not an aircraft. The chief engineer of a turbine manufacturer would have to complete the competition analysis for each sector.

The five forces are *in-sector competition* (direct competition), *substitute products, new entrants, supplier power* and *buyer power* (channel and consumer).

In-sector or *direct competition* means those organizations that supply the same goods or services in the same way. This does not necessarily mean that they have the same competitive position. Exit and entry costs greatly affect the flexibility and potential for competitive response. For example, three companies produce a similar product. One makes all the components and has a high investment in automated production lines (high entry costs). Another has a simple assembly factory and buys in its components and subassemblies (modest entry costs). The third subcontracts out all its manufacture and concentrates wholly on marketing and distribution. In a recession the latter two companies have greater flexibility, for they can reduce their output with much less impact than the fully integrated company, which has to operate at high capacity to cover its fixed and capital costs.

Indirect competition can come from industry competitors and from outside an industry in the form of *substitute products* which could be used in place of the offered product or service.

New entrants are organizations that have the potential to supply goods or services. A continual consideration is the international dimension, where many companies are looking for new markets; and the corporate strategist will be continually alert to such expansionist moves into a given market under appraisal. Shifts and convergence of technology may allow new entrants. Amazon, which identified the Internet as a mechanism for the distribution of music and books, slowly extended its product range, introducing gifts and strategic affiliations to travel and tourism operators. Established high-street music and book distributors followed it online. However, Amazon had defined its competitive advantage as price and high-street retailers have problems justifying differing pricing strategies for their different channels of distribution. The Internet provides a low-cost entry option for organizations and it enables organizations to disguise their size and scale of operations, so a one-man band can compete with an international market leader. This has resulted in a large number of new entrants across all market sectors – predictably, this dot-com boom was followed by many dot-com crashes in the late 1990s (see Chapter 23).

In evaluating *supplier power*[1] the relative bargaining power of suppliers vis-à-vis the organization is appraised. If the organization is large and there are many suppliers, then the leverage of the supplier is reduced. If there are only a few suppliers and the organization is a small user, then the positions are reversed. Care must be taken that all the goods, licences (patents, copyrights etc.) and services (including labour) required by the organization are taken into account. Consider food retailing. When the distribution and retailing of food was highly fragmented, the large food processors of branded foods had great supplier power. With the high concentration of retailing power in a few supermarket groups, the roles are now effectively reversed, with the supplier power much diminished and the power lying with the retailers.

In evaluating *buyer power* in the engineering context, it is usual to consider the power of the channels of supply and the ultimate customers. The direct customers of a manufacturer of consumer electronics are the high-street or Internet retailers (in the UK Dixons, Comet etc.). However, consumers will only purchase a company's products if they provide distinctive benefits (high quality, more useful features and so on) and value for money in a competitive market.

It is also wise to consider indirect channel power. A supplier of car engine and transmission lubricants may not enjoy major sales to Ford, but will be most concerned if its product is not approved for topping up Ford vehicles. A supplier of UPS (uninterruptible power supplies) will have direct sales to computer manufacturers (large customers with high buyer power). The computer manufacturers will also have indirect buyer power, since it will be necessary for the UPS supplier to be included in the list of approved power suppliers for customers configuring their own systems. It will be necessary to demonstrate to agents and computer distributors that the system is effective, that it will give them an acceptable profit margin and that supply will be prompt. The ultimate buyers of these units are knowledgeable and will be able to make accurate performance and cost comparisons. As this analysis demonstrates, this is a competitive market where the engineer and technologist ensure extremely good performance with strict control of manufacturing costs for the product to be successful. The strategist must consider the external environment in conjunction with the internal environment of the organization.

---

[1] This aspect of Porter's analysis is inconsistent with current concepts of supply system management. As shown in Chapter 21, successful engineering companies now seek to establish alliances and partnerships with suppliers. Porter's perspective applied more frequently in the early 1980s when his analysis was developed, but an engineering company's ability to develop partnerships continues to be influenced by the factors identified by Porter.

Example 22(1)

PEST and Porter's analysis for a Division of British Telecom (BT)

PEST

Political issues: The deregulation of the telecoms market in the early 1990s allowed new competition. Industry regulator Ofcom has undertaken a review of the British telecoms sector. Ofcom has the ability to impose price controls and to limit the charges for network use and broadband provision to the 500 companies that BT provides with products, services and network capacity. BT provides a near-universal service that its competitors threaten with recourse to legislation. Ofcom is interested in whether access costs are stifling competition in the broadband and fast Internet market. The government has pledged to turn the UK into 'Broadband Britain' by 2007.

Economic: There are substantial costs of maintaining 48 000 staff, including 17 000 engineers to service 21 million customers in the UK alone. Rapid change in the industry makes staying apace with new technology expensive, as the need to diversify into new areas such as business communications, ISP (Internet service provision) and entertainment takes BT away from its core business. The costs of bit rate (cost of transmission of a unit of information) are rapidly decreasing compared with consumers' disposable income. There are also decreasing costs of technology and automation compared with labour costs.

Social: BT has a commitment to environment, society and sustainability – in 2003 it won a Queen's Award for Enterprise. There were cultural problems in converting from a public service to a profit organization, magnified by the need to shed large numbers of staff. It took advantage of the opportunity to exploit teleworking with part-time staff. BT has a presence in all communities – relationships exist with the majority of UK households.

Technical: Fast-moving global technical development requires resources and staff to be constantly updated. BT has made significant investments in technology during the past decade. 300 million calls are made every day and 350 Internet connections made every month through BT. BT Wholesale has 500 customers to whom it supplies products, services and network capacity – some sell this on. BT supplies capacity to its competitors. BT Global Services provides information communications technology to 10 000 customers in 200 countries. BT Open World services 1.7 million customers as an Internet service provider utilizing modem, broadband and ADSL options. BT Exact is BT's research, technology and IT operations business. BT is well positioned to benefit and lead future technology innovations.

Competition analysis

Direct: BT operates in 200 countries and has 10 000 global business customers and 21 million UK residential customers across five strategic business units. As a result, BT has literally thousands of competitors in each market sector. In the UK it supplies 500 of these with products, services and network capacity. In the residential market direct competitors in the UK include ntl, Telewest and Cable & Wireless – all of which base their differentiation on price and invest heavily in advertising and brand building. In the UK there are 700 different Internet service providers (ISPs), mostly reliant on the BT network – which they pay for. Cable companies are developing their own networks, but only cover small sections of the country.

New entrants: Local networks have been established by cable TV companies. Mobile telephone companies are moving into domestic markets offering significantly cheaper calls to mobiles used in the home. International entry is occurring by other telephone companies.

Substitute products: E-mail and text are rapidly increasing media for both business and domestic use; however, both utilize BT technology. There has been a growth in the amount of communication over the past decade in line with the growth in communication technologies. A substitute product could result in or provide for a decline in time dedicated to communications – with less time spent at work in the future and more time spent pursuing leisure activities.

Note: In most cases of the use of a model some aspects will not be applicable. It is important to recognize this and not to force the model to fit where it does not have application. Some strategists might prefer to consider mobile telephone technology as a substitute product rather than a new entrant. It is more important to ensure that all potential competition has been identified than to enter long debates as to the most appropriate classification.

Supplier power: Suppliers to BT have relatively little supplier power, while BT has exceedingly strong supplier power – so strong that Ofcom and its competitors are looking at it closely to see if it has too much supplier power and is actually a threat to the development of new technologies.

Note: A key source of supplier power in the industry in general was the ownership of the long-distance land lines. However, changing technology and the development of fibreoptic networks, which offer significantly improved capacity, could potentially change that. Currently, installation of fibreoptic networks is prohibitively expensive, but technology may soon offer an alternative cableless network that would open up the market to competitors.

Buyer power: Domestic consumers are not used to price and service competition for telephones. However, once a consumer is connected to cable TV the costs of switching to the new supplier for package services incorporating telephony may be very low (even zero). There is enormous latent consumer buying power once consumers become more educated.

Note: Since BT owns the local distribution network, there is no channel power to BT. However, a new-entry telephone company that wanted to use a local TV company's fibre network for local connection or a cableless alternative would find that the supplying company had considerable channel bargaining power.

General competition notes: This is a highly regulated environment and who can compete with what, when and where is still subject to a vast amount of political pressure (both nationally and internationally). The establishment of local fibreoptic networks is at enormous entry cost, and therefore such communications organizations are under considerable pressure to maximize their return as soon as possible. Conversely, the establishment of a substantial long-distance network, in proportional terms, has allowed organizations that had the rights of way (railways and power supply companies) to establish networks at a modest entry cost and with government subsidies. Clearly, in the space of a decade BT has moved from an effective monopoly position to one of extreme competitive pressure and turbulence. The next decade should provide even greater dynamism as alternatives to network technology are developed and BT will no longer be able to rely on the dominance provided by its existing infrastructure. It has invested heavily in a range of strategic directions that may prove costly if its core income stream (cash cow) is threatened.

22.5   THE BUSINESS PLANNING PROCESS: THE INTERNAL ENVIRONMENT

Figure 22.4 shows two tools for considering the internal environment, Porter's value chain and the 7Ss.

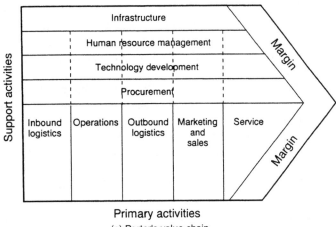

Primary activities

(a) Porter's value chain

Source: M.E.Porter (1990) *The Competitive Advantage of Nations*, The Free Press, London

Three Hard Ss

Strategy Structure Systems

Shared values

Skills   Staff   Style

Three Soft Ss

(b) McKinsey's 7Ss

Figure 22.4   McKinsey's 7Ss and Porter's value chain

22.5.1   The value chain

The *value chain* is in two parts: primary activities and support activities. The effective and efficient conduct of all the value chain activities, especially when supported by a *total quality management* orientation, produces added value, the margin. The model can be used to audit an existing business or can be used as a framework to consider the facilities that will be needed in a new production location.

Inbound logistics are the activities concerned with the shipping, storage and retrieval of materials for the productive process. Given the need to control working capital in inventory and the high cost of labour, this has been an intensive area of engineering management activity with the introduction of just-in-time production methods and of materials resource planning computer systems, as described in Chapter 20. In any production operation, competitive advantage comes from using superior process technologies that quickly produce first-time quality every time with minimum downtime. Outbound logistics should have the ability to gain advantage by superior packaging and timely distribution to customers. Sales and marketing activities are considered in more detail later in this chapter.

Efficient sales systems are vital and efficiency definitions are dependent on what the customer perceives is the difference between the actual time between ordering and receiving the service or product and what it should be. It is of no value to have an efficient production system if the *customer's order* takes several weeks to be processed. (Chapter 8 describes how those companies that have 'managed to succeed' have paid attention to reducing the order–produce–deliver cycle.) The manager should carefully audit the complete system from initial customer contact through manufacture to eventual customer receipt of the goods and payment (something that is often neglected). What is the value of a super-efficient distribution system if customers do not receive their bill for weeks? (See Chapter 14, section 14.8 for management of the working capital cycle and also Chapter 23.) Many engineering products, both in the consumer and industrial sector, require support and field service. The customer will not be very impressed with prompt delivery of a system if installation and commissioning are slow and inefficient.

These primary systems will not remain in good order and adapt to changing conditions without support. An organization should have an appropriate financial structure, good-quality physical facilities (buildings and equipment) and, of increasing importance for competitive advantage, total quality management and effective and efficient information systems. Other elements of the support activities have both general and specific applications, hence the dotted lines. A company will need to run general quality awareness programmes for all staff from the managing director to the new starter. It will also need to provide specialist training – advanced CAM (computer-aided management) training for the production department, product training for the sales staff and so on.

Technology development is obviously concerned with the development of the technology of the product and its manufacture. However, all aspects of the value chain system can benefit from the application of engineering-style innovation.

For example, in the service phase, remote diagnostics in a system may allow earlier and more rapid clearing of field failures with a consequent reduction in downtime for the customer. In some ways this may be more apparent to the customer than super-efficient production (which, of course, is vital) because it is visible, whereas the production process is not. Procurement also applies to all aspects of an organization's purchasing. Effective sourcing of raw materials must include consideration of cost, quality, delivery and technology. However, such a strategy can be completely negated if the same quality, delivery, technology and cost controls are not exercised on the procurement of distribution and communication.

Example 22(2)

Outline value chain analysis for a car repair garage

Primary activity – inbound logistics: Good delivery bay for components and good store facilities for the rapid retrieval of anything from a bulb to a complete gear box. Part of the process is the collection of the vehicles for repair, so good parking with clearly marked bays and good recovery vehicles for the collection of broken-down and accident-damaged vehicles are required.

Primary activity – operations: Good systems for rapid inspection and fault diagnostics, good equipment for the removal and fitting of parts.

Primary activity – outbound logistics: Reflection of inbound logistics with good and clearly marked parking bays for completed vehicles. In addition, a good system for the disposal of parts and materials is vital, for some of these may be hazardous if disposed of badly (e.g. spent oil).

Primary activity – marketing and sales: Efficient sales system, good telephone booking process, good and clear collection of customer fault reports, on-line checking of customer credit status, rapid checkout and billing, good electronic point of sale with funds transfer for credit cards, provision of detailed documentation of good quality (e.g. estimates for insurance claims, certificates and itemized bills).

Primary activity – service: Replacement car during service period, possibly additional services such as car valeting, insurance etc.

Support activity – infrastructure: Good buildings, equipment and workshops. Good computer systems to provide effective parts management and customer service.

Human resource management: Customer care training, recruitment of well-qualified staff, continual specialist training (e.g. updates for fitters as new models are released, computer training with new software for sales staff, etc.). Effective reward systems to recognize exceptional contributions.

Technology development: Environmental and safety issues important, better systems for the control of exhaust fumes during testing, solvent vapours, etc. New methods of fault detection, systems to maintain advanced auto electronics. Continual attention to improvements to information support systems.

Procurement: Effective purchase of maintenance materials including parts. Good buying for own vehicles, communications and computers. Note: other heavy expenses include insurance, etc. and these will need attention.

22.5.2   McKinsey's 7S framework

The 7Ss provide an additional framework for evaluating the 'software' of a company, its people. (See Chapter 3 for a full explanation of the 7Ss.) The company must have a sense of direction, a strategy with mission and objectives, and implementation procedures and policies. These must be supported with an organizational structure and systems. This is only possible if the correct staff are recruited and provided with the appropriate range and depth of skills. Two other elements are vital for long-term success. All organizations develop their own style and this must be congruent with the industry needs. The other elements can only be successfully orchestrated once a spirit of shared values has been developed. It is no use the production and sales staff having highly customer service-oriented attitudes if the rest of the organization does not. Organizations only succeed if people pull in the same direction. In a fast-moving and complex environment it is not possible to codify all eventualities into a manual. Shared styles and values should ensure that staff will make the right decisions, as the shared culture would mean that all understand what the company demands.

22.6   SWOT – FROM ANALYSIS TO DECISION AND ACTION

External and internal analyses involve the consideration of a vast range of issues that may have a potentially critical impact on the organization. This expansive process needs to focus on the key issues that, without prompt management action, will result in a loss of competitive advantage. The *strengths, weaknesses, opportunities* and *threats* (SWOT) analysis is the mechanism for this [3, 4]. Strengths and weaknesses are internal to the organization. This is not to say that they are taken without reference to the external environment. If staff have been trained to a certain level this can be either a strength or a weakness depending on how well the competitors' staff are trained. Threats and opportunities are external to the organization. Threats can come from changes in the environment such as legislation. Opportunities are areas where the organization can extend its operations to achieve a profitable expansion. (See Chapter 2 for a more complete review of global threats and opportunities, and Chapter 23 for an example of SWOT applied in a start-up technology company.)

This form of analysis focuses on the organization's key competencies and strategic advantages. This involves a detailed examination of the marketing issues, which are covered in the remaining sections of this chapter.

22.7   MARKET SEGMENTATION

Not all consumers are the same and to be successful it is essential to match the products or services to the precise needs of different groups of people. A company manufacturing jeans will be concerned with age, sex, height, weight and income to create a range that caters for needs from everyday use to designer products, for the poor student and the affluent fashion conscious. Segmentation is just as important in industrial markets. An analytical instrument manufacturer will be concerned with the type of industry, geography and size of customer as possible variables. The precise variables that should be used depend on the nature of the market and the decisions to be made. In general there need to be six conditions satisfied if segmentation is to be useful:

(1) The segment should be *accessible*. A company could make computer keyboards for left-handed accountants, but just how would these people be found?

(2) The segment should be *unique*. The population within the segment is different to the populations outside the segment.

(3) The segment should be *measurable*. It must be possible to profile the population in the segment and determine its characteristics and the total size and potential value of the segment.

(4) The segment should be *substantial*. The value of the segment (transaction profit × number of potential customers) should be of sufficient size to make the development of a specific strategy cost-effective. If this is not so, the segmentation is of no practical value.

(5) The segmentation should be *appropriate* to the decision in hand. There are any number of segmentation variables for people (language, colour of eyes, education, etc.). It is necessary to decide which ones are key to the specific product. There are no set rules for this, it is a skill that has to be developed. It may take a page to explain the concept of segmentation, but years of practice to acquire the skill to apply it well.

(6) The segment should be *stable and predictable*, so that it is possible to develop and implement a complete marketing plan.

The precise segmentation variables depend on the decisions to be made. In consumer products, age, sex, lifestyle, income, language and geography might be appropriate. In industrial markets, the size of customer, markets served and manufacturing technologies used are potential variables. Tables 22.1 and 22.2 illustrate typical variables for industrial and consumer markets.

Table 22.1 Major segmentation variables for industrial markets

Variable	Typical breakdown
Type of industry	Manufacturing, service, chemical, etc.
Technology used	Robots, high pressure, clean room operations, etc.
Geographical location	Country, region
Purchasing procedures	Contract arrangements, tendering procedures
Number of sites	1; 2–5; 5–10; 10–50; 50+; global; multinational
Stage of development	New to technology, established user
Size of purchases	Value of product purchased, number of orders, size of delivery (e.g. drums, palettes, small package)
Buying criteria	Price, quality, service, technology
State of R & D	Leader or follower
Organization size	Number of employees, sales revenue, turnover, etc.

Table 22.2   Major segmentation variables for consumer markets

Variable	Typical breakdown
Demographic	
Age	Under 6, 6–11, 12–19, 20–34, 35–49, 50–64, 65 +
Sex	male, female
Family size	1–2, 3–4, 5+
Family life cycle	Young, single; young, married, no children; young, married, youngest child under 6; young, married, youngest child 6 or over; older, married, with children; older married, no children under 18; older, single; other
Income	Under £7500; £7500–£12 000; £12 000–£15 000; £15 000–£18 000; £18 000–£25 000; £25 000–£50 000; £50 000 and over
Occupation	Professional and technical; managers, officials and proprietors; clerical, sales; craftspeople, foremen; operatives; farmers; retired; students; housewives; unemployed; military
Education	Fifth form; sixth form; college of further education; technical college; university; NVQs; professional qualifications
Religion	Protestant; Catholic; Muslim; Buddhist; other
Race	White, black, oriental, other
Nationality	British, Chinese, Indian, Pakistani, Japanese, Other Far East, West Indian, French, German, Other European, American, other
Psychographic	
Social class	Lower lowers, upper lowers, working class, middle class, upper middles, lower uppers, upper uppers
Personality	Compulsive, gregarious, authoritarian, ambitious
Behavioural	
Occasions	Regular occasion, special occasion
Benefits	Quality, service, economy
User status	Non-user, ex-user, potential user, first-time user, regular user
Usage rate	Light user, medium user, heavy user
Loyalty status	None, medium, strong, absolute
Readiness stage	Unaware, aware, informed, desirous, intending to buy
Attitude towards product	Enthusiastic, positive, indifferent, negative, hostile
Geographic	
Region	South West, Wales, South, London, Scotland, Northern Ireland, Home Counties, Midlands, North, West
City or town size	Under 5000; 5000–20 000; 20 000–50 000; 50 000–100 000; 100 000–250 000; 250 000–500 000; 500 000–1 million; 1 million–4 million; 4 million or over
Density	Urban, suburban, rural

22.8   PRODUCT LIFE CYCLE (PLC)

Products have their day, they grow, have their period of ascendancy and then decline. The time for this can vary from a few months – often shorter life cycles belong to a fashion such as drainpipe trousers, or a fad such as the Rubik's cube or

Tamagotchi – to decades for industrial products (e.g. steam turbines for powering electrical generation). The reproduction of music shows a number of such life cycles: the introduction of the 78 rpm record, the growth of the micro groove record, an extension and extra impetus with the enhancement of stereo and the ultimate decline of records with the introduction of CDs, mini discs and MP3. Part of this product life cycle is shown in Figure 22.5.

In the introduction stage the product may be expensive and may appeal to only a select market. The first group of customers will require much information and advice. This will be provided by specialist outlets. At this stage competition may be rather limited, but the cash flow will be negative as the product will have had high development costs and will require extensive launch marketing communications. In the growth phase many more customers will be drawn to the product, as will some competition. Profits will start to be made, but often the cash flow may still be negative. Heavy marketing is still demanded to attract new customers and extra capital will be needed for the new production facilities (and to finance stocks if 'lean' production and logistics have not been introduced). Once the product has reached maturity the competition will become more intense, but the demands for capital will become stabilized and the customer base becomes established. Moreover, the learning curve effect will have sharply reduced manufacturing and distribution costs. Thus there will be good profits and strong positive cash flows.

All good things come to an end and, as new technologies challenge and the competition becomes intense, the weaker producers will exit as the product enters the decline stage. This decline may be fast, given the rapid advancement of new technologies. For example, the postal system – the forerunner of which was developed by King Louis XI of France in 1464 – after nearly 600 years of being the only method of communicating at a distance, has lost significant market share to e-mail in less than a decade. The development of the telephone was the first threat to post, with Bell's initial experiments in 1875. Then 50 years ago, the fax enabled the transmission of data, but neither phone nor fax made a significant impact on the lucrative business postal traffic. However, within the last 10 years e-mail has established itself as a way of communicating globally due to its cost-effectiveness,

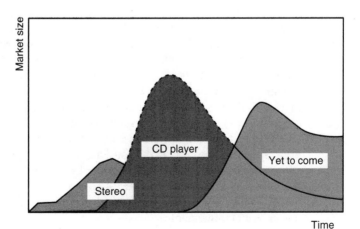

Figure 22.5   Product life cycle

versatility and reliability. In technological and scientific markets it is vital to scan the environment for issues that can bring an abrupt decline to the product life cycle.

In the product life cycle different customer groups enter the market as the cycle develops; this is called the diffusion of innovation. Innovators are the first customer group to enter a market and are normally expected to be a small percentage of the total lifetime market for the product. In the case of the postal system the innovators would have been Louis XI and the friends with whom he chose to communicate. Innovators are followed by the early adopters, such as the Pony Express in the USA in 1860, and then by a much larger group, the early majority. At this point the product has theoretically reached approximately 50% of its total lifetime customer base. However, it is now established with a steady customer base and has decreased margins and increased revenue. Prices will probably drop to open the market to more customers and the late majority, another large group, follow into the market. The final group to enter the market are the laggards, who commence purchase as the cycle is coming to an end – unless a way is found to extend it.

22.9   PORTFOLIO ANALYSIS

It is a major concern to the organization that only in the maturity phase can strong positive cash flows and profits be expected. During the introduction and growth stages profits may be small and cash flow under great pressure. Failure to exit appropriately may mean that the 'decline' products may also become a drain. However, it should be noted that an orderly exit should allow the recovery of working capital. At first sight the organization needs a collection of mature products, but this may only be obtained in a dynamic system by a continual flow of products. This balancing of the products, or business areas, is called portfolio analysis. First the company has to classify its products and decide which to promote and which to drop or exit. The simplest such model is the Boston matrix [3] shown in Figure 22.6.

In this model the two key parameters are the relative market share measured on a log scale and the market growth. This yields a 2×2 matrix. A product characterized by a low share and high growth is represented by a 'question mark'. (This would be a product in the introduction phase of the product life cycle.) One with a high growth and a high market share is represented by a 'star' (a product in the growth phase of the cycle). This is typical for a new technology where the first entry companies have a strong market share position (there is little or no competition). The 'cash cows' are for mature products where there are strong profits and cash flows as long as manufacturing, distribution and marketing expenses are strictly controlled. In the decline phase the 'dogs' can be a continual drain on company resources. 'Dogs' must be driven back to 'cash cows' (sharp reduction in manufacturing costs to recover market share through cost leadership) or an orderly exit. 'Dogs' are a trap for many technical companies. A company may be built up on a product by a founder and it takes strong will to drop a product once it has become equated with the company. At one time IBM made typewriters; it does not do so now. Exit requires organizational decisions and skilled management.

The Boston matrix is much discussed in first-level marketing books, but is a poor model except in the sphere of fast-moving consumer goods where the key

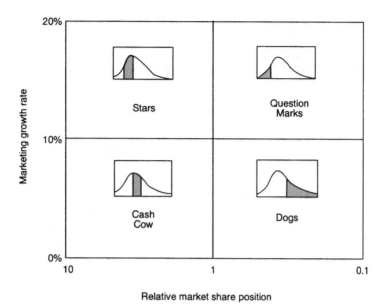

Figure 22.6   Boston matrix
Source: Reprinted from *Long Range Planning*, Volume 10, B. Hedley, 'Strategy and the "business portfolio"', pp. 9–15, Copyright (1977), with permission from Elsevier

parameters for success are market growth and market share. For industrial and technological markets it fails. The GE matrix [4] shown in Figure 22.7 provides a robust and more general framework.

The two key parameters are market attraction (vertical axis) and competitive position (horizontal axis). (The Boston matrix represents a degenerate case of this matrix where competitive advantage is simply measured by market share, and attraction by market growth.) For any particular organization, factors such as technology fit, market size, profit margins and risk can be evaluated and given weighting factors. The fit of products to these factors can also be evaluated and a weighting matrix constructed. Similarly, the factors for competitive position for the market can be judged (e.g. strength of product line, cost structure, brand image, patent position etc.) and the individual product positions judged and another weighting matrix constructed. It is then possible to construct the GE matrix and position the products. The size of the market is shown by the area of the circle. An organization's market share is indicated by the size of the wedge in the circle.

Such a matrix shows the position at a particular time, but markets are dynamic and both market attraction and competitive position will be changing. The direction and speed of this change are indicated by the length and direction of the arrows.

The power of this model is that it allows the consolidation of a range of key parameters and incorporates the best judgements of the marketing engineers and scientists for not only the present situation but for the dynamics of the situation. Clearly, products with high attraction and a good competitive position should be developed; where there is little attraction and little competitive advantage, exit is the recommended strategy.

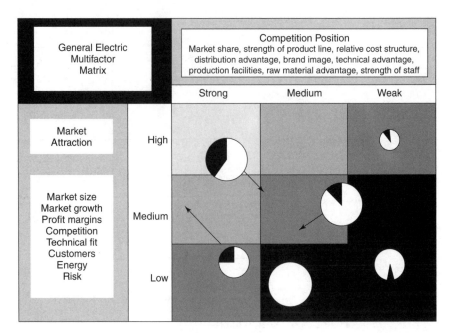

Figure 22.7 The General Electric multifactor matrix
Source: Reprinted by permission from page 202 of *Analysis for Strategic Market Decisions* by George S. Day; © 1986 by West Publishing Company. All rights reserved

Clearly, both the Boston and the GE models show that the company is in an unstable position and unable to sustain long-term growth and success, unless the designers can provide a stream of new successful products. A decision support model is required to support this activity.

The Ansoff matrix, Figure 22.8, provides a model for this [5]. All the manager has to decide is what the existing products are and what new potential products the organization could make. To complement this, a decision is required as to what are the existing markets and the potential new markets. This yields the 2 × 2 matrix and four potential strategies.

Strategy 1: market penetration (existing markets – existing customers). All forms of marketing communications must be expanded to increase market share, encourage more frequent usage and/or alternative usage of the product by existing customers. Given that the marketing engineers and scientists understand the product and their customers, this represents a strategy of modest risk. Clearly, if there is a high market share or the product is in the decline stage, this strategy will not be sufficient and alternatives are needed.

Strategy 2: product development (existing markets – new products). Here there are two issues: what new products the company can competitively produce and which of these will be of value to existing customers. This is a more risky strategy than simple market penetration. Establishment of new products can be difficult, even in existing markets, where current customers may be resistant to innovation or may have loyalty to an existing supplier.

Strategy 3: market development (new markets – existing products). The product may be of value in new markets, for example by the diffusion of existing

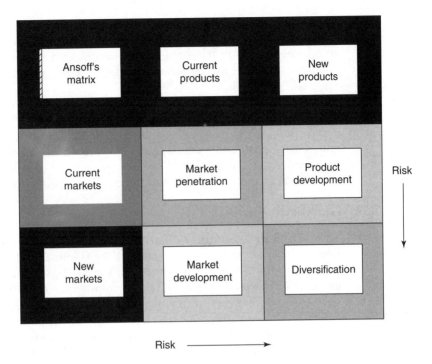

Figure 22.8 Ansoff's matrix

technologies into new markets or by moving into new geographical markets. In a mature product life cycle it may be a good strategy to export products to an area where the cycle has not reached a decline stage. Again, one parameter will be fully understood, in this case the product, but much less will be known about the other, the markets. So again, this is a strategy of increased risk.

Strategy 4: diversification (new products – new markets). If Strategies 1, 2 and 3 do not provide the expansion required, the company might consider new markets and new products. This is the area of maximum risk where extensive market research is required to reduce that risk. Where conditions are right this can be very profitable. Diversification can be concentric, horizontal or conglomerate.

In the telecommunications field in the UK companies are using all these strategies. BT has developed its market by going global with its business communications products. It has extended its product range, providing mobile phones, selling spare network capacity to competitors and developing a large research centre to enable it to be a market leader in product innovation. BT has stuck to its core communications provision, while other telecommunications providers have diversified into entertainment, delivering cable TV as a package with telephone services. It is not so clear when diversification strategies may be appropriate for BT. However, with 48 000 staff, consider the immense training tasks faced by BT during all these changes. The skills developed in training may be a marketable skill in their own right in areas outside traditional BT markets and could represent a profitable diversification. Likewise, the entry of the electrical power grid companies into fibreoptic super-highways represents a potentially profitable diversification strategy.

Successful Ansoff analysis for technical markets needs imaginative interpretation of both the technological possibilities and the developing market benefit needs. It is only the engineer and scientist with marketing knowledge who will have a sufficiently complete understanding to conduct this structured but very creative process.

There is no set formula for devising new products or for providing the insight to recognize a new attractive market.

Once these business ideas have been evaluated they can be fed into the GE matrix for a decision as to which product to accept. In an innovative company the above process should generate far more opportunities than can be commercially realized. A rational decision process to focus on the best prospects is vital.

22.10   RELATIONSHIP MARKETING

The marketing process can be defined as a series of micro and macro environment relationships, networks and interactions. Evert Gummesson [6] identifies 30 such relationships, which he claims underpin the marketing process. The strategic manager is aware that there are relationships with customers, suppliers, stakeholders and competitors, all of which affect the effectiveness of the organization, while the nature of those relationships defines the culture, values and policies of the organization. Relationships with customers determine the profitability of the organization. Sometimes organizations require basic transitory relationships where no repeat purchase is required and the generation of customer loyalty is not required. In these instances the cost of maintaining the relationship is low and these tend to be price-led markets where demand is variable and price elasticity is high.

Some organizations require complex relationships with their customers. In some instances the customer is a co-producer of the end product so a very close partnership is required, such as with Business Angels, a business financier that provides money and support to enable business development. Differing levels of relationship are aspired to, depending on the industry sector, and differing levels of success in developing relationships are achieved depending on the strategic and operational approach of the organization. The main steps in the customer development process are shown in Figure 22.9.

The starting point is your market segment – the suspects who may potentially become prospects. Prospects have a potential interest in the product and the ability to purchase, and the organization aims to convert prospects into first-time customers. For most organizations a simple rule of thumb is that retaining existing customers and generating increased sales revenue from existing customers is cheaper than attracting new ones, so the simple aim is to utilize marketing tools to develop them into repeat customers. Developing a customer into a client requires a relationship strategy that enables the organization to recognize and know clients. Some companies may further develop their benefits package supplied to customers in order to develop a membership scheme, and then take the step of turning the customer into an advocate, which has obvious benefits to marketing and to revenue. The final step in the process is to turn the customer into a partner. Relationships underpin marketing in terms of customer satisfaction, value, retention and profit. In addition to increasing revenue, relationships influence the value of the communications process.

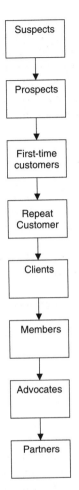

Figure 22.9   Relationship marketing
Source: Adapted from Philip Kotler (2000) *Marketing Management,* Millennium edn, Prentice Hall International, Upper Saddle River, N.J.

The aim of the process is to develop loyalty and generate increased revenue from that loyalty. Some customers will inevitably be lost along the way, some who choose to leave and some for whom the organization finds the cost of keeping them exceeds the gain. The organization must make some difficult decisions with regard to the profitability of relationships and implement them if the model is to be profitable.

The investment worth is determined by the product type and industry sector. For example, a hotel chain requiring an internal refit across all the hotels in its chain is more likely to develop a partnership with its supplier (see also Chapter 17, section 17.8) and to benefit from an intensive long-term partnership, while a small grocery shop is unlikely to consider developing a relationship strategy for the ten-year-old who purchases a sherbet dibdab. However, culture and attitude play a part, for if the grocery store implements a relationship strategy with the child it could result in lifelong custom with the child, the child's friends and family.

Building relationships with customers before these relationships become profitable is a strategy that UK banks have developed, providing youth and student accounts as loss leaders with the aim of developing lifelong relationships worth thousands of pounds of cross-sales income, such as credit card income and mortgages.

All elements of the marketing mix underpin the relationships that organizations develop with their customers and the speed and success with which they take them through the customer development process.

22.11   THE MARKETING MIX

The marketing mix is the lever of marketing power; it is the way an organization reacts and positions its offerings in the marketplace. The marketing mix structure is given in Figure 22.10. It is divided into two parts – the traditional marketing mix:

- Product (consumer good).
- Place (channel of distribution).
- Price (cost).
- Promotion (marketing communication).

and the service extended marketing mix:

- People.
- Process.
- Physical evidence.
- Period (time).

Services differ from products in that they are intangible, inseparable, variable and perishable. If an engineer or scientist consults a patent agent on how to protect an invention, the advice is critical but no tangible object changes hands. The advice is provided at the same time as it is consumed, while physical goods are produced and consumed in separate timeframes. The type, quality and location of advice are variable and the advice cannot be stored – if the appointment is missed it no longer exists. This is completely different from the purchase of a new computer, for example, where a physical object changes ownership. However, the purchasers of physical goods have intangible feelings, and these are important factors in the purchase decision. There can be issues of separability, variability, perishability and tangibility with physical products.

Most products incorporate attributes of a service, for example the purchase of a banana from a supermarket, where the environment, attitude of staff and ease of purchase will affect the likelihood of purchase and of satisfaction. Likewise, most services incorporate tangible elements. The hairdresser or barber shop provides a tangible environment and equipment such as scissors and brushes. In addition there may be an extended range of products on sale, such as shampoo, conditioner, gel or wax.

Feelings of security and confidence are intangible, but apply as much to industrial purchases as consumer products. A safety engineer installing a new fire detection system will have these feelings. In analysing the product offering required, the marketing engineer must consider the intangible benefits the

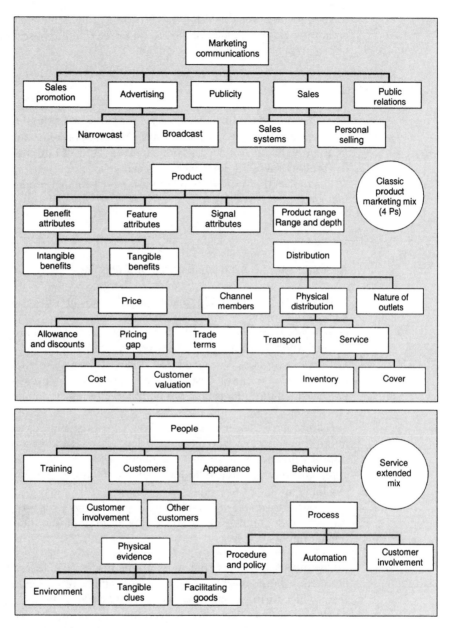

Figure 22.10   Marketing mix

customers may have. These intangibles will be supported by tangible features. For a fire detection system, it might be self-checking diagnostics to report fault conditions. This list of tangible benefits provides the marketing engineer with a 'wish list' that the design process must develop in terms of specific features. In many cases the resulting product can be relatively featureless in appearance. Alongside functionality, it may be essential to provide some way of communicating the product's appeal. So, with consumer products functionality must be reflected in an innovative style of design that is not only functional but attractive, and also communicates the quality of the product. (See Chapter 6, section 6.3.4 for more information about this aspect of design.)

22.11.1 Product

The collection of products or services a company offers should not be an accident but a specific strategic decision resulting from the portfolio analysis and the Ansoff analysis of the market needs. There are two vectors to this decision: the depth of the product range (the number of variants of a given product) and the breadth of the range (the number of different types of related products that should be included in the portfolio).

Example 22(3)

Attribute analysis for a small office photocopier

Target market: Small offices such as solicitors, architects, sole traders where there is little or no technical support.

Intangible benefits: Feeling of being up to date, security, empowerment (copies when you want them, how you want them).

Tangible benefits: Low cost (life-cycle costs, purchase, maintenance, power, toner, paper, etc. must cost less than bureau service), ease of use, low heat output, small footprint (space often limited), low service needs, short warm-up, jam free etc.

Feature attributes: The precise design features such as simple paper path, micro toner in easy-to-replace cartridge, not-in-use detection with switch-to-stand-by power levels, photo-quality copy etc.

Signal attributes: Stylish design to fit in with modern office furniture, possibly available in different colours.

Product range and depth: Product depth – range of speeds, A4 only, A3 and A4, etc. Product range – plain-paper fax, laser printers, colour (all you need for office printing?).

One aspect of product that must not be neglected is *packaging*. For industrial products it should meet legal requirements and be robust enough to fully protect the product during extended transport and handling. In many countries it must be easy to recycle. Packaging is even more important for consumer goods that have to be sold off the shelf, and in such situations should be considered as part of the marketing communication aspect of the marketing mix.

There are two key elements of *distribution*, the physical distribution of the goods or services and the nature of outlets and intermediaries the organization needs to use. Physical distribution is where marketing links in with the subject of operations. The best possible configuration for an electricity grid or for the most economical distribution of frozen chickens is approachable with techniques such as linear programming. Wholesalers, distributors and retailers provide essential value chain links. Typically they break bulk and provide an assortment of goods. However, physical goods can now be supplied via virtual outlets on the Internet.

A supermarket takes deliveries of pallet loads of goods from a wide range of manufacturers, and consumers can then buy small quantities of a wide assortment of products that they require. However, in the marketing of technical products, distributors and agents also provide other functions such as the supply of consumable items (paper, oil, etc.), installation, field service and staff training. The value chain links between the organization and its outlets must be excellent. For consumer goods, electronic point-of-sale systems provide an aid to excellent stock control and vital feedback on marketing performance of a product. Future plans include the microchip tagging of all products where the chip replaces the bar code to further develop this facility. Similar links for improving stock control of industrial goods are discussed in Chapter 20.

The Internet is an intangible distribution system that enables significant changes in the traditional distribution chain. Industries that utilize brokers, agents or intermediaries can now sell direct to the customer using a website. The benefit to the customer is the reduced price, but there is also the convenience – the ability to order the product from the comfort of your home or anywhere else and to have it delivered to you exactly when and where you want. A pre-ordered DVD, for example, can arrive with the customer on the day of its release. The Internet is significantly affecting industry sectors that trade as intermediate distributors, such as estate agents or holiday brokers.

22.11.2　Price

Price is a vital part of the market positioning of the product. Again, the company should view the customer's perception of pricing in benefit-oriented terms. Often in technology-driven (product-oriented) organizations there is a temptation for an accountant to say the cost of a product is x and so the price should be $(x + y\%)$, where y is the typical industry mark-up. This process will work for a stable industry with free competition, but most definitely will not work for rapidly changing markets. (See Chapter 13 for further discussion on pricing and costing.)

Certainly, cost control is vital to ensure maximum competitive advantage. However, given a strategic cost advantage, the organization may decide to maintain high profit margins at small volumes, or pass the benefits on to the customers and gain market share. Whatever the policy, it is essential that the organization has a clear understanding of the customer's benefit valuation. The Porter five forces of competition analysis is one tool for appraising this situation. Further information can be gained by market research. The difference between the organization's cost structures and the market benefit valuation is known as the *strategic pricing gap*. If it is large the organization has many options, if it is small it has few, and if it is negative it does not have a business!

Given that there is a good pricing gap, the organization may operate a number of different pricing policies to reflect different objectives. If the objective is to maximize the short-term profits, the marketing engineers and scientists need to estimate the volumes that would be sold at different prices. At low prices the volumes may be high, but the margins will be small and the total profit will be small. If the price is high the margins will be very good, but the small volumes will again yield a small total profit. What must be found is the combination of margin and volume that will yield the maximum profit in the period (see Exercise

2 of the Chapter 13 exercises provided on the website). This is only one policy. If the objective is to build market share, low margins may be acceptable. A low pricing policy is a market penetration policy, a high pricing policy is a market skimming policy.

In technical markets a variation of the skimming strategy may be used. New technology tends to be expensive at first, with costs dropping in later years through the learning curve effect. (The use of simultaneous engineering, as described in Chapter 6, may compress the learning period to months rather than years.) The technology will have applications in differing market areas with different benefit valuations. In such a situation the company can skim the first market and then, once the profits have been maximized in that market, drop the prices to bring in the next level of benefit valuation and repeat this process. This strategy will maximize total profit over the product life cycle and is particularly applicable where the company has some substantial patented technical advantage, such as a rewritable DVD.

The other aspects of pricing are the normal ones of deciding what discounts should be given to the trade, what price breaks will be given to volume purchasers and what trade-in allowances may be appropriate. In fast-moving areas where equipment may become obsolete before its physical life is over, the organization may find it desirable to offer 'migration packages' to encourage earlier upgrades. Increasingly customers are not simply concerned with the purchase costs of items but with life-cycle costs. Life-cycle costs include the costs of capital purchase, installation, training, servicing, insurance, energy, consumables (e.g. toner, paper), labour costs in operation and costs of ultimate decommissioning and disposal (see also Chapter 15, section 15.5.2). In pricing policy it is often only the engineer with an understanding of the technology and with marketing and accountancy skills to calculate paybacks and so on, who can make the appropriate strategic judgements. These considerations also have to be weighed by the designer, as discussed in Chapter 6.

22.11.3 Promotion

Promotion is the aspect of marketing with which people are most familiar and, of this communication mix, advertising in its many forms surrounds us all. Promotion is also known as communications and it is important to consider that communications is a two-way process. Before considering the actual methods of communication, it is essential to evaluate the underlying strategies. Maslow's and Herzberg's ideas of motivation (as described in Chapter 9) are applicable to purchase motivation. Moreover, the subject has to be taken through a number of stages. One general model for this process is the 'hierarchy of effects' model where the consumer has to be taken through six stages (awareness, knowledge, liking, preference, conviction and purchase). In the industrial situation the model should be extended to cover the post-purchase experience (e.g. training and servicing), as the ultimate objective is to become a preferred supplier with many repeat purchases.

In industrial buying, as discussed in Chapter 21, more than one person may be involved in the decision. In marketing some new software, the computer managers (the users) as well as the buyer will be involved. The great skill in marketing is to decide who is in this decision-making unit (DMU), what their agendas are, and devising communications that answer their different needs (e.g. one person may be more concerned with ease of use, another with total costs).

It is possible to promote products either by a push strategy, a pull strategy or a combination push–pull strategy. In a push strategy, the organization uses its marketing budget to promote and communicate with the distributors and outlets and relies on these to complete the process with the customer. In the pull policy, the company advertises and communicates with the customers to create a general market demand that distributors need to satisfy (i.e. a pull through the channels strategy). The most generally used strategy is the combination strategy, where the channels are primed with a 'push' communication strategy before a 'pull' advertising campaign.

Sales promotion may be of value in both strategies. Competitions can be run for agents with the best results and incentive offers given to customers (e.g. 10% extra free). Sales promotion efforts are usually intended to stimulate purchase in a short period. So, in the launch of a new product, the producer might offer free installation and service for the first three months of introduction to encourage rapid take-up and induce trial.

Advertising is in some ways the simplest method of communication in the sense that the media are paid for and are largely under the control of the organization. Advertising needs to be efficient and effective. To be effective, advertising needs the right message delivered to all the desired customers at the right time. To be efficient, advertising must be provided at a minimum cost and with the minimum of wastage. This is why direct marketing is becoming so important. No matter what selection of media is made (TV, press), the chances are that advertising will not reach all potential customers. In attempting to reach them, the company could waste much of its effort in placing the advertisement before people who are of no interest to the organization at all. A simple measure is a cost per thousand of potential customers. While consumer advertising is all around us, many engineers and scientists will be concerned with company-to-company marketing of technical products. Here the 'narrowcast' (as opposed to 'broadcast') media is much more important. This is most commonly in the form of specialist journals. However, as shown below, advertising by any medium is a relatively minor part of industrial marketing.

Advertising is where the media earn revenue. Publicity is where a company gets the media to print information on its products but does not pay for it. This is why companies issue press releases on innovative products and why trade journalists are provided with prototype products to evaluate. *Public relations* may be used to provide events that will attract the press. A specialist aspect of public relations is damage limitation where adverse publicity needs to be managed. This aspect is of special relevance to engineers, scientists and technologists, as in the event of an accident it is often the technically competent staff that must meet the media. Contingency planning is the process of pre-planning and risk assessment to avoid possible disasters and to prepare responses for risks that cannot be avoided.

In consumer markets the organization has to communicate with large numbers of customers and therefore broadcast advertising is effective and efficient. In industrial markets complex products are being sold to complex decision-making units. Here advertising is much less important and the key role is that of the sales manager who can enter into direct contact with the individuals in the decision-making unit. The process of personal selling is simple but needs skill to implement. The stages are:

(1) Research to identify the targets and assemble information.
(2) Relationship building through listening.
(3) Presentation (show and tell).
(4) Meeting objections (e.g. 'this will need a lot of training' etc.).
(5) Closing the sale.
(6) Aftersales follow-up to generate repeat sales.

Sales systems are vital to customer satisfaction. A customer may be delighted with a capital purchase; but if the ordering system for spare parts is slow this all may be lost. In many markets with customers operating electronic data interchange (EDI) and just-in-time (JIT) policies, the organization's sales systems may become an outright source of competitive advantage. (See Chapter 8 for ABB's use of Industrial IT in this way.)

22.11.4   Service extended mix

In services, the benefits to be communicated are sometimes intangible. In these cases people are the most important element of the marketing effort. They must be well trained, have an appropriate appearance and behave in a customer-responsive way. The way a company delivers a service may be important and clear policy guidance on process should be given to staff. The degree of service delivery automation should be part of the strategic marketing policy. Where the customer comes into the organization's premises for training, the physical environment is physical evidence of the quality of the service. The quality of manuals will provide tangible clues to service quality. For many services the sale of supporting goods (e.g. textbooks for a professional institution) will not only be a statement of the service quality, but also a source of additional revenue. (These aspects of quality are discussed under total quality management in Chapter 4.)

22.12   MARKETING INFORMATION SYSTEMS

To develop and market a product, a continual flow of information is needed. This is provided by the marketing information system, which has four elements: *research, intelligence, internal data* and *analytic systems*.

22.12.1   Market research

In the launch of a new product, the organization will need to know the size of the potential market and how people will find out about the proposed product. There are two types of marketing research: expensive secondary research, and very expensive primary research. Secondary research is where the marketing engineers and scientists review published information from industry surveys and so on about the products and markets. Then, and only then, the much more expensive market research can be considered. Here, marketing staff actively go out into the marketplace and conduct experiments and surveys. This is a very expensive process and even a modest and restricted survey may cost £15 000. Typical research methods are customer preference trials, questionnaires, surveys and interviews. The information collected can be divided into quantitative and qualitative (e.g. what price will a consumer be prepared to pay – quantitative – and their attitudes – qualitative). Both types of information are required to underpin the development of an effective marketing strategy.

22.12.2 Marketing intelligence and internal data

Market intelligence is vital to long-term success. For technical products it involves tracking competitors' activities such as new promotional campaigns and product launches. Competitors' products may be purchased for retro-engineering to see how they are built and to evaluate their technical performance, strengths and weaknesses. An organization has vast amounts of information that arise in its normal day-to-day business. This should be examined for its marketing information content. So, sales figures should be analysed to see which accounts are more profitable, which areas have smaller margins and so on.

Long-term trends should be looked for – declining sales may indicate new competition in a given sector. Apart from the accountancy information, care should be taken to capture and use other sources of data such as field service reports and customer complaints. These can often provide useful insights that will help to develop competitive new products. This area is vital in developing market-driven quality (MDQ) programmes, and provides marketing input for the simultaneous/concurrent engineering process.

22.12.3 Analytic methods

Much of the information is hidden in vast amounts of data points. An organization can use statistical and modelling systems to convert a flood of data into specific information. Sales may have seasonal characteristics and month-by-month comparisons may not be helpful. Various methods of forecasting from basic (simple moving average) to more complex seasonally adjusted trend-detecting models (exponential smoothing) may be appropriate.

With these four elements (research, intelligence, internal and analytic) the organization has the ability to adjust its marketing plans and marketing mix to changing market conditions – an essential requirement for long-term competitive advantage.

22.13 SUMMARY

This chapter began with the business planning process, as described in Figure 22.1. The figure shows that marketing must form an integral part of the management of any organization. The secret of long-term competitive advantage is not just excellence in the achievement of any single parameter in the business development process. Technological leadership may be essential, but if integration with the planning process is lacking, the battle in the market may be lost.

Integrated plans start with clear statements of mission, aims and objectives, which must be shared with all within the business team. It is not possible to move to action without completing the marketing audit. If you are lost in a strange city, the first thing you have to do is to find where you are. The external analysis (PEST and Porter competition analysis) and the internal analysis (value chain) will provide the raw data, and the SWOT analysis will focus it into information and provide the bridge from analysis to the action agenda. At this stage the integrated plan can be outlined giving operating objectives for each element of the plan.

Production targets can be set in terms of volumes, timescales and cost structures. The market needs should be organizationally linked to the product design specification through product attribute analysis for each of the target market segments. Other elements of the marketing strategy need to be evolved. What will be the strategic pricing gap? Will this be exploited with a penetration or a skimming pricing strategy? What field and distribution service will be required (what outlets will be needed, what physical methods of distribution will be appropriate)? Will the communication strategy be push, pull or push–pull? Who are the communication publics, what is the nature of the decision-making units in the market segments? What media will be efficient and effective in making targets aware and thence converting these targets into active, long-term accounts?

Once answers to these questions have been agreed within the team, the integrated plan can be drawn up in critical path planning (see Chapter 18) format with milestones and budgets set. These two elements provide the mechanism of control and feedback. Continual effort is needed to make certain that each element is on schedule and within budget and that where deviations occur corrective actions are taken. Communications are most effective when they are integrated and can be used to take the customer from suspect through to partner. The marketing information management system is vital in providing the feedback and control information from the marketplace. Profits flow by satisfying customer benefit needs and wants more effectively and efficiently than the competition. Only with a strategically integrated marketing approach from the whole organization will this be achieved consistently.

REFERENCES

1. Curtis, T. (1994) *Business and Marketing for Engineers and Scientists*, McGraw-Hill, Maidenhead. Chapter 2 for further discussion on PEST analysis.
2. Porter, M. (1985) *Competitive Advantage: Creating and Sustaining Superior Performance*, Free Press, London.
3. Hedley, B. (1977) 'Strategy and the "business portfolio" ', *Long Range Planning*, Volume 10, pp. 9–15.
4. Day, G. S. (1986) *Analysis for Strategic Marketing Decisions*, West Publishing, St Paul, MN.
5. Ansoff, I. H. (1957) *Strategies for Diversification*, McGraw-Hill, Maidenhead.
6. Gummesson, E. (1999) *Total Relationship Marketing*, Butterworth-Heinemann, Oxford.

BIBLIOGRAPHY

Cowell, D. (1989) *The Marketing of Services*, Heinemann, London.

Crawford, M. and Benedetto, A. D. (2003) *New Products Management*, 7th edn, McGraw-Hill, New York.

Johnson, J. and Scholes, K. (2002) *Exploring Corporate Strategy*, 6th edn, Prentice Hall, Harlow.

Kotler, P. (1991) *Marketing Management, Analysis, Planning, Implementation and Control*, 7th edn, Prentice-Hall International, London.

Kotler, P (2000) *Marketing Management*, Millennium edn, Prentice Hall International, Upper Saddle River, NJ.

Kotler, P. (2003) *Marketing Management*, 11th edn, Prentice Hall, Upper Saddle River, NJ.

Lancaster, G. and Messingham, L. (1993) *Marketing Management*, McGraw-Hill, Maidenhead.

Lynch, R. (2003) *Corporate Strategy*, 3rd edn, Prentice Hall, Harlow.

McDonald, M. (1990) *Marketing Plans: How to Prepare Them, How to Use Them*, 2nd edn, Heinemann, London.

Moore, W. L. (1993) *Product Planning and Management: Design and Delivering Value*, McGraw-Hill, Maidenhead.

Trott, P. (2002) *Innovation Management and New Product Development*, 2nd edn, Prentice Hall, Harlow.

23

A Case Study in Starting an SME

23.1   INTRODUCTION

The early days of Ford Motor Company were typical of small and medium enterprise (SME) experience. Henry Ford, the company's founder, was a maintenance engineer at a Detroit power plant, and worked on his version of a horseless carriage in his 'leisure' time. The vehicle he assembled functioned satisfactorily, once he had got it out of the shed where it was built (by demolishing the shed). However, each of the first two companies that Mr Ford put together, with help from local businessmen, were failures within a year. With the determination that every successful entrepreneur needs, he tried again and the rest is history – without being bunk.

Two failures out of three are fairly 'normal' for business start-ups in the UK. The third one that succeeds rarely becomes a global giant, but there are more than 2m SMEs in this country and they account for 56% of the UK's employment, and 52% of sales turnover. By tracing the inchoate history of one SME start-up over a 10-year period, this chapter identifies the major challenges involved in setting up a new business – and shows how one entrepreneur overcame them. The story proceeds on a broad front, with changes of pace and direction, but running a small company is rather like that. The main character, the entrepreneur, is an electrical engineer, identified as 'Mr Red', who, with limited work experience when he set up on his own, applied the sort of management skills and tools described in this book (though that is not where he acquired them) and became a successful businessman. In addition to his degree, this enterprising engineer had two years' experience installing and maintaining high-tech diagnostic equipment in UK and Irish hospitals. He gave up that job to gain a PhD with a thesis on 'The interpretation of data in intensive care medicine: an application of knowledge-based techniques'. The opportunity for 'crossover' – applying advanced technology to new fields – that sparked the PhD thesis was recognized and seized at several stages in Mr Red's career as a business manager, and is key to the success of his company. Crossovers are also the basis of the continuous corporate reinvention advocated by Gary Hamel in *Leading the Revolution* (see Chapter 3), but Hamel is ten years behind Mr Red, whose adventures are described in two parts. The first part deals with how Mr Red dealt with seven problems common to most SMEs. The second describes his attempts to step up the scale of his company's operations.

The narrative style of this chapter is informal rather than academic, in keeping with the atmosphere in most SMEs. It may be difficult to believe from the earlier parts of this book, but business management can be fun. It can also be exciting. This happens in organizations of any size, but it is particularly true in SMEs. It is hoped that some of the fun and excitement of Mr Red's adventure will be better conveyed by the change of style.

23.2   PART 1: SMALL BEGINNINGS

23.2.1   Task 1: The idea

As shown in Chapter 1, every business starts with an idea. The idea may be a completely new product or service, or the adaptation of an existing product or service for a new use or market.

Mr Red's idea was in the latter category – another crossover. In the early 1990s, while putting the finishing touches to his PhD (note the hard work ethic), Mr Red worked in a not-for-profit (NFP) organization in Oxford that developed analytical software, and was tasked with applying a (then) new tool to the organization's techniques, which used object-level information management systems – itself a somewhat esoteric field. The tool was Standard Generalized Mark-up Language (SGML), which later spawned HTML and XML, two key technologies for the representation of information on the World Wide Web (WWW). At the time there were very few SGML experts in the UK, or elsewhere, and Mr Red spotted the opportunity to apply his newly learned skills in other fields, in a maybe-for-profit organization of his own.

23.3.2   Task 2: The Plan

The entrepreneur needs a plan to turn their idea into a business. Its first use is to clear the entrepreneur's own mind about what they intend to do and how they are going to do it. At this stage little detail is required, but enough should be available so that the mind-clearing process can be assisted by discussion with a few, well-chosen advisers or potential customers. Later, particularly if the plan is to be used to help to raise funds (Task 7), more detail will be needed. Much later (if it survives), the plan becomes the basis of running the organization and yet more detail is required. Appendix 1 gives a guide to writing a business plan, but the essential elements are:

(1)   A brief executive summary.
(2)   A description of the proposal.
(3)   Market information.
(4)   Resource requirements.
(5)   Marketing strategy.
(6)   Development strategy/exit plan.
(7)   Risk analysis.
(8)   Summary financial information.

Mr Red's entire initial plan could have been written on the back of an envelope, but it may not have been so formal. The proposal was to resign from his job, having persuaded his employer to contract with him to complete the SGML project as a consultant. He had little knowledge of other market applications but, rightly, judged the potential to be enormous. The resources required were himself and a computer, working from home. His marketing strategy was to negotiate a win/win deal with his employer, which was that its project would be completed on time, at lower cost than if Mr Red were still on its payroll; Mr Red would have a secured income for six months while he developed other opportunities. The opportunities were the development strategy. The exit plan was: if successful, take on staff, build up a profitable business and maybe sell it one day; if unsuccessful, return to paid employment. Risk analysis? With no dependants and only a small mortgage there was little to lose. Financial summary? The monthly fee for working on the project was enough to live on.

That was sufficient to get started in 1993. The 2001/2002 Business Plan for the group that Mr Red developed was more comprehensive. It comprised 28 pages on 10 topics: group activities; business development; financial information; markets; sales and marketing plan; competition; staff; infrastructure and organization; managing growth; strategic partnerships – plus an appendix of 20 financial schedules. But a lot of water and cash flowed before that degree of sophistication was required and there were many tasks ahead.

23.2.3   Task 3: Dealing with the authorities

A business in the UK has to deal with the Inland Revenue [1] for payment of income tax, National Insurance (NI) and corporation tax, and possibly with HM Customs and Excise [2] for payment of Value Added Tax (VAT). A company is also required to register with Companies House [3] and its premises have to comply with requirements of the Health and Safety Executive. New businesses may choose to seek help from Business Link [4], a service providing 'advice, support and information on setting up and running your own business', which is government-backed through the Department of Trade and Industry (DTI).

Though not essential at the outset, a new business would probably benefit from the appointment, or at least the identification, of suitable professional advisers: a solicitor for legal advice, and an accountant for financial advice. Small, SME-friendly practices are better than big-name firms, initially. Administration is simplified by having a separate business bank account and telephone line.

Mr Red chose to start his new, independent career as an employee of a company that he had set up some years earlier, together with two fellow postgraduate students, which will be identified here as OURCO. The registered office of the company was moved from a 'convenience' City of London address to Mr Red's private address, and Companies House was notified.

He approached his High Street bank to move the company's business account from a branch in London SE1 to its Oxford branch, and asked for its new business start-up pack that was advertised on television and in the national press. According to the advertisements, the pack came with the services of a business adviser. The bank agreed to move the account, but refused to provide the start-up pack because the business was already started. This turned out to be fairly typical of its attitude – there were other occasions when advertised services, help or facilities for SMEs were not forthcoming. However, the bank did assign a very helpful business manager. Although he had over 200 SME accounts to deal with, this manager took a special interest in OURCO, got to know its plans and people, and was supportive in times of financial stress. Unfortunately, when annual turnover reached £1m, OURCO was moved out of his portfolio into the care of a more senior manager, with fewer clients but a less helpful disposition.

Mr Red calculated his own monthly income tax (PAYE) using Inland Revenue help cards and tables. He used the same process for part-time staff employed in the first few months of 'going it alone', and for the first few full-time employees a year or so later. By including the part-timers, Mr Red was able to say, in all honesty, 'We have eight people on the payroll.' Combined with being registered for VAT, this gave OURCO a bit more status in the eyes of some potential clients. Such stratagems may be necessary for start-up companies to be taken seriously.

Mr Red developed a manual monthly bookkeeping routine to record sales invoices, sales receipts, purchases and travel expenses in general-purpose accounting ledgers. A problem with these monthly calculations, and the filing of documents to support them, was that they used some of Mr Red's most valuable resource – his own *time*. After a year of struggling entirely on his own, Mr Red was able to *delegate* the bookkeeping upwards to an unpaid family member, though he continued to handle the PAYE. This allowed Mr Red to devote more of his time and talents to fee-earning activities. Adam Smith would have approved – division of labour increases productivity. This was achieved through delegation, which is an important management skill!

Initially, annual accounts for the company were prepared by a small London practice, which did all that was legally required but no more. After a year, the account was moved to a one-man practice that had several SMEs as clients, some of them outside the UK, and that was itself an SME. Prior to setting up on his own, the 'one man' had been finance director of a substantial international trading company. This business experience, combined with his portfolio of SMEs, made him a valuable source of guidance – a *mentor* – as well as an auditor. (Entrepreneurs who do not have someone in their circle of friends and family with the business background enabling them to act as a mentor may find one who helps free of charge through the National Federation of Enterprise Agencies [5], or one for whom a small charge is made through Business Link [4].)

23.2.5 Task 4: Marketing

The early marketing of OURCO was by *networking* – the use of personal contacts who could be given some idea of the new company's capabilities, and who themselves had contacts with business or other activities where those abilities might be applied. This led to work in two areas that proved to be fundamental to OURCO's success. One area was the automotive industry, where OURCO worked as a subcontractor to the IT division of a medium SME that had several vehicle maker clients. The work involved management of databases for multilingual service manuals, where the new technique of SGML was particularly effective, giving the automotive companies faster, cheaper updates of their manuals. The other area was health – OURCO won business with a Regional Health Authority and with the National Health Service (NHS) Information Management Centre.

Together with the work from Mr Red's former employer, these contracts set OURCO on the road to growth:

	Sales (£K)	PBT*(£K)
1993/4 (8 months)	7	–
1994/5	49	4

*Profit before tax

The next phase of OURCO's growth was secured through another marketing or promotional ploy – *participation in conferences*. Mr Red participated in the NHS annual IT conference in Harrogate, the European SGML Users' Conferences at various continental locations and (not strictly a conference, but a useful promotional opportunity) a three-day SGML Training Course, where he was one of three expert speakers. Prior to the conferences, Mr Red identified other delegates that he thought might be the source of new business and developed plans to meet them – a proactive form of networking. At the training courses he had the opportunity of introducing his company as well as himself.

Within two years, by which time the company consisted of three young men and one young woman (see Section 23.2.6), OURCO had its own exhibition stand at the NHS conference and a shared stand at the SGML Users' conference. Later, Mr Red presented papers at the European and World SGML Users' conferences, and took over some of the organization of the training courses. All these activities helped to raise the profile of OURCO and to generate new business. Successful execution of the new business and repeat engagements from existing clients provided more opportunities to present papers (as well as welcome revenue) and gradually established OURCO's international reputation as SGML experts.

Mr Red recognized that OURCO lacked marketing expertise and sought help through the DTI's Business Link scheme, which offered three days' free consultancy in a range of subjects, including marketing. The expert assigned to OURCO confined his efforts to merely suggesting links with other SMEs in the Thames Valley area, demonstrating a surprising inability to see the true potential of OURCO's capabilities, or to recognize that SGML was geared to handling large-scale databases unlikely to be found in SMEs. In retrospect, it can be seen that working at the leading edge of new technology increases the risk that the potential of an SME will not be recognized by civil servants or bank managers. This lack of recognition was also evident when OURCO applied for DTI help to finance its participation in overseas SGML Users' conferences. It was told that funds could only be released to members of a recognized trade association. SGML was too new, and its practitioners too few, to have generated a trade association, so OURCO pressed on alone. In later years Mr Red was an invited speaker at the conferences, and OURCO developed its own annual International Summer School in Oxford – a business activity in itself, but also a low-cost, effective form of *sales promotion*.

Business with the automotive industry as a subcontractor grew steadily and helped to establish OURCO's reputation as a capable IT consultancy company. Contacts cultivated through the SGML conferences earned OURCO contracts with divisions of Xerox. These included 'body shopping' (assigning OURCO personnel to work in Xerox-led teams), software product development to enhance Xerox products, and postsales customer training on behalf of Xerox. OURCO also began to work with 'blue-chip' clients such as Reed-Elsevier and the British Medical Journal in publishing, and Towers Perrin and Kleinwort-Benson in financial services.

Through Xerox, OURCO was introduced into a new technical field – business process management (workflow). By being quicker to react than some of Xerox's bigger solutions partners, OURCO landed a key role in developing a workflow management system to control the engineering drawings for the construction of the Heathrow Express rail link. This project provided a steady income stream for OURCO over several years, but ultimately came to a close when the customer decided to make the system that OURCO had developed into a product that it could use in other areas of its business. Although OURCO was considered to undertake this work, the contract was awarded to a larger company that had better quality control and the type of software development procedures necessary to develop a product. OURCO learned that if it wanted to compete for this type of business, it would need to make its whole software engineering process more formalized. It started looking at how it might achieve this through *ISO 9001 certification*.

ISO 9000 certification was to become an important part of OURCO's marketing platform. (See Chapter 4 and Appendix 2 for details of this quality management system or QMS.) Companies in the automotive industry require compliance with their own or the ISO 9000 QMS, and for supply to government purchasing agencies, such as the National Health Service, ISO 9000 qualification is a mandatory requirement under EU regulations. Mr Red recognized the marketing value of ISO 9000 registration, but his reasons for pursuing it went deeper, and the way he went about achieving it was typical of his management style.[1]

Combined with the ability to deliver what it promised, OURCO's marketing helped to keep the company growing:

	Sales (£K)	PBT (£K)
1995/6	70(a)	3
1996/7	140(b)	15
1997/8	298(c)	39
1998/9	675(d)	64

(a)   £40k with auto industry through the medium-sized IT company.

---

[1] Mr Red wanted a QMS to help him manage the company, and saw that a development of ISO 9000 for use by software and IT systems suppliers, known as 'ISO 9000 with TickIT', would bring improved project management and consistent standards for software writing. A Quality Manual and ISO 9000 certification can be acquired by telephone for a few hundred pounds, but Mr Red knew that this approach had little value. He wanted everyone in the company to 'buy in' to the QMS and make it part of 'the way we do business here'. So *he* led the programme. The first step was a weekend meeting off-site for *everyone* in the company, where the reasons for a QMS were explained and the implementation plan was developed in a *participative* way. Working sessions in the mornings were followed by a range of shared leisure activities in the afternoons. Overnight accommodation and dinner were provided for staff and their families – all part of making OURCO a good place to work. When it came to implementation, Mr Red wrote a substantial part of the quality manual and the supporting procedures, *leading by example* and *participating*.

OURCO selected Lloyds Registration Quality Assurance (LRQA), one of the most respected authorities, as its QMS assessor. This was partly to make sure that the QMS was of the highest standard, but also because LRQA approval would be particularly good for OURCO's corporate image. Approval was obtained within less than a year of the weekend launch meeting, which is unusually fast. The completed and constantly updated QMS is accessible by all employees on the OURCO intranet. Introduction to it forms part of each new employee's initial training, with the aim that every member of the company will always know how to respond to any customer contact – so it *is* part of OURCO's marketing.

(b)   £14k with auto industry, £75k with divisions of Xerox.

(c)   £180k with divisions of Xerox.

(d)   £244k with divisions of Xerox, £70k with auto industry.

The danger of relying on two or three clients for a large proportion of sales was recognized, but this was a risk that had to be taken. OURCO's expertise was most valuable to large organizations with massive databases, and as a small start-up company it needed a portfolio of 'blue-chip' clients to gain credibility. Its risk management policy was to expand the portfolio and enter new business sectors.

23.2.6   Task 5: Hiring the right people

Finding and keeping good staff is a 'key success factor' and there are lessons to be learned from the way in which OURCO has managed to do this from the outset.

After a year of working alone, Mr Red took on his first full-time employee. The new recruit was an analyst/programmer whom Mr Red had worked with at the NFP organization. He also worked from home initially, and was paid the same as Mr Red. In fact, at one stage the employee was paid more than his boss: when new business was hard to win, the employee was asked to take a pay cut and Mr Red stopped paying himself altogether. When business picked up, *Mr Red kept his promise* and the previous level of pay was restored, though Mr Red delayed resumption of payments to himself.

A year later, a second analyst/programmer was invited to join – also drawn from the ranks of the Oxford NFP organization where Mr Red had worked, and also paid the same as Mr Red. Recruitment of someone whose skills and personality were known through years of acquaintance was preferred to the use of an agency or advertisement. This helped to ensure that everyone 'got on' – particularly important in a small team. At this point it was necessary to find 'proper' offices and OURCO was fortunate to locate low-cost accommodation in a privately funded 'starter' development; see Task 6. It was also time to recruit someone to deal with day-to-day administration. A young woman was taken on as a temp through an agency, which allowed both her and OURCO to see whether they suited each other. They did and she was subsequently hired as a full-time employee at half the salary of the three SGML experts – more division of labour and more delegation. Although adept at the immediate tasks to hand, her ambitions were more concerned with marketing than administration, and she was encouraged in this direction through a *sponsored training* course. Four years later she was marketing manager, and in 2004, like the first two recruits, she was still with OURCO.

At an early stage, when there were just four full-time staff, bonuses, based on profit sharing, and a staff share purchase scheme were introduced. Together with the egalitarian pay structure, these helped to set the *style* of the company (see Chapter 3), which was reinforced by *all* employees attending the European SGML Users' conference. They travelled, with their exhibit, in a hired van and stayed in something like a youth hostel, but they did it as a team.

More sophisticated versions of the bonus and share purchase schemes were developed and still exist, though there was a period when profits were not there to be shared. A voluntary, contributory pension plan was introduced and all employees were covered by travel and permanent health insurance. By these means OURCO offered competitive remuneration, but more important to hiring and keeping the best people was the 'buzz' of working at the leading edge of new technology, the excitement of dealing with big-name clients, and the opportunity to share success.

This total package enabled OURCO to add to the first few recruits some other very talented people, through local and national press advertisements, through agencies and by postings on the electronic noticeboards of selected universities. It also enabled Mr Red to recruit another colleague from his former NFP employer – the company secretary. Networking also produced a key hiring. It is not clear whether Mr Red recognized Mr Orange from one of the NHS conferences or vice versa, but the two got talking at a bus stop in London. By the time the bus reached Oxford, they had discovered a common interest in web technology and it transpired that Mr Orange was doing some ground-breaking IT work setting up an intranet at the Oxford Radcliffe Hospital (ORH). A short-term effect was that OURCO became consultants to ORH. A medium-term result was that Mr Orange joined OURCO, continued working on ORH projects, and introduced two other clinical IT experts (Mr Yellow and Mrs Green) from another NHS Trust, who also joined OURCO. By these means, staff numbers increased to 18 by the end of the 1999/2000 financial year and the business continued to grow:

	Sales (£K)	PBT (£K)
1999/2000	850	60

Accommodating 18 people and providing dedicated space for training on behalf of Xerox was a problem – and part of an opportunity (see Section 23.3).

23.2.7   Task 6: Finding the right place, and space, to grow

Working from home is a low-cost way to get started in business, but it does not project the right image, or generate confidence with buyers when trying to sell to global corporations like Ford or General Motors, or state bodies such as the National Health Service and the Royal Air Force. Organizations like these have routines for evaluating potential new suppliers that examine financial status, staff numbers and facilities, among other things. Also, when there are two or three employees all working from home, communication and cooperation become difficult. So, OURCO needed office accommodation that it could afford, that re-flected its new-technology sales offering, and that would allow it to grow. It found this in the Oxford Centre for Innovation, which provided office and workshop space in various sizes, at low rental and on short leases. It was located within walking distance of the rail and bus stations, provided a shared receptionist, telephone answering in the tenant company name, meeting rooms, and duplicating and fax services. Although it had close links with government-sponsored schemes, the Centre was originally the result of a private initiative, not part of Business Links ('ill-focused and poor value for money' [6]), or a Regional Development Authority scheme ('failing to deliver' [7]), or any one of the 183 Department of Trade and Industry initiatives designed to help small firms.

OURCO took one of the smallest offices initially, when there were just three staff, and later moved across the hall to a two-office suite, which served until there were eight employees. It then took half the top floor, which served it well until it was time to grow, and time to go.

23.2.8   Task 7: Finding funding

As shown in Chapter 1, section 1.4, and Chapter 14, section 14.8, a business needs enough in the bank to cover its outgoings until these are matched or exceeded by proceeds from sales, its income. Chapters 1 and 14 indicated where these start-up funds may come from.

In the case of OURCO, initial funding requirements for its Oxford start-up were small. The first two computers were bought on Mr Red's father's charge card, and he was paid back later. In the first three years operating costs were generally funded by sales, with occasional small shortfalls covered by short-term family loans.

Once the company was operating profitably, retained profits were the principal source of funds. In Year 4, 1996, and in 1998 authorized share capital was increased, reaching £20 000 at the second step, of which £15 500 was issued and taken up by the directors and employees. This provided sufficient funds to meet the capital costs of taking on more staff – just some furniture and a computer – and the lag between taking someone on and their generating revenues to more than cover their salary and benefits. OURCO could have ticked over quite nicely, growing at 50% a year, but there were opportunities for major change. The business strategy to seize these opportunities entailed a new organization, new offices and a new form of funding. All these changes are described below.

23.3   PART 2: GOING FOR IT!

23.3.1   A turning point

Section 2.5, External factors, opened with:

> 'To determine a business strategy, or any other strategy, it is necessary to consider the environment within which the manager's unit is performing.'

In the late 1990s there were fundamental changes in the environment in which OURCO was performing, which led its manager, Mr Red, to develop a complete new strategy. He went through the phases of SMEAC (see section 2.5) – Situation, Mission, Execution, Administration, Communication – and repeated most of the tasks described in Part 1 of this chapter. He came up with a new idea, a new plan, a new organization, new hirings, new premises and new funding.

The changes in OURCO's environment included:

- On a global basis, SGML had spawned Extensible Mark-up Language (XML), which provided the basis of the next generation of the World Wide Web.
- The Internet was established, opening the way to e-commerce.
- Forecasts of e-commerce growth showed a requirement for services beyond the capability of the then current Internet service providers.
- Some of OURCO's clients, particularly in the automotive industry, were 'going global' in their information management.

- One major client, Xerox, was planning to spin off or close down some of its divisions.
- At national level, in October 1998 the UK government published its new *Information in Health* strategy, as a central theme of its Modernising Britain programme. A key feature was the use of web-based technology for the delivery and exchange of healthcare information – exactly the field where OURCO excelled. An additional £1bn was committed to NHS clinical IT over a five-year period.
- Locally, OURCO had outgrown the largest available OCFI office unit.
- The industrial and geographical spread of OURCO's client portfolio and the scale of operations required organizational change in order to maintain excellent customer service.

23.3.2 SWOT analysis, leading to a bigger idea

Anticipating the effects of these changes, OURCO's SWOT analysis (see Chapter 22) showed:

Strengths

- International reputation as a high-quality solutions provider.
- Expertise in SGML/XML.
- Portfolio of blue-chip clients.
- Three top UK clinical informatics experts employed.
- Other talented staff on board.
- Electronic health records (EHR) developed.
- Two managed customer websites operational.

Weaknesses

- Management team too small (just Mr Red and the company secretary, who was also director of operations).
- Lack of focus due to wide, flat organization.
- Skills shortage.
- No space for expansion.
- Overreliance on a few large customers.
- Small player on the global stage.

Threats

- Fast-moving technology.
- Fast-moving competition.
- Enforced relocation.

Opportunities

- Global e-commerce.
- NHS IT programme.

23.3.3 A new strategic plan

In order to exploit the two big opportunities, OURCO would need to change from a niche consultancy company (which had begun to develop software solutions for some of its customers) to both a product company and an application service provider (ASP). Mr Red had seen other companies in OURCO's sector try and fail to make the move from consultancy company to product or solutions provider. To successfully move the company into these new areas would require a detailed plan and proper financial backing. Two things were certain: making the move would be very risky, but if it didn't go for it, OURCO would have to watch someone else cash in on the big opportunities.

The strategic plan to exploit the opportunities, counter the threats, address the weaknesses and build on the strengths involved:

- *A new corporate structure*, comprising two new subsidiary companies of OURCO:
    (a) OURCO Online to develop and host innovative web applications and services;
    (b) OURCO Health to focus on developing a product for electronic health records (EHR) for the NHS.
- *Strengthening the management team*, by recruiting
    - a group finance director;
    - managing directors for OURCO Online and OURCO Health.
- *Raising capital* from private investors.
- *Forming alliances* with global IT organizations.
- *Moving to modern offices* on the Oxford Business Park.

According to the financial model, OURCO would continue to be profitable, OURCO Online would be profitable within two years, and OURCO Health within three years. The exit plan included a possible trade sale or initial public offering (IPO) for one or more of the companies when its trading performance commanded an acceptable price.

This was a comprehensive plan, too big for the back of an envelope. The whole exercise – market research, SWOT analysis, risk assessment, consultation, actually writing the plan – was performed by Mr Red while he continued to run the company, make presentations, close deals, manage projects, lead (and sometimes drive) the quality programme, spot talent, and communicate with employees, clients and potential business partners. He also found time to get married, move house and run the New York marathon. (Entrepreneurs, like everyone else, have to find the life–work balance that suits them, and, while starting an SME can be fascinating, it should be remembered that 100% work means 0% life, which has adverse consequences after just a few months.)

23.3.4   Moving out, moving up

The previous two sections dealt with the situation and mission of OURCO's repositioning. This section deals with execution, administration and communication, but not in that order.

Having developed the new business plan and decided that the Oxford Business Park offered the best relocation opportunity, Mr Red arranged a two-day weekend meeting to *communicate* the plans to all employees. This followed a similar format to the quality programme weekend described in the earlier footnote in this chapter. Having heard and discussed the plans, staff were encouraged to identify which of the three OURCO companies they would prefer to work in. In almost every case their choice could be met, and the opportunity for later change of preference was available. That covered internal communication. Externally, customers and suppliers had to be informed, as well as the Inland Revenue, Customs and Excise and Companies House – a mixture of administration and communication, all handled by the two-man management team. The company's bankers also had to be kept up to date.

Administration included a rewrite of the quality management system, since the accreditation to ISO 9001 standards covered only the original company, OURCO.

Another item on the management agenda was fundraising. Funds were needed to launch the two new companies, just as described in Chapter 1, and to meet the costs of 'extraordinary items' such as the office move and the reorganization. OURCO's bank balance was at about the right level for 'business as usual', but was not sufficient to finance the necessary expansion of the business. It was planned to seek private investments in OURCO Online initially, and later in OURCO Health. The process involved communication to 'those individuals who had expressed an interest in the business of OURCO Online Limited as an investment opportunity' of 'information upon which a decision to buy shares may be based'. To generate 'expressions of interest' an outline of OURCO's business plans was presented to a Business Angel network, and the opportunity was spread among staff, friends, family and business contacts.

Sufficient funds were raised to meet the objectives for launching OURCO Online, but there were other demands on OURCO's finances. The Oxford Business Park was not OURCO's first choice, but when long negotiations to secure another property were abruptly ended by the owner's change of plans it was the only viable option. It had drawbacks: the lease was for a minimum of 14 years, which was longer than OURCO would have liked, and the terms required deposit of one year's rent, or a bank's guarantee to pay that amount. OURCO did not have the spare cash to deposit and the bank would not give a guarantee without security. Eventually, the security was provided by a charge on the homes of the directors. How lucky that this was possible! (George Washington, Napoleon Bonaparte and Winston Churchill, amongst others, all recognized that luck is a major determinant of human destiny.)

Execution of the office move was completed over a long (holiday) weekend. Execution of the business plan was designed to take place over a period of years. That was the plan, but the reality was different.

Having formed the two new companies and started operations, Mr Red decided to have a professional appraisal of OURCO's business strategy in preparation for a further round of fundraising, and he commissioned a study by Pricewater-houseCoopers (PwC). The study endorsed the overall plan, but PwC made several suggestions for improvement, aimed at relationships between the three companies that would be more tax efficient, simplifying the task of further fundraising by making the group more attractive to larger private investors, and better preparing the group for an IPO or trade sale. The bottom line was that PwC advised the company to restructure into a single group holding company so that further funding could be brought into that company, rather than the subsidiaries; without the restructuring it was unlikely that larger investors could be attracted.

The PwC observations were consistent with what OURCO was experiencing with the new set-up. A lot of time was spent doing business between the three companies and allocating resources from each of the companies to deal with outside customers. The result of these difficulties and PwC's appraisal was a plan to restructure, but meanwhile, business continued.

An operations director (Mr Blue), with experience of managing complex global IT projects, was recruited for OURCO Online. This turned out to be a key appointment that strengthened the management of the whole group. Contracts for OURCO Online were secured from Jaguar Cars, The European Foundation (an agency of the European Union, which placed its business in ECUs, the precursor of euros, which involved currency risks for OURCO), Stemcor (one of the world's largest steel traders) and the British Medical Journal. OURCO Health installed electronic patient records (EPR) at 'Beacon' sites[2] that were leading the IT revolution in the NHS, but had to do deals on special introductory terms. OURCO continued to provide systems consultancy, training and body-shopping to Ford, BMW/Rover, Rathbone, Xerox and others. However, the automotive industry was going through hard times and reverting to type: buying at the lowest daily rate, rather than the best-quality and lowest-cost total solution, and delaying payments to suppliers to ease their own cash-flow problems. Operating costs for the three companies were higher: annual office rents doubled; the group finance director's salary was more than double that of the part-time, amateur bookkeeper previously employed; capital expenditure on equipment and Internet connections quadrupled. And there were one-off costs of the office move and reorganization. The end result was OURCO's first loss:

	Sales (£K)	PBT/(Loss)(£K)
2000/2001	1867	(353)

The loss was covered by the funding that had been brought in from external investors, but it signalled a new 'mindset' for OURCO, which had previously expanded through reinvesting its own profits. The positive result was that the sales growth put OURCO into the *Sunday Times* listing of the 100 Fastest Growing UK Technology Companies.

23.3.5   The long road to success

Execution of the business strategy continued through 2001, particularly the formation of alliances, which were agreed with major international IT companies such as Oracle, Arbortext and Software AG.

In April 2001, OURCO embarked on the restructuring recommended by PwC. This involved the formation of a new company, OURCO Group Ltd, which, after a complex valuation process, would acquire all the shares of OURCO, OURCO Health and OURCO Online. Shareholders accepted the proposal unanimously and it was implemented in May. Besides making the group easier to manage, the restructuring aimed to make it more attractive for further private funding, and the management (the finance director and Mr Red) began a round of presentations to potential investors among venture capital funds. As an interim measure, OURCO approached its bank for help, which would include 'factoring' of outstanding

---

[2]The sites included South Staffordshire Health Community, a development of one of OURCO's first customers.

invoices, a higher overdraft limit, and assistance in applying for a small firm loan or development grant under schemes operated by banks on behalf of the DTI. The bank did agree to the bigger overdraft, but only if it was secured against charges on residential property. The overdraft guarantee was added to the rent guarantee in return for further charges on the homes of directors.

Quite quickly an investment fund was found that was willing to invest a significant amount in OURCO, and technical and financial due diligence was performed to determine a fair share price. Some existing shareholders, including directors, also contributed to the fundraising and the total injection of almost £1m would have been sufficient to meet all OURCO's cash requirements for business expansion, and leave management free to concentrate on the twin tasks of building sales and developing the product to capture a major share of the health records market. Then came 11 September 2001. The investment fund put all new business on hold and the deal was off. (Does anyone remember the introduction to Chapter 2? It includes: 'even at the start, the new company is influenced by events and developments on . . . the other side of the world'). Demand for IT services, already at a reduced level through 2001, also collapsed, along with many other industry sectors.

Up to that point recruitment had continued, mainly for OURCO Online, including in July a managing director, who came with a strong sales background – the first senior OURCO manager with that experience. Through some additions to OURCO Health and with OURCO continuing to provide consultancy services, the total payroll had reached 50 by October. With a lag of something like six months before a new hiring becomes 'cash flow positive', OURCO needed that £1m of new investment to continue its expansion. In fact it had raised only a quarter of that amount.

Despite the collapse of stock markets around the world and sharp falls in all business activities, OURCO's search for funds had to continue. So did its sales efforts. In July, it had secured its first order with a value in excess of £500 000, which was from a car manufacturer for an engineering drawings management system. Two orders for the product of similar value had followed, and there were other positive developments. In another of Mr Red's crossovers, the electronic health records technology had been adapted and applied for the first time outside the healthcare sector, to provide an information management portal for the BBC.

One major incentive to stick to the expansion strategy came from the healthcare sector. The UK government had clearly indicated that it intended to stand by its targets to provide an electronic health record to every citizen in England, and that a new streamlined procurement process would be put in place to achieve this. Although the details of the process had not been worked out, there was a clear political imperative to commit significant funding to the health records sector; the opportunities for companies with the right products could be immense.

23.3.6   Funds raised, hopes dashed

One of the PwC recommendations was the appointment of nonexecutive directors, who would help with strategic development and grooming of OURCO for a trade sale or IPO. Accepting this advice, Mr Red made an extensive search (more networking), which led to the appointment of a nonexecutive chairman for OURCO Group, in January 2002.

One of the first pieces of advice given by the chairman was for the company to focus on its EHR product; this was the area of greatest opportunity and the chairman considered that by simultaneously developing the second strand of opportunities as an application service provider, OURCO risked falling short of both objectives. In addition, product companies are always more attractive to investors than services companies.

The new chairman had a finance background and, more importantly, brought with him a major new investor (a 'unique selling point' during the selection process). The investor completed technical and financial due diligence assessments of OURCO, and in January 2002 was ready to commit the remaining investment that OURCO needed. Half this sum would be through direct investment, and up to a further half through underwriting a one for five rights issue to existing shareholders.

By the end of March this investment was in place, and OURCO was poised for the Health Minister's announcement of new EHR contracts in the NHS, which was expected at the NHS IT Conference. There was no such announcement. Instead, it was revealed that the Department of Health was working on a new centrally driven plan for transforming the NHS through IT, which would be far greater in its scope, but would take much longer to procure.

23.3.7 The long wait

The announcement of the larger National Programme for IT in the NHS presented OURCO with an even greater opportunity than before, but in the short term the extended timescale for procurement left it seriously overstaffed. Sales projections were revamped and plans developed to enable the company to break even with a lower group turnover. That would be a reduction in sales from the year just ended (the first time the company had not planned to increase sales), but the company had been making losses as it invested in its product development, and the year-end results showed that those losses could not continue without further investment:

	Sales (£K)	PBT/(Loss)(£K)
2001/2002	2205	(510)

A process of staff redundancies and redeployment had to be undertaken. All directors and some senior staff took salary reductions. Every affected employee was interviewed and counselled, which was painful for everyone involved. Headcount reduction, 'downsizing', 'delayering', however it is labelled, is difficult enough in large organizations. In SMEs it is even more stressful, since those involved are probably friends as well as colleagues. But it has to be done. Having dealt with Statutory Sick Pay, Statutory Maternity Pay and Student Loan Refunds, OURCO had to learn about Statutory Redundancy Pay. It took until the end of October 2002 to complete the reduction to a staff level of 30.

The remaining staff regrouped to run the company more efficiently and to intensify their sales efforts. The smaller team quickly recovered its morale and set about chasing the ever-growing opportunity in the NHS.

Resource allocation for the whole group was assigned to Mr Blue, who became director of operations for the whole company. His experience and project management skills helped to improve customer satisfaction by completing more work on time – and on budget, with the reduced staff numbers. This was good for OURCO, but Mr Blue's was not a back-room job. He met regularly with clients at project review meetings, listening to their needs, building confidence in OURCO. Rewards came in repeat business from satisfied customers. Mr Blue had complete support across the company, which showed up in the performance indicators: almost every month, resource utilization was more than 100% of theoretical capacity (despite complying with the EU Working Time Directive).

Leadership of nonhealth sales was assumed by Mr Red. He developed new business in the automotive and publishing sectors, and secured a contract with the Food and Agricultural Organization (FAO) of the United Nations for a multilingual information management system. This FAO project was the subject of a keynote address, presented jointly with the FAO chief of information dissemination management, at the XML 2002 Convention in Boston, USA – the largest annual gathering of XML users and developers in the world. Quite a publicity coup for OURCO!

Healthcare sales were pursued by OURCO's clinical informatics experts Mr Orange, Mr Yellow and Mrs Green. There were still contracts to be won at Regional Health Authority and Hospital Trust level and these were used to listen to and learn from users about ways to enhance the product. As a consultancy company, Ourco had originally been technology driven; as a product company it needed to become customer or market driven. With the continued push from its customers, product development was one area where there was no cutback. A team of OURCO's top analyst/programmers continued development of the product to match customers' expressed and anticipated needs, but at considerable 'opportunity cost'. These same star performers had to be withdrawn from fee-earning activities – a tough decision when sales were vital for cash flow, but one that was fully supported by the major shareholders.

Other activities not cut back were the OURCO XML Summer School, which was reoriented to give more emphasis to information management in healthcare (with special terms for delegates from the NHS) and work on the OURCO QMS. There was no choice on the QMS, since the ISO standard had been revised from ISO 9000:1994 to ISO 9000:2000, which would apply from 2003 (see Chapter 4) and entailed a new accreditation process.

Throughout 2002, the general business environment was difficult, but it was particularly bad for IT companies. The dot-com bubble had burst, the collapse of Enron had given 'big business', particularly commodity trading, a bad name, and the forecast benefits of business-to-business e-trading were not being realized. Sales were hard to come by, and when they were won, like many SMEs, OURCO suffered from slow payment by some of its clients. But OURCO continued to believe in its product and that the market for it was about to grow substantially – the major risk was that larger companies, forced out of other areas of IT by the collapse of those markets, would start to encroach on OURCO's healthcare business.

23.3.8 The home straight

In January 2003 the Department of Health announced:

'The National Programme for IT in the NHS is now ready to embark on the formal procurement process to secure a wide range of products and services from the private sector and deliver the national applications as outlined in the paper, *Delivering 21st Century IT Support for the NHS.*'

The programme related only to England. Scotland and Wales were developing their own plans.

Details of the NHS programme emerged slowly, but by April 2003 it was known that £2.3bn would be spent in the first three years to create an Integrated Care Records Service. The concept outlined was a national data 'spine' providing a central health records service to support five regional local service providers (LSPs). 'Expressions of interest' were sought from companies wishing to be involved. Originally almost 100 companies put themselves forward and this number was gradually reduced at each stage of the procurement. Many of the original companies grouped themselves together into partnerships and consortia to move forward with joint bids.

OURCO positioned itself to team up with larger service providers and systems integrators, which were large enough to be considered as prime contractors in the bidding. Also, OURCO 'expressed interest' in bidding on its own for pilot EHR projects in Wales and Scotland, and teamed up again with larger partners to be considered for a bid to provide the health records system for the Armed Forces. OURCO faced a four-stage sales effort:

- Continue to promote installations of its EHR product, to establish it as the 'product of choice' among NHS users.
- Convince companies bidding as prime contractors that they should tender OURCO's EHR product.
- Help the prime contractors develop and sell their proposals to the NHS IT procurement team.
- Pursue the separate bids for Scotland and Wales using its own resources.

At the same time, OURCO was negotiating with IT service providers in France and Finland for the installation of the EHR product under licence in their healthcare markets.

Mr Red recognized that OURCO did not have the ability to sustain such an effort, nor to handle contract negotiations with the bigger players involved. He proposed that a team of experienced commercial and sales specialists should be assembled, to work on short-term contracts on terms that were heavily weighted for payment by results. He again got full backing from shareholders and unstinting support from employees.

By mid-2003 there were about a dozen consortia pursuing each of the six major health record contracts for England, and smaller numbers lining up to compete for the Wales and Armed Forces projects (which were for concept design and development, rather than implementation). OURCO was eliminated from the bid for Scotland, on the grounds of being too small – clearly it should have teamed up with a larger service provider for this bid. The mistake was duly noted for the future, but OURCO did not dwell on this setback for too long.

OURCO not only had to work flat out, it also had to prepare a hiring plan that would allow it to ramp up for implementation of any tenders that it won, and to continue with 'business as usual'. The 'business as usual' effort was important as insurance against *not* winning *any* of the major tenders, so sales efforts continued in the automotive and publishing sectors, as well as in healthcare outside the major tenders.

To provide some relief to the almost overstretched existing staff, and to prepare for some success in the bidding process, the OURCO board approved hiring of new staff and contractors. The company had already committed about a quarter of its fee-earning staff on development work in preparation for winning some of the tenders, with no guarantee of a return, and so detailed planning and mitigation of the financial risks were needed.

Once again, OURCO's major investor agreed to underwrite a rights issue, at a time when venture capitalists were 'preparing to turn their backs on small companies' [8]. This was not an act of blind faith. OURCO's investor had made his own independent assessment of the EHR product and the market position and was prepared to believe that OURCO would win at least some of the tenders it was involved with.

Through the remaining months of the bidding process OURCO remained optimistic, but mindful of the need to prepare staff and investors for all eventual outcomes – including not winning anything! With so much focus on product development for a market that might well dry up if all tenders were lost, OURCO embarked on establishing the 'crossover' of the EHR product as a knowledge management system (KMS) for applications in the publishing and automotive sectors. This completed a product development cycle: OURCO's early work on multilingual service manuals for the automotive industry, using SGML technology, led to the use of SGML/XML in healthcare IT, and in its developed form this was fed back to multilingual authoring and remote editing in other sectors.

Development of the KMS product meant that all customer projects would now use OURCO's own products, so completing the journey from consultancy company, through solutions provider, to product company. Recognizing this last step, the management team sought to focus the operations of the company on product development and support for sales channels through third-party systems-integrator partners. New processes and procedures were formalized and written into a new quality system. Using this new system, OURCO obtained first-time accreditation to the ISO 9001:2000 standard.

23.3.9   The finishing line

On 8 December 2003 there were 500 000 people celebrating on the streets of London, to greet England's Rugby World Cup champions. There were also 40 people celebrating some other champions in offices in Oxford. At the end of a procurement process that had taken almost a year, OURCO's product was chosen to form the basis of the National Care Records Service, the 'data spine' in England.

Two weeks later, OURCO learned that it had been selected as the sole preferred supplier for the development of a proof of concept for the single integrated health records in Wales. A month later, the prime contractor bidding OURCO's product was chosen as one of two developers of the first prototypes for health records for the UK Armed Forces.

23.4   EXCELSIOR

Having cooperated with some of the giants of the IT industry, competed against others and held its own on technical and commercial merit, OURCO raised its sights. The new strategic plan for 2004 to 2007 included the Strategic Ambition:

'To become a *global* market leader in e-records management, using open standards.'

Sloan's advice about the dangers of success, quoted in Chapter 7, section 7.3, is particularly apposite for SMEs. If a successful SME rests on its laurels or, to mix metaphors, stops swimming against the currents of indifference from institutions and competition from larger, established organizations, it will sink or be swept away. The only way to go from one success is on towards the next.

23.4   SUMMARY

This chapter traced the progress of an entrepreneur from a one-man consultancy to the head of a successful IT service provider. Starting as salesman, accountant, administrator, analyst, programmer, buyer and planner, he recruited a team to take on some of these tasks, and more. He used, perhaps instinctively, management tools that are described in this book, such as SWOT and SMEAC analysis and Ansoff's matrix, to develop business plans and marketing strategies that worked. He was a 'true believer' in quality management systems, and used ISO 9000 with TickIT for project management and as part of a marketing plan. His management style was consistent with the 'lessons from America's best-run companies' [9] and Deming's '14 obligations of management' [10].

As a small SME, the entrepreneur was able to fund expansion through retained profits. To make the leap to being a medium SME, he had to look for outside investors and was successful in finding them among Business Angels and through a nonexecutive chairman – people who had 'been there, done that'.

Guidance was obtained from professional advisers, such as PwC, and from nonexecutive directors, rather than through government schemes intended to assist SMEs or from the banks' small business advisers.

The importance of cash-flow management was illustrated frequently, and some of the risks of dealing with large organizations, governmental or private, were identified. (As the late Professor Ernest Grebenik used to tell his students: 'Dealing with government departments is like making love with an elephant. Any result takes a very long time, and there is a constant danger of being crushed.' He never revealed how he knew this.)

The critical success factors for this SME can be summarized as:

- Spotting opportunities to apply scarce knowledge.
- Choosing and keeping excellent people.
- Listening to customers and anticipating their needs.
- Developing the right products and processes.
- Belief in the importance of quality in all its forms.
- Managing to a plan.

- Constancy of purpose (one of Deming's 14 points).
- Leading by example.
- Keeping promises.
- Luck.

The last factor is frequently linked with Gary Player's response to a spectator who shouted 'Lucky!' when Player sank a 30-foot putt. 'You're right, it was lucky. And you know what? The harder I practise, the luckier I get.' A lot of OURCO people practise very hard.

23.5   CAST LIST

It may encourage student readers, and perhaps their tutors, to see how many engineering, science and technology qualifications are held by some of OURCO's key personnel. The list was compiled when OURCO had fewer than 30 employees and may not be comprehensive.

Mr Red	BA (Oxon) (Engineering), PhD (City University, London) (Medical Informatics)
Mr Orange	MSc (University of Wales College of Medicine), Fellowship in Clinical Biochemistry (Portsmouth Polytechnic)
Mr Blue	BSc (York) (Chemistry and the Environment)
Mr Indigo	BSc (Southampton)(Physics), PhD (Astrophysics)
Mr Violet	BSc (Mathematics and Computer Science), MSc (Cranfield)(Knowledge-based Systems in Manufacturing)
Mr Black	HND (Computer Science)
Mr White	BM (Medicine), MSc (Computer Science), MSc (Medical Informatics)
Mr Grey	HND (Computing)
Mr Brown	BSc (Information and Computing)
Ms Beige	BSc (Computer Science)
Ms Purple	BA (Natural Sciences), MSc (Software Systems Technology)
Mr Puce	BSc (Physics)
Mr Pink	BSc (Computer Software Technology)
Mr Lime	BSc (Computer Science)

Not all those listed have been mentioned in the text, but they have all made a contribution to OURCO's success.

23.6   EXERCISES

On the John Wiley website, at www.wileyeurope.com/go/chelsom, there are two cash-flow management exercises based on OURCO's experience. Also on the site there is a description of a customer relations situation that could form the basis of group discussions.

REFERENCES

1. www.inlandrevenue.gov.uk/employers/download.htm.
2. www.hmce.gov.uk.
3. www.companieshouse.gov.uk.
4. www.businesslink.org.
5. www.nfea.com.
6. Report by Richard Brooks of the Institute for Public Policy Research, quoted in *The Times*, Business section, 14 October 2003.
7. Report by Kevin Morgan, Cardiff University for the National Audit Office, quoted in *The Times*, Business section, 1 April 2003.
8. 'Investment firms shun start-ups', *The Times*, Business section, 16 January 2004.
9. Peters, T. and Waterman, R. (1982) *In Search of Excellence*, McGraw-Hill, New York.
10. Peratec Ltd (1994) *Total Quality Management*, 2nd edn, Chapman and Hall, London, gives a list of Deming's 14 Points. See, also, W. E. Deming (1982) *Out of the Crisis*, Cambridge University Press, Cambridge.

BIBLIOGRAPHY

The Sunday Times recommends:

Stone, P. (2001) *Your Own Business: The Complete Guide to Succeeding with a Small Business*, How to Books, London.

Reuvid, J. and Millar, R. (2003) *Start Up and Run Your Own Business*, 2nd edn, Kogan Page, London.

Bailey, A. (2003) *The Which? Guide to Starting Your Own Business*, Which? Consumer Guides, London.

Appendix 1

A Guide to Writing a Business Plan

A1.1 INTRODUCTION

A business plan is an important tool to facilitate management of an organization, whether it is just starting up or has been established for several years. The most important 'customer' of the plan is the person who is running the organization, but a well-prepared plan will also be vital when seeking financial backing from banks or venture capitalists. It will be useful when seeking assistance or advice from any one of the many government agencies that exist to help small and medium enterprises (SMEs), or when seeking help from lawyers or accountants.

The plan helps managers to focus on where the business is now, what it is trying to achieve and how it is going to reach its objectives. When implementation is underway, the plan provides benchmarks against which actual performance may be measured, and variances from plan will identify the need for corrective action if things are going wrong, or opportunities to reinforce success when things are going right.

The process of preparing and using a business plan is similar to the SMEAC and SWOT analysis referred to in Chapter 2. It is also not unlike running a scientific experiment, or using the scientific method of inductive/deductive reasoning. An idea or theory is formed, the plan or experiment is set up and started; performance is measured and interpreted; conclusions are drawn and (perhaps) actions are taken. A formal overview of the business planning process is given in Chapter 22, Figure 22.1.

The following two sections cover the basic content of a business plan. The first section deals with the framework of the text, and the second with the financial schedules. Together they comprise the output from the process of writing a plan, but a look at the desired end result is not a bad way to start. Having gained some idea of the finished product, the compiler of the plan has to *think* and then write about each section, starting with 'Scenario' and working through to 'Business development'. Then the financial schedules should be prepared and the first three sections retrofitted in reverse order: summary, index, title page.

A business plan, like all written communications, will benefit from the advice given in Chapter 11, section 11.4.2.

A1.2 BASIC STRUCTURE

A1.2.1 Contents

As a minimum, the plan should contain the following sections:

- Title page.
- Index.
- Executive summary.
- Scenario.
- The offering.
- The market.
- The organization and its management.
- Business development.
- Financial data.

A1.2.2 Title page

This page should show the organization and the period to which the plan relates, for example:

PQR Ceramics

Business Plan 2004–2006

There is an opportunity here for some eye-catching graphics. It is good practice to mark the cover page 'Confidential', to show restricted circulation, and to display the copyright sign.

A1.2.3 Index

It is helpful to give the reader a list of contents, with page numbers. Busy people like to go straight to the parts that interest them most. Some will go to 'the offering', others to the financial summary, others to 'the market'. Some may even go to the executive summary. In each case, what they see should make them want to read the rest of the plan.

A1.2.4 Executive summary

The executive summary should cover not more than one page. At the outset it should show the purpose of the document – whether it is the framework for managing the organization, or whether the plan has been written to help raise funds.

The summary should contain the whole story: what the big idea is; what the objectives are; how they are to be met; who is going to do it; what success looks like; and a timetable for achieving it.

This section is sometimes called 'the elevator pitch' – what would be said in a few minutes to 'sell' the plan, should the entrepreneur find himself or herself sharing a short ride in an elevator (lift) with a venture capitalist or other influential person with the power to make the dream a reality.

A1.2.5   The scenario

This section is a combination of 'Situation' and 'Mission' from SMEAC. It describes the organization (its size and structure), where it is now in business terms (fledgling or well established, aimed at certain business sectors or already engaged with a sound client portfolio, stable or growing etc.), where it is planning to go (enter new markets, introduce new products or services, achieve X% sales increase, achieving break-even or profitability) and the resources it requires (people, funding, facilities). Chapter 22, sections 22.3 to 22.6 describes some tools that can be used in thinking about and writing a business scenario.

A1.2.6   The offering

Here the plan outlines the products or services that form, or will form, the basis of the business. They should be described in nontechnical terms, avoiding trade jargon and obscure acronyms. The unique selling points (USPs – there's an acronym) should be highlighted – those features that will make the product or service successful. Future development potential of the product or service should be outlined. Use Ansoff's matrix: same product to new customers; new product to existing customers and so on. If the plan is being used to raise funds, the product or service description should give sufficient detail to be convincing, but should withhold data that could be useful to competitors. In some cases, for example if finance is being sought for a truly revolutionary product, it may be necessary to obtain a nondisclosure agreement from the other party.

A1.2.7   The market

The business planner has to determine what is to be sold, to whom, in what volumes, at what price and through which channels. Chapter 22, sections 22.7 and 22.10 can be helpful here. The size and structure of target markets should be outlined and competition assessed. The basis for beating competition needs to be established – will it be on price, quality, technology, delivery, service or a combination of all these? The resultant sales plan should be the starting point for all other elements of the business plan. It will determine resource requirements, product development, cost budgets and capital expenditure budgets.

A1.2.7   The organization and its management

If the plan is being developed as a management tool, the organization and its personnel – particularly its management – has to be subjected to 'gap analysis'. This identifies the gap between what exists and what is required in the organizational structure, manning levels and skill sets of the business. The plan then has to show how and when the gap will be filled. (On different scales, Chapters 8 and 23 describe how organizations were changed and appointments made in order to meet business plans.)

If the plan is to be used for fundraising, this section of the business plan must describe the existing organization and give names and 'thumbnail sketches' of its management team. If some of the new funds are to be applied to recruitment, the additional numbers and types of employee will need to be shown.

A1.2.8   Business development

For most businesses, particularly SMEs, a three-year business plan is sufficient. The first year should be planned in more detail, with monthly figures for sales, costs, profits (or losses) and headcount, and a diary of new product launches, sales or promotional events. For the second year, quarterly summaries may be adequate, and the third year may be just one set of annual figures and highlights. Year-on-year comparisons should be provided.

Over the three-year period, information should be set out for *product development*, including targets for new product introductions, and *market development*, including target changes in share of existing markets, and any new geographical or industry sectors to be entered. Any changes to organization, staff levels and facilities that are required to meet the product and market developments should be covered.

A1.3   FINANCIAL DATA

A1.3.1   Introduction

The business plan should provide summary profit and loss accounts and balance sheets covering the starting point and each year-end in the period covered by the plan. These two 'primary statements' are described in Chapter 14, section 14.5. A cash flow statement for the planning period should also be prepared. These are the outputs of this part of the planning process. The inputs are a sales budget and a cost budget. This part of the appendix gives guidelines for the preparation of these two budgets and the cash flow forecast, on a scale suitable for a start-up SME.

The financial data attachments to the business plan deserve their own index. A plan to be used for presentation to potential investors might contain:

- Summary profit and loss accounts, Years 1, 2 and 3.
- Year-end balance sheets, Years 1, 2 and 3.
- Summary cash flow forecast.

For internal management purposes there are other schedules that may be useful:

- Sales budget, Years 1, 2 and 3.
- Staff (headcount) budget, Years 1,2 and 3.
- Capital expenditure budget, Years 1,2 and 3.
- Cost budget, Years 1,2 and 3.

Illustrations of each of these schedules are provided in sections A1.3.2 to A1.3.5 to conclude this appendix.

A1.3.2   The sales budget

After a few years, a company may find it useful to have two sales budgets. One would be based on sales planning volumes (SPV), which are realistic projections, but containing an element of 'stretch' for the sales force. These volumes should be used for resource planning – the personnel and facilities necessary to support that level of sales. The other sales budget would be based on financial planning volumes, set, say, 10% below the SPV. It is prudent to use this lower level of sales to develop projected profit and loss statements. The sales budgets should be prepared following dialogue between the head of sales and the head of the organization – which for a start-up SME would probably be the same person.

The budget should be constructed with sufficient detail that it can be used as a management tool. This requires identification of sales by product (or service) line, possibly by major customer, shown on a monthly basis for Year 1, and possibly just as annual figures for Years 2 and 3. Management could then compare actual monthly sales performance with the budgeted figure for each product or service, in order to determine what corrective or reinforcing action is necessary.

The summary might look this:

PQR Ceramics Sales Budget 200X–200Y(£K)

Product/Service	Month 1	Month 2	Month 3 etc.	Total Year 1	Total Year 2	Total Year 3
Product/service A	3	3	4	44	65	103
Product/service B	2	4	5	62	82	148
Product/service C	2	3	4	56	66	122
etc.						
Total sales	20	45	55	630	840	1250

A1.3.3   The staff (or headcount) budget

Development of the staff budget is an important part of resource planning, and generates one of the major elements of the cost budget. It is developed, like a bill of materials or bill of quantities (see Chapter 17), as a list of each type of employee, the cost per unit of each type, and a schedule of numbers employed – by month for Year 1, by quarter, possibly, for Year 2, and as an average figure for Year 3. The cost of each type of employee should include salary, employer's National Insurance and the cost of 'fringe benefits' such as holiday pay, pension fund contributions, health insurance and travel insurance. The staff budget might look like this:

PQR Ceramics Staff Budget 200X–200Y

Job title	Salary and FB (£kpa)	200X				200X + 1				200Y	
		Month 1		Month 2 etc.		Total Y1		Q1		Q2 etc.	
		No.	£k	No.	£k	No.	£k	No.	£k	No.	£k
CEO	48	1	4.0	1	4.0	1	48.0	1	12.0	1	48
Finance director	36	1	3.0	1	3.0	1	36.0	1	9.0	1	36
Adm. asst.	18	1	1.5	1	1.5	1	18.0	1	4.5	1	18
Production mgr	24	1	2.0	1	2.0	1	24.0	1	6.0	1	24
Technicians	18	10	15.0	10	15.0	10	180.0	10	45.0	12	216
Warehouseman	12	1	1.0	1	1.0	1	12.0	1	3.0	1	12
Chief designer	36	1	3.0	1	3.0	1	36.0	1	9.0	1	36
Designers	24	2	4.0	2	4.0	2	48.0	2	12.0	2	48
Salesman	24	1	2.0	1	2.0	1	24.0	1	6.0	1	24
Total		19	35.5	19	35.5	19	426.0	19	106.5	21	462

The numbers of each type of employee will be one input to the capital expenditure (capex) budget, since every member of staff will require a place to work and some equipment. The total costs of staff salaries and fringe benefits will be one of the items in the cost budget. Simple examples of capital expenditure and cost budgets are shown below.

A1.3.4   The capital expenditure budget

As explained in Chapter 14, the purchase of things like furniture, computers, plant and machinery and buildings, which provide a benefit over an extended period, is accounted for in the balance sheet. These items are entered in the balance sheet as *fixed assets* from the time that they are installed, and their value is written down over their useful life by *depreciation* (see Chapter 14). That is, they are *capitalized*. Hence, the plan or programme to purchase such items is the *capital expenditure budget*. Generally, fixed assets have to be paid for on delivery or, in the case of buildings and some major pieces of equipment, in stage payments during construction, manufacture or installation. Each payment following delivery, and each stage payment, is an item in the capital expenditure budget. A simple budget for the first half-year of operation of a small manufacturing company might look like this:

PQR Ceramics Capital Expenditure Budget 200X (£k)

Item	Month 1	Month 2	Month 3	Month 4	Month 5	Month 6	Half Yr Total
Office equipment	20			10			30
Office fixtures	5						5
Office furniture	10			5			15
Total office	35			15			50
Machinery	40		10				50
Factory equipment	12						12
Materials handling	8						8
Total factory	60		10				70
Total company	95		10	15			120

The monthly totals, £95 000, £10 000 etc., would appear as outflows in the cash flow forecast. Depreciation at the appropriate rate would start when the equipment was installed, and would become an item in the cost budget.

A1.3.5 The cost budget

This budget covers all the operating costs of the enterprise. For a consultancy, or other form of service company, the major element will be staff costs. Besides salaries and wages, staff costs include items such as employer's National Insurance contributions, holiday pay and pension contributions. For a manufacturing company materials will be the most important item, and distribution costs may be significant. A production company that also offers design services would have a more balanced cost structure. An example cost budget is shown below.

PQR Ceramics Operating Cost Budget 200X (£k)

Item	Month 1	Month 2	Month 3	etc.	Month 12	Total Year
Personnel	35.5	35.5	35.5		35.5	426.0
Rent & rates	3.0	3.0	3.0		3.0	36.0
Utilities	1.0	1.0	0.8		1.0	10.0
Phone, fax etc.	1.0	1.0	1.0		1.0	12.0
Stationery	0.5	0.0	0.1		0.6	4.2
Travel expenses	0.5	2.0	1.0		2.6	18.4
Materials	8.2	10.6	11.4		9.1	104.4
Packing & shipping	0.6	1.1	1.4		1.8	18.6
Depreciation	1.2	1.2	1.7		2.4	16.6
Adverts/promotions	0.0	0.0	2.0		1.5	5.7
Professional fees	0.0	0.0	4.0		1.5	7.7
Finance costs	0.3	0.3	0.5		0.5	5.2
Miscellaneous	1.0	1.0	1.5		1.0	10.0
Total	52.8	56.7	63.9		61.5	674.8

If PQR Ceramics were to meet its sales budget of £630 000 in Year 1, it would show a small operating loss in its profit and loss account (see Chapter 14). With only slight increases in personnel costs in Year 2, and sales possibly rising to £840 000, the company might make profits before tax in that year. If all the example budgets were met, the company would be making quite good returns in Year 3. This scenario is fairly typical of SME start-up business plans. With more than half of start-ups failing to survive Year 3, such optimistic plans are viewed with scepticism by banks and venture capitalists. To prepare for the financiers' questioning, and to satisfy himself or herself that the plan is sound, the entrepreneur who is seeking funds should do 'sensitivity analysis' on the business model. This involves evaluating the effect on profits of, say, a 20% shortfall in sales in Years 2 and 3, and development of contingency plans to survive in those circumstances. (Chapter 15, section 15.11, gives more information about sensitivity analysis.)

A1.3.6   The cash flow forecast

The cash flow forecast is one of the most vital pieces of management informa-tion. It is important because if the company runs out of cash, it may cease to be a 'going concern' and go out of business. The cash flow forecast combines elements from the revenue accounts and the capital accounts. The revenue accounts include inflows, such as payments for what has been sold, and out-flows, such as payments for purchases and salaries. The capital accounts also include both inflows such as share subscriptions, and outflows such as capital investments.

The timing of some cash flow transactions is fixed by legislation or contracts. Salaries, for example, may have to be paid every month end, and the income tax and National Insurance on them by the middle of the following month. Corporation Tax and Value Added Tax have to be paid by set dates. There can be some flexibility in other outflows, such as payments for supplies, but continual late payment is bad practice. It can lead to delays in supply or demands for payment with order, and may damage the company's credit rating. The major source of inflow – that is, receipts from sales – is the greatest area of uncertainty. The selling company may specify 'Net 30 days' on its invoices, meaning that they should be paid within 30 days of receipt, but most customers will have a monthly payment cycle that does not fit neatly with receipt of the seller's invoices. That may mean a delay of 60 days and, until a pattern of payment times is established for all major clients, it would be prudent to base the cash flow forecast on an even longer period, say 90 days. The seller can minimize the payment delay by ensuring that the sales invoices are properly prepared, showing exactly what has been sold, when it was supplied, and the purchase order or other authorization for supplying, and by sending the invoice to the correct address – which may not be the location supplied, nor the address from which the purchase originated.

A cash flow forecast for the first six months of a start-up company might appear as follows:

PQR Ceramics Six Months Cash Flow, 200X (£k)

	Month1	Month2	Month3	Month4	Month5	Month6
Cash inflow from						
Billings (net 90 days)	–	–	18	35	52	48
Cash outflow						
Operating costs	53	57	64	58	55	62
VAT payment			10			14
Fixed asset purchases	95	–	10	15	–	–
Net cash in(out)flow	(148)	(57)	(66)	(38)	(3)	(28)
New funding						
Opening cash balance	500	352	295	229	191	188
Closing cash balance	352	295	229	191	188	156
Memo:						
Overdraft facility	20	20	20	20	20	20

The cash inflow from sales billings is net of VAT, but the company has to plan for payment to Customs and Excise of the VAT added to sales, less the VAT paid on purchases.

The opening cash balance of £500k may be the entrepreneur's own money, or may have been borrowed, or subscribed by shareholders or venture capitalists. It appears that this initial funding was sufficient to see the company through the first six to nine months, but, unless sales receipts start to increase in the second half-year, it may be necessary to obtain new funding by the end of the year.

A1.4 COMMENT

At various points in this book it has been recommended that engineers, scientists and technologists should think of a business as a system. If they do so, they will recognize that the business plan is a control mechanism. This perspective may help them in the challenge of managing their own business.

Appendix 2

Quality Management Tools

A2.1   INTRODUCTION

This appendix presents some of the forms and processes that are used in 'planning for quality' and 'managing for quality'. Two of them are 'generic' – the eight disciplines (8D) or team oriented problem solving (TOPS) approach and the cause-and-effect (fishbone) diagram can be used in many problem-solving situations, as well as in the context of quality management.

The descriptions of the three 'tools' have been provided by the Ford Motor Company, and are reproduced with their permission from publications that the company provides free to all suppliers.

Information is given about a fourth tool: ISO 9000:2000. This revised standard now qualifies as a quality management tool, and incorporates principles that are applicable to management generally. The standard is too lengthy to be included here. The Appendix merely lists the major changes from ISO 9000:1994 and the subject headings against which companies are assessed when seeking recognition of compliance with the BS/ISO 9000 quality standard.

A2.2   8D TEAM ORIENTED PROBLEM SOLVING

The eight disciplines (8D) approach, also known as team oriented problem solving (TOPS), is the Ford method for addressing concerns including those issues concerning capability indices that are below desired values.

The 8D approach to problem solving:

- Provides an orderly team-oriented method for solving problems using facts rather than personal bias. Creative, permanent solutions usually require input from many activities.
- Applies to any problem or activity and assists in achieving effective communication between departments which share a common objective.
- Requires documentation through the concern analysis report. A typical report form is shown on the following pages.
- Provides the missing link between SPC and realized quality improvement.

The eight disciplines are identified and defined on the following page, and diagrammed with a flow chart.

While the documentation of each problem is essential, the order of steps taken for resolution may vary depending on the degree of difficulty or complexity of a particular problem. For example, by the time a problem is reported and a team formed, interim action may have been taken already by the manufacturing personnel but the permanent solution may require subsequent team involvement. The eight disciplines are:

(1) Use team approach
Establish a small group of people with the process/product knowledge, allocated time, authority and skill in the required technical disciplines to solve the problem and implement corrective actions. The group must have a designated champion.

(2) Describe the problem
Specify the internal/external customer problem by identifying in quantifiable terms the who, what, when, where, why, how, how many (5 W2H) for the problem.

(3) Implement and verify interim (containment) actions
Define and implement containment actions to isolate the effect of problem from any internal/external customer until corrective action is implemented. Verify the effectiveness of the containment action.

(4) Define and verify root causes
Identify all potential causes which could explain why the problem occurred. Isolate and verify the root cause by testing each potential cause against the problem description and test data. Identify alternative corrective actions to eliminate root cause.

(5) Verify corrective actions
Through pre-production test programmes quantitatively confirm that the selected corrective actions will resolve the problem for the customer, and will not cause undesirable side effects. Define contingency actions, if necessary, based on risk assessment.

(6) Implement permanent corrective actions
Define and implement the best permanent corrective actions. Choose ongoing controls to ensure the root cause is eliminated. Once in production, monitor the long term effects and implement contingency actions, if necessary.

(7) Prevent recurrence
Modify the management systems, operating systems, practices and procedures to prevent recurrence of this and all similar problems.

(8) Congratulate your team
Recognize the collective efforts of the team.

These steps do not have to be followed in the order given. They can vary with each problem. For example, by the time a problem is reported and a team formed, the interim action may have been taken by the foreman or operator.

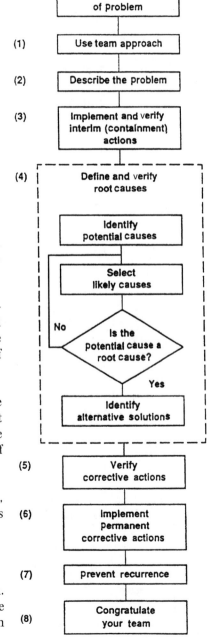

PROBLEM SOLVING PROCESS

PROBLEM SOLVING DISCIPLINES — CONCERN ANALYSIS REPORT

Status Date:

Concern/RFR No.	CS No	Concern Title		Date Opened	Assigned To
(2) Describe Concern		Date Completed	Concern Code	Vehicle	Engine/Trans/Axle
				Build Date	Build Date
				(1) Team/Activity—Phone	

(3) Containment/ (6) Corrective Actions (Des Manut B&A Ser)	% Effect	Effective Date	
		Supply	B&A

PCR SREA TSL/TSB Number—

(4) Define Root Causes	Root Cause Code	Commit Date	Completion Date
		Transfer Code	% Contribution (each cause)

If additional analysis is required indicate completion dates.

(5) Verification of Containment/Corrective Actions

(7) Action to Prevent Recurrence

	Containment Action Date			Corrective Action Dates	
	Committed	Completion		Committed	Completion
Define			Define		
Verify			Verify		
Implement			Implement		

Reporting Engineer	Date Status/Closed		Concurrence Supervisor	Date

(8) CONGRATULATE YOUR TEAM

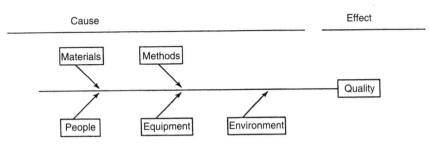

Figure A2.1 Cause and effect

A2.3 CAUSE AND EFFECT DIAGRAMS

The cause and effect diagram was developed to represent the relationship between some 'effect' and all possible 'causes' influencing it. This relationship is shown in Figure A2.1.

A2.3.1 Making cause and effect diagrams

The possible causes of dispersion in the quality characteristic (effect) are arranged in the cause and effect diagram in such a way that all relationships are clearly shown.

Step 1. Decide the quality characteristic (e.g. wobble during machine rotation). This is something you may want to improve and control. In this case you may have found that most of the defectives were due to wobble during rotating. To eliminate this wobble, you must identify its causes.

Step 2. Construct the diagram by boxing in the quality characteristic (e.g. wobble) on the right side. Draw an arrow extending from the left to the box. Then, write the main factors which may be causing the wobble directing a branch arrow to the main arrow. Group the main possible causes of dispersion into such items as materials, equipment, methods, people and environment. Each individual group will form a branch. See Figure A2.2.

Step 3. Onto each of these branches, write in the detailed factors which may be regarded as the causes. And onto each of these, write in even more detailed causes. See Figure A2.3.

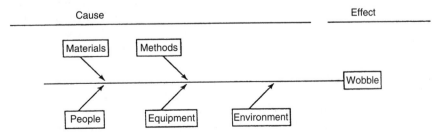

Figure A2.2 Cause and effect with 'wobble'

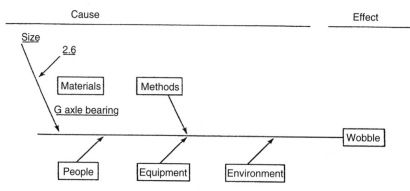

Figure A2.3   Cause and effect with more detail

If you pursue the following thought process, you will eventually identify all possible causes of the problem.

(1)   Why do production process defects occur?
Because of machine wobble (dispersion), therefore, machine wobble is a quality characteristic (effect).

(2)   Why does the machine wobble (dispersion) occur?
Because of the dispersion in the materials, 'Materials' is illustrated on the diagram as a branch.

(3)   Why does dispersion in the materials occur?
Because of the dispersion in the G axle bearing, the G axle bearing becomes a cause.

(4)   Why does the dispersion in the G axle bearing occur?
Because of the dispersion in the size of the G axle bearing, size becomes a cause.

(5)   Why does the dispersion in the size of the G axle bearing occur?
Because of the dispersion at the 2.6 mm point, the 2.6 mm point becomes a cause.

In this way you add to a cause and effect diagram until it clearly shows the causes of the dispersion.

Step 4.   Finally check to make certain that all factors that may be causing dispersion are included in the diagram. If they are, and the relationships of causes to effect are properly illustrated, then the diagram is complete.

Step 5.   From this well-defined diagram of possible causes, identify and select the most likely causes for further analysis. When examining each cause, search for things that have changed, and deviations from the norm or patterns. Seek to cure the causes and not the symptoms of the problem.

A2.3.1   SUMMARY

Your aim is to get results. A cause and effect diagram is an aid to clearly show the relationship between the causes and effect of a problem. After the causes of the problem have been identified, corrective actions can be taken.

A2.4   POTENTIAL FAILURE MODE AND EFFECTS ANALYSIS

POTENTIAL FAILURE MODE AND EFFECTS ANALYSIS (FMEA)

AN INSTRUCTION MANUAL
Revised September 1988

DESIGN FMEA

INTRODUCTION

A Design potential FMEA is an analytical technique utilized by Product Engineers as a means to assure that, to the extent possible, potential failure modes and their associated causes have been considered and addressed. End items, along with every related subassembly and detail part, should be evaluated. In its most rigorous form, an FMEA is a summary of the engineer's thoughts (including an analysis of items that could go wrong based on experience and past concerns) as a component or system is designed. This systematic approach parallels and formalizes the mental discipline that an engineer normally goes through in any design process.

The Design potential FMEA supports the design process in reducing the risk of failures by:

- Aiding in the objective evaluation of design requirements and design alternatives.

- Increasing the probability that potential failure modes and their effects on system operation have been considered in the design/development process.

- Providing additional information to aid in the planning of thorough and efficient design test and development programs.

- Developing a list of potential failure modes ranked according to their effect on the "customer," thus establishing a priority system for design improvements and development testing.

- Providing an open issue format for recommending and tracking risk reducing actions.

- Providing future reference to aid in analyzing field concerns, evaluating design changes, and developing advanced designs.

Customer Defined

The definition of "CUSTOMER" for a Design FMEA should normally be seen as the "END USER." However, the design engineer's customers are also the design engineers of the vehicle or higher level assemblies, and/or the manufacturing process engineers in activities such as Manufacturing, Assembly, and Service, who require a clearly defined, manufacturable and service-friendly design.

When fully implemented, the FMEA discipline requires a Design FMEA for all new parts, changed parts, and carryover parts in new applications. It is initiated by an engineer from the responsible design activity, which for a proprietary design (black/gray box) may be the supplier.

Team Effort

During the preparation of the Design potential FMEA, the responsible engineer is expected to seek input from such areas as Manufacturing, Quality, and Service, as well as from the Design area responsible for next assembly. In addition, for any black/gray box items, the responsible Ford Product Engineer should be consulted. The FMEA should be a catalyst to stimulate the interchange of ideas between the functions affected and thus promote a team approach.

The Design FMEA is a living document and should be initiated at or by design concept finalization, be continually updated as changes occur throughout the phases of product development, and be fundamentally completed along with the final drawings.

DESIGN FMEA

INTRODUCTION (Continued)

Team Effort (cont'd)

The Design FMEA addresses the design intent and assumes the design will be manufactured/assembled to this intent. Potential failure modes/causes which can occur during the manufacturing or assembly process should not be included in a Design FMEA, as their identification, effect and control are covered by the Process FMEA. However, potential manufacturing/assembly concerns known by the design engineer should be conveyed to the manufacturing/assembly source, using means such as team meetings.

The Design FMEA does not rely on process controls to overcome potential weaknesses in the design, but it does take the technical/physical limits of a manufacturing/assembly process into consideration, e.g.:

- necessary mold drafts
- limited surface finish
- assembling space/access for tooling
- limited hardenability of steels
- process capability

DEVELOPMENT OF A DESIGN FMEA

The design engineer has at his or her disposal a number of documents that will be of use in preparing the Design potential FMEA. The process begins by developing a listing of what the design is expected to do, and what it is expected not to do, i.e., the design intent. Customer wants and needs, as may be determined from sources such as Quality Function Deployment (QFD), Corporate Vehicle Requirements Manual (CVRM), known product requirements and/or manufacturing wants, should be incorporated. The better the definition of the wanted characteristics, the easier it is to identify potential failure modes for corrective action.

In order to facilitate documentation of the analysis of potential failures and their consequences, form No. 1695, shown in the Design FMEA Appendix, was developed for use in Ford Motor Company.

Application of the form is described below; points are numbered according to the numbers encircled on the form (as shown in Appendix).

1) Subsystem/Name

Enter the number and name of the subsystem.

2) Design Responsibility

Enter the name of the area responsible for the design of the component, assembly or system.

3) Other Areas Involved

Enter any areas/departments or organizations affected by or involved in the design or function of the component(s).

DESIGN FMEA

DEVELOPMENT OF A DESIGN FMEA (Continued)

4) Suppliers and Plants Affected

List any supplier(s) or manufacturing plants involved in the design or manufacture of components or assemblies being analyzed.

5) Model Years/ Vehicle(s)

Enter the model year and all car lines that will utilize the design being analyzed.

6) Scheduled Engineering Release Date

Indicate the date the component or assembly is scheduled to be released.

7) Prepared By

Indicate the name, telephone number, address and company of the engineer preparing the FMEA.

8) FMEA Date

Show the date the original FMEA was compiled, and then show the latest FMEA revision date.

9) Part Name and Number/Function

Enter the name and number of the part or assembly being analyzed. Use suffixes, change letters and/or Concern Report/Change Request (CR/CR) numbers, as appropriate. Prior to initial release, experimental part numbers should be used. In the space below the part name and number, indicate as concisely as possible the function of the part or assembly being analyzed. Where the assembly has numerous functions with different potential modes of failure, it may be desirable to list the functions separately.

10) Potential Failure Mode

Potential Failure Mode is defined as the manner in which a part or assembly could potentially fail to meet the design intent, performance requirements, and/or customer expectations. The potential failure mode may also be the cause of a potential failure mode in a higher level assembly, or be the effect of one in a lower level part.

List each potential failure mode for the particular part and part function. The assumption is made that the failure could occur, but will not necessarily occur. A recommended starting point is a review of past FMEAs, test reports, quality, warranty, durability and reliability concerns, things-gone-wrong, concern reports, and group "brainstorming" on similar components.

Potential failure modes that would only occur under certain operating conditions (i.e., hot, cold, wet, dry, dusty, etc.) and under certain usage conditions (i.e., above average mileage, rough terrain, only city driving, etc.) shall be considered. Typical failure modes could be:

Cracked	Sticking
Deformed	Short Circuited (electrical)
Worn	Open Circuited (electrical)
Corroded	Oxidized
Loosened	Vibrating
Leaking	Fractured

Note: Potential failure modes should be described in "physical" or technical terms, not as a symptom noticeable by the customer.

DESIGN FMEA

DEVELOPMENT OF A DESIGN FMEA (Continued)

11) Potential Effect(s) of Failure

Potential Effects of Failure are defined as the effects of the failure mode on the customer.

Describe the effects of the failure in terms of what the customer might notice or experience. These should always be stated in terms of vehicle or system performance. Typical failure effects could be:

Noise	Rough
Erratic Operation	Excessive Effort Required
Inoperative	Unpleasant Odor
Unstable	Operation Impaired
Intermittent Operation	Draft
Vehicle Control Impaired	Poor Appearance

If the effect of failure could potentially affect safe vehicle operation, or involves potential noncompliance with government regulations, it must be so indicated, e.g., "may not comply with FMVSS #XXX."

12) Severity

Severity is an assessment of the seriousness of the effect (listed in the previous column) of the potential failure mode to the next assembly, the vehicle, or the customer. Severity applies to the effect and to the effect only. A reduction in Severity Ranking index can be effected only through a design change. Severity should be estimated on a "1 to 10" scale.

Evaluation Criteria:

Severity of Effect	Ranking
Minor: Unreasonable to expect that the minor nature of this failure would cause any real effect on the vehicle or system performance. Customer will probably not even notice the failure.	1
Low: Low severity ranking due to nature of failure causing only a slight customer annoyance. Customer will probably only notice a slight deterioration of the system or vehicle performance.	2 3
Moderate: Moderate ranking because failure causes some customer dissatisfaction. Customer is made uncomfortable or is annoyed by the failure (e.g., engine misfire, compressor rumble, sunroof leak). Customer will notice some subsystem or vehicle performance deterioration.	4 5 6
High: High degree of customer dissatisfaction due to the nature of the failure such as an inoperable vehicle (e.g., engine fails to start) or an inoperable convenience subsystem (e.g., air conditioning system, power sunroof). Does not involve vehicle safety or noncompliance to government regulations.	7 8
Very High: Very high severity ranking when a potential failure mode affects safe vehicle operation and/or involves noncompliance with government regulations.	9 10

13) Critical Characteristics (∇)

Critical Characteristics should be identified by entering an inverted delta (∇) in this column. Determine if the inverted delta should be assigned by following the flow chart in Engineering Practice 5 (shown in Reference section of this manual) whenever the severity ranking is 9 or 10 and the occurrence (Step 15) and detection (Step 17) are both greater than 1.

DESIGN FMEA

DEVELOPMENT OF A DESIGN FMEA (Continued)

14) Potential Cause(s) of Failure

Potential Cause of Failure is defined as an indication of a design weakness, the consequence of which is the failure mode.

List, to the extent possible, every conceivable failure cause assignable to each failure mode. The causes should be listed as concisely and completely as possible so that remedial efforts can be aimed at pertinent causes. Typical failure causes could be:

Incorrect Material Specified	Inappropriate Material Specified
Incorrect Assembling Instruction	Inadequate Design Life Assumption
Incorrect Torque Specified	Over-stressing
Insufficient Lubrication Capability	Overload
Permissible Material Impurity Level	Inadequate Maintenance
Poor Mold Form	Instructions
Incorrect Material Thickness	Imbalance
Specified	Poor Environment Protection

15) Occurrence

Occurrence is the likelihood that a specific cause (listed in the previous column) will result in the failure mode. The occurrence ranking number has a meaning rather than a value. Removing or controlling one or more of the causes of the failure mode through a design change is the only way a reduction in the occurrence ranking can be effected.

Estimate the likelihood of the occurrence of potential failure modes on a "1 to 10" scale. In determining this estimate, questions such as the following should be considered:
- How adequate is the proposed Design Verification (DV) program?
- Is part carryover or similar to previous level part or assembly?
- How significant are changes from previous level part or assembly?
- Is part radically different from previous level part?
- Is part completely new?
- What are the environmental changes?
- What is the service history/field experience with similar parts or assemblies?

The following occurrence ranking system should be used to ensure consistency. The "Design Life Possible Failure Rates" are based on the number of failures which are anticipated during the design life of the part or assembly.

Evaluation Criteria:

Probability of Failure	Ranking	Design Life Possible Failure Rates
Remote: Failure is unlikely.	1	<1 in 10^6
Low: Relatively few failures.	2	1 in 20000
	3	1 in 4000
Moderate: Occasional failures.	4	1 in 1000
	5	1 in 400
	6	1 in 80
High: Repeated failures.	7	1 in 40
	8	1 in 20
Very High: Failure is almost inevitable.	9	1 in 8
	10	1 in 2

DESIGN FMEA

DEVELOPMENT OF A DESIGN FMEA (Continued)

16) Design Verification (DV)

List all current DVs which are intended to prevent the design cause(s) of potential failure from occurring or are intended to detect the design cause(s) of the potential failure or the resultant failure mode.

Current DVs (e.g., road testing, design reviews, mathematical studies rig/lab testing, feasibility reviews, prototype tests, fleet testing) are those that have been or are being used with the same or similar designs. The initial occurrence and detection rankings will be based on these DV controls, considering the representatives of the prototypes and models being used. The DV controls listed should be directly related to the prevention or detection of specific causes of failure.

If any other specific DVs, such as those for a radically new design, are necessary, they should be listed in the Recommended Action column.

17) Detection

Detection is an assessment of the ability of the proposed design program (listed in the previous column) to identify a potential design weakness before the part or assembly is released for production. In order to achieve a lower ranking, generally the planned verification program has to be improved.

Evaluation Criteria:

Likelihood of Detection by D V Program	Ranking
Very High: D V Program will almost certainly detect a potential design weakness	1
	2
High: D V Program has a good chance of detecting a potential design weakness	3
	4
Moderate: D V Program may detect a potential design weakness	5
	6
Low: D V Program not likely to detect a potential design weakness	7
	8
Very Low: D V Program probably will not detect a potential design weakness	9
Absolute Certainty of Non-Detection: D V Program will/can not detect a potential design weakness, or there is no D V Program	10

18) Risk Priority Number (RPN)

The Risk Priority Number is the product of the occurrence, severity and detection rankings. This value should be used to rank order the concerns in the design (e.g., in Pareto fashion). In themselves, RPNs have no other value or meaning.

DESIGN FMEA

DEVELOPMENT OF A DESIGN FMEA (Continued)

19) Recommended Action(s)

When the failure modes have been rank ordered by RPN, corrective action should be first directed at the highest ranked concerns and critical items. The intent of any recommended action is to reduce any one or all of the occurrence, severity and/or detection rankings. An increase in design verification actions will result in a reduction in the detection ranking only. A reduction in the occurrence ranking can only be effected by removing or controlling one or more of the causes of the failure mode through a design revision. Only a design revision can bring about a reduction in the severity ranking. Actions such as the following could be considered.

- Design of Experiments (particularly when multiple or interactive causes are present)
- Revised Test Plan
- Revised Design
- Revised Material Specification

If no actions are recommended for a specific cause, then this should be indicated.

20) Area/Engineer Responsible (for the Recommended Action)

Enter the area and engineer responsible for the recommended action as well as the target completion date.

21) Actions Taken:

After an action has been completed, enter a brief description of the actual action and effective or completion date.

22) Resulting RPN:

After the corrective action has been identified, estimate and record the resulting occurrence, severity, and detection rankings. Calculate and record the resulting RPN. If no actions are taken, leave the "Resulting RPN" and related ranking columns blank.

All Resulting RPN(s) should be reviewed and if further action is considered necessary repeat Steps 19 through 22.

Follow-Up: The design engineer is responsible for assuring that all actions recommended have been implemented or adequately addressed. The FMEA is a living document and should always reflect the latest design level, as well as the latest relevant actions, including those occurring post Job #1.

The design engineer has several means of assuring that concerns are identified and that recommended actions are implemented. They include the following:

- Engineering drawings and specifications: These show design changes, critical characteristics and supplier and/or manufacturing test requirements.
- Sign-off responsibility for Manufacturing Installation Drawings: Installation Drawings specify such items as critical torques, assembly sequences and part positioning. Verify that assembly concerns identified by the Design FMEA are addressed by the installation drawings.
- Review of Process FMEAs and Manufacturing Control Plans.

POTENTIAL FAILURE MODE AND EFFECTS ANALYSIS (DESIGN FMEA)

Page 1 of 8

(1) Subsystem/Name: 01.03 Body Closures
(2) Design Responsibility: Body Engineering
(3) Other Areas Involved: Car Product Dev., Manufacturing, B&A
(4) Supplier and Plants Affected: Dalton, Fraser, Henley Assembly Plants
(5) Model Year/Vehicle(s): 199x /Lion 4dr/Wagon
(6) Engineering Release Date: 9x 03 01
(7) Prepared By: A. Tate — x64 12 — Ford Body Eng.
(8) FMEA Date (Org): 8x 03 22 (Rev) 8x 08 14

(22) Action Results

(9) Part Name & Number / Part Function	(10) Potential Failure Mode	(11) Potential Effect(s) of Failure	(12) S	(13) Class	(14) Potential Cause(s) of Failure	(15) O	(16) Design Verification	(17) D	(18) RPN	(19) Recommended Action(s)	(20) Area/Individual Responsible & Completion Date	(21) Actions Taken	S	O	D	RPN
Front door L.H. H8HX-0000-A • Ingress to and egress from vehicle • Occupant protection from weather, noise, and side impact • Support/anchorage for door hardware including mirror, hinges, latch and window regulator • Provide proper surface for appearance items — paint and soft trim	Corroded interior lower door panels	• Deteriorated life of door leading to: • Unsatisfactory appearance due to rust through paint over time • Impaired function of interior door hardware	7	Y	• Upper edge of protective wax application specified for inner panels is too low	6	Vehicle general durability tests veh. T-118 T-109 T-301	7	294	Add laboratory accelerated corrosion testing	A. Tate — B&CE 8x 09 30	Based on test results (Test No. 1481) upper edge spec raised 125mm	7	2	2	28
					• Insufficient wax thickness specified	4	Vehicle general durability testing — as above	7	112	• Add laboratory accelerated corrosion testing • Conduct Design of Experiments (DOE) on wax thickness	Combine w/test for wax upper edge verification; A. Tate — B&CE 9x 01 15	Test results (Test No. 1481) show specified thickness is adequate	7	2	2	28
					• Inappropriate wax formulation specified	2	Physical and Chem Lab test — Report No. 1265	2	28	None						
					• Entrapped air prevents wax from entering corner/edge areas	5	Design aid investigation with non-functioning spray head	8	280	Team evaluation using production spray equipment and specified wax	B&CE & B&A 8x 11 15	Based on test, additional holes will be provided in affected area	7	1	3	21
					• Wax application plugs door drain holes	3	Laboratory test using "worst case" wax application and hole size	1	21	None						
					• Insufficient room between panels for spray head egress	4	Drawing evaluation of spray head access	4	112	Team evaluation using design aid buck and spray head	B&CE & B&A 8x 09 15	Evaluation showed adequate access	7	1	1	7

NOTE Reference — xxxxxxx xxxxxx xxxx xxxxxxxxxxxxxxx xxxxxx xxxxx Instruction Manual xxxxxx xxxxx

xxx 1695 xxxx

A2.5   ISO 9000:2000

The complete standard is available from the British Standards Institute [1]. This Appendix identifies the most significant changes introduced with ISO 9000:2000, and lists the section headings of the revised standard. There is a brief review of ISO 9000:2000 in Chapter 4, section 4.5.

A2.5.1   Changes from ISO 9000:1994

The British Standards Institution (BSI) CD entitled *Quality Management 2000 TOOLKIT* [2] identifies, among others, the following changes from version 1994 to version 2000:

- ISO 9000:2000 replaces ISO 9001:1994 *and* ISO 9002:1994 *and* ISO 9003:1994. Organizations requiring only the reduced coverage of ISO 9002/3 can secure this by excluding sections of the new standard.
- Previously there was a 'shopping list' of 20 clauses, with rather tenuous links. There are now five sections that show clear links with each other to provide a logical sequence of requirements that are in line with generic organizational structures. (Numbers 4, 5, 6, 7 and 8 in the Contents list shown below.)
- The term 'quality assurance' is no longer used, and is replaced by 'quality management'.
- The scope statement is now more positive and speaks of achieving customer satisfaction through application of the system as opposed to preventing nonconformity.
- There is a clear indication of the process nature of the standard, and that it is applicable to all organizations no matter what size or industry sector.
- There is a consistent message that top management is responsible for the system.
- The management review clause has been considerably expanded. It asks for action plans to be considered to ensure that outputs from the review are acted on.
- There is a completely new section on resource management, which includes requirements dealing with staff competencies.
- Documentation requirements are more flexible. Procedures are not essential when those carrying out an activity are deemed to be competent.
- Preventive actions, to stop problems arising, are stressed, rather than corrective actions to be taken when problems have arisen.
- Quality objectives are now related to continual improvement and must be measurable.
- A new section on planning of realization processes makes the user think about what processes they need and how they will be controlled.
- Customer-related processes now replace 'contract review', and include a requirement for the organization to understand the implied needs of the customer.

A2.5.2   Contents of ISO 9000:2000

1. Scope
    1.1   General
    1.2   Application
2. Normative reference
3. Terms and definitions
4. Quality management system
    4.1   General requirements
    4.2   Documentation requirements
5. Management responsibility
    5.1   Management commitment
    5.2   Customer focus
    5.3   Quality policy
    5.4   Planning
    5.5   Responsibility, authority and communication
    5.6   Management review
6. Resource management
    6.1   Provision of resources
    6.2   Human resources
    6.3   Infrastructure
    6.4   Work environment
7. Product realization
    7.1   Planning of product realization
    7.2   Customer-related processes
    7.3   Design and development
    7.4   Purchasing
    7.5   Production and service provision
    7.6   Control of monitoring and measuring devices
8. Measurement, analysis and improvement
    8.1   General
    8.2   Monitoring and measurement
    8.3   Control of non-conforming product
    8.4   Analysis of data
    8.5   Improvement

REFERENCES

1. The standard can be ordered from BSI Customer Services, 389 Chiswick High Road, London W4 4AL, telephone: 020 8996 9001, or through the website shown below.
2. The CD is available from British Standards Institution. Contact details are on www.bsi-global.com.

Appendix 3

Case Study: Developing a Network

A3.1 INTRODUCTION

The objective of this case study is to illustrate the manner in which a *precedence list* can be developed for the purposes of producing a first-draft network and carrying out *critical path analysis*. The analysis itself is not carried out. However, for the precedence list that is eventually developed there is only one possible solution, and the duration of the project for that solution and some comments on the solution are given at the end of the case study for those who wish to complete the whole exercise. The methods of critical path analysis are described in Chapter 18.

A3.2 GENERAL DATA

An irrigation scheme is to be built. It will consist of a pump house on a river and 17 km of 500 mm diameter steel pipeline to which are connected, at 4 km intervals, 3 km lengths of 100 mm diameter PVC pipeline, which feed surface networks of 25 mm diameter PVC pipes. The 500 mm and 100 mm pipeline are placed in trenches that are then backfilled. The 25 mm surface pipes are connected by means of standpipes to the 100 mm pipes.

At one location the 500 mm pipeline crosses a valley. This is achieved by means of a 1 km long bridge that will be built as part of the project. The bridge will also provide a single-lane vehicle crossing for local traffic.

Figure A3.1 shows a schematic diagram of the project, Figure A3.2 provides details of the pump house and Figure A3.3 shows details of the bridge.

A3.3 ACTIVITIES LISTS

The main activities are listed in Tables A3.1 to A3.3. Activities are not necessarily listed in chronological order.

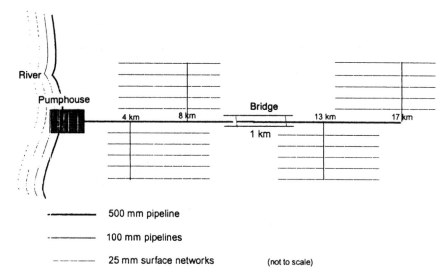

500 mm pipeline

100 mm pipelines

25 mm surface networks   (not to scale)

Figure A3.1   Irrigation scheme

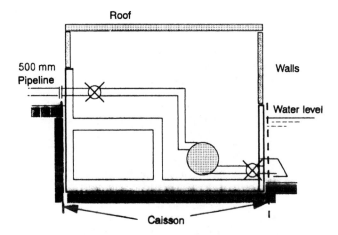

Figure A3.2   Pump house

A3.4   ACTIVITIES DESCRIPTIONS

Some of the activities may require a little explanation. This is provided as briefly as possible.

Pump house

The pumps are to be installed below river surface level. This is achieved by constructing in the river a watertight caisson from which the water is pumped. A concrete substructure chamber is then built in which the pumps are installed. When the caisson is dismantled the pump chamber can be flooded and the system

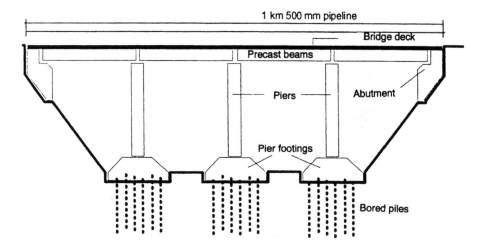

Figure A3.3   Bridge

Table A3.1   Pump house activities

Pump house	Duration (days)
PH1 Order pumps and deliver	85
PH2 Order pipework and valves and deliver	50
PH3 Construct caisson	10
PH4 Excavate within caisson	10
PH5 Concrete substructure	15
PH6 Install pumps, pipework and valves	20
PH7 Dismantle caisson	5
PH8 Build walls	10
PH9 Build roof	10
PH10 Test and commission	10

Table A3.2   Bridge activities

Bridge	Duration (days)
B1 Order precast beams and deliver	50
B2 Bore one set of piles	20
B3 Construct one footing	10
B4 Construct one pier	20
B5 Construct one abutment	15
B6 Place one span of precast beams	5
B7 Lay bridge deck	7

Table A3.3   Pipeline activities

Pipelines	Duration (days)	Work rate (km/day)
500 mm		
P51 Order 17 km of pipes and fittings and deliver	50	
P52 Clear 16 km route for pipeline		1
P53 Dig trench		0.5
P54 Place pipe in trench and join		0.5
P55 Backfill around pipe, and complete to surface		1
P56 Place 1 km length across bridge	2	
100 mm		
P11 Order 12 km of pipes and fittings and deliver	40	
P12 Clear route for 3 km pipeline (×4)		1
P13 Dig trench for 3 km pipeline (×4)		0.5
P14 Place pipe in trench and join (includes all standpipes for connecting to 25 mm pipes) (×4)		0.25
P15 Backfill around pipe, and complete to surface (×4)		1
P16 Join 100 mm pipe to 500 mm pipe (each)	3	
25 mm		
PN1 Order all pipes and fittings, and deliver	30	
PN2 Clear area for each of the four surface networks (×4)	20	
PN3 Install each surface network (×4)	60	
PC1 Commission and hand over all pipelines	10	

tested. The pumps have to be installed before the roof is constructed as they are lifted into position using a crane.

Bridge

The construction of each footing and pier is an independent operation. A span of precast beams cannot be placed until the piers and abutments for supporting that span are complete. The bridge deck cannot be laid until all precast beams have been placed.

Pipelines

Work on the different pipelines and on the surface networks are independent operations that can be carried out in parallel or series. At each connection between the 500 mm and 100 mm pipes a short length of trench will be left open until the pipes have been joined. The activity 'Join 100 mm pipe to 500 mm pipe' includes backfilling around the completed connection. For the 500 mm and 100 mm pipe lines, *backfilling* will follow *placing* (and joining) the pipes at a practical distance, and *placing* will follow *trenching* in a similar fashion.

A3.5   DEVELOPING THE PRECEDENCE LIST

The task is to determine the linkages between activities. Two questions will need to be asked of each activity in turn. They are:

(1)   Which activities must be completed/started to allow the given activity to proceed and finish?
(2)   Should there be any minimum duration prescribed for the linkage?

The linkages between activities are determined by:

- logic;
- resources;
- method.

Logic is the most important determinant of precedence. A pier cannot be constructed until its foundation is complete. That is logic. There is no logical linkage between one pier and another. A decision that one pier should precede another can only be justified on the grounds of *resources*, for example because only one team of workers and/or equipment is available for pier construction, or *method*, because, for example, the equipment used for the construction of one pier would interfere with the equipment and operations on the other pier.

It should be borne in mind that the objective of the precedence list is to produce a *first-draft network*. The fewer conditions that are imposed in the form of dependencies (linkages) between activities, the greater the freedom for the contractor in planning the operations. Links that are determined on the basis 'it would be a good idea to complete *X* before *Y*' should be avoided because the consequence of this is that, logically, *Y* cannot proceed until *X* is complete. What would happen if *X* was delayed indefinitely because of supply problems? Obviously, in practice, the contractor would ignore the network linkage and precede with *Y*, but a properly constructed network should work just as well when activities are delayed as when work proceeds according to programme. Indeed, an essential purpose of a network is to provide information on new critical paths and on which activities might need to be accelerated when activities are held up, so each link should be an expression of necessary dependency not of arbitrary choice. This should lead to a greater degree of float on a greater number of activities, allowing the contractor more options when producing the programme bar chart. In developing the precedence list, Question 1 and Question 2 will be asked of each activity in turn, in the order presented. Where the answer to Question 1 is that no activity needs be completed/started to allow the activity to proceed, a preceding activity, 'START', of zero duration, will be entered.

The analyses are presented in Tables A3.4 to A3.7.

Pump house

Table A3.4

Activity	Answer to Question 1	Answer to Question 2
PH1 Order pumps etc.	START	nil
PH2 Order pipework etc.	START	nil
PH3 Construct caisson	START	nil
PH4 Excavation in caisson	PH3 – logic	nil
PH5 Concrete substructure	PH4 – logic	nil
PH6 Install pumps etc.	PH1 and PH2 – logic	nil
	PH5 – logic	7 days to cure concrete
PH7 Dismantle caisson	PH5 and PH6 – method	nil
PH8 Build walls	PH5 – logic	nil
PH9 Build roof	PH8 – logic	nil
	PH6 – method	nil
PH10 Test etc.	PH6 – logic	nil
	PH9 – optional	nil

Bridge

Before a precedence list can be developed for the bridge certain decisions have to be made. Should it be assumed that there are sufficient resources for working on the piers concurrently or not? In practice this would be determined by a number of factors, including the anticipated size of the labour force, the types of

Table A3.5

Activity	Answer to Question 1	Answer to Question 2
Bridge		
B1 Order beams etc.	START	nil
B21 Bore piles 1	START	nil
B22 Bore piles 2	B21 – resources	nil
B23 Bore piles 3	B22 – resources	nil
B31 Construct Footing 1	B21 – logic	nil
B32 Construct Footing 2	B22 – logic	nil
B33 Construct Footing 3	B23 – logic	nil
B41 Construct Pier 1	B31 – logic	nil
B42 Construct Pier 2	B32 – logic	nil
B43 Construct Pier 3	B33 – logic	nil
B51 Abutment 1	START	nil
B52 Abutment 2	START	nil
B61 Precast span 1	B1, B51 and B41 – logic	nil
B62 Precast span 2	B61 – method	nil
	B41 and B42 – logic	nil
B63 Precast span 3	B62 – method	nil
	B42 and B43 – logic	nil
B64 Precast span 4	B63 – method	nil
	B43 and B52 – logic	nil
B7 Bridge deck	B61, B62, B63, B64 – method	nil

equipment to be used, the materials lead times and delivery rates and the ground conditions. However, maintaining the view that the fewer dependencies that are imposed the better, it is assumed for the purpose of this case study that sufficient resources could be provided to allow independent and parallel operations at any of the pier locations, except for boring the piles, where it is assumed that a single subcontractor will carry out this work using a single team that will move across the site from Pier 1 to Pier 2 to Pier 3. It may be the case, once the whole network has been developed, that there will be sufficient float on the independent pier operations to allow them to be constructed in series. Instead of predetermining the position, this approach allows network analysis to provide a solution. For the precast beams it is assumed that sequential working will take place from span to span.

In Table A3.5, B21 indicates Activity B2 at Pier 1, B32 indicates Activity B3 at Pier 2, and so on.

Pipelines

Similarly, decisions need to be made with respect to the methods of constructing the pipelines. The question of resources must be considered.

For the construction of the 16 km of 500 mm pipeline, there are many alternatives:

- starting from one end and proceeding to the other;

Table A3.6

Activity	Answer to Question 1	Answer to Question 2
P51 Order 500 mm	START	nil
P52 Clear 500 mm	START	nil
P53 Dig 500 mm	P52 – logic	S–S (1 day) this allows P52 a 1 km start
		F–F (2 days) this keeps P53 1 km behind P52
P54 Place 500 mm	P51 – logic	nil
	P53 – logic	S–S (2 days) this allows P53 a 1 km start
		F–F (2 days) this keeps P54 1 km behind P53
P55 Backfill 500 mm	P54 – logic	S–S (2 days) this allows P54 a 1 km start
		F–F (1 day) this keeps P55 1 km behind P54
P56 500 mm on bridge	B7 – logic	7 days – to cure concrete
	P54 – method	S–S (16 days) assumes 8 km of pipeline have been placed
P11 Order 100 mm	START	nil
P12 Clear 100 mm	START	nil
P13 Dig 100 mm	P12 – logic	S–S (1 day) this allows P12 a 1 km start
		F–F (2 days) this keeps P13 1 km behind P12
P14 Place 100 mm	P11 – logic	nil
	P13 – logic	S–S (2 days) this allows P13 a 1 km start
		F–F (4 days) this keeps P14 1 km behind P13
P15 Backfill 100 mm	P14 – logic	S–S (4 days) this allows P14 a 1 km start
		F–F (1 day) this keeps P15 1 km behind P14
P16 Join 100 mm to 500 mm	P14 and P54 – method (assumes that a separate team does this work)	nil

Table A3.7

Activity	Answer to Question 1	Answer to Question 2
PN1 Order 25 mm	START	nil
PN21 Clear 25 mm	START	nil
PN22 Clear 25 mm	START	nil
PN23 Clear 25 mm	START	nil
PN24 Clear 25 mm	START	nil
PN31 Place 25 mm	PN1	nil
	PN21	for PN3 following PN2 S–S (5 days) and F–F (10 days), this allows PN3 to start after one quarter of PN2 has been completed and keeps PN3 behind PN2 by one quarter of the area
PN32 Place 25 mm	PN1	nil
	PN22	S–S (5 days), F–F (10 days)
PN33 Place 25 mm	PN1	nil
	PN23	S–S (5 days), F–F (10 days)
PN34 Place 25 mm	PN1	nil
	PN24	S–S (5 days), F–F (10 days)
PC1	PN31 to PN34	nil
	P15, P16	nil
	P55, P56	nil

- working on the sections either side of the bridge concurrently;
- having three or more operational sections concurrently.

For the construction of the four 3 km lengths of 100 mm pipeline, the alternatives could be:

- using one team to proceed from one 3 km length to another;
- working concurrently on all four lengths;
- using two teams, each to work on two 3 km lengths.

For the four 25 mm surface networks, similar alternatives to those above present themselves. For the purposes of this case study the following assumptions have been made:

- 500 mm pipeline – one team.
- 100 mm pipeline – one team.
- 25 mm networks – four teams could be available if necessary.

In Table A3.7, PN21 indicates Activity PN2 at Network 1, PN32 indicates Activity PN3 at Network 2, and so on.

A3.6 FINALIZING THE PRECEDENCE LISTS

One further check must be made. Which activities in the left-hand columns of Tables A3.4 to A3.7 do not appear in the middle column? All activities except those that are the final tasks of the project must appear at least once as a preceding

Table A3.8   Finalized precedence list

Activity	Duration days	Precursor activity	Link type and duration
PH1	85	START	
PH2	50	START	
PH3	10	START	
PH4	10	PH3	
PH5	15	PH4	
PH6	20	PH1, PH2	
		PH5	F–S (7)
PH7	5	PH5, PH6	
PH8	10	PH5	
PH9	10	PH8, PH6	
PH10	10	PH6, PH9	
B1	50	START	
B21	20	START	
B22	20	B21	
B23	20	B22	
B31	10	B21	
B32	10	B22	
B33	10	B23	
B41	20	B31	
B42	20	B32	
B43	20	B33	
B51	15	START	
B52	15	START	
B61	5	B1, B51, B41	
B62	5	B61, B41, B42	
B63	5	B62, B42, B43	
B64	5	B63, B43, B52	
B7	7	B61, B62, B63, B64	
P51	50	START	
P52	16	START	
P53	32	P52	S–S(1), F–F(2)
P54	32	P51	
		P53	S–S(2), F–F(2)
P55	16	P54	S–S(2), F–F(1)
P56	2	B7	
		P54	S–S(16)
P11	40	START	
P12	12	START	
P13	24	P12	S–S(1), F–F(2)
P14	48	P11	
		P13	S–S(2), F–F(4)
P15	12	P14	S–S(4), F–F(1)
P16	12	P14, P54	
PN1	30	START	
PN21	20	START	
PN22	20	START	
PN23	20	START	
PN24	20	START	
PN31	60	PN1	
		PN21	S–S(5), F–F(10)
PN32	60	PN1	
		PN22	S–S(5), F–F(10)
PN33	60	PN1	
		PN23	S–S(5), F–F(10)
PN34	60	PN1	
		PN24	S–S(5), F–F(10)
PC1	10	PN31, PN32, PN33, PN34	
		P15, P16, P55, P56	
FINISH		PH7, PH10, PC1	

activity; that is, in the middle column. The following activities have not been found in the middle column:

- PH7 Dismantle caisson.
- PH10 Test and commission pumps.
- PC1 Commission and handover pipelines.

It is arguable that PH7 should have preceded PH10, but it has been left as shown. Thus the final activity on the network, FINISH, is preceded by PH7, PH10, PC1.

The finalized precedence list is given in Table A3.8. All links are F–S (0) unless otherwise shown.

Network

Analysis of the network shows a project duration of 125 days. The critical path passes through activities PH1–PH6–PH9–PH10. What makes the pumphouse work critical is the long lead time required for ordering and delivering the pumps (PH1).

Operations on the bridge piers were set up to be independent of one another. In fact an examination of the network shows that construction of the bridge footings B31 to B33 and of the bridge piers B41 to B43 could be done sequentially without putting them on the critical path. The project manager is likely to select this option.

Work on the surface networks was also assumed to be a series of independent operations. Clearing can take place in the period from Day 0 to Day 105; the network shows that this could be done sequentially for the first three areas (Activities PN21 to PN23) but that the final area (Activity PN24) must not start later than Day 50, which would be halfway through Activity PN23, so, unless the project manager decides that the project should take 135 days instead of 125 days, there will have to be some overlap between PN23 and PN24 and extra resources will be needed.

Installation of the surface pipe is a different matter. The four surface networks must be installed in the period from Day 30 until Day 115, each takes 60 days; these activities cannot be programmed sequentially and all four surface networks will be worked at the same time, although their start dates could be staggered by seven or eight days.

Appendix 4

DCF Tables

Table A4.1   Present value tables

n \ i	0.5	1	1.5	2	2.5	3	3.5	4	4.5	5
1	0.9950	0.9901	0.9852	0.9804	0.9756	0.9709	0.9662	0.9615	0.9569	0.9524
2	0.9901	0.9803	0.9707	0.9612	0.9518	0.9426	0.9335	0.9246	0.9157	0.9070
3	0.9851	0.9706	0.9563	0.9423	0.9286	0.9151	0.9019	0.8890	0.8763	0.8638
4	0.9802	0.9610	0.9422	0.9238	0.9060	0.8885	0.8714	0.8548	0.8386	0.8227
5	0.9754	0.9515	0.9283	0.9057	0.8839	0.8626	0.8420	0.8219	0.8025	0.7835
6	0.9705	0.9420	0.9145	0.8880	0.8623	0.8375	0.8135	0.7903	0.7679	0.7462
7	0.9657	0.9327	0.9010	0.8706	0.8413	0.8131	0.7860	0.7599	0.7348	0.7107
8	0.9609	0.9235	0.8877	0.8535	0.8207	0.7894	0.7594	0.7307	0.7032	0.6768
9	0.9561	0.9143	0.8746	0.8368	0.8007	0.7664	0.7337	0.7026	0.6729	0.6446
10	0.9513	0.9053	0.8617	0.8203	0.7812	0.7441	0.7089	0.6756	0.6439	0.6139
11	0.9466	0.8963	0.8489	0.8043	0.7621	0.7224	0.6849	0.6496	0.6162	0.5847
12	0.9419	0.8874	0.8364	0.7885	0.7436	0.7014	0.6618	0.6246	0.5897	0.5568
13	0.9372	0.8787	0.8240	0.7730	0.7254	0.6810	0.6394	0.6006	0.5643	0.5303
14	0.9326	0.8700	0.8118	0.7579	0.7077	0.6611	0.6178	0.5775	0.5400	0.5051
15	0.9279	0.8613	0.7999	0.7430	0.6905	0.6419	0.5969	0.5553	0.5167	0.4810
16	0.9233	0.8528	0.7880	0.7284	0.6736	0.6232	0.5767	0.5339	0.4945	0.4581
17	0.9187	0.8444	0.7764	0.7142	0.6572	0.6050	0.5572	0.5134	0.4732	0.4363
18	0.9141	0.8360	0.7649	0.7002	0.6412	0.5874	0.5384	0.4936	0.4528	0.4155
19	0.9096	0.8277	0.7536	0.6864	0.6255	0.5703	0.5202	0.4746	0.4333	0.3957
20	0.9051	0.8195	0.7425	0.6730	0.6103	0.5537	0.5026	0.4564	0.4146	0.3769
21	0.9006	0.8114	0.7315	0.6598	0.5954	0.5375	0.4856	0.4388	0.3968	0.3589
22	0.8961	0.8034	0.7207	0.6468	0.5809	0.5219	0.4692	0.4220	0.3797	0.3418
23	0.8916	0.7954	0.7100	0.6342	0.5667	0.5067	0.4533	0.4057	0.3634	0.3256
24	0.8872	0.7876	0.6995	0.6217	0.5529	0.4919	0.4380	0.3901	0.3477	0.3101
25	0.8828	0.7798	0.6892	0.6095	0.5394	0.4776	0.4231	0.3751	0.3327	0.2953
26	0.8784	0.7720	0.6790	0.5976	0.5262	0.4637	0.4088	0.3607	0.3184	0.2812
27	0.8740	0.7644	0.6690	0.5859	0.5134	0.4502	0.3950	0.3468	0.3047	0.2678
28	0.8697	0.7568	0.6591	0.5744	0.5009	0.4371	0.3817	0.3335	0.2916	0.2551
29	0.8653	0.7493	0.6494	0.5631	0.4887	0.4243	0.3687	0.3207	0.2790	0.2429
30	0.8610	0.7419	0.6398	0.5521	0.4767	0.4120	0.3563	0.3083	0.2670	0.2314
31	0.8567	0.7346	0.6303	0.5412	0.4651	0.4000	0.3442	0.2965	0.2555	0.2204
32	0.8525	0.7273	0.6210	0.5306	0.4538	0.3883	0.3326	0.2851	0.2445	0.2099
33	0.8482	0.7201	0.6118	0.5202	0.4427	0.3770	0.3213	0.2741	0.2340	0.1999
34	0.8440	0.7130	0.6028	0.5100	0.4319	0.3660	0.3105	0.2636	0.2239	0.1904
35	0.8398	0.7059	0.5939	0.5000	0.4214	0.3554	0.3000	0.2534	0.2143	0.1813
36	0.8356	0.6989	0.5851	0.4902	0.4111	0.3450	0.2898	0.2437	0.2050	0.1727
37	0.8315	0.6920	0.5764	0.4806	0.4011	0.3350	0.2800	0.2343	0.1962	0.1644
38	0.8274	0.6852	0.5679	0.4712	0.3913	0.3252	0.2706	0.2253	0.1878	0.1566
39	0.8232	0.6784	0.5595	0.4619	0.3817	0.3158	0.2614	0.2166	0.1797	0.1491
40	0.8191	0.6717	0.5513	0.4529	0.3724	0.3066	0.2526	0.2083	0.1719	0.1420
50	0.7793	0.6080	0.4750	0.3715	0.2909	0.2281	0.1791	0.1407	0.1107	0.0872
60	0.7414	0.5504	0.4093	0.3048	0.2273	0.1697	0.1269	0.0951	0.0713	0.0535
80	0.6710	0.4511	0.3039	0.2051	0.1387	0.0940	0.0638	0.0434	0.0296	0.0202
90	0.6383	0.4084	0.2619	0.1683	0.1084	0.0699	0.0452	0.0293	0.0190	0.0124
100	0.6073	0.3697	0.2256	0.1380	0.0846	0.0520	0.0321	0.0198	0.0123	0.0076
110	0.5777	0.3347	0.1944	0.1132	0.0661	0.0387	0.0227	0.0134	0.0079	0.0047
120	0.5496	0.3030	0.1675	0.0929	0.0517	0.0288	0.0161	0.0090	0.0051	0.0029

Table A4.1   Present value tables (cont'd)

n \ i	5.5	6	6.5	7	7.5	8	8.5	9	9.5	10
1	0.9479	0.9434	0.9390	0.9346	0.9302	0.9259	0.9217	0.9174	0.9132	0.9091
2	0.8985	0.8900	0.8817	0.8734	0.8653	0.8573	0.8495	0.8417	0.8340	0.8264
3	0.8516	0.8396	0.8278	0.8163	0.8050	0.7938	0.7829	0.7722	0.7617	0.7513
4	0.8072	0.7921	0.7773	0.7629	0.7488	0.7350	0.7216	0.7084	0.6956	0.6830
5	0.7651	0.7473	0.7299	0.7130	0.6966	0.6806	0.6650	0.6499	0.6352	0.6209
6	0.7252	0.7050	0.6853	0.6663	0.6480	0.6302	0.6129	0.5963	0.5801	0.5645
7	0.6874	0.6651	0.6435	0.6227	0.6028	0.5835	0.5649	0.5470	0.5298	0.5132
8	0.6516	0.6274	0.6042	0.5820	0.5607	0.5403	0.5207	0.5019	0.4838	0.4665
9	0.6176	0.5919	0.5674	0.5439	0.5216	0.5002	0.4799	0.4604	0.4418	0.4241
10	0.5854	0.5584	0.5327	0.5083	0.4852	0.4632	0.4423	0.4224	0.4035	0.3855
11	0.5549	0.5268	0.5002	0.4751	0.4513	0.4289	0.4076	0.3875	0.3685	0.3505
12	0.5260	0.4970	0.4697	0.4440	0.4199	0.3971	0.3757	0.3555	0.3365	0.3186
13	0.4986	0.4688	0.4410	0.4150	0.3906	0.3677	0.3463	0.3262	0.3073	0.2897
14	0.4726	0.4423	0.4141	0.3878	0.3633	0.3405	0.3191	0.2992	0.2807	0.2633
15	0.4479	0.4173	0.3888	0.3624	0.3380	0.3152	0.2941	0.2745	0.2563	0.2394
16	0.4246	0.3936	0.3651	0.3387	0.3144	0.2919	0.2711	0.2519	0.2341	0.2176
17	0.4024	0.3714	0.3428	0.3166	0.2925	0.2703	0.2499	0.2311	0.2138	0.1978
18	0.3815	0.3503	0.3219	0.2959	0.2720	0.2502	0.2303	0.2120	0.1952	0.1799
19	0.3616	0.3305	0.3022	0.2765	0.2531	0.2317	0.2122	0.1945	0.1783	0.1635
20	0.3427	0.3118	0.2838	0.2584	0.2354	0.2145	0.1956	0.1784	0.1628	0.1486
21	0.3249	0.2942	0.2665	0.2415	0.2190	0.1987	0.1803	0.1637	0.1487	0.1351
22	0.3079	0.2775	0.2502	0.2257	0.2037	0.1839	0.1662	0.1502	0.1358	0.1228
23	0.2919	0.2618	0.2349	0.2109	0.1895	0.1703	0.1531	0.1378	0.1240	0.1117
24	0.2767	0.2470	0.2206	0.1971	0.1763	0.1577	0.1412	0.1264	0.1133	0.1015
25	0.2622	0.2330	0.2071	0.1842	0.1640	0.1460	0.1301	0.1160	0.1034	0.0923
26	0.2486	0.2198	0.1945	0.1722	0.1525	0.1352	0.1199	0.1064	0.0945	0.0839
27	0.2356	0.2074	0.1826	0.1609	0.1419	0.1252	0.1105	0.0976	0.0863	0.0763
28	0.2233	0.1956	0.1715	0.1504	0.1320	0.1159	0.1019	0.0895	0.0788	0.0693
29	0.2117	0.1846	0.1610	0.1406	0.1228	0.1073	0.0939	0.0822	0.0719	0.0630
30	0.2006	0.1741	0.1512	0.1314	0.1142	0.0994	0.0865	0.0754	0.0657	0.0573
31	0.1902	0.1643	0.1420	0.1228	0.1063	0.0920	0.0797	0.0691	0.0600	0.0521
32	0.1803	0.1550	0.1333	0.1147	0.0988	0.0852	0.0735	0.0634	0.0548	0.0474
33	0.1709	0.1462	0.1252	0.1072	0.0919	0.0789	0.0677	0.0582	0.0500	0.0431
34	0.1620	0.1379	0.1175	0.1002	0.0855	0.0730	0.0624	0.0534	0.0457	0.0391
35	0.1535	0.1301	0.1103	0.0937	0.0796	0.0676	0.0575	0.0490	0.0417	0.0356
36	0.1455	0.1227	0.1036	0.0875	0.0740	0.0626	0.0530	0.0449	0.0381	0.0323
37	0.1379	0.1158	0.0973	0.0818	0.0688	0.0580	0.0489	0.0412	0.0348	0.0294
38	0.1307	0.1092	0.0914	0.0765	0.0640	0.0537	0.0450	0.0378	0.0318	0.0267
39	0.1239	0.1031	0.0858	0.0715	0.0596	0.0497	0.0415	0.0347	0.0290	0.0243
40	0.1175	0.0972	0.0805	0.0668	0.0554	0.0460	0.0383	0.0318	0.0265	0.0221
50	0.0688	0.0543	0.0429	0.0339	0.0269	0.0213	0.0169	0.0134	0.0107	0.0085
60	0.0403	0.0303	0.0229	0.0173	0.0130	0.0099	0.0075	0.0057	0.0043	0.0033
80	0.0138	0.0095	0.0065	0.0045	0.0031	0.0021	0.0015	0.0010	0.0007	0.0005
90	0.0081	0.0053	0.0035	0.0023	0.0015	0.0010	0.0006	0.0004	0.0003	0.0002
100	0.0047	0.0029	0.0018	0.0012	0.0007	0.0005	0.0003	0.0002	0.0001	0.0001
110	0.0028	0.0016	0.0010	0.0006	0.0004	0.0002	0.0001	0.0001	0.0000	0.0000
120	0.0016	0.0009	0.0005	0.0003	0.0002	0.0001	0.0001	0.0000	0.0000	0.0000

Table A4.1   Present value tables (cont'd)

n \ i	10.5	11	11.5	12	12.5	13	13.5	14	14.5	15
1	0.9050	0.9009	0.8969	0.8929	0.8889	0.8850	0.8811	0.8772	0.8734	0.8696
2	0.8190	0.8116	0.8044	0.7972	0.7901	0.7831	0.7763	0.7695	0.7628	0.7561
3	0.7412	0.7312	0.7214	0.7118	0.7023	0.6931	0.6839	0.6750	0.6662	0.6575
4	0.6707	0.6587	0.6470	0.6355	0.6243	0.6133	0.6026	0.5921	0.5818	0.5718
5	0.6070	0.5935	0.5803	0.5674	0.5549	0.5428	0.5309	0.5194	0.5081	0.4972
6	0.5493	0.5346	0.5204	0.5066	0.4933	0.4803	0.4678	0.4556	0.4438	0.4323
7	0.4971	0.4817	0.4667	0.4523	0.4385	0.4251	0.4121	0.3996	0.3876	0.3759
8	0.4499	0.4339	0.4186	0.4039	0.3897	0.3762	0.3631	0.3506	0.3385	0.3269
9	0.4071	0.3909	0.3754	0.3606	0.3464	0.3329	0.3199	0.3075	0.2956	0.2843
10	0.3684	0.3522	0.3367	0.3220	0.3079	0.2946	0.2819	0.2697	0.2582	0.2472
11	0.3334	0.3173	0.3020	0.2875	0.2737	0.2607	0.2483	0.2366	0.2255	0.2149
12	0.3018	0.2858	0.2708	0.2567	0.2433	0.2307	0.2188	0.2076	0.1969	0.1869
13	0.2731	0.2575	0.2429	0.2292	0.2163	0.2042	0.1928	0.1821	0.1720	0.1625
14	0.2471	0.2320	0.2178	0.2046	0.1922	0.1807	0.1698	0.1597	0.1502	0.1413
15	0.2236	0.2090	0.1954	0.1827	0.1709	0.1599	0.1496	0.1401	0.1312	0.1229
16	0.2024	0.1883	0.1752	0.1631	0.1519	0.1415	0.1318	0.1229	0.1146	0.1069
17	0.1832	0.1696	0.1572	0.1456	0.1350	0.1252	0.1162	0.1078	0.1001	0.0929
18	0.1658	0.1528	0.1409	0.1300	0.1200	0.1108	0.1023	0.0946	0.0874	0.0808
19	0.1500	0.1377	0.1264	0.1161	0.1067	0.0981	0.0902	0.0829	0.0763	0.0703
20	0.1358	0.1240	0.1134	0.1037	0.0948	0.0868	0.0794	0.0728	0.0667	0.0611
21	0.1229	0.1117	0.1017	0.0926	0.0843	0.0768	0.0700	0.0638	0.0582	0.0531
22	0.1112	0.1007	0.0912	0.0826	0.0749	0.0680	0.0617	0.0560	0.0508	0.0462
23	0.1006	0.0907	0.0818	0.0738	0.0666	0.0601	0.0543	0.0491	0.0444	0.0402
24	0.0911	0.0817	0.0734	0.0659	0.0592	0.0532	0.0479	0.0431	0.0388	0.0349
25	0.0824	0.0736	0.0658	0.0588	0.0526	0.0471	0.0422	0.0378	0.0339	0.0304
26	0.0746	0.0663	0.0590	0.0525	0.0468	0.0417	0.0372	0.0331	0.0296	0.0264
27	0.0675	0.0597	0.0529	0.0469	0.0416	0.0369	0.0327	0.0291	0.0258	0.0230
28	0.0611	0.0538	0.0475	0.0419	0.0370	0.0326	0.0288	0.0255	0.0226	0.0200
29	0.0553	0.0485	0.0426	0.0374	0.0329	0.0289	0.0254	0.0224	0.0197	0.0174
30	0.0500	0.0437	0.0382	0.0334	0.0292	0.0256	0.0224	0.0196	0.0172	0.0151
31	0.0453	0.0394	0.0342	0.0298	0.0260	0.0226	0.0197	0.0172	0.0150	0.0131
32	0.0410	0.0355	0.0307	0.0266	0.0231	0.0200	0.0174	0.0151	0.0131	0.0114
33	0.0371	0.0319	0.0275	0.0238	0.0205	0.0177	0.0153	0.0132	0.0115	0.0099
34	0.0335	0.0288	0.0247	0.0212	0.0182	0.0157	0.0135	0.0116	0.0100	0.0086
35	0.0304	0.0259	0.0222	0.0189	0.0162	0.0139	0.0119	0.0102	0.0087	0.0075
36	0.0275	0.0234	0.0199	0.0169	0.0144	0.0123	0.0105	0.0089	0.0076	0.0065
37	0.0249	0.0210	0.0178	0.0151	0.0128	0.0109	0.0092	0.0078	0.0067	0.0057
38	0.0225	0.0190	0.0160	0.0135	0.0114	0.0096	0.0081	0.0069	0.0058	0.0049
39	0.0204	0.0171	0.0143	0.0120	0.0101	0.0085	0.0072	0.0060	0.0051	0.0043
40	0.0184	0.0154	0.0129	0.0107	0.0090	0.0075	0.0063	0.0053	0.0044	0.0037
50	0.0068	0.0054	0.0043	0.0035	0.0028	0.0022	0.0018	0.0014	0.0011	0.0009
60	0.0025	0.0019	0.0015	0.0011	0.0009	0.0007	0.0005	0.0004	0.0003	0.0002
80	0.0003	0.0002	0.0002	0.0001	0.0001	0.0001	0.0000	0.0000	0.0000	0.0000
90	0.0001	0.0001	0.0001	0.0000	0.0000	0.0000	0.0000	0.0000	0.0000	0.0000
100	0.0000	0.0000	0.0000	0.0000	0.0000	0.0000	0.0000	0.0000	0.0000	0.0000
110	0.0000	0.0000	0.0000	0.0000	0.0000	0.0000	0.0000	0.0000	0.0000	0.0000
120	0.0000	0.0000	0.0000	0.0000	0.0000	0.0000	0.0000	0.0000	0.0000	0.0000

Table A4.1   Present value tables (cont'd)

n \ i	15.5	16	16.5	17	17.5	18	18.5	19	19.5	20
1	0.8658	0.8621	0.8584	0.8547	0.8511	0.8475	0.8439	0.8403	0.8368	0.8333
2	0.7496	0.7432	0.7368	0.7305	0.7243	0.7182	0.7121	0.7062	0.7003	0.6944
3	0.6490	0.6407	0.6324	0.6244	0.6164	0.6086	0.6010	0.5934	0.5860	0.5787
4	0.5619	0.5523	0.5429	0.5337	0.5246	0.5158	0.5071	0.4987	0.4904	0.4823
5	0.4865	0.4761	0.4660	0.4561	0.4465	0.4371	0.4280	0.4190	0.4104	0.4019
6	0.4212	0.4104	0.4000	0.3898	0.3800	0.3704	0.3612	0.3521	0.3434	0.3349
7	0.3647	0.3538	0.3433	0.3332	0.3234	0.3139	0.3048	0.2959	0.2874	0.2791
8	0.3158	0.3050	0.2947	0.2848	0.2752	0.2660	0.2572	0.2487	0.2405	0.2326
9	0.2734	0.2630	0.2530	0.2434	0.2342	0.2255	0.2170	0.2090	0.2012	0.1938
10	0.2367	0.2267	0.2171	0.2080	0.1994	0.1911	0.1832	0.1756	0.1684	0.1615
11	0.2049	0.1954	0.1864	0.1778	0.1697	0.1619	0.1546	0.1476	0.1409	0.1346
12	0.1774	0.1685	0.1600	0.1520	0.1444	0.1372	0.1304	0.1240	0.1179	0.1122
13	0.1536	0.1452	0.1373	0.1299	0.1229	0.1163	0.1101	0.1042	0.0987	0.0935
14	0.1330	0.1252	0.1179	0.1110	0.1046	0.0985	0.0929	0.0876	0.0826	0.0779
15	0.1152	0.1079	0.1012	0.0949	0.0890	0.0835	0.0784	0.0736	0.0691	0.0649
16	0.0997	0.0930	0.0869	0.0811	0.0758	0.0708	0.0661	0.0618	0.0578	0.0541
17	0.0863	0.0802	0.0746	0.0693	0.0645	0.0600	0.0558	0.0520	0.0484	0.0451
18	0.0747	0.0691	0.0640	0.0592	0.0549	0.0508	0.0471	0.0437	0.0405	0.0376
19	0.0647	0.0596	0.0549	0.0506	0.0467	0.0431	0.0398	0.0367	0.0339	0.0313
20	0.0560	0.0514	0.0471	0.0433	0.0397	0.0365	0.0335	0.0308	0.0284	0.0261
21	0.0485	0.0443	0.0405	0.0370	0.0338	0.0309	0.0283	0.0259	0.0237	0.0217
22	0.0420	0.0382	0.0347	0.0316	0.0288	0.0262	0.0239	0.0218	0.0199	0.0181
23	0.0364	0.0329	0.0298	0.0270	0.0245	0.0222	0.0202	0.0183	0.0166	0.0151
24	0.0315	0.0284	0.0256	0.0231	0.0208	0.0188	0.0170	0.0154	0.0139	0.0126
25	0.0273	0.0245	0.0220	0.0197	0.0177	0.0160	0.0144	0.0129	0.0116	0.0105
26	0.0236	0.0211	0.0189	0.0169	0.0151	0.0135	0.0121	0.0109	0.0097	0.0087
27	0.0204	0.0182	0.0162	0.0144	0.0129	0.0115	0.0102	0.0091	0.0081	0.0073
28	0.0177	0.0157	0.0139	0.0123	0.0109	0.0097	0.0086	0.0077	0.0068	0.0061
29	0.0153	0.0135	0.0119	0.0105	0.0093	0.0082	0.0073	0.0064	0.0057	0.0051
30	0.0133	0.0116	0.0102	0.0090	0.0079	0.0070	0.0061	0.0054	0.0048	0.0042
31	0.0115	0.0100	0.0088	0.0077	0.0067	0.0059	0.0052	0.0046	0.0040	0.0035
32	0.0099	0.0087	0.0075	0.0066	0.0057	0.0050	0.0044	0.0038	0.0033	0.0029
33	0.0086	0.0075	0.0065	0.0056	0.0049	0.0042	0.0037	0.0032	0.0028	0.0024
34	0.0075	0.0064	0.0056	0.0048	0.0042	0.0036	0.0031	0.0027	0.0023	0.0020
35	0.0065	0.0055	0.0048	0.0041	0.0035	0.0030	0.0026	0.0023	0.0020	0.0017
36	0.0056	0.0048	0.0041	0.0035	0.0030	0.0026	0.0022	0.0019	0.0016	0.0014
37	0.0048	0.0041	0.0035	0.0030	0.0026	0.0022	0.0019	0.0016	0.0014	0.0012
38	0.0042	0.0036	0.0030	0.0026	0.0022	0.0019	0.0016	0.0013	0.0011	0.0010
39	0.0036	0.0031	0.0026	0.0022	0.0019	0.0016	0.0013	0.0011	0.0010	0.0008
40	0.0031	0.0026	0.0022	0.0019	0.0016	0.0013	0.0011	0.0010	0.0008	0.0007
50	0.0007	0.0006	0.0005	0.0004	0.0003	0.0003	0.0002	0.0002	0.0001	0.0001
60	0.0002	0.0001	0.0001	0.0001	0.0001	0.0000	0.0000	0.0000	0.0000	0.0000
80	0.0000	0.0000	0.0000	0.0000	0.0000	0.0000	0.0000	0.0000	0.0000	0.0000
90	0.0000	0.0000	0.0000	0.0000	0.0000	0.0000	0.0000	0.0000	0.0000	0.0000
100	0.0000	0.0000	0.0000	0.0000	0.0000	0.0000	0.0000	0.0000	0.0000	0.0000
110	0.0000	0.0000	0.0000	0.0000	0.0000	0.0000	0.0000	0.0000	0.0000	0.0000
120	0.0000	0.0000	0.0000	0.0000	0.0000	0.0000	0.0000	0.0000	0.0000	0.0000

Table A4.1   Present value tables (cont'd)

n \ i	21	22	23	24	25	30	35	40	50	100
1	0.8264	0.8197	0.8130	0.8065	0.8000	0.7692	0.7407	0.7143	0.6667	0.5000
2	0.6830	0.6719	0.6610	0.6504	0.6400	0.5917	0.5487	0.5102	0.4444	0.2500
3	0.5645	0.5507	0.5374	0.5245	0.5120	0.4552	0.4064	0.3644	0.2963	0.1250
4	0.4665	0.4514	0.4369	0.4230	0.4096	0.3501	0.3011	0.2603	0.1975	0.0625
5	0.3855	0.3700	0.3552	0.3411	0.3277	0.2693	0.2230	0.1859	0.1317	0.0313
6	0.3186	0.3033	0.2888	0.2751	0.2621	0.2072	0.1652	0.1328	0.0878	0.0156
7	0.2633	0.2486	0.2348	0.2218	0.2097	0.1594	0.1224	0.0949	0.0585	0.0078
8	0.2176	0.2038	0.1909	0.1789	0.1678	0.1226	0.0906	0.0678	0.0390	0.0039
9	0.1799	0.1670	0.1552	0.1443	0.1342	0.0943	0.0671	0.0484	0.0260	0.0020
10	0.1486	0.1369	0.1262	0.1164	0.1074	0.0725	0.0497	0.0346	0.0173	0.0010
11	0.1228	0.1122	0.1026	0.0938	0.0859	0.0558	0.0368	0.0247	0.0116	0.0005
12	0.1015	0.0920	0.0834	0.0757	0.0687	0.0429	0.0273	0.0176	0.0077	0.0002
13	0.0839	0.0754	0.0678	0.0610	0.0550	0.0330	0.0202	0.0126	0.0051	0.0001
14	0.0693	0.0618	0.0551	0.0492	0.0440	0.0254	0.0150	0.0090	0.0034	0.0001
15	0.0573	0.0507	0.0448	0.0397	0.0352	0.0195	0.0111	0.0064	0.0023	0.0000
16	0.0474	0.0415	0.0364	0.0320	0.0281	0.0150	0.0082	0.0046	0.0015	0.0000
17	0.0391	0.0340	0.0296	0.0258	0.0225	0.0116	0.0061	0.0033	0.0010	0.0000
18	0.0323	0.0279	0.0241	0.0208	0.0180	0.0089	0.0045	0.0023	0.0007	0.0000
19	0.0267	0.0229	0.0196	0.0168	0.0144	0.0068	0.0033	0.0017	0.0005	0.0000
20	0.0221	0.0187	0.0159	0.0135	0.0115	0.0053	0.0025	0.0012	0.0003	0.0000
21	0.0183	0.0154	0.0129	0.0109	0.0092	0.0040	0.0018	0.0009	0.0002	0.0000
22	0.0151	0.0126	0.0105	0.0088	0.0074	0.0031	0.0014	0.0006	0.0001	0.0000
23	0.0125	0.0103	0.0086	0.0071	0.0059	0.0024	0.0010	0.0004	0.0001	0.0000
24	0.0103	0.0085	0.0070	0.0057	0.0047	0.0018	0.0007	0.0003	0.0001	0.0000
25	0.0085	0.0069	0.0057	0.0046	0.0038	0.0014	0.0006	0.0002	0.0000	0.0000
26	0.0070	0.0057	0.0046	0.0037	0.0030	0.0011	0.0004	0.0002	0.0000	0.0000
27	0.0058	0.0047	0.0037	0.0030	0.0024	0.0008	0.0003	0.0001	0.0000	0.0000
28	0.0048	0.0038	0.0030	0.0024	0.0019	0.0006	0.0002	0.0001	0.0000	0.0000
29	0.0040	0.0031	0.0025	0.0020	0.0015	0.0005	0.0002	0.0001	0.0000	0.0000
30	0.0033	0.0026	0.0020	0.0016	0.0012	0.0004	0.0001	0.0000	0.0000	0.0000
31	0.0027	0.0021	0.0016	0.0013	0.0010	0.0003	0.0001	0.0000	0.0000	0.0000
32	0.0022	0.0017	0.0013	0.0010	0.0008	0.0002	0.0001	0.0000	0.0000	0.0000
33	0.0019	0.0014	0.0011	0.0008	0.0006	0.0002	0.0001	0.0000	0.0000	0.0000
34	0.0015	0.0012	0.0009	0.0007	0.0005	0.0001	0.0000	0.0000	0.0000	0.0000
35	0.0013	0.0009	0.0007	0.0005	0.0004	0.0001	0.0000	0.0000	0.0000	0.0000
36	0.0010	0.0008	0.0006	0.0004	0.0003	0.0001	0.0000	0.0000	0.0000	0.0000
37	0.0009	0.0006	0.0005	0.0003	0.0003	0.0001	0.0000	0.0000	0.0000	0.0000
38	0.0007	0.0005	0.0004	0.0003	0.0002	0.0000	0.0000	0.0000	0.0000	0.0000
39	0.0006	0.0004	0.0003	0.0002	0.0002	0.0000	0.0000	0.0000	0.0000	0.0000
40	0.0005	0.0004	0.0003	0.0002	0.0001	0.0000	0.0000	0.0000	0.0000	0.0000
50	0.0001	0.0000	0.0000	0.0000	0.0000	0.0000	0.0000	0.0000	0.0000	0.0000
60	0.0000	0.0000	0.0000	0.0000	0.0000	0.0000	0.0000	0.0000	0.0000	0.0000
80	0.0000	0.0000	0.0000	0.0000	0.0000	0.0000	0.0000	0.0000	0.0000	0.0000
90	0.0000	0.0000	0.0000	0.0000	0.0000	0.0000	0.0000	0.0000	0.0000	0.0000
100	0.0000	0.0000	0.0000	0.0000	0.0000	0.0000	0.0000	0.0000	0.0000	0.0000
110	0.0000	0.0000	0.0000	0.0000	0.0000	0.0000	0.0000	0.0000	0.0000	0.0000
120	0.0000	0.0000	0.0000	0.0000	0.0000	0.0000	0.0000	0.0000	0.0000	0.0000

Table A4.2 Present value of annuity tables

n \ i	0.5	1	1.5	2	2.5	3	3.5	4	4.5	5
1	0.9950	0.9901	0.9852	0.9804	0.9756	0.9709	0.9662	0.9615	0.9569	0.9524
2	1.9851	1.9704	1.9559	1.9416	1.9274	1.9135	1.8997	1.8861	1.8727	1.8594
3	2.9702	2.9410	2.9122	2.8839	2.8560	2.8286	2.8016	2.7751	2.7490	2.7232
4	3.9505	3.9020	3.8544	3.8077	3.7260	3.7171	3.6731	3.6299	3.5875	3.5460
5	4.9259	4.8534	4.7826	4.7135	4.6458	4.5797	4.5151	4.4518	4.3900	4.3295
6	5.8964	5.7955	5.6972	5.6014	5.5081	5.4172	5.3286	5.2421	5.1579	5.0757
7	6.8621	6.7282	6.5982	6.4720	6.3494	6.2303	6.1145	6.0021	5.8927	5.7864
8	7.8230	7.6517	7.4859	7.3255	7.1701	7.0197	6.8740	6.7327	6.5959	6.4632
9	8.7791	8.5660	8.3605	8.1622	7.9709	7.7861	7.6077	7.4353	7.2688	7.1078
10	9.7304	9.4713	9.2222	8.9826	8.7251	8.5302	8.3166	8.1109	7.9127	7.7217
11	10.6770	10.3676	10.0711	9.7868	9.5142	9.2526	9.0016	8.7605	8.5289	8.3064
12	11.6189	11.2551	10.9075	10.5753	10.2578	9.9540	9.6633	9.3851	9.1186	8.8633
13	12.5562	12.1337	11.7315	11.3484	10.9832	10.6350	10.3027	9.9856	9.6829	9.3936
14	13.4887	13.0037	12.5434	12.1062	11.6909	11.2961	10.9205	10.5631	10.2228	9.8986
15	14.4166	13.8651	13.3432	12.8493	12.3814	11.9379	11.5174	11.1184	10.7395	10.3797
16	15.3399	14.7179	14.1313	13.5777	13.0550	12.5611	12.0941	11.6523	11.2340	10.8378
17	16.2586	15.5623	14.9076	14.2919	13.7122	13.1661	12.6513	12.1657	11.7072	11.2741
18	17.1728	16.3983	15.6726	14.9920	14.3534	13.7535	13.1897	12.6593	12.1600	11.6896
19	18.0824	17.2260	16.4262	15.6785	14.9789	14.3238	13.7098	13.1339	12.5933	12.0853
20	18.9874	18.0456	17.1686	16.3514	15.5892	14.8775	14.2124	13.5903	13.0079	12.4622
21	19.8880	18.8570	17.9001	17.0112	16.1845	15.4150	14.6980	14.0292	13.4047	12.8212
22	20.7841	19.6604	18.6208	17.6580	16.7654	15.9369	15.1671	14.4511	13.7844	13.1630
23	21.6757	20.4558	19.3309	18.2922	17.3321	16.4436	15.6204	14.8568	14.1478	13.4886
24	22.5629	21.2434	20.0304	18.9139	17.8850	16.9355	16.0584	15.2470	14.4955	13.7986
25	23.4456	22.0232	20.7196	19.5235	18.4244	17.4131	16.4815	15.6221	14.8282	14.0939
26	24.3240	22.7952	21.3986	20.1210	18.9506	17.8768	16.8904	15.9828	15.1466	14.3752
27	25.1980	23.5596	22.0676	20.7069	19.4640	18.3270	17.2854	16.3296	15.4513	14.6430
28	26.0677	24.3164	22.7267	21.2813	19.9649	18.7641	17.6670	16.6631	15.7429	14.8981
29	26.9330	25.0658	23.3761	21.8444	20.4535	19.1885	18.0358	16.9837	16.0219	15.1411
30	27.7941	25.8077	24.0158	22.3965	20.9303	19.6004	18.3920	17.2920	16.2889	15.3725
31	28.6508	26.5423	24.6461	22.9377	21.3954	20.0004	18.7363	17.5885	16.5444	15.5928
32	29.5033	27.2696	25.2671	23.4683	21.8492	20.3888	19.0689	17.8736	16.7889	15.8027
33	30.3515	27.9897	25.8790	23.9886	22.2919	20.7658	19.3902	18.1476	17.0229	16.0025
34	31.1955	28.7027	26.4817	24.4986	22.7238	21.1318	19.7007	18.4112	17.2468	16.1929
35	32.0354	29.4086	27.0756	24.9986	23.1452	21.4872	20.0007	18.6646	17.4610	16.3742
36	32.8710	30.1075	27.6607	25.4888	23.5563	21.8323	20.2905	18.9083	17.6660	16.5469
37	33.7025	30.7995	28.2371	25.9695	23.9573	22.1672	20.5705	19.1426	17.8622	16.7113
38	34.5299	31.4847	28.8051	26.4406	24.3486	22.4925	20.8411	19.3679	18.0500	16.8679
39	35.3531	32.1630	29.3646	26.9026	24.7303	22.8082	21.1025	19.5845	18.2297	17.0170
40	36.1722	32.8347	29.9158	27.3555	25.1028	23.1148	21.3551	19.7928	18.4016	17.1591
50	44.1428	39.1961	34.9997	31.4236	28.3623	25.7298	23.4556	21.4822	19.7620	18.2559
60	51.7256	44.9550	39.3803	34.7609	30.9087	27.6756	24.9447	22.6235	20.6380	18.9293
80	65.8023	54.8882	46.4073	39.7445	34.4518	30.2008	26.7488	23.9154	21.5653	19.5965
90	72.3313	59.1609	49.2099	41.5869	35.6658	31.0024	27.2793	24.2673	21.7992	19.7523
100	78.5426	63.0289	51.6247	43.0984	36.6141	31.5989	27.6554	24.5050	21.9499	19.8479
110	84.4518	66.5305	53.7055	44.3382	37.3549	32.0428	27.9221	24.6656	22.0468	19.9066
120	90.0735	69.7005	55.4985	45.3554	37.9337	32.3730	28.1111	24.7741	22.1093	19.9427

Table A4.2   Present value of annuity tables (cont'd)

i / n	5.5	6	6.5	7	7.5	8	8.5	9	9.5	10
1	0.9479	0.9434	0.9390	0.9346	0.9302	0.9259	0.9217	0.9174	0.9132	0.9091
2	1.8463	1.8334	1.8206	1.8080	1.7956	1.7833	1.7711	1.7591	1.7473	1.7355
3	2.6979	2.6730	2.6485	2.6243	2.6005	2.5771	2.5540	2.5313	2.5089	2.4869
4	3.5052	3.4651	3.4258	3.3872	3.3493	3.3121	3.2756	3.2397	3.2045	3.1699
5	4.2703	4.2124	4.1557	4.1002	4.0459	3.9927	3.9406	3.8897	3.8397	3.7908
6	4.9955	4.9173	4.8410	4.7665	4.6938	4.6229	4.5536	4.4859	4.4198	4.3553
7	5.6830	5.5824	5.4845	5.3893	5.2966	5.2064	5.1185	5.0330	4.9496	4.8684
8	6.3346	6.2098	6.0888	5.9713	5.8573	5.7466	5.6392	5.5348	5.4334	5.3349
9	6.9522	6.8017	6.6561	6.5152	6.3789	6.2469	6.1191	5.9952	5.8753	5.7590
10	7.5376	7.3601	7.1888	7.0236	6.8641	6.7101	6.5613	6.4177	6.2788	6.1446
11	8.0925	7.8869	7.6890	7.4987	7.3154	7.1390	6.9690	6.8052	6.6473	6.4951
12	8.6185	8.3838	8.1587	7.9247	7.7353	7.5361	7.3447	7.1607	6.9838	6.8137
13	9.1171	8.8527	8.5997	8.3577	8.1258	7.9038	7.6910	7.4869	7.2912	7.1034
14	9.5896	9.2950	9.0138	8.7455	8.4892	8.2442	8.0101	7.7862	7.5719	7.3667
15	10.0376	9.7122	9.4027	9.1079	8.8271	8.5595	8.3042	8.0607	7.8282	7.6061
16	10.4622	10.1059	9.7678	9.4466	9.1415	8.8514	8.5753	8.3126	8.0623	7.8237
17	10.8646	10.4773	10.1106	9.7632	9.4340	9.1216	8.8252	8.5436	8.2760	8.0216
18	11.2461	10.8276	10.4325	10.0591	9.7060	9.3719	9.0555	8.7556	8.4713	8.2014
19	11.6077	11.1581	10.7347	10.3356	9.9591	9.6036	9.2677	8.9501	8.6496	8.3649
20	11.9504	11.4699	11.0185	10.5940	10.1945	9.8181	9.4633	9.1285	8.8124	8.5136
21	12.2752	11.7641	11.2850	10.8355	10.4135	10.0168	9.6436	9.2922	8.9611	8.6487
22	12.5832	12.0416	11.5352	11.0612	10.6172	10.2007	9.8098	9.4424	9.0969	8.7715
23	12.8750	12.3034	11.7701	11.2722	10.8067	10.3711	9.9629	9.5802	9.2209	8.8832
24	13.1517	12.5504	11.9907	11.4693	10.9830	10.5288	10.1041	9.7066	9.3341	8.9847
25	13.4139	12.7834	12.1979	11.6536	11.1469	10.6748	10.2342	9.8226	9.4376	9.0770
26	13.6625	13.0032	12.3924	11.8258	11.2995	10.8100	10.3541	9.9290	9.5320	9.1609
27	13.8981	13.2105	12.5750	11.9867	11.4414	10.9352	10.4646	10.0266	9.6183	9.2372
28	14.1214	13.4062	12.7465	12.1371	11.5734	11.0511	10.5665	10.1161	9.6971	9.3066
29	14.3331	13.5907	12.9075	12.2777	11.6962	11.1584	10.6603	10.1983	9.7690	9.3696
30	14.5337	13.7648	13.0587	12.4090	11.8104	11.2578	10.7468	10.2737	9.8347	9.4269
31	14.7239	13.9291	13.2006	12.5318	11.9166	11.3498	10.8266	10.3428	9.8947	9.4790
32	14.9042	14.0840	13.3339	12.6466	12.0155	11.4350	10.9001	10.4062	9.9495	9.5264
33	15.0751	14.2302	13.4591	12.7538	12.1074	11.5139	10.9678	10.4644	9.9996	9.5694
34	15.2370	14.3681	13.5766	12.8540	12.1929	11.5869	11.0302	10.5178	10.0453	9.6086
35	15.3906	14.4982	13.6870	12.9477	12.2725	11.6546	11.0878	10.5668	10.0870	9.6442
36	15.5361	14.6210	13.7906	13.0352	12.3465	11.7172	11.1408	10.6118	10.1251	9.6765
37	15.6740	14.7368	13.8879	13.1170	12.4154	11.7752	11.1897	10.6530	10.1599	9.7059
38	15.8047	14.8460	13.9792	13.1935	12.4794	11.8289	11.2347	10.6908	10.1917	9.7327
39	15.9287	14.9491	14.0650	13.2649	12.5390	11.8786	11.2763	10.7255	10.2207	9.7570
40	16.0461	15.0463	14.1455	13.3317	12.5944	11.9246	11.3145	10.7574	10.2472	9.7791
50	16.9315	15.7619	14.7245	13.8007	12.9748	12.2335	11.5656	10.9617	10.4137	9.9148
60	17.4499	16.1614	15.0330	14.0392	13.1594	12.3766	11.6766	11.0480	10.4809	9.9672
80	17.9310	16.5091	15.2848	14.2220	13.2924	12.4735	11.7475	11.0998	10.5189	9.9951
90	18.0350	16.5787	15.3315	14.2533	13.3135	12.4877	11.7571	11.1064	10.5233	9.9981
100	18.0958	16.6175	15.3563	14.2693	13.3237	12.4943	11.7613	11.1091	10.5251	9.9993
110	18.1315	16.6392	15.3695	14.2773	13.3287	12.4974	11.7632	11.1103	10.5258	9.9997
120	18.1524	16.6514	15.3766	14.2815	13.3311	12.4988	11.7640	11.1108	10.5261	9.9999

Table A4.2 Present value of annuity tables (cont'd)

n \ i	10.5	11	11.5	12	12.5	13	13.5	14	14.5	15
1	0.9050	0.9009	0.8969	0.8929	0.8889	0.8850	0.8811	0.8772	0.8734	0.8696
2	1.7240	1.7125	1.7012	1.6901	1.6790	1.6681	1.6573	1.6467	1.6361	1.6257
3	2.4651	2.4437	2.4226	2.4018	2.3813	2.3612	2.3413	2.3216	2.3023	2.2832
4	3.1359	3.1024	3.0696	3.0373	3.0056	2.9745	2.9438	2.9137	2.8841	2.8550
5	3.7429	3.6959	3.6499	3.6048	3.5606	3.5172	3.4747	3.4331	3.3922	3.3522
6	4.2922	4.2305	4.1703	4.1114	4.0538	3.9975	3.9425	3.8887	3.8360	3.7845
7	4.7893	4.7122	4.6370	4.5638	4.4923	4.4226	4.3546	4.2883	4.2236	4.1604
8	5.2392	5.1461	5.0556	4.9676	4.8820	4.7988	4.7177	4.6389	4.5621	4.4873
9	5.6463	5.5370	5.4311	5.3282	5.2285	5.1317	5.0377	4.9464	4.8577	4.7716
10	6.0148	5.8892	5.7678	5.6502	5.5364	5.4262	5.3195	5.2161	5.1159	5.0188
11	6.3482	6.2065	6.0697	5.9377	5.8102	5.6869	5.5679	5.4527	5.3414	5.2337
12	6.6500	6.4924	6.3406	6.1944	6.0535	5.9176	5.7867	5.6603	5.5383	5.4206
13	6.9230	6.7499	6.5835	6.4235	6.2698	6.1218	5.9794	5.8424	5.7103	5.5831
14	7.1702	6.9819	6.8013	6.6282	6.4620	6.3025	6.1493	6.0021	5.8606	5.7245
15	7.3938	7.1909	6.9967	6.8109	6.6329	6.4624	6.2989	6.1422	5.9918	5.8474
16	7.5962	7.3792	7.1719	6.9740	6.7848	6.6039	6.4308	6.2651	6.1063	5.9542
17	7.7794	7.5488	7.3291	7.1196	6.9198	6.7291	6.5469	6.3729	6.2064	6.0472
18	7.9451	7.7016	7.4700	7.2497	7.0398	6.8399	6.6493	6.4674	6.2938	6.1280
19	8.0952	7.8393	7.5964	7.3658	7.1465	6.9380	6.7395	6.5504	6.3701	6.1982
20	8.2309	7.9633	7.7098	7.4694	7.2414	7.0248	6.8189	6.6231	6.4368	6.2593
21	8.3538	8.0751	7.8115	7.5620	7.3256	7.1016	6.8889	6.6870	6.4950	6.3125
22	8.4649	8.1757	7.9027	7.6446	7.4006	7.1695	6.9506	6.7429	6.5459	6.3587
23	8.5656	8.2664	7.9845	7.7184	7.4672	7.2297	7.0049	6.7921	6.5903	6.3988
24	8.6566	8.3481	8.0578	7.7843	7.5264	7.2829	7.0528	6.8351	6.6291	6.4338
25	8.7390	8.4217	8.1236	7.8431	7.5790	7.3300	7.0950	6.8729	6.6629	6.4641
26	8.8136	8.4881	8.1826	7.8957	7.6258	7.3717	7.1321	6.9061	6.6925	6.4906
27	8.8811	8.5478	8.2355	7.9426	7.6674	7.4086	7.1649	6.9352	6.7184	6.5135
28	8.9422	8.6016	8.2830	7.9844	7.7043	7.4412	7.1937	6.9607	6.7409	6.5335
29	8.9974	8.6501	8.3255	8.0218	7.7372	7.4701	7.2191	6.9830	6.7606	6.5509
30	9.0474	8.6938	8.3637	8.0552	7.7664	7.4957	7.2415	7.0027	6.7778	6.5660
31	9.0927	8.7331	8.3980	8.0850	7.7923	7.5183	7.2613	7.0199	6.7929	6.5791
32	9.1337	8.7686	8.4287	8.1116	7.8154	7.5383	7.2786	7.0350	6.8060	6.5905
33	9.1707	8.8005	8.4562	8.1354	7.8359	7.5560	7.2940	7.0482	6.8175	6.6005
34	9.2043	8.8293	8.4809	8.1566	7.8542	7.5717	7.3075	7.0599	6.8275	6.6091
35	9.2347	8.8552	8.5030	8.1755	7.8704	7.5856	7.3193	7.0700	6.8362	6.6166
36	9.2621	8.8786	8.5229	8.1924	7.8848	7.5979	7.3298	7.0790	6.8439	6.6231
37	9.2870	8.8996	8.5407	8.2075	7.8976	7.6087	7.3390	7.0868	6.8505	6.6288
38	9.3095	8.9186	8.5567	8.2210	7.9089	7.6183	7.3472	7.0937	6.8564	6.6338
39	9.3299	8.9357	8.5710	8.2330	7.9191	7.6268	7.3543	7.0997	6.8615	6.6380
40	9.3483	8.9511	8.5839	8.2438	7.9281	7.6344	7.3607	7.1050	6.8659	6.6418
50	9.4591	9.0417	8.6580	8.3045	7.9778	7.6752	7.3942	7.1327	6.8886	6.6605
60	9.5000	9.0736	8.6830	8.3240	7.9932	7.6873	7.4037	7.1401	6.8945	6.6651
80	9.5206	9.0888	8.6942	8.3324	7.9994	7.6919	7.4071	7.1427	6.8964	6.6666
90	9.5226	9.0902	8.6952	8.3330	7.9998	7.6922	7.4073	7.1428	6.8965	6.6666
100	9.5234	9.0906	8.6955	8.3332	7.9999	7.6923	7.4074	7.1428	6.8965	6.6667
110	9.5236	9.0908	8.6956	8.3333	8.0000	7.6923	7.4074	7.1429	6.8965	6.6667
120	9.5237	9.0909	8.6956	8.3333	8.0000	7.6923	7.4074	7.1429	6.8966	6.6667

Table A4.2 Present value of annuity tables (cont'd)

n \ i	15.5	16	16.5	17	17.5	18	18.5	19	19.5	20
1	0.8658	0.8621	0.8584	0.8547	0.8511	0.8475	0.8439	0.8403	0.8368	0.8333
2	1.6154	1.6052	1.5952	1.5852	1.5754	1.5656	1.5560	1.5465	1.5371	1.5278
3	2.2644	2.2459	2.2276	2.2096	2.1918	2.1743	2.1570	2.1399	2.1231	2.1065
4	2.8263	2.7982	2.7705	2.7432	2.7164	2.6901	2.6641	2.6386	2.6135	2.5887
5	3.3129	3.2743	3.2365	3.1993	3.1629	3.1272	3.0921	3.0576	3.0238	2.9906
6	3.7341	3.6847	3.6365	3.5892	3.5429	3.4976	3.4532	3.4098	3.3672	3.3255
7	4.0988	4.0386	3.9798	3.9224	3.8663	3.8115	3.7580	3.7057	3.6546	3.6046
8	4.4145	4.3436	4.2745	4.2072	4.1415	4.0776	4.0152	3.9544	3.8950	3.8372
9	4.6879	4.6065	4.5275	4.4506	4.3758	4.3030	4.2322	4.1633	4.0963	4.0310
10	4.9246	4.8332	4.7446	4.6586	4.5751	4.4941	4.4154	4.3389	4.2647	4.1925
11	5.1295	5.0286	4.9310	4.8364	4.7448	4.6560	4.5699	4.4865	4.4056	4.3271
12	5.3069	5.1971	5.0910	4.9884	4.8892	4.7932	4.7004	4.6105	4.5235	4.4392
13	5.4605	5.3423	5.2283	5.1183	5.0121	4.9095	4.8104	4.7147	4.6222	4.5327
14	5.5935	5.4675	5.3462	5.2293	5.1167	5.0081	4.9033	4.8023	4.7047	4.6106
15	5.7087	5.5755	5.4474	5.3242	5.2057	5.0916	4.9817	4.8759	4.7738	4.6755
16	5.8084	5.6685	5.5342	5.4053	5.2814	5.1624	5.0479	4.9377	4.8317	4.7296
17	5.8947	5.7487	5.6088	5.4746	5.3459	5.2223	5.1037	4.9897	4.8801	4.7746
18	5.9695	5.8178	5.6728	5.5339	5.4008	5.2732	5.1508	5.0333	4.9205	4.8122
19	6.0342	5.8775	5.7277	5.5845	5.4475	5.3162	5.1905	5.0700	4.9544	4.8435
20	6.0902	5.9288	5.7748	5.6278	5.4872	5.3527	5.2241	5.1009	4.9828	4.8696
21	6.1387	5.9731	5.8153	5.6648	5.5210	5.3837	5.2524	5.1268	5.0065	4.8913
22	6.1807	6.0113	5.8501	5.6964	5.5498	5.4099	5.2763	5.1486	5.0264	4.9094
23	6.2170	6.0442	5.8799	5.7234	5.5743	5.4321	5.2964	5.1668	5.0430	4.9245
24	6.2485	6.0726	5.9055	5.7465	5.5951	5.4509	5.3134	5.1822	5.0569	4.9371
25	6.2758	6.0971	5.9274	5.7662	5.6129	5.4669	5.3278	5.1951	5.0685	4.9476
26	6.2994	6.1182	5.9463	5.7831	5.6280	5.4804	5.3399	5.2060	5.0783	4.9563
27	6.3198	6.1364	5.9625	5.7975	5.6408	5.4919	5.3501	5.2151	5.0864	4.9636
28	6.3375	6.1520	5.9764	5.8099	5.6518	5.5016	5.3588	5.2228	5.0932	4.9697
29	6.3528	6.1656	5.9883	5.8204	5.6611	5.5098	5.3661	5.2292	5.0989	4.9747
30	6.3661	6.1772	5.9986	5.8294	5.6690	5.5168	5.3722	5.2347	5.1037	4.9789
31	6.3775	6.1872	6.0073	5.8371	5.6758	5.5227	5.3774	5.2392	5.1077	4.9824
32	6.3875	6.1959	6.0149	5.8437	5.6815	5.5277	5.3818	5.2430	5.1111	4.9854
33	6.3961	6.2034	6.0214	5.8493	5.6864	5.5320	5.3854	5.2462	5.1139	4.9878
34	6.4035	6.2098	6.0269	5.8541	5.6905	5.5356	5.3886	5.2489	5.1162	4.9898
35	6.4100	6.2153	6.0317	5.8582	5.6941	5.5386	5.3912	5.2512	5.1182	4.9915
36	6.4156	6.2201	6.0358	5.8617	5.6971	5.5412	5.3934	5.2531	5.1198	4.9929
37	6.4204	6.2242	6.0393	5.8647	5.6996	5.5434	5.3953	5.2547	5.1212	4.9941
38	6.4246	6.2278	6.0423	5.8673	5.7018	5.5452	5.3969	5.2561	5.1223	4.9951
39	6.4282	6.2309	6.0449	5.8695	5.7037	5.5468	5.3982	5.2572	5.1233	4.9959
40	6.4314	6.2335	6.0471	5.8713	5.7053	5.5482	5.3993	5.2582	5.1241	4.9966
50	6.4468	6.2463	6.0577	5.8801	5.7125	5.5541	5.4043	5.2623	5.1275	4.9995
60	6.4505	6.2492	6.0600	5.8819	5.7139	5.5553	5.4052	5.2630	5.1281	4.9999
80	6.4515	6.2500	6.0606	5.8823	5.7143	5.5555	5.4054	5.2632	5.1282	5.0000
90	6.4516	6.2500	6.0606	5.8823	5.7143	5.5556	5.4054	5.2632	5.1282	5.0000
100	6.4516	6.2500	6.0606	5.8824	5.7143	5.5556	5.4054	5.2632	5.1282	5.0000
110	6.4516	6.2500	6.0606	5.8824	5.7143	5.5556	5.4054	5.2632	5.1282	5.0000
120	6.4516	6.2500	6.0606	5.8824	5.7143	5.5556	5.4054	5.2632	5.1282	5.0000

Table A4.2 Present value of annuity tables (cont'd)

i / n	21	22	23	24	25	30	35	40	50	100
1	0.8264	0.8197	0.8130	0.8065	0.8000	0.7692	0.7407	0.7143	0.6667	0.5000
2	1.5095	1.4915	1.4740	1.4568	1.4400	1.3609	1.2894	1.2245	1.1111	0.7500
3	2.0739	2.0422	2.0114	1.9813	1.9520	1.8161	1.6959	1.5889	1.4074	0.8750
4	2.5404	2.4936	2.4483	2.4043	2.3616	2.1662	1.9969	1.8492	1.6049	0.9375
5	2.9260	2.8636	2.8035	2.7454	2.6893	2.4356	2.2200	2.0352	1.7366	0.9688
6	3.2446	3.1669	3.0923	3.0205	2.9514	2.6427	2.3852	2.1680	1.8244	0.9844
7	3.5079	3.4155	3.3270	3.2423	3.1611	2.8021	2.5075	2.2628	1.8829	0.9922
8	3.7256	3.6193	3.5179	3.4212	3.3289	2.9247	2.5982	2.3306	1.9220	0.9961
9	3.9054	3.7863	3.6731	3.5655	3.4631	3.0190	2.6653	2.3790	1.9480	0.9980
10	4.0541	3.9232	3.7993	3.6819	3.5705	3.0915	2.7150	2.4136	1.9653	0.9990
11	4.1769	4.0354	3.9018	3.7757	3.6564	3.1473	2.7519	2.4383	1.9769	0.9995
12	4.2784	4.1274	3.9852	3.8514	3.7251	3.1903	2.7792	2.4559	1.9846	0.9998
13	4.3624	4.2028	4.0530	3.9124	3.7801	3.2233	2.7994	2.4685	1.9897	0.9999
14	4.4317	4.2646	4.1082	3.9616	3.8241	3.2487	2.8144	2.4775	1.9931	0.9999
15	4.4890	4.3152	4.1530	4.0013	3.8593	3.2682	2.8255	2.4839	1.9954	1.0000
16	4.5364	4.3567	4.1894	4.0333	3.8874	3.2832	2.8337	2.4885	1.9970	1.0000
17	4.5755	4.3908	4.2190	4.0591	3.9099	3.2948	2.8398	2.4918	1.9980	1.0000
18	4.6079	4.4187	4.2431	4.0799	3.9279	3.3037	2.8443	2.4941	1.9986	1.0000
19	4.6346	4.4415	4.2627	4.0967	3.9424	3.3105	2.8476	2.4958	1.9991	1.0000
20	4.6567	4.4603	4.2786	4.1103	3.9539	3.3158	2.8501	2.4970	1.9994	1.0000
21	4.6750	4.4756	4.2916	4.1212	3.9631	3.3198	2.8519	2.4979	1.9996	1.0000
22	4.6900	4.4882	4.3021	4.1300	3.9705	3.3230	2.8533	2.4985	1.9997	1.0000
23	4.7025	4.4985	4.3106	4.1371	3.9764	3.3254	2.8543	2.4989	1.9998	1.0000
24	4.7128	4.5070	4.3176	4.1428	3.9811	3.3272	2.8550	2.4992	1.9999	1.0000
25	4.7213	4.5139	4.3232	4.1474	3.9849	3.3286	2.8556	2.4994	1.9999	1.0000
26	4.7284	4.5196	4.3278	4.1511	3.9879	3.3297	2.8560	2.4996	1.9999	1.0000
27	4.7342	4.5243	4.3316	4.1542	3.9903	3.3305	2.8563	2.4997	2.0000	1.0000
28	4.7390	4.5281	4.3346	4.1566	3.9923	3.3312	2.8565	2.4998	2.0000	1.0000
29	4.7430	4.5312	4.3371	4.1585	3.9938	3.3317	2.8567	2.4999	2.0000	1.0000
30	4.7463	4.5338	4.3391	4.1601	3.9950	3.3321	2.8568	2.4999	2.0000	1.0000
31	4.7490	4.5359	4.3407	4.1614	3.9960	3.3324	2.8569	2.4999	2.0000	1.0000
32	4.7512	4.5376	4.3421	4.1624	3.9968	3.3326	2.8569	2.4999	2.0000	1.0000
33	4.7531	4.5390	4.3431	4.1632	3.9975	3.3328	2.8570	2.5000	2.0000	1.0000
34	4.7546	4.5402	4.3440	4.1639	3.9980	3.3329	2.8570	2.5000	2.0000	1.0000
35	4.7559	4.5411	4.3447	4.1644	3.9984	3.3330	2.8571	2.5000	2.0000	1.0000
36	4.7569	4.5419	4.3453	4.1649	3.9987	3.3331	2.8571	2.5000	2.0000	1.0000
37	4.7578	4.5426	4.3458	4.1652	3.9990	3.3331	2.8571	2.5000	2.0000	1.0000
38	4.7585	4.5431	4.3462	4.1655	3.9992	3.3332	2.8571	2.5000	2.0000	1.0000
39	4.7591	4.5435	4.3465	4.1657	3.9993	3.3332	2.8571	2.5000	2.0000	1.0000
40	4.7596	4.5439	4.3467	4.1659	3.9995	3.3332	2.8571	2.5000	2.0000	1.0000
50	4.7616	4.5452	4.3477	4.1666	3.9999	3.3333	2.8571	2.5000	2.0000	1.0000
60	4.7619	4.5454	4.3478	4.1667	4.0000	3.3333	2.8571	2.5000	2.0000	1.0000
80	4.7619	4.5455	4.3478	4.1667	4.0000	3.3333	2.8571	2.5000	2.0000	1.0000
90	4.7619	4.5455	4.3478	4.1667	4.0000	3.3333	2.8571	2.5000	2.0000	1.0000
100	4.7619	4.5455	4.3478	4.1667	4.0000	3.3333	2.8571	2.5000	2.0000	1.0000
110	4.7619	4.5455	4.3478	4.1667	4.0000	3.3333	2.8571	2.5000	2.0000	1.0000
120	4.7619	4.5455	4.3478	4.1667	4.0000	3.3333	2.8571	2.5000	2.0000	1.0000

Table A4.3 Annual value tables
(for intermediate values use the inverse of Table A4.2 values)

n \ i	1	2	3	4	5	6	7	8	9	10
1	1.0100	1.0200	1.0300	1.0400	1.0500	1.0600	1.0700	1.0800	1.0900	1.1000
2	0.5075	0.5150	0.5226	0.5302	0.5378	0.5454	0.5531	0.5608	0.5685	0.5762
3	0.3400	0.3468	0.3535	0.3603	0.3672	0.3741	0.3811	0.3880	0.3951	0.4021
4	0.2563	0.2626	0.2690	0.2755	0.2820	0.2886	0.2952	0.3019	0.3087	0.3155
5	0.2060	0.2122	0.2184	0.2246	0.2310	0.2374	0.2439	0.2505	0.2571	0.2638
6	0.1725	0.1785	0.1846	0.1908	0.1970	0.2034	0.2098	0.2163	0.2229	0.2296
7	0.1486	0.1545	0.1605	0.1666	0.1728	0.1791	0.1856	0.1921	0.1987	0.2054
8	0.1307	0.1365	0.1425	0.1485	0.1547	0.1610	0.1675	0.1740	0.1807	0.1874
9	0.1167	0.1225	0.1284	0.1345	0.1407	0.1470	0.1535	0.1601	0.1668	0.1736
10	0.1056	0.1113	0.1172	0.1233	0.1295	0.1359	0.1424	0.1490	0.1558	0.1627
11	0.0965	0.1022	0.1081	0.1141	0.1204	0.1268	0.1334	0.1401	0.1469	0.1540
12	0.0888	0.0946	0.1005	0.1066	0.1128	0.1193	0.1259	0.1327	0.1397	0.1468
13	0.0824	0.0881	0.0940	0.1001	0.1065	0.1130	0.1197	0.1265	0.1336	0.1408
14	0.0769	0.0826	0.0885	0.0947	0.1010	0.1076	0.1143	0.1213	0.1284	0.1357
15	0.0721	0.0778	0.0838	0.0899	0.0963	0.1030	0.1098	0.1168	0.1241	0.1315
16	0.0679	0.0737	0.0796	0.0858	0.0923	0.0990	0.1059	0.1130	0.1203	0.1278
17	0.0643	0.0700	0.0760	0.0822	0.0887	0.0954	0.1024	0.1096	0.1170	0.1247
18	0.0610	0.0667	0.0727	0.0790	0.0855	0.0924	0.0994	0.1067	0.1142	0.1219
19	0.0581	0.0638	0.0698	0.0761	0.0827	0.0896	0.0968	0.1041	0.1117	0.1195
20	0.0554	0.0612	0.0672	0.0736	0.0802	0.0872	0.0944	0.1019	0.1095	0.1175
21	0.0530	0.0588	0.0649	0.0713	0.0780	0.0850	0.0923	0.0998	0.1076	0.1156
22	0.0509	0.0566	0.0627	0.0692	0.0760	0.0830	0.0904	0.0980	0.1059	0.1140
23	0.0489	0.0547	0.0608	0.0673	0.0741	0.0813	0.0887	0.0964	0.1044	0.1126
24	0.0471	0.0529	0.0590	0.0656	0.0725	0.0797	0.0872	0.0950	0.1030	0.1113
25	0.0454	0.0512	0.0574	0.0640	0.0710	0.0782	0.0858	0.0937	0.1018	0.1102
26	0.0439	0.0497	0.0559	0.0626	0.0696	0.0769	0.0846	0.0925	0.1007	0.1092
27	0.0424	0.0483	0.0546	0.0612	0.0683	0.0757	0.0834	0.0914	0.0997	0.1083
28	0.0411	0.0470	0.0533	0.0600	0.0671	0.0746	0.0824	0.0905	0.0989	0.1075
29	0.0399	0.0458	0.0521	0.0589	0.0660	0.0736	0.0814	0.0896	0.0981	0.1067
30	0.0387	0.0446	0.0510	0.0578	0.0651	0.0726	0.0806	0.0888	0.0973	0.1061
31	0.0377	0.0436	0.0500	0.0569	0.0641	0.0718	0.0798	0.0881	0.0967	0.1055
32	0.0367	0.0426	0.0490	0.0559	0.0633	0.0710	0.0791	0.0875	0.0961	0.1050
33	0.0357	0.0417	0.0482	0.0551	0.0625	0.0703	0.0784	0.0869	0.0956	0.1045
34	0.0348	0.0408	0.0473	0.0543	0.0618	0.0696	0.0778	0.0863	0.0951	0.1041
35	0.0340	0.0400	0.0465	0.0536	0.0611	0.0690	0.0772	0.0858	0.0946	0.1037
36	0.0332	0.0392	0.0458	0.0529	0.0604	0.0684	0.0767	0.0853	0.0942	0.1033
37	0.0325	0.0385	0.0451	0.0522	0.0598	0.0679	0.0762	0.0849	0.0939	0.1030
38	0.0318	0.0378	0.0445	0.0516	0.0593	0.0674	0.0758	0.0845	0.0935	0.1027
39	0.0311	0.0372	0.0438	0.0511	0.0588	0.0669	0.0754	0.0842	0.0932	0.1025
40	0.0305	0.0366	0.0433	0.0505	0.0583	0.0665	0.0750	0.0839	0.0930	0.1023
50	0.0255	0.0318	0.0389	0.0466	0.0548	0.0634	0.0725	0.0817	0.0912	0.1009
60	0.0222	0.0288	0.0361	0.0442	0.0528	0.0619	0.0712	0.0808	0.0905	0.1003
80	0.0182	0.0252	0.0331	0.0418	0.0510	0.0606	0.0703	0.0802	0.0901	0.1000
90	0.0169	0.0240	0.0323	0.0412	0.0506	0.0603	0.0702	0.0801	0.0900	0.1000
100	0.0159	0.0232	0.0316	0.0408	0.0504	0.0602	0.0701	0.0800	0.0900	0.1000
110	0.0150	0.0226	0.0312	0.0405	0.0502	0.0601	0.0700	0.0800	0.0900	0.1000
120	0.0143	0.0220	0.0309	0.0404	0.0501	0.0601	0.0700	0.0800	0.0900	0.1000

Table A4.3   Annual value tables (cont'd)
(for intermediate values use the inverse of Table A4.2 values)

i / n	11	12	13	14	15	20	25	30	40	50
1	1.1100	1.1200	1.1300	1.1400	1.1500	1.2000	1.2500	1.3000	1.4000	1.5000
2	0.5839	0.5917	0.5995	0.6073	0.6151	0.6545	0.6944	0.7348	0.8167	0.9000
3	0.4092	0.4163	0.4235	0.4307	0.4380	0.4747	0.5123	0.5506	0.6294	0.7105
4	0.3223	0.3292	0.3362	0.3432	0.3503	0.3863	0.4234	0.4616	0.5408	0.6231
5	0.2706	0.2774	0.2843	0.2913	0.2983	0.3344	0.3718	0.4106	0.4914	0.5758
6	0.2364	0.2432	0.2502	0.2572	0.2642	0.3007	0.3388	0.3784	0.4613	0.5481
7	0.2122	0.2191	0.2261	0.2332	0.2404	0.2774	0.3163	0.3569	0.4419	0.5311
8	0.1943	0.2013	0.2084	0.2156	0.2229	0.2606	0.3004	0.3419	0.4291	0.5203
9	0.1806	0.1877	0.1949	0.2022	0.2096	0.2481	0.2888	0.3312	0.4203	0.5134
10	0.1698	0.1770	0.1843	0.1917	0.1993	0.2385	0.2801	0.3235	0.4143	0.5088
11	0.1611	0.1684	0.1758	0.1834	0.1911	0.2311	0.2735	0.3177	0.4101	0.5058
12	0.1540	0.1614	0.1690	0.1767	0.1845	0.2253	0.2684	0.3135	0.4072	0.5039
13	0.1482	0.1557	0.1634	0.1712	0.1791	0.2206	0.2645	0.3102	0.4051	0.5026
14	0.1432	0.1509	0.1587	0.1666	0.1747	0.2169	0.2615	0.3078	0.4036	0.5017
15	0.1391	0.1468	0.1547	0.1628	0.1710	0.2139	0.2591	0.3060	0.4026	0.5011
16	0.1355	0.1434	0.1514	0.1596	0.1679	0.2114	0.2572	0.3046	0.4018	0.5008
17	0.1325	0.1405	0.1486	0.1569	0.1654	0.2094	0.2558	0.3035	0.4013	0.5005
18	0.1298	0.1379	0.1462	0.1546	0.1632	0.2078	0.2546	0.3027	0.4009	0.5003
19	0.1276	0.1358	0.1441	0.1527	0.1613	0.2065	0.2537	0.3021	0.4007	0.5002
20	0.1256	0.1339	0.1424	0.1510	0.1598	0.2054	0.2529	0.3016	0.4005	0.5002
21	0.1238	0.1322	0.1408	0.1495	0.1584	0.2044	0.2523	0.3012	0.4003	0.5001
22	0.1223	0.1308	0.1395	0.1483	0.1573	0.2037	0.2519	0.3009	0.4002	0.5001
23	0.1210	0.1296	0.1383	0.1472	0.1563	0.2031	0.2515	0.3007	0.4002	0.5000
24	0.1198	0.1285	0.1373	0.1463	0.1554	0.2025	0.2512	0.3006	0.4001	0.5000
25	0.1187	0.1275	0.1364	0.1455	0.1547	0.2021	0.2509	0.3004	0.4001	0.5000
26	0.1178	0.1267	0.1357	0.1448	0.1541	0.2018	0.2508	0.3003	0.4001	0.5000
27	0.1170	0.1259	0.1350	0.1442	0.1535	0.2015	0.2506	0.3003	0.4000	0.5000
28	0.1163	0.1252	0.1344	0.1437	0.1531	0.2012	0.2505	0.3002	0.4000	0.5000
29	0.1156	0.1247	0.1339	0.1432	0.1527	0.2010	0.2504	0.3001	0.4000	0.5000
30	0.1150	0.1241	0.1334	0.1428	0.1523	0.2008	0.2503	0.3001	0.4000	0.5000
31	0.1145	0.1237	0.1330	0.1425	0.1520	0.2007	0.2502	0.3001	0.4000	0.5000
32	0.1140	0.1233	0.1327	0.1421	0.1517	0.2006	0.2502	0.3001	0.4000	0.5000
33	0.1136	0.1229	0.1323	0.1419	0.1515	0.2005	0.2502	0.3001	0.4000	0.5000
34	0.1133	0.1226	0.1321	0.1416	0.1513	0.2004	0.2501	0.3000	0.4000	0.5000
35	0.1129	0.1223	0.1318	0.1414	0.1511	0.2003	0.2501	0.3000	0.4000	0.5000
36	0.1126	0.1221	0.1316	0.1413	0.1510	0.2003	0.2501	0.3000	0.4000	0.5000
37	0.1124	0.1218	0.1314	0.1411	0.1509	0.2002	0.2501	0.3000	0.4000	0.5000
38	0.1121	0.1216	0.1313	0.1410	0.1507	0.2002	0.2501	0.3000	0.4000	0.5000
39	0.1119	0.1215	0.1311	0.1409	0.1506	0.2002	0.2500	0.3000	0.4000	0.5000
40	0.1117	0.1213	0.1310	0.1407	0.1506	0.2001	0.2500	0.3000	0.4000	0.5000
50	0.1106	0.1204	0.1303	0.1402	0.1501	0.2000	0.2500	0.3000	0.4000	0.5000
60	0.1102	0.1201	0.1301	0.1401	0.1500	0.2000	0.2500	0.3000	0.4000	0.5000
80	0.1100	0.1200	0.1300	0.1400	0.1500	0.2000	0.2500	0.3000	0.4000	0.5000
90	0.1100	0.1200	0.1300	0.1400	0.1500	0.2000	0.2500	0.3000	0.4000	0.5000
100	0.1100	0.1200	0.1300	0.1400	0.1500	0.2000	0.2500	0.3000	0.4000	0.5000
110	0.1100	0.1200	0.1300	0.1400	0.1500	0.2000	0.2500	0.3000	0.4000	0.5000
120	0.1100	0.1200	0.1300	0.1400	0.1500	0.2000	0.2500	0.3000	0.4000	0.5000

Index

ABB (Asea Brown Boveri)
  customer expectations   411
  delayering   117
  go for growth   135
  matrix organization   115
  organizational transformation   117–19
  use of Ansoff's matrix   135, 136
  use of IT   141
ABC inventory management   392, 396
accounting standard   272
accounts   270–83
  financial accounts   270, 273
  an introduction   10–11
  management accounts   273
  profit and loss account   274–6
  trading account   275–6
accruals   283
activity sampling   243–5
activity-on-arrow diagrams   364
activity-on-node diagrams   364–75
ACWP (actual cost of work performed)   346–8
Adair, J.   184, 203
advertising   460–2
A/E/PM in construction contracts   380–6
aggression   155, 156
alliances   361–2
  see also partnership sourcing
annual value *see* equivalent annual value
annuity   301–4
Ansoff's matrix   135–6, 452–3
anthropologists   152
arbitration   384
assets   276–82
  current assets   281–2
  fixed assets   278–81
automobile industry
  organizational innovation   34
  vehicle producer operating systems   68

BAe (British Aerospace)
  concurrent engineering   91, 92, 100

  design change cost effects   54
balance sheets   274–83
  an introduction   10, 11
bar charts   339–42, 364, 370–1
Barnevik, Percy   138, 141, 142
Basic Mean Time (BMT)   245–8
Basic Time (BT)   245–8
Bass, B. M.   176, 203
bathtub curve   326
BCWP (budgeted cost of work performed)   346–8
BCWS (budgeted cost of work scheduled)   346–8
Beer, M.   196, 203
Beer, Stafford   5
behaviour   154
behaviour – constructive   155, 156
behaviour – destructive   155, 156
behavioural sciences   150, 151
Belbin – eight-role team model   190
Belbin – nine-role team model   191
Belbin, R. M.   190–4, 203
benchmarking   124
benefit cost (B/C) ratio   296, 310–11
Berne, E.   166, 167, 168, 169, 170, 171, 172
bills of quantities (BoQ)   357–8
Blake, R. R.   184, 203
blanket orders   397, 416, 417
BOMs (bills of material)   67, 70
book-keeping   283–4
Boston matrix   450–1
break-even   250, 259–63
British Steel
  use of SPC   48, 49
budgets   263–4, 273–4, 346
business chain   6, 7
business functions   6
Business Link   468
business mentors   469
business performance measurement
  an introduction   10–12

business plan
  elements 467
  a guide 487–95
business planning 435–46
business process reengineering (BPR) 67
  at Chrysler 112, 113, 132–5

capability index (Cp, Cpk) 49
capital
  working capital cycle 285–6
capital expenditure 276
capital rationing 315–16
cash 284–5
cash flow diagrams 9, 299–308, 313
Cattell, R. B. 158, 172
Cattell 16PF questionnaire 158
cause and effect (fishbone) diagrams
    47, 499–500
change control 348–9
change orders *see* variation orders
Channel Tunnel 351
Chrysler Corporation
  merger with Daimler 139
  purchasing policies 425
  reengineering 112, 113, 132–5
  reorganization 113
civil engineering management *see* construction
    management
Clarke, C. 187, 203
COBA program 312
collective responsibility 179
committees 177
communication 164, 165, 204
  code 208
  channel 205, 206, 207
  conceptual model 206
  encoding/decoding 209
  elevator speech 219
  feedback 206, 208, 212
  in practice 207
  medium 205, 206, 208
  maintenance problems
    attenuation 211, 212
    distortion 211, 212
    noise 211, 212
    overload 211, 212
    redundancy 211, 212
  multichannel
    multiple 222
    one-to-one 221
    one-to-more-than-one 221
  multimedia 230
  nonverbal 225

  one-to-one 205
  patterns 222–35
    all channel 222, 224
    chain 222, 223
    circle 222, 223
    wheel 222, 223
    Y network 222, 223
  presentations 231–4
    basic principles 231
    good practice 233, 234
    preparation 231, 232
    speaker 233
    visual aids 232, 233
  receiver 205, 206, 212
  senses 221
  transmitter 205, 206, 212
  verbal 207
  visual 205–29
    body language 207, 229
    charts 228
    cues 227
    impact 227
    tables 228
  written 215–20
    abstracts 219
    content 216
    clarity 218
    essays 217
    fog factor 218
    readability 218
    reports 217
    structure 216
    summaries 219, 220
    syntax 216
communication barriers
  accent 209, 212
  culture 210, 212
  dialect 210, 212
  education 210, 212
  halo effect 210, 212
  jargon 209, 212, 217, 219
  linguistic style 209, 212
  personal 210, 212
  receiver behaviour 211, 212
  transmitter behaviour 211, 212
complexity, hard 149
complexity, soft 149
concurrent engineering (simultaneous engineering,
    SE) 84–96, 188, 335, 350, 354–7
construction management 353–62, 376–86,
    512–21
  see also contract management
construction procurement 376–86

Consumer Price Index (CPI)   317
contract
    formal   189
    informal   189
    psychological   189
contract management (CM)   355, 356–7
contracts
    cost reimbursement   359–60, 383
    design-and-build *see* design-and-construct
    design-and-construct   355, 382
    fixed price   351–2
    lump sum   357, 382
    measurement   357–9, 381–2
    target contracts   359, 383
    turnkey   355
contribution   260–1
control charts   48
conversation   207
corporation *see* limited company
cost-benefit analysis (CBA)   311–13
cost centres   251
cost, life-cycle *see* life-cycle costing
cost of quality
    elements   45
    TQM effects   52
Cost Performance Index (CPI)   346–8
cost plan   339–40
costing   250–8
costs
    direct costs   250–1, 258
    fixed costs   258–63
    indirect costs   250
    initial cost   250–1
    labour costs   254–5
    materials costs   251–4
    standard costs   257–8
    variable costs   258–63
CPA *see* critical path planning and analysis
creative cycle model   176, 177
creditors   283, 285–6
critical path planning and analysis   341, 365–86,
    464
Crosby, Philip   46, 55, 219, 234
cultural system   152
culture   23, 24
customer care   59

DCF (discounted cash flow)   297–323
DCF tables   522–33
debtors   277, 282–3
Deming, W. Edwards   xiii, xiv, 43, 46, 55, 56,
    398, 424
demography   21–3

depreciation   276, 278–81
    reducing balance method   279–81
    straight line method   279
    sum-of-the-digits method   280
    usage method   281
design
    dilemma   56, 77
    pressures to change   79–82
    using simultaneous engineering   84–101
design-and-construct contracts *see* contracts
discounted cash flow *see* DCF
division of work   194–6
    common processes   195
    customer basis   195
    location basis   195
    major purpose or function   195
    product or service   195
    staff basis   195
    time basis   195
DuPont
    links with General Motors   34, 106–8
    organization   105

earned value   346
earned value analysis   346–8
ECC2 *see* Engineering and Construction Contract
economic life   278–81
Egan Report   362, 385
employee involvement
    in problem solving teams   496–9
    in simultaneous engineering teams   88
empowerment   117
engineer, role of in construction contracts *see*
    A/E/PM
Engineering and Construction Contract   383–4
EOQ (economic order quantity)   391
equity theory   163
equivalent annual value (EAV)   303–7, 532–3
estimating costs   263–6
exchange rates   20, 137
expectancy theory   163

failure mode and effects analysis *see* FMEA
failure rate   325–6
failure rate data   327
fast-track engineering   335, 355
Fiedler, F. E. A.   187, 203
financial accounts *see* accounts
financial performance   268–95
financial ratios   287–94
    gearing ratios   292–3
        debts/assets employed ratios   293
        debt/equity ratio   292–3

financial ratios (*continued*)
  interest cover   293
  investor ratios   293–4
    earnings per share (EPS)   293–4
    price/earnings (P/E) ratio   294
  liquidity ratios   289–90
    current ratio   289–90
    quick ratio   290
  operating ratios   287–9
    asset turnover   288–9
    gross profit margin   287–8
    return on capital employed (ROCE)   288
    working capital turnover   289
  working capital ratios   290–2
    creditors turnover ratio   292
    debtors turnover ratio   291–2
    stock turnover ratios   290–1
first in first out (FIFO)   252–3
fixation   155, 156
float (in critical path analysis)   368
flow charts   236–8
flow diagrams   236–8
flow process charts/diagrams   236–8
FMEA (failure mode and effects analysis)   54, 325, 501–9
Ford, Henry I   4, 124, 466
Ford Motor Company
  2000 programme   126
  Q101/Q1 427
  QS9000 58
  rediscover SE   88
  staff recruitment policy   109, 110
  supply base management   426–9
  and Taylorism   33, 34
Fridrich, Heinz K.   128
frustration   155, 156
frustration model   156

Gantt chart *see* bar charts
GDP (gross domestic product)   20
GE matrix   451–2
Geneen, Harold   36
General Motors
  management model   34–6
  organization   106–8
  reorganization   111, 112
GNP (gross national product)   19
goal setting theory   162
group behaviour   173
group development   176
  Bass and Ryterband model   176
  Tuckman model   176
group effectiveness   174

individual objectives   175
maturity of group   175
member characteristics   175
nature of task   175
size   175
group interactions   220
group purpose   174
  individual   174
  organizational   174
group type   174
groups   174
groups – mature   176
growth rates in leading economies   19
Gunning, R.   218, 234

Hamel, Gary   39, 40, 97, 115
Harvard model   196
heavyweight programme management   87, 112
Hellrreigel, D.   181, 203
Hendry, C.   197, 203
Herzberg, F.   149, 159, 161, 162, 163, 172, 460
  two factor theory   161, 162
  see also hygiene factors, motivators
Hewlett-Packard   130–1
hidden agenda   175
Hofstede, G.   200, 203
human resource management (HRM)   196–200
  development   198
  forecasting   197
  HRM models   196
  performance appraisal   199
  personnel movement   199
  recruitment   197
  remuneration   199
  selection   197
  training   198
  see also international HRM
hygiene factors (Herzberg)   162, 163

IBM
  centres of excellence   116
  decline and recovery   128, 129
ICE7 381–2
IChemE Green Book   383
incentive schemes   243–4
individual   152
individuals in pairs   164
individuality   156, 157
inflation   264–5, 316–20, 350–1
interactive management   96
interest rate   294–322
  money interest rate   316–18
  real interest rate   316–18

internal rate of return (IRR)   307–10
international HRM   200–2
   economic development   201
   economic system   202
   legislative framework   201
   national culture   200
   political system   202
interpersonal relationships   171
interrogative method   241–3
interviews   212–15
   appraisal   213
   counselling   214
   discipline   214
   induction   213
   progress   213
   recruitment   213
   selection   213
   termination   214
inventory   72, 73, 388–408
investment decisions   296–323
ISO 9000   57, 471, 510, 511
   with TickIT   57
ISO 14000   58, 137
ITT   36, 38, 39

JCT98 381
JIT *see* just-in-time
job enlargement   163
job enrichment   163
job scope   163
Juran, Joseph   43, 46, 55
just-in-time   324, 401–7, 462

kanban   403
Krech, D.   185, 203

last in first out (LIFO)   253
Latham Report   336, 385
leadership
   in ABB   141, 142
   in Chrysler   139
   functional model   184
   functions   184
   qualities   183
   in Shell   143
   situation   183
   styles   185, 186
   traits   183
leadership power
   organizational   182
   personal   182
   in scientific/technical environment   182

   sources   182
leadership theories
   business maturity   187
   contingency   187
   technical context   187
lean manufacturing   324
Leavitt, H. J.   152, 157, 172
liabilities (in accounting)   276–8, 282–3
life-cycle costing (LCC)   296, 299, 311–13
lightweight programme management   87
Likert, Rensis   194
limited company   269–270
liquidated damages   361
logistics   407
Lutz, Robert A.   39, 87, 112, 113, 124, 132, 135

maintenance management   324–34
   maintenance cost   324
   maintenance planning   330–4
   maintenance strategy   328–30
management contracting (MC)   355–6
management – participative   156
managerial grid   184
market research   250, 462–3
market segmentation   438, 446–8
marketing   433–65
marketing information systems   462–3
marketing mix   434, 456–62
Martin, J.   165, 166, 167, 170, 172
Maslow, A. H.   149, 159, 160, 161, 172, 460
   hierarchy of human needs   159, 160, 163
matching principle in accounting   270
materials
   advanced materials   65, 97
   cost contribution   33
   profit influence   72
   selection   64
matrix organization   115, 116
McClelland, D. C.   162, 172
McKinsey & Co.   36
McKinsey's 7Ss   36, 37, 434, 443, 446
meetings   177–80, 225, 345
   communications   225
   functions   178
   management of   179
   minutes   225
   types   177
method statements   338
method study   235–43
MF/1 382
Mintzberg, Henry   5, 6, 185
mission   433–5
motivation   149, 154

motivation model   155
motivation theories   159
motivators (Herzberg)   162
MRO (maintenance, repair and operating supplies)
    74, 395
MRPI/MRPII   68
Mouton, J. S.   184, 203
Mullins, L. J.   156, 172, 182, 203
multiple activity charts   238–41
Myers-Briggs   158, 172
Myers-Briggs test (MBTI)   158

negotiated contracts   352, 360–2
net present value (NPV)   296–323
networks and network analysis   341, 364–75,
    512–21
New Engineering Contract *see Engineering and*
    Construction Contract
Nissan Manufacturing UK
    lean logistics   407
    purchasing policy   424
    use of SPC   49, 50
nomination of subcontractors   353
NPV and NPW *see* net present value

over-the-walls (OTW) engineering   77, 78, 354
overall plant effectiveness (OPE)   325
overheads   251, 255–7, 275, 286

Pareto analysis   287, 394–5
Pareto, Vilfredo   394
partnering   362, 385–6, 455
partnership sourcing   423–32
partnerships   269
payback   296–7
perfect capital market   314–15
performance measurement   10–12, 345–8, 389
performance rating factor   245–6
performance ratios   346–8
personality   156–9
    dynamics   165
    system   151
    types   157
personality theory
    ideographic approach   158
    monothetic approach   158
personal relationships   164
PERT   372–5
PEST   436–8, 441
PFI/PPP   380
PLC *see* product life cycle
Porter's five forces   436, 438–42

Porter's value chain   443–5
portfolio analysis   450–4
power
    connection   182
    coercive   182
    expert   182
    within engineering environment   182
    information   182
    legitimate   182
    referent   182
    reward   182
    scientific/technical   182
PPP *see* PFI/PPP
Pratt, S.   187, 203
precedence lists   364–75, 512–21
predetermined motion-time systems (PMTS)
    244–5
present value   296–323
present worth *see* present value
pricing   263–7, 459–60
Private Finance Initiative *see* PFI/PPP
privity of contract   353, 382
procurement/purchasing   414–31
product life cycle (PLC)   448–50
product support   251
profit   250, 258–63, 284–7
profit and loss account   268, 274–6
    an introduction   10,11
programme evaluation and review technique *see*
    PERT
project definition   336–7
project management   335–63, 376–87
project management software   337
project manager   336
    in construction projects *see* A/E/PM
project planning   337, 338–44
prudence concept in accounting   271
psychological energy   153
psychological needs   155
psychological success   153
psychologists   151
psychology, interpersonal   165
Public Private Partnership *see* PFI/PPP
purchasing *see* procurement

QFD (quality function deployment)   93–6

rating in time study   245–8
ratios *see* financial ratios
Raven, S.   215, 234
reinforcement theory   163
regression   155, 156
relationship marketing   454–6

reliability   326–7
reliability-centred maintenance (RCM)   331
resource charts   340–3
resource smoothing   341
Retail Price Index (RPI)   264–5, 316–18, 351
revenue expenditure   276
ROCE (return on capital employed)
   an introduction   10
   materials management effects   70–4
   see also financial ratios
Rover Group
   purchasing policy   424
Ryterband, E. C.   176, 203

sales revenue   275, 276, 288, 289, 291
Saturn Car Company
   formation   114
   inventory performance   408
SBU (strategic/separate business unit)   105
schedule performance index (SPI)   346–7
schedule variance   346–7
S-curves   343, 345
self   152–4
   self-concept   152, 153
   self-esteem   153, 154
   self-ideal   152, 153
sensitivity analysis   322
seven Ss of management *see* McKinsey's 7Ss
Shell
   planning as learning   126
   performance decline   143
Siemens   97, 98
simultaneous engineering *see* concurrent
      engineering
simultaneous engineering in construction   354–7
Six Sigma   58
Sloan, Alfred P.
   joins GM   34, 107
   and marketing   34
   and organization   35
   and success   112, 123
SMART targets   199
social system   151
sociologists   151
sole traders   268
SPC (statistical process control)   46, 47
standard costs   257–8
standard rating *see* rating in time study
standard times   243–8
stock valuation   252–4
stocks   282, 286
stretch activity   372
strokes *see* transactional analysis

subcontracting   353, 355–6, 381
supply and demand   266–7
supply base management   429
supply chain   15, 16
supply system
   description   411, 412, 413
   engineers' involvement   64–7
   replacing supply chain   98, 411
SWOT analysis   434, 446, 475
   an introduction   18

target contracts *see* contracts
tasks   236, 244
Taylor, Frederick W.   33, 34, 235
Taylorism   33, 235
team membership   187
team models
   Belbin eight-role model   190
   Belbin nine-role model   191
team role behaviour   192
team roles   190, 191
teams
   flexible   194
   member characteristics   190
   motivation   189
   project   188
   technical   193
teamworking   244
tendering   264, 352, 354, 355, 360–2
three needs theory   162
time study   245–8
TOPS (team oriented problem solving)   496–9
total productive maintenance (TPM)   325
TQM (total quality management)   59–61
transactional analysis (TA)   165–71
   ego substates   168
   extended model   167
   games   170
   major ego states   167
   strokes   166
   subpersonality characteristics   169
   theoretical model   166
   transactions   165–71
      complementary   167, 168, 169
      crossed   168, 170
      rituals   170
      ulterior   170
Tuckman, B. W.   176, 192, 203
turnkey contracts *see* contracts

unconscious self   153

variation orders *see* change control

wear-in   325–6
wear-out   326
WIP (work-in-process/progress)   74
withdrawal   156
work breakdown structure (WBS)   338–9, 344–5
work measurement   234, 243–8
work relationships   165
work study   234–8
working capital cycle *see* capital

Xerox
   cooperative contracting   425
   corporate turnaround   127, 128
   leadership through quality   127

yield *see* internal rate of return

ZETA engine – Ford's return to SE   88–91